DEFECTS AND TRANSPORT IN OXIDES

BATTELLE INSTITUTE MATERIALS SCIENCE COLLOQUIA

Published by Plenum Press

1972: Interatomic Potentials and Simulation of Lattice Defects
Edited by Pierre C. Gehlen, Joe R. Beeler, Jr., and Robert I. Jaffee

1973: Deformation and Fracture of High Polymers
Edited by H. Henning Kausch, John A. Hassell, and Robert I. Jaffee

1974: Defects and Transport in Oxides
Edited by Martin S. Seltzer and Robert I. Jaffee

In preparation:

1975: The Physical Basis for Heterogeneous Catalysis

DEFECTS AND TRANSPORT IN OXIDES

Edited by

MARTIN S. SELTZER
Materials Science Department
Battelle Memorial Institute, Columbus Laboratories

ROBERT I. JAFFEE
Materials Science Department
Battelle Memorial Institute, Columbus Laboratories

BATTELLE INSTITUTE
MATERIALS SCIENCE COLLOQUIA

Columbus and Salt Fork, Ohio
September 17-22, 1973

Robert I. Jaffee, Chairman

PLENUM PRESS • New York – London

Library of Congress Cataloging in Publication Data

Battelle Institute Materials Science Colloquia, 8th, Columbus and Salt Fork, Ohio, 1973.
Defects and transport in oxides.

Includes bibliographical references.
1. Crystals–Defects–Congresses. 2. Mass transfer–Congresses. 3. Oxides–Congresses.
I. Seltzer, Martin S., ed. II. Jaffee, Robert Isaac, 1917- ed. III. Title.
QD921.B35 1973 548'.842 74-19162
ISBN 0-306-30824-X

A Division of Plenum Publishing Corporation
227 West 17th Street, New York, N.Y. 10011

United Kingdom edition published by Plenum Press, London
A Division of Plenum Publishing Company, Ltd.
4a Lower John Street, London W1R 3PD, England

Printed in the United States of America

To the late **PROFESSOR JOHN E. DORN**

whose contributions to the science of deformation of solids via thermal activation and climb of defects has been part of the broad foundation of transport processes by defects described for oxides in this volume.

PARTICIPANTS

J. S. ANDERSON *University of Oxford, Oxford, England*

P. CARO *CNRS, Bellevue-Meudon, France*

T. D. CHIKALLA *Battelle, Pacific Northwest Laboratories, Richland, Washington, U.S.A.*

A. H. CLAUER *Battelle, Columbus Laboratories, Columbus, Ohio, U.S.A.*

J. B. COHEN *Northwestern University, Evanston, Illinois, U.S.A.*

R. CONDIT *University of California, Livermore, California, U.S.A.*

J. M. COWLEY *Arizona State University, Tempe, Arizona, U.S.A.*

A. L. DRAGOO *National Bureau of Standards, Washington, D.C., U.S.A.*

L. EYRING *Arizona State University, Tempe, Arizona, U.S.A.*

P. L. FARNSWORTH *Exxon Nuclear Company, Richland, Washington, U.S.A.*

A. D. FRANKLIN *National Bureau of Standards, Washington, D.C., U.S.A.*

J. B. GOODENOUGH *Massachusetts Institute of Technology, Cambridge, Massachusetts, U.S.A.*

H. C. GRAHAM *Aerospace Research Laboratories, Wright-Patterson Air Force Base, Ohio, U.S.A.*

K. HAGEMARK *3M Company, St. Paul, Minnesota, U.S.A.*

P. E. HART *Battelle, Pacific Northwest Laboratories, Richland, Washington, U.S.A.*

K. HAUFFE *Institut für Physikalische Chemie der Universität Göttingen, Göttingen, West Germany*

A. H. HEUER *Case Western Reserve University, Cleveland, Ohio, U.S.A.*

J. M. HONIG *Purdue University, Lafayette, Indiana, U.S.A.*

R. A. HUGGINS *Stanford University, Stanford, California, U.S.A.*

J. C. HURT *Army Research Office, Durham, North Carolina, U.S.A.*
B. ILSCHNER *Universität Erlangen-Nürnberg, Erlangen, West Germany*
R. I. JAFFEE *Battelle, Columbus Laboratories, Columbus, Ohio, U.S.A.*
D. L. JOHNSON *Northwestern University, Evanston, Illinois, U.S.A.*
P. J. JORGENSEN *Stanford Research Institute, Menlo Park, California, U.S.A.*
P. KOFSTAD *Central Institute for Industrial Research, Blindern-Oslo, Norway*
F. A. KRÖGER *University of Southern California, Los Angeles, California, U.S.A.*
R. A. LAD *NASA Lewis Research Center, Cleveland, Ohio, U.S.A.*
G. G. LIBOWITZ *Allied Chemical Corporation, Morristown, New Jersey, U.S.A.*
A. B. LIDIARD *Atomic Energy Research Establishment, Harwell, Didcot, Berks., England*
P. LINDENMEYER *National Science Foundation, Washington, D.C., U.S.A.*
B. LUX *Battelle, Geneva Research Centre, Geneva, Switzerland*
A. J. MARKWORTH *Battelle, Columbus Laboratories, Columbus, Ohio, U.S.A.*
F. A. NICHOLS *Westinghouse Electric Corporation, West Mifflin, Pennsylvania, U.S.A.*
A. S. NOWICK *Columbia University, New York, New York, U.S.A.*
M. O'KEEFFE *Arizona State University, Tempe, Arizona, U.S.A.*
R. A. RAPP *The Ohio State University, Columbus, Ohio, U.S.A.*
H. SCHMALZRIED *Technische Universität Clausthal, Zellerfeld, West Germany*
M. S. SELTZER *Battelle, Columbus Laboratories, Columbus, Ohio, U.S.A.*
P. G. SHEWMON *National Science Foundation, Washington, D.C., U.S.A.*
W. W. SMELTZER *McMaster University, Hamilton, Ontario, Canada*
H. G. SOCKEL *Institut für Werkstoffwissenschaften I der Universität Erlangen-Nürnberg, Erlangen, West Germany*
J. STRINGER *University of Liverpool, Liverpool, England*
V. S. STUBICAN *The Pennsylvania State University, University Park, Pennsylvania, U.S.A.*
N. M. TALLAN *Aerospace Research Laboratories, Wright-Patterson Air Force Base, Ohio, U.S.A.*
D. S. TANNHAUSER *Technion-Israel Institute of Technology, Haifa, Israel*
R. P. TURCOTTE *Battelle, Pacific Northwest Laboratories, Richland, Washington, U.S.A.*
H. J. VINK *Philips Research Laboratories, Eindhoven, Netherlands*
J. B. WAGNER, JR. *Northwestern University, Evanston, Illinois, U.S.A.*
D. D. WELSH *Brookhaven National Laboratory, New York, New York, U.S.A.*
J. E. WERTZ *University of Minnesota, Minneapolis, Minnesota, U.S.A.*
J. M. WIMMER *Aerospace Research Laboratories, Wright-Patterson Air Force Base, Ohio, U.S.A.*
I. G. WRIGHT *Battelle, Columbus Laboratories, Columbus, Ohio, U.S.A.*

PREFACE

DEFECTS AND TRANSPORT IN OXIDES is the proceedings of the eighth Battelle Colloquium in the Materials Sciences, held in Columbus and Salt Fork, Ohio, September 17-22, 1973. It took as its theme the relationship between defects and transport of both mass and charge in oxides. Applications of defect-controlled transport to a number of important processes in oxides also were covered. In selecting this topic, the Organizing Committee thought that 1973 was timely to bring together the leading theoretical and experimental researchers in the oxide transport field to review its status in a critical way, and to consider current major research directions and how research in the future might be guided into fruitful areas.

The meeting was highlighted by the presentation of several papers which suggest that major advances in our understanding of transport in oxides appear to be imminent. These papers dealt with the results of new theoretical approaches whereby the energies and configurations of defects may be calculated, and with new experimental techniques for indirectly observing these defects, previously thought to be below the limits of experimental resolving power. Other papers, dealing with the application of defect chemistry to technological processes, served to demonstrate the successes and to point out yet unresolved problems associated with understanding the chemistry of imperfect crystals.

As the usual custom, the Colloquium started with a first day of introductory lectures by major contributors to the field. In the evening of the first day an address was given by Dr. H. J. Vink, one of the important early contributors. The first introductory lecture by Dr. F. A. Kröger provided the Colloquium with an excellent historical review of defect thermodynamics and the various means for characterizing defects. Dr. J. S. Anderson then introduced the topic of the real structure of defect solids. Dr. John B. Goodenough introduced charge transport, and Dr. H. Schmalzried discussed mass transport involved in solid-solid reactions. Finally, Dr. A. B. Lidiard introduced the theoretical calculation of defect configurations through the use of interatomic potentials.

On the following days, the overall field was divided into the following sessions: Theory and Characterization, Transport of Mass and Charge, and Defect-Dependent Processes (in two groups). Each session was introduced by a morning presentation of research papers and, after a free afternoon, ended with an evening agenda-discussion session covering the critical issues in each of the subtopics. Finally, a concluding agenda discussion covering issues in the overall field was led by Dr. A. S. Nowick. The reader is commended to these discussions for an up-to-date statement of the status of research in oxide transport.

Each Colloquium customarily honors one of the distinguished participants by inscribing the proceedings in his name. At Salt Fork, we deviated from this custom by honoring the late Professor John E. Dorn of the University of California at Berkeley. Although Professor Dorn had not worked on defects and transport in oxides, he was an important contributor to the field of defect transport in solids. He was a great friend of Battelle, and had participated in many of the early Colloquia. His untimely death in September of 1971, several months after contracting pneumonia while participating in the June 1971 Colloquium on "Interatomic Potentials and Simulation of Lattice Defects", prevented us from honoring him as a participant during his lifetime. The Colloquium on "Defects and Transport in Oxides" is the first opportunity Battelle has had to acknowledge its great admiration for Professor Dorn and his many contributions to the science of materials. We were happy that Mrs. Dorn was able to attend the concluding banquet on September 22 held in Professor Dorn's honor and memory.

We also wish to thank our Battelle colleagues who have supported the Colloquia. In particular, we wish to thank Dr. Sherwood L. Fawcett, President of Battelle Memorial Institute, and Dr. William H. Jones, Jr., Director of Research in the Physical Sciences of Battelle Institute, under whose auspices these Colloquia are a continuing activity. We are grateful to Dr. John M. Batch, Director of Battelle's Columbus Laboratories, for his participation in the opening and closing events of the Colloquium, and for the excellent support provided by members of the Battelle-Columbus staff: Duane Yothers, in charge of local arrangements; the secretariat headed by Darlene Weaver, including Rose Bowman and Dahlene Walnoha, who provided excellent secretarial assistance; Phillip La Mori, responsible for coordinating meals and wines, bringing *haut cuisine* to central Ohio; Philip Mc Rury, in charge of sound and lighting arrangements; and Lew De Walt, in charge of transportation and slide projection. Edna Jaffee was in charge of the Ladies Program and hospitality. To all these, and many others, we wish to extend our sincere thanks.

The Organizing Committee is grateful to J. S. Anderson, I. G. Austin, L. Eyring, P. Kofstad, F. A. Kröger, G. G. Libowitz, A.-H. Heuer, H. Schmalzried, and C. Wagner for their advice and the choice of subject and participants for this Colloquium.

This book dealing with the transport behavior of oxides is the first in the Battelle Colloquia series that has been concerned with oxides. We trust it will not be the last, and that it will be useful to old and new workers in the field. If new advances in oxides result from having held this Colloquium, our purposes will have been served well.

The Organizing Committee
R. I. Jaffee, Chairman
T. D. Chikalla
P. L. Farnsworth
R. A. Rapp
M. S. Seltzer
N. M. Tallan
B. A. Wilcox

CONTENTS

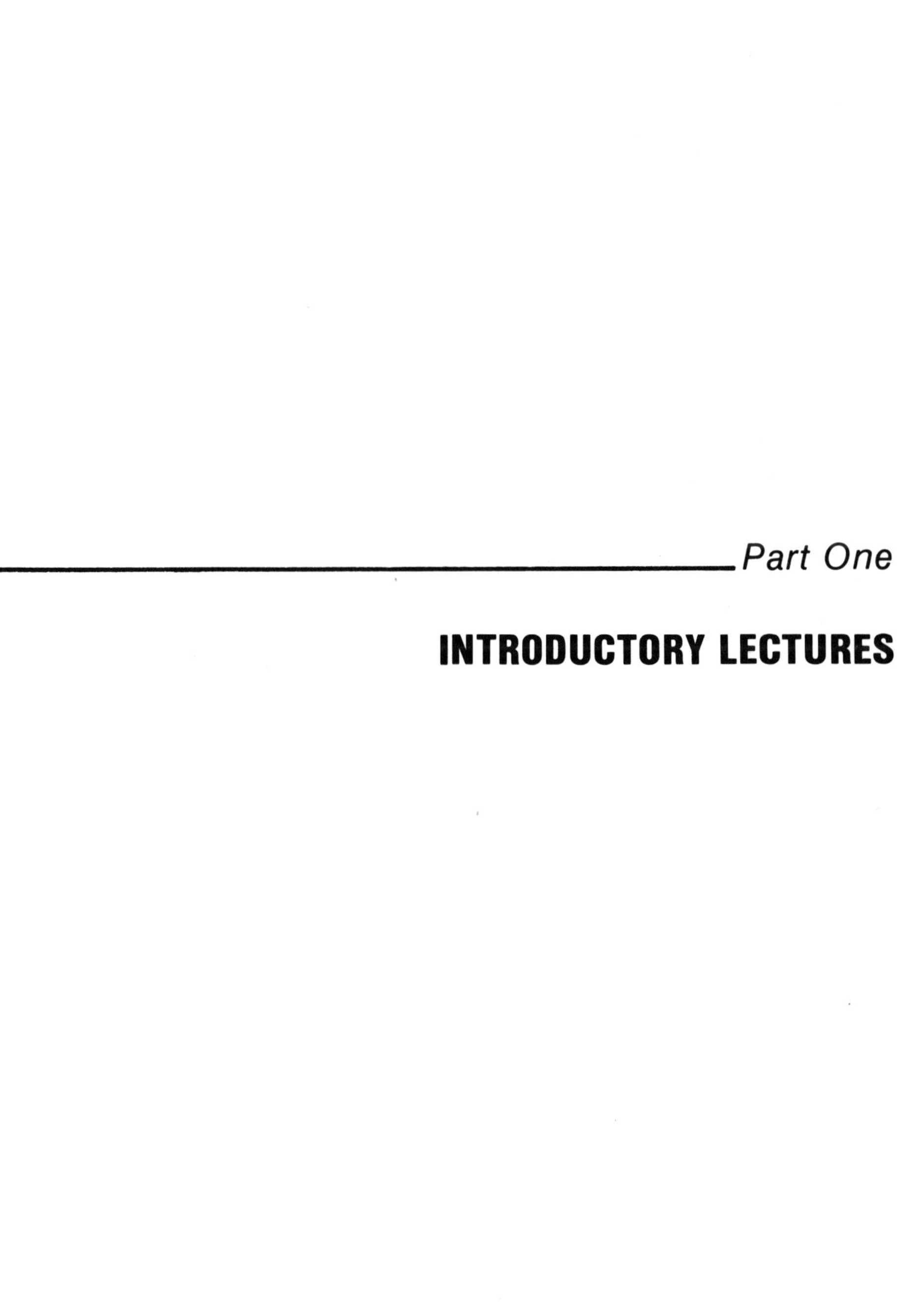

Part One

INTRODUCTORY LECTURES

DEFECT THERMODYNAMICS—HISTORICAL

F. A. Kröger

Department of Materials Science
University of Southern California
Los Angeles, California

ABSTRACT

Ionic conduction in crystals was initially attributed to dissociation. In 1926 Frenkel introduced the concept of atomic disorder. Schottky and Wagner (1931-1935) introduced further types of disorder and nonstoichiometry. Quasi-chemical-defect reactions and the corresponding mass-action relations were introduced in 1933. Medium- and short-range interaction between defects is accounted for by Debye-Hückel theory (Dünwald and Wagner, 1933) and pairing (Schottky, 1939). Regulation of defect concentrations by doping with aliovalent foreign atoms began with the work on $AgBr + CdBr_2$ by Koch and Wagner (1937), followed by semiconductor work after 1942. Application of the Schottky-Wagner approach to all possible defects, combined with a graphical method of representation using log-log or log-1/T plots and approximation of the neutrality condition (Brouwer, 1954), has become known as defect chemistry (Kröger and Vink, 1954, 1958). The method was extended to ternary compounds

(Schmalzried and Wagner, 1962). Near surfaces (Frenkel, 1946), grain boundaries, solid-liquid interfaces (Grimley and Mott, 1947), and dislocations (Eshelby et al., 1958), local neutrality is not maintained: we have planar or linear charges of one type balanced by a space charge in the crystal nearby. This makes it possible to determine the chemical potential of individual defects. Magnéli (1950) observed extended defects in βWO_3 giving rise to shear structures (Wadsley, 1955).

1 BULK PROPERTIES

1.1 Undoped Binary Compounds

Earliest indications of the presence of defects in crystals were obtained from ionic conduction of solid salts. Joffé (1923)[1] observed that the conductivity of complex salts such as sulfates, nitrates, and chlorates decreased with purification, approaching a limit which he explained by dissociation – comparable to the well-known dissociation of water. This dissociation was found to increase with increasing temperature. Dissociation was also introduced to explain the fact that solidus curves at the melting point of binary compounds usually have a horizontal tangent.[2]

Frenkel (1926)[3,4] gave the concept of dissociation a more precise interpretation, introducing the concept of *inner evaporation* which involves the displacement of an atom (ion) from a lattice site to an interstitial site, and thus forming a vacancy and an interstitial (called by him a dislocated atom). Evidently this is what we now call Frenkel disorder. He recognized the difference between the situation in which the interstitial occupies the site next to the vacancy and that in which it occupies a site far away. Only in the latter case does the disorder contribute to ionic conduction and self-diffusion – with both the vacancy and the interstitial contributing to the current. He showed that the product of the concentration vacancy to interstitials and vacancies must be a constant, dependent on temperature only. Additional disorder processes, again seen as analogues of the H^+, OH^- ion formation in H_2O, were introduced by Schottky and Wagner (1931)[5], Wagner (1931, 1933)[6,7] and Schottky (1935)[8], viz., disorder in which only vacancies are formed (now usually called Schottky disorder), in which only interstitials are formed (interstitial disorder), and in which only atoms occupying the wrong type of lattice sites occur (substitutional disorder, presently also called antistructure disorder). These investigators included Frenkel disorder of both components in their discussion. Two further possibilities for atomic disorder, viz., disorder involving one type of vacancy and one misplaced atom – a hybrid of the Schottky and antistructure disorder, and disorder involving an interstitial and a misplaced atom – a hybrid of interstitial and antistructure disorder, were proposed much later.[9,10] All possible atomic disorder processes involving only two types of defects are shown in Fig. 1. The discussion given by Schottky and Wagner[5], under the title "Theory of ordered mixed phases", is very general. It is based on statistical mechanics, assuming a random distribution of defects and native atoms over individual sublattices. It includes a discussion of equilibria between phases, nonstoichiometry, and the way in which component activities vary with the deviation from stoichiometry. Wagner's paper[7] concentrates on semiconductors and the way in which electronic conduction depends on stoichiometry and component activities. It introduces electrons as excess ("überschuss") defects and missing electrons as deficiency defects (holes), and two classes of semiconductors are introduced: electron-excess conductors such as ZnO, CdO, BaO, and electron-deficiency conductors such as CuI, Cu_2O, NiO. He also introduced the concept of electronic disorder through formation of free electrons and holes by intrinsic excitation. Following an earlier paper (1931)[11], the holes are seen as high-valency ions present in a row of low-valency ions, e.g., Ni^{3+} in a row of Ni^{2+}, Cu^{2+} in a row of Cu^{+}, etc., conduction being formulated by $Cu^{\cdot} + Cu^{\cdot\cdot} \rightarrow Cu^{\cdot\cdot} + Cu^{\cdot}$[12]. It was recognized that holes of this type are in essence similar to holes in band states[13-15] as introduced at about the same time (1931) by Heisenberg[14] for metals and by Wilson[16] for semiconductors; the latter also introduced formation of electrons and holes by excitation across a band gap and

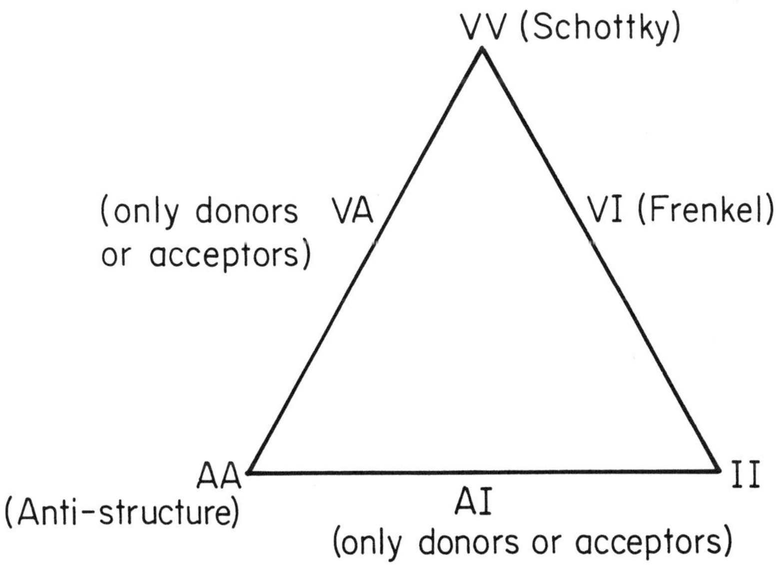

Fig. 1. Atomic disorder processes in binary crystals according to various investigators.

showed that the product of the concentrations of electrons and holes is proportional to exp (- E_{gap}/kT). Wagner's concept of electronic defects as ions of a changed valency was later extended to quasi free electrons, these being equivalent to ions of a low valency in a row of ions of a high valency (Verwey et al., 1950).[17,19] The representation of electrons or holes as ions of a changed valency is related to the concept of conduction by hopping electrons.[19-21]

The discussions mentioned so far were based on statistical mechanics: expressions for the Gibbs free energy of the crystal, G, are written in terms of enthalpies and vibrational entropies of lattice atoms and defects, plus configurational entropies due to random distribution of defects and atoms over particular sublattices, and equilibrium relations are derived by differentiating G for the number of particular defects. At the same time also the formation of defect formation by quasi-chemical reactions makes its appearance, first in words, by Wagner (1933)[7]:

$$\text{perfect crystal} \rightleftharpoons \text{quasi-free electron} + \text{electron hole} \qquad (1)$$

$$\text{perfect crystal} \rightleftharpoons \text{cation at interstitial site} + \text{cation vacancy.} \qquad (2)$$

Defect symbols introduced in the same paper to characterize the defects dominating the neutrality conditions were not yet used to formulate the actual formation of defects:

e^- (z) = quasi free electron
$Me^{\cdot}$ (z) = interstitial Me^+ ion
$\square$ (X') = X^- vacancy
$\square$ ($Me^{\cdot}$) = M^+ vacancy.

The first full-fledged quasi-chemical reactions, with application of the law of mass action to find quantitative relations between the defect concentrations and partial pressures of components in adjacent phases, appear in papers on ZnO and NiO by von Baumbach and Wagner in 1933[22],

$$O_2(\text{gas}) + 2\,Zn^{\cdot\cdot}\,(z) + 4e^-\,(z) \rightleftharpoons 2\,ZnO \quad , \qquad (3)$$

and in 1934[23],

$$O_2\,(\text{gas}) + 4\,e^-\,(g) \rightleftharpoons 2\,O^{2-}(g) + 4\square(e^-) + 2\square(Ni^{2+}) \quad . \qquad (4)$$

Using the ionic notation for electrons and holes, intrinsic electronic excitation is formulated (1933)[24]:

$$2\,Cu^{\cdot\cdot} \rightleftharpoons Cu^{\cdot} + Cu^{\cdot\cdot\cdot} \quad . \qquad (5)$$

It is no accident that these quasi-chemical formulations and the corresponding mass-action relations are introduced by the physical chemist

(Wagner); the physicists (Schottky[8], Frenkel[3]) prefer the statistical-mechanics approach. This dichotomy still exists today. The relations between defect concentrations arrived at by the two methods are the same if the statistical treatment is based on random distribution of defects and lattice constituents over individual sublattices. In chemistry, the use of the mass-action law in combination with chemical equations has a long history (Ostwald and Arrhenius, 1880-1890) and has led to a branch of chemistry known as analytical chemistry. As in electrolyte theory, in solids nonideal behavior is accounted for in two ways: by Debye-Hückel interaction[25-27] and by ion pair formation[27]. Schottky and Wagner[5], and Wagner in various papers, arrive at solutions for individual defects in terms of reaction constants partial pressures by solving the set of equilibrium relations obtained either from the statistical treatment or from mass-action relations combined with a complete neutrality condition. From the general results, solutions corresponding to specific simple cases in which only two species dominate the neutrality condition can be obtained. While the simpler solutions are often sufficient to explain experimental results, it was recognized that sometimes a change in dominant defects occurs. Thus, in CuI at high p_{I_2} there is a large excess of iodine, and the hole concentration is equal to the copper-vacancy concentration. At low p_{I_2}, however, the compound is almost stoichiometric and copper vacancies and interstitials formed by Frenkel disorder are the major species.[28]

In semiconductors, nonstoichiometry leads to changes in the electronic conduction by creation of free electrons and/or holes. In the alkali halides, temperature variation causes a variation of ionic conduction, which is explained by atomic disorder. Work carried out by Pohl and his co-workers in Göttingen, reviewed in 1937[29], shows that in the alkali halides nonstoichiometry gives rise to color centers, excess metal giving so-called F-centers and excess halogen V-centers, with the strength of the absorption bands proportional to the deviation from stoichiometry,[30,31] which in turn is a simple function of the pressure of metal or halogen with which the crystals are equilibrated during preparation (Mollwo[32], Rögener[33]). Owing to Pohl's hesitancy to accept anything that is not strictly proven, models are not proposed.

1.2 Doped Binary Compounds

So far, pure compounds have been considered. Though the existence of the effects of foreign atoms, present as impurities, on the physical properties were known[1], these were not explained. This changes dramatically with the work of Koch and Wagner (1937)[34] who extended the defect theory for AgCl and AgBr to include effects of aliovalent foreign atoms, and derived expressions for the concentrations of native ionic defects as a function of the foreign-atom concentration and showed how these must be expected to affect ionic conductivity for various ratios (ϕ) of the mobility of individual defects. Dependent on the mobility ratio, the ionic conduction must be expected to go through a minimum for addition of either a high-valency cation (Cd, Pb) or a high-valency anion (S) (Fig. 2a). As shown in Fig. 2b, experiments on the conductivity of AgCl doped with Cd or Pb and AgBr + Pb reveal the expected minimum, which proves that the mobility of interstitial Ag is larger than that of Ag vacancies. This paper constitutes a milestone in solid state chemistry.

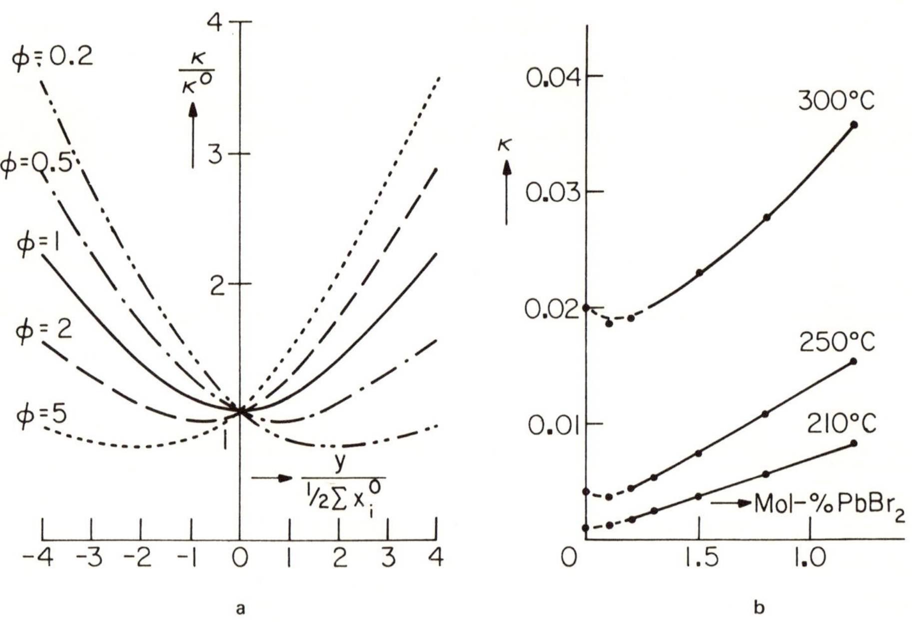

Fig. 2. (a) Predicted variations of the conductivity of a binary ionic conductor upon doping with aliovalent atoms with valencies larger ($y > 0$) or smaller ($y < 0$) than the atoms they replace for various values of the mobility ratio (ϕ) of individual defects; (b) Experimental results for the conductivity (η) as f[$PbBr_2$] for AgBr + $PbBr_2$ (Koch and Wagner[34]).

The dopant effect on the neutrality condition was introduced initially[34] through a parameter y, defined by

$$y = \frac{\text{number of cations - number of anions}}{\text{number of cation sites}} ,$$

with $y < 0$ for Cd, Pb, $y >$ for S. Schottky[27] realized that y can be identified with the site fraction of (nonassociated) foreign ions.

A first example of the effect of foreign solutes on the electronic conduction of crystalline solids is found in the work of Hägg (1935)[35] on WO_3 + Li, Na, K, and Ca. It was several years before the concept of doping to deliberately change the concentrations of electrons or holes in semiconductors became accepted. Doping of the elemental semiconductors Ge and Si is described by Seitz (1942)[36], of Ge by Lark-Horowitz et al. (1945)[37], of Si by Pearson and Shockley (1947)[38], of Pearson and Bardeen (1948 and 1949)[39] and Scaff et al. (1949)[40]. The elements As, P, and Sb substituting for Ge or Si give rise to excess electron (n-type) semiconductivity; Al or B give rise to p-type (hole) semiconductivity. Doping of compounds is described for ZnSb + Ag or Sn by Telkes (1947)[41], for NiO + Li by Verwey et al. (1948)[42], and for NiO + Li or Ag, and TiO_2 plus many dopants by Johnson and Weyl (1949)[43]. In all these, the concentrations of free electrons or holes are increased by doping with aliovalent atoms. Stabilization of elements in unusual valencies – corresponding to trapping of electrons or holes at the aliovalent dopant is described for Mn^{3+} in Al_2O_3, where $Mn^{3+} \equiv Mn^{2+}$ + trapped hole (Selwood, 1949).[44]

The dopants affecting electronic conduction are, in general, species having a normal valency differing from the species of the base crystal for which they substitute, higher in the case of a donor and lower in that of an acceptor. In Koch and Wagner's experiment on AgBr + $PbBr_2$ where aliovalent atoms also are involved, an effect on ionic conduction was observed.

What is it that determines whether an aliovalent dopant affects electronic rather than ionic defects – or vice versa? The clue to this lies in the fact that ionic defects are formed if the aliovalent atoms are incorporated together with their normal number of counter ions, e.g., Cd as $CdBr_2$. Formation of free electrons or holes, on the other hand, requires incorporation with an amount of counter ions corresponding to a compound fitting the base crystal: LiO in NiO, AlO in ZnO. This means that high-valency atoms have to be incorporated under reducing conditions to produce electrons, and low-valency atoms under oxidizing conditions to produce holes.[42,45]

This is demonstrated by CdS doped with Ga, which shows both mechanisms of incorporation[46]: when prepared under reducing condition (high p_{Cd}), a solid solution CdS-GaS is formed containing one free electron per gallium atom: $[Ga_{Cd}^{\cdot}] = [e']$. When prepared under oxidizing conditions (high p_{S_2}) we have CdS-GaS_2, with $[Ga_{Cd}^{\cdot}] = [V_{Cd}']$, or CdS-$Ga_2S_3$ with $[Ga_{Cd}^{\cdot}] = 2[V_{Cd}'']$. The latter is obviously equivalent to the Koch-Wagner mechanism; in semiconductor parlance it is called self-compensation[47,48].

Incorporation under the two conditions is formulated by

$$2\ Ga_{Cd}^{\cdot} + V_{Cd}'' + 3\ S_S^x \rightleftharpoons Ga_2S_3 \rightleftharpoons 2\ Ga_{Cd}^{\cdot} + 2\ e' + 3\ S_S^x - Cd(g) \quad . (6)$$

High Cd pressures tend to fill the cadmium vacancies and thereby shift the equilibrium to the right:[45,47,49]

$$Cd(g) + V_{Cd}'' \rightarrow Cd_{Cd}^x + 2\ e';\ K_{V,e} \quad . \qquad (7)$$

The transition occurs at the point where $[e'] \cong 2[V_{Cd}''] = \frac{1}{2}\ [Ga_{Cd}^{\cdot}]$,

$$(p_{Cd})_{tr} = \frac{[e']^2}{[V_{Cd}'']} K_{V,d}^{-1} = [Ga_{Cd}^{\cdot}]\ K_{V,e}^{-1} \quad .$$

Thus, a changeover from one mechanism to the other can occur either by a change of p_{Cd} at constant $[Ga_{Cd}^{\cdot}]$ and temperature, or by a change of $[Ga_{Cd}^{\cdot}]$ at constant p_{Cd} and temperature. For singly charged vacancies, the latter is not possible; $(p_{Cd})_{tr} = \frac{[e']}{[V_{Cd}']} K^{-1} = K^{-1}$, independent of doping.

The species dominating one range are present as a minority in the other range, with a gradual increase in their concentration when the transition point is approached. The concentrations of minority as well as majority species can be expressed in terms of equilibrium constants, component activities, and dopant activities or concentrations on the basis of the Schottky-Wagner defect theory – and all the concentrations can be represented in a single figure. The paper on CdS-Ga[46] mentioned above is the first one in which this was done. Log-log plots are used both to extend the concentration and pressure range and because large line sections in such plots turn out to be straight lines (Fig. 3). Brouwer[50] pointed out that the work involved in carrying out the calculations can be reduced considerably if the neutrality condition and/or the dopant balance is approximated by its dominant members. Then linear solutions are obtained which are the asymptotes of the accurate solutions. The combination of quasi-chemical reactions to describe defect reactions, mass-action law to find the corresponding relation between defect concentrations and component activities, Brouwer's approximation, and representation of results in log-log or log-1/T plots[49,51,52] has since found general acceptance and is known as *defect chemistry*[53].

Figure 4 shows such approximate isotherms for CdS doped with indium, according to a recent paper.[54] Such figures are useful because they relate the defect structure present under one condition to that present under another condition and make it possible to check for consistency; too often results under one condition have been discussed

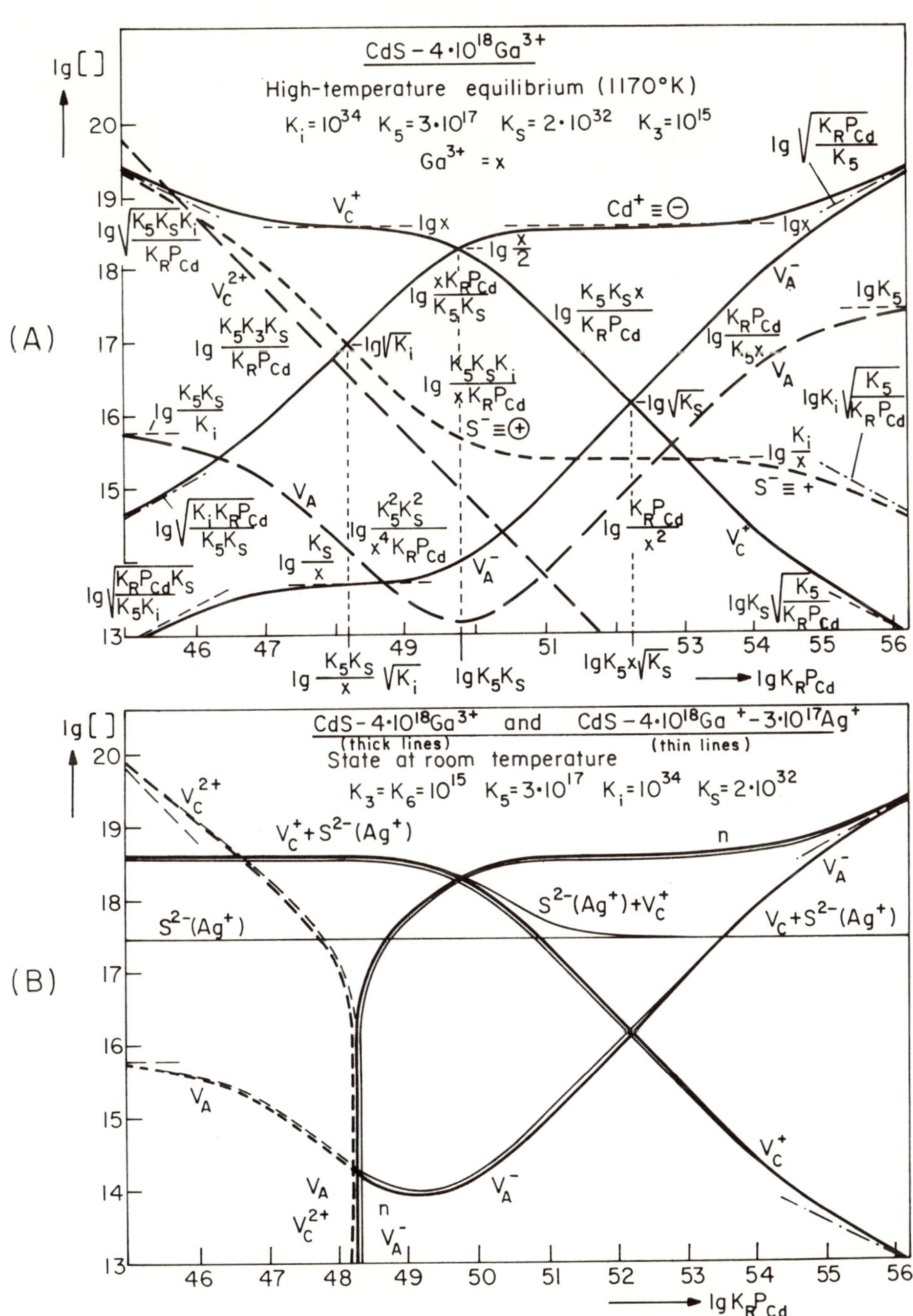

Fig. 3. Defect isotherms for CdS-Ga (Kröger et al.[46]).

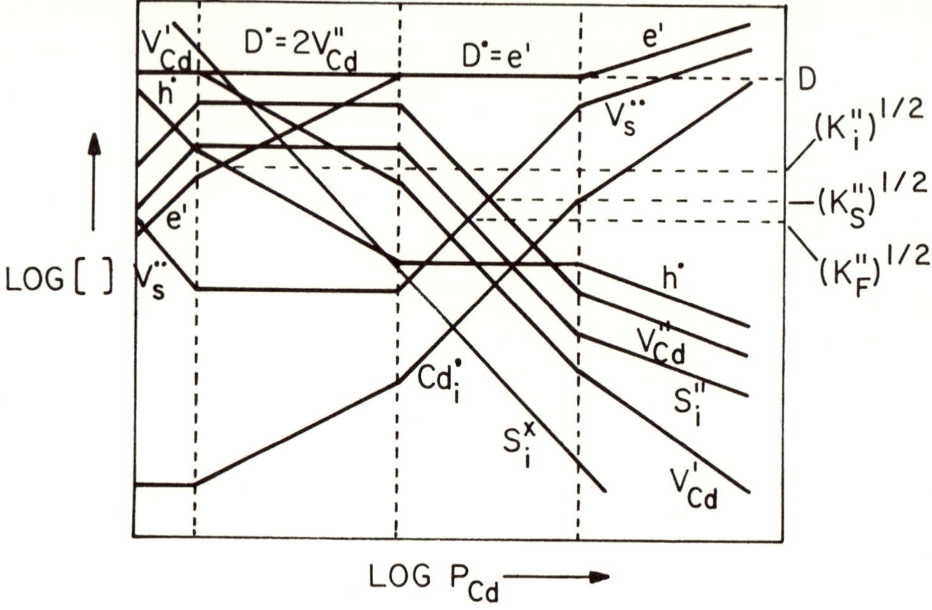

Fig. 4. Approximate isotherms for CdS-In at ≈ 800°C.[54]

without regard for results obtained under other conditions. It also helps to discuss within one framework all types of different experiments such as semiconductor properties, ionic conduction, diffusion, optical properties, and magnetic properties. Investigations along the lines followed for CdS-Ga interpreted by the methods of defect chemistry were subsequently carried out for PbS, pure and doped with Ag or Bi[55], and for CdTe, pure and doped with In, Ag, Cu, or Au[56]. Earlier work on PbS[57,58] and PbSe[59,60] had shown that these substances are amphoteric semiconductors, showing n-type electron conduction for an excess of lead and p-type (hole) conduction for an excess of chalcogen. Doping with high-valence foreign elements (Bi) now is seen to extend the n-type region at the cost of the p-type region, with the low valency dopants (Ag) having the opposite effect. The PbS system was the first for which the relation between nonstoichiometry in the solid and the P-T-x phase diagrams was indicated[61] (Fig. 5). Subsequently, the relation between the defect isotherms and the G(x) curve was also determined.[62] Figure 6 shows G(x) as calculated on the basis of results obtained from the study of PbS as a semiconductor.

Many of the studies involving nonstoichiometric crystalline solids were carried out on crystals cooled after equilibration at some high temperature. Interpretation of the results then involves the uncertainties of the effects occurring during cooling. It is preferable to measure the

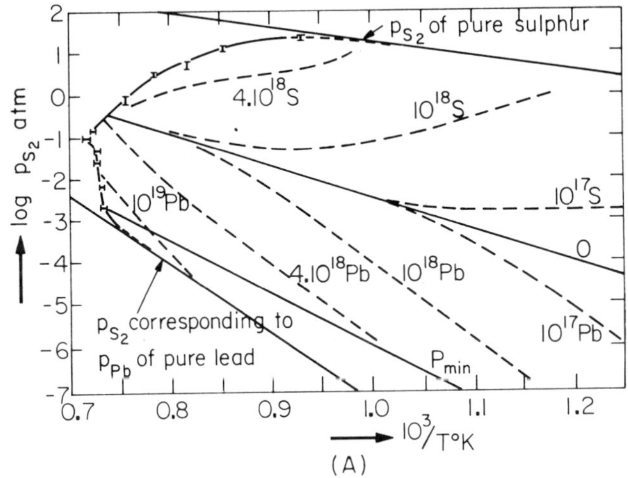

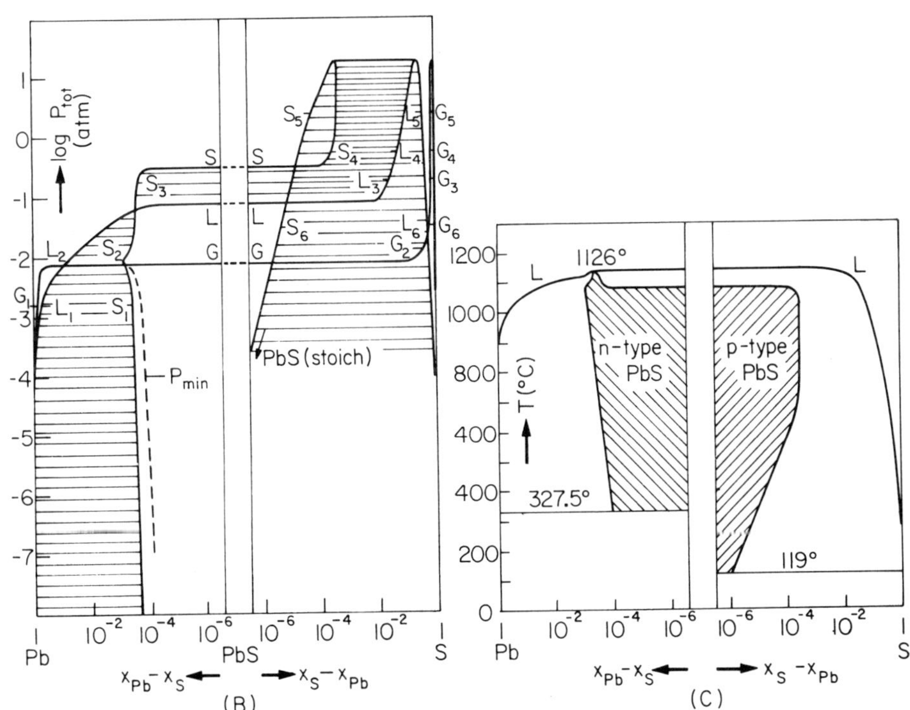

Fig. 5. P-T, P-x, and T-x projections of the three-phase line PbS(s) + liquid + vapor in the system Pb-S.[61]

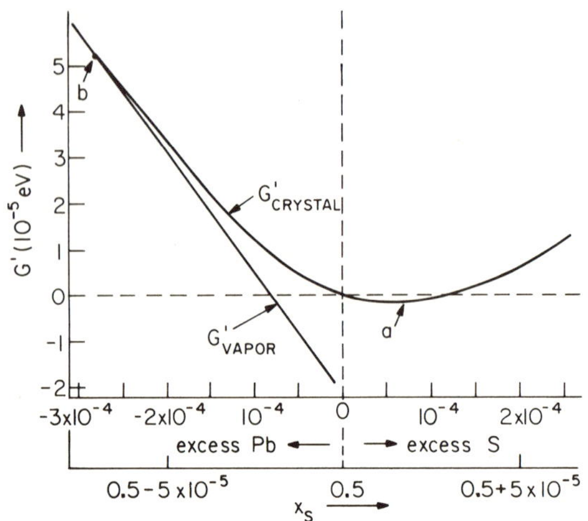

Fig. 6. G(x) for PbS(s) calculated from Hall data.[62]

properties of crystals at high temperature while they are at equilibrium with the outer phase. Such measurements were already done in 1933. They include measurement of conductivity (CdO, ZnO[22], NiO[23], Cu_2O[12], CoO, FeO, Fe_2O_3[63], CuI[28]), weight change (Cu_2O[64], CuI[65]), and Hall effect (CuI[65], ZnSe[66], CdSe[67], CdTe[68], CdS[69]). The rates of diffusion and of diffusion-limited processes such as oxidation may be determined either by weight change at temperature or by analysis after cooling. Anderson and Richards (1946)[70] studied Pb diffusion in PbS, using radioactive tracers. It was only much later, however, that the dependence of the self-diffusion on nonstoichiometry, predicted by Wagner (1950)[71], was studied systematically for PbS[72] and PbSe[73]. Other cases in which such a dependence has been studied are CoO[74], ZnO[75], CdO[76] and CdS[54].

Tarnishing processes and chemical diffusion in nonmetals generally involve migration of charged species. In order to prevent the setting up of a space charge, at least two species must be involved: counter diffusion of defects with charges of the same sign, or parallel diffusion of defects with charges of opposite sign (ambipolar diffusion). Wagner[77] has shown that for tarnishing, both atomic and electronic defects must be involved; he derived expressions for the tarnishing rate on this basis, and discussed the effects of nonstoichiometry and doping.

2 TERNARY COMPOUNDS

As we saw earlier[1], the earliest conduction measurements included experiments on ternary compounds. On the other hand, ternary compounds (spinels, perovskites) have become of considerable importance as magnetic materials, dielectrics, and resistors. In spite of this, a discussion of the defect structure in such materials was for a long time limited to the cation distribution in spinels[78], the effect of aliovalent dopants on the conduction of perovskites[78], and the electronic conductivity in the system Co_3O_4-Fe_3O_4 near the compound $CoFe_2O_4$ resulting from the presence of ions with 2+ and 3+ charges. This was remedied in 1962, when Schmalzried and Wagner[81-83] presented a basic discussion of the defect thermodynamics of such systems, using the quasi-chemical approach. Such systems, when compared with binary compounds, have an additional degree of freedom. Variation in the ratio of subcompounds (AX, B_2X_3 in AB_2X_4) gives rise to atomic defects; variation in the ratio of metal to nonmetal gives rise to electronic as well as atomic defects. Figure 7 shows defect isotherms for AB_2X_4 as a function of the B_2X_3 activity.[83] Driessens[84] has used the same approach in discussing the activity of component compounds AB, NM in the system (A, N) (B, M). The defect structure of the ternary compound NH_4Cl has been discussed taking into account ionic, electronic, and protonic defects.[85]

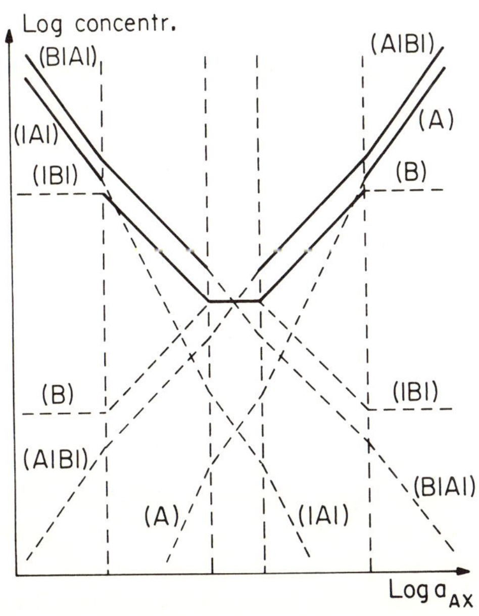

Fig. 7. Concentrations of point defects in AB_2X_4 as functions of the activity of AX according to Schmalzried.[83]

3 DEFECTS – SYMBOLS AND DEFINITIONS; CHEMICAL POTENTIALS

Defect chemistry – as normal chemistry – requires symbols for the indication of defects as well as normal crystal constituents. Several types of symbols have already been presented above. Various types found in the literature are shown in Table I. Apart from the actual symbols used, the symbols differ basically in two other ways:

1. In some, actual charges are given; others are characterized by effective charges, defined relative to the ideal unperturbed crystal.
2. Some are structure elements – items present at a certain site, defined with respect to empty space.[52] Others are building units – units that may be added to a crystal without upsetting the basic site ratio.[86]

There are two building-unit systems, as can best be demonstrated by giving the equivalent in structure elements[86]:

a. additive – subtractive (Schottky[87])

$$MX \equiv M_M^x + X_X^x$$

$$|M| \equiv V_M^x - M_M^x$$

b. additive – additive (Albers[88], Ong Jr.[89])

$$MX \equiv M_M^x + X_X^x$$

$$V_X^{\cdot\cdot}\,^{88} \text{ or } M^{2+}\,^{89} \equiv M_M^x + V_X^{\cdot\cdot} \,.$$

It is of course extremely confusing if a building-unit system uses the symbols generally accepted for structure elements, as is the case in ref. 88. The type of definition used can be seen from the way in which quasi-chemical equations are written. So Eqs. (4) and (6) are written in structure-element language; Eq. (4) in terms of the additive-subtractive building-unit system would read

$$O_2(\text{gas}) \rightleftharpoons 2\,\text{NiO} + 2\,|\text{Ni}|'' + 4\,|e|^{\cdot} \quad , \qquad (4')$$

and in the additive-additive building-unit system

$$O_2(\text{gas}) \rightleftharpoons 2V_{Ni}'' + 4\,h \qquad (4'')$$

It was originally believed that the use of structure elements is objectionable because they have no well-defined chemical potential[87]; this led to the introduction of *virtual* chemical potentials[86], proper combinations of structure elements having true chemical potentials. However, as we shall see, processes upsetting the site ratio of a crystal may occur near dislocations and at surfaces and interfaces. As a result, structure elements have

Table I. Symbols for Defects and Structure Elements in $M^{2+}X^{2-}$ According to Various Authors

	Wagner[7]	Baumbach and Wagner[23]	Schottky[8]	Rees[90]	Kröger et al.[46]	Besson[92]	Kröger and Vink[51]	Schottky and Stöckmann[91]	Lander[93]	Thomas[99]	Schottky[87]	Albers[88]	Ong Jr.[89]
Cation Interstitials	$M^{\cdot\cdot}(z)$			$\|M^{2+}\|\Delta$		$(M^{2+})^{2+}$	$M_i^{\cdot\cdot}$	$Mo^{\cdot\cdot}$	M_I^{++}	M_I^{++}	$M^{\cdot\cdot}$	$M_i^{\cdot\cdot}$	M_i^{++}
Anion Interstitials	$X''(z)$			$\|X^{2-}\|\Delta$		$(X^{2-})^{2-}$	X_i''	Xo''	$X_I^{=}$	$X_I^{=}$	X''	X_i''	X_i^{--}
Cation Vacancy	$\square(M^{\cdot\cdot})$	$\square(M^{2+})$	$\boxed{M^{2+}}$	$\square^{+}$	V_c	$<>^{2-}_{2+}$	V_M''	$M\square''$	$V_M^{=}$	$V_M^{=}$	$\|M\|''$	V_M''	$X^{=}$ or $(V_MX_X)^{=}$
Anion Vacancy	$\square(X'')$		$\boxed{X^{2-}}$	$\square^{-}$	V_A	$<>^{2+}_{2-}$	$V_X^{\cdot\cdot}$	$X\square^{\cdot\cdot}$	V_X^{++}	V_X^{++}	$\|X\|^{\cdot\cdot}$	$V_X^{\cdot\cdot}$	M^{++} or $(M_MV_X)^{++}$
Interstitial M Atom	M (z)	M(x)		$e_2\|M^{2+}\|\Delta$		$(M)^{o}$	M_i^{x}	Mo^{x}	M_I	M_I	M^{x}	$M_{i=}^{x}$	M_i
Interstitial X Atom	X (z)	X (z)		$p_2\|X^{2-}\|\Delta$		$(X)^{o}$	X_i^{x}	Xo^{x}	X_I	X_I	X^{x}	X_i^{x}	X_i
M Atom Vacancy	$\square(M)$			$p_2\square^{+}$	V_c^{++}	$<>^{o}_{2+}$	V_M^{x}	$M\square^{x}$	V_M	V_M	$\|M\|^{x}$	V_M^{x}	X or $(V_M^{2+}X_X^{2-})$
X Atom Vacancy	$\square(X)$			$e_2\square^{-}$	$V_A^{\cdot}$	$<>^{o}_{2-}$	V_X^{x}	$X\square^{x}$	V_X	V_X	$\|X\|^{x}$	V_M^{x}	M or $(M_M^{2+}V_X^{2-})$
Normal M^{2+} Ion		$M^{2+}(g)$	(M^{2+})			$<M^{2+}>^{o}_{2+}$	M_M^{x}		N.O.	M_M	–	–	
Normal X^{2-} Ion		$X^{2-}(g)$	(X^{2-})			$<X^{2-}>^{o}_{2-}$	X_X^{x}		N.O.	X_X	–	–	
Electrons in Valence Band		$e^{-}(g)$											
Electrons in Conduction Band	$e^{-}(z)$			e	M^{+}	$(e^{-})^{-}$	e'	⊖	e^{-}	e^{-}	e'	e	e
Hole in Valency Band		(e^{-})		p	X^{-}		$h^{\cdot}$	⊕	e^{+}	e^{+}	$\|e\|^{\cdot}$	h	M^{+}
Substitutional Foreign Ion													
F^{3+} at M^{2+} site				$F^{3+}\|\square^{+}$		$<F^{3+}>^{+}_{2+}$	$F_M^{\cdot}$	$F\bullet^{\cdot}(M)$	F_S^{+}	F_c^{+}	$F\|M\|^{\cdot}$	$F_M^{\cdot}$	
F^{+} at M^{2+} site				$F^{+}\|\square^{+}$		$<F^{+}>^{-}_{2+}$	F_M'	$F\bullet'(M)$	F_S^{-}	F_c^{-}	$F\|M\|'$	F_M'	
e^{-} trapped at Fe^{3+} at M^{2+} site				$e\|F^{3+}\|\square^{+}$			F_M^{x}	$F\bullet^{x}(M)$	F_S	F_c	$F\|M\|^{x}$	F_M^{x}	F_M
← Structure elements → (Wagner to Thomas)											← Building units → (Schottky to Ong Jr.)		

determinable chemical potentials: objections against their use are invalid, and the concept of virtual chemical potentials can be dropped.

All systems in which symbols with effective charge zero represent atomic defects (Kröger-Vink[51], Lander[93], Thomas[94], Schottky[87], Schottky and Stöckmann[91], and Albers[88]) have the advantage of making it possible to describe elemental crystals and compounds in one system – thus showing the basic similarity in many phenomena[52].

4 DEFECT EQUILIBRIA NEAR SURFACES, GRAIN BOUNDARIES, AND DISLOCATIONS

As indicated by Frenkel (1946)[4], Grimley and Mott[95], and Lehovec (1953)[96], the formation of different atomic defects in binary compounds will, in general, involve different amounts of energy. As a result, there will be a tendency to form more of the type whose formation costs the least energy. For charged defects in the bulk of the crystal, this tendency cannot be followed: electroneutrality has to be maintained; near the surface, however, a deviation from neutrality is allowed. At a gas-crystal interface, we get a planar surface charge consisting of the difference of the charges of positive and negative ions at the surface, compensated by an equal space charge in the subsurface region of opposite sign.[4,96] There is a potential difference between the surface and the bulk equal to $\Delta V = \Delta G/q$, ΔG being the difference in the free energy of formation of individual defects. The variation of the potential and the space charge with distance from the surface is arrived at by Debye-Hückel theory: the potential varies approximately exponentially, and the concentrations of charged defects depend exponentially on the potential, with defects of one charge increasing, and those of the other decreasing, toward the surface (Fig. 8).

For doped crystals, the dopant takes part in forming the space charge (Bassani and Thompson, 1956)[98]; see also ref. 97. Similar considerations apply to solid-liquid interfaces[95], grain boundaries, and dislocations[97]; in the latter case, there is a linear charge of one sign surrounded by a cylindrical space charge with an excess of defects having a charge of the opposite sign.

In semiconductors, the surface or dislocation charge is usually formed by electrons or holes trapped at surface states, the space charge being formed by electrons or holes and/or atomic defects[99]. The potential variation gives rise to band bending.

If complete defect equilibrium is not established, the space charge is made up of a fixed concentration, up to a certain depth, of ionized donors or acceptors (Schottky, 1938).[100] At solid-liquid interfaces we may have space charges of opposite signs in the two phases[95] or a planar-interface charge of one sign with space charges in the two phases of the same, opposite sign.[101]

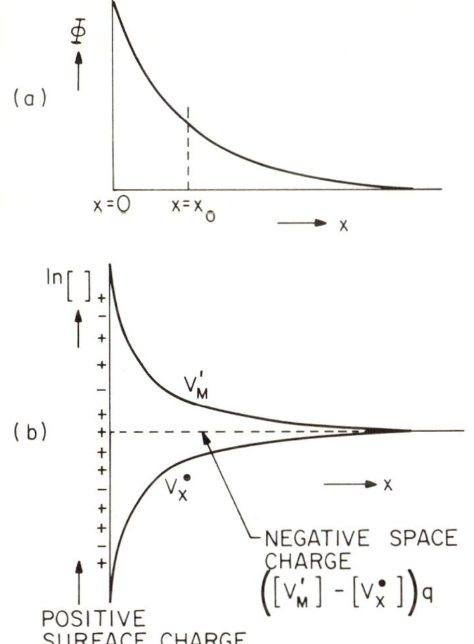

Fig. 8. Variation with distance from the surface (x) of the potential (Φ) and the concentrations of defects V_M' and $V_X^{\bullet}$ for a compound MX for the case that there is an excess of M^+ ions over X^- ions at the surface (Schematical).

For dislocations in ionic crystals, the sign and magnitude of the charge on the dislocation has been determined from the charge developed on the surfaces of a bent crystal. The dislocations temporarily escape from their defect clouds, setting up a potential corresponding to the dislocation charge at the side of the crystal to which they move (Caffyn and Goodfellow, 1962).[102] In NaCl, the positive dislocation charge was increased by doping with oxygen, and decreased or even inverted by doping with cadmium (Rueda and Dekeyser, 1963).[103] The concentration of Cd at which the dislocation changes sign (the so-called flat-band situation) has been used to quantitatively determine the free energy of formation of individual defects (Davidge, 1963).[104] Herewith the need for virtual chemical potentials disappears. The formation of space charges is not limited to interfaces between grains of the same substance (grain boundaries) but also occurs at boundaries between different solids. Potentials at interfaces Ag/AgCl and Ag/AgBr have been determined by polarization techniques (Sonoike et al., 1959).[105] The space charge around precipitates in semiconductors may markedly change their properties (Wagner, 1972).[106] In a complete analysis, coulombic as well as elastic effects have to be taken into account, with the elastic energy varying with the distance to the surface or the dislocation (Bacon, 1972).[107] In metals, the elastic effect is dominant, in nonmetals the coulombic term dominates.

5 EXTENDED DEFECTS

Up to 1950, deviations from stoichiometry were believed to be accommodated by the formation of point defects as outlined above. Since that time, however, various systems have been found in which larger deviations from stoichiometry are established in a different way. The first example was found in βWO_3 of the exact composition $W_{20}O_{58}$ (Magnéli, 1950).[108] Excess W over the stoichiometric composition was found to give rise to a structure in which slabs of the normal WO_3 structure twin with each other in such a way that what are lattice sites in one slab are interstitial sites in the next slab. The structure consists of a repetition of slabs of equal thickness, the thickness being determined by the deviation from stoichiometry; it has been called *shear structure* (Wadsley, 1955).[109] A series of structures with the general composition W_nO_{3n-1} and W_nO_{3n-2} have been found.[110] It has not been definitely established that each shear structure represents a new phase. The observation of nonuniformly spaced shear planes at intermediate compositions rather than a set of two uniformly spaced situations with different spacings seem to indicate the opposite.[111] However, a strong case has been made for the acceptance of shear structures of uniform spacing as stable, separate phases.[112]

REFERENCES

1. Joffé, A., *Ann. Physik*, **72**, 461 (1923).
2. Schottky, W., Ulich, H., and Wagner, C., *Thermodynamik*, J. Springer, Berlin (1949), pp. 400-409; Zernike, J., *Chemical Phase Theory* Uitg. Mij A.E.E. Kluwer, Deventer (1955), p. 229.
3. Frenkel, J. I., *Z. Physik* **35**, 652 (1926).
4. Frenkel, J. I., *Kinetic Theory of Liquids*, Oxford University Press, New York (1946).
5. Schottky, W., and Wagner, C., *Z. Phys. Chem.*, **B11**, 163 (1931).
6. Wagner, C., *Z. Phys. Chem. Bodenstein Festband* (1931), p. 177.
7. Wagner, C., *Z. Phys. Chem.*, **B22**, 181 (1933).
8. Schottky, W., *Z. Phys. Chem.*, **B29**, 335 (1935).
9. Néel, L., *Compt. Rend.*, **230**, 190 (1950).
10. Kröger, F. A., *J. Phys. Chem. Solids*, **23**, 1342 (1962).
11. Wagner, C., *Physik Z.*, **32** 641 (1931).
12. Dünwald, H., and Wagner, C., *Z. Phys. Chem.*, **B17**, 467 (1932).
13. Heisenberg, W., *Ann. Phys. Ser.*, **5**, **10**, 888 (1931).
14. Bloch, F., *Z. Phys.*, **52**, 555 (1928).
15. Peierls, R., *Ergebn. Exakt. Naturwissenschaften*, **11**, 264 (1932).

16. Wilson, A. H., *Proc. Roy. Soc. (London)*, **A133**, 485 (1931).
17. Verwey, E.J.W., Haaijman, P. W., Romejn, F. C. and van Oosterhout, G. W., *Philips Res. Rep.*, **5**, 173 (1950).
18. Kröger, F. A., Vink, H. J., and van den Boomgaard, J., *Z. Phys. Chem.*, **203**, 1 (1954).
19. de Boer, J. H., and Verwey, E.J.W., *Proc. Phys. Soc. (London)*, **49** 66 (1937).
20. Mott, N. F., *Proc. Phys. Soc. (London)*, **A62**, 416 (1949); *Advances in Metal Physics*, **3** 76 (1952).
21. Joffé, A. F., *Soviet Phys. Solid State*, **1**, 139 (1959).
22. von Baumbach, M. H., and Wagner, C., *Z. Phys. Chem.*, **B22**, 199 (1933).
23. von Baumbach, H. F., and Wagner, C., *Z. Phys. Chem.*, **B24**, 59 (1934).
24. von Baumbach, H. H., Dünwald, H., and Wagner, C., *Z. Phys. Chem.*, **B22**, 226 (1933).
25. Dünwald, H., and Wagner, C., *Z. Phys. Chem.*, **B22**, 222 (1933).
26. Wagner, C., and Hammen, H., *Z. Phys. Chem.*, **B40**, 197 (1938).
27. Schottky, W., *Z. Elektrochem.*, **45**, 33 (1939).
28. Nagel, K., and Wagner, C., *Z. Phys. Chem.*, **B25**, 71 (1934).
29. Pohl, R. W., *Proc. Phys. Soc. (London)*, **49**, extra part, p. 1 (1937).
30. Stasiw, O., *Gött. Nachr. Math. Physik Klasse* p. 261 (1932).
31. Kleinschrodt, F. G., *Ann Phys. Ser. 5*, **27**, 97 (1936).
32. Mollwo, E., *Ann. Phys. (Leipzig), Ser. 5*, **29**, 394 (1937).
33. Rögener, H., *Ann. Phys. (Leipzig), Ser. 5*, **29**, 386 (1937).
34. Koch, E., and Wagner, C., *Z. Phys. Chem.*, **B38**, 295 (1937).
35. Hägg, G., *Z. Phys. Chem.*, **B29**, 192 (1935).
36. Seitz, F., *The Electrical Conductivity of Si and Ge*, NDRC 14-110, University of Pennsylvania, November, 1942.
37. Lark-Horowitz, K., Final Report NDRC, 14-585, Purdue University, 1945; Lark-Horowitz, K., Middleton, A. E., Miller, E. R., Scanlon, W. W., and Walerstein, I., *Phys. Rev.*, **69**, 258 (1946).
38. Pearson, G. L., and Shockley, W., *Phys. Rev.*, **71**, 142A (1947).
39. Pearson, G. L., and Bardeen, J., *Phys. Rev.*, **73**, 1256A (1948); **75**, 865 (1949).
40. Scaff, J. H., Theuerer, H. C., and Schumacher, E. E., *J. Metals*, **185** 383 (1949).
41. Telkes, M., *J. Appl. Phys.*, **18**, 1116 (1947).
42. Verwey, E.J.W., Haaijman, P. W., and Romeyn, F. C., *Chem. Weekblad.*, **44**, 705 (1948).
43. Johnson, G. H., and Weyl, W. A., *J. Am. Ceram. Soc.*, **32**, 398 (1949).
44. Selwood, P. W., Moore, T. E., Ellis, M., and Wethington, W., *J. Am. Chem. Soc.*, **71**, 693 (1949).
45. Hauffe, K., *Reaktionen in und an Festen Stoffen*, (J. Springer, Berlin, Göttingen, Heidelberg (1955).
46. Kröger, F. A., Vink, H. J., and van den Boomgaard, J., *Z. Phys. Chem. (Leipzig)*, **203**, 1 (1954).
47. Prener, J. S., and Williams, F. E., *J. Phys. Chem. Solids*, **8**, 461 (1959).
48. Mandel, G., *Phys. Rev.*, **134**, A1073 (1964); Mandel, G., Morehead, F. F., and Wagner, P. R., *Phys. Rev.*, **136**, A826 (1964).
49. Kröger, F. A., and Vink, H. J., *J. Phys. Chem. Solids*, **5**, 208 (1958).
50. Brouwer, G., *Philips Res. Rep.*, **9**, 366 (1954).
51. Kröger, F. A., and Vink, H. J., *Solid State Physics* (F. Seitz and D. Turnbull, Eds.), Academic Press, New York), (1956), Vol. 3, p. 307.
52. Kröger, F. A., *Chemistry of Imperfect Crystals*, North-Holland Publishing Company, Amsterdam (1964).
53. van Gool, W., *Principles of Defect Chemistry of Crystalline Solids*, Academic Press, New York (1966).
54. Kumar, V., and Kröger, F. A., *J. Solid State Chemistry*, **3**, 382 (1971).
55. Bloem, J., Kröger, F. A., and Vink, H. J., *Report on Conference on Defects in Crystalline Solids*, The Physical Society, London (1955), p. 273; Bloem, J., *Philips Res. Rep.*, **11**, 273 (1956).
56. de Nobel, D., *Philips Res. Rep.*, **14**, 361, 430 (1959).
57. Hintenberger, H., *Z. Physik*, **119**, 1 (1942); *Z. Naturforsch*, **1**, 13 (1946).

58. Morton, M. C., *Trans. Faraday Soc.*, **43**, 194 (1946).
59. Bauer, K., *Ann. Physik (Leipzig) Ser. 5*, **38**, 84 (1940).
60. Schottky, W., *Schweiz. Arch.*, No. 1, p. 141 (1941).
61. Bloem, J., and Kröger, F. A., *Z. Phys. Chem (Frankfort)*, **7**, 1 (1956).
62. Kröger, F. A., *The Chemistry of Imperfect Crystals*, North-Holland Publishing Company, Amsterdam (1964), Section 15.2.
63. Wagner, C., and Koch, E., *Z. Phys. Chem.*, **B32**, 439 (1936).
64. Wagner, C., and Hammen, H., *Z. Phys. Chem.*, **B40**, 197 (1938).
65. Maurer, R. J., *J. Chem. Phys.*, **13**, 321 (1945).
66. Smith, F.T.J., *Solid State Comm.*, **7**, 1757 (1969).
67. Smith, F.T.J., *Solid State Comm.*, **8**, 263 (1970).
68. Smith, F.T.J., *Metallurg. Trans.*, **1**, 617 (1970).
69. Hershman, G. H., and Kröger, F. A., *J. Solid State Chem.*, **2**, 483 (1970).
70. Anderson, J. S., and Richards, R., *J. Chem. Soc. (London)* (1946), p. 537.
71. Wagner, C., *J. Chem. Phys.*, **18**, 62 (1950).
72. Simkovich, G., and Wagner, J. B., Jr., *J. Chem. Phys.*, **38**, 1368 (1963).
73. Seltzer, M., and Wagner, J. B., Jr., *J. Chem. Phys.*, **36**, 130 (1962).
74. Carter, R. E., and Richardson, F. D., *Trans. AIME*, **200**, 1244 (1954).
75. Birchenall, C. E., *Met. Revs.*, **3**, 325 (1958).
76. Haul, R., and Just, D., *J. Appl. Phys.*, **33S**, 487 (1962).
77. Wagner, C., *Z. Phys. Chem.*, **B21**, 25 (1932); **B32**, 447 (1936).
78. Neel, L., *Compt. Rend.*, **230**, 190 (1950); for a survey see ref. 52, Chapter 18.
79. Jonker, G. H., *Physica*, **20**, 1118 (1954).
80. Jonker, G. H., *J. Phys. Chem. Solids*, **9**, 165 (1959).
81. Schmalzried, H., and Wagner, C., *Z. Phys. Chem. (Frankfurt)*, **31**, 198 (1962).
82. Schmalzried, H., *Proc. 5th Symposium on Reactivity of Solids* (G. M. Schwab, Ed.), Elsevier Publishing Co., (1965), p. 204.
83. Schmalzried, H., *Progress in Solid State Chemistry* (H. Reiss Ed.), Pergamon Press, New York (1965), Vol. 2, p. 265.
84. Driessens, F.C.M., *Ber. Bunsenges. Phys. Chem.*, **72**, 754 (1968).
85. Kröger, F. A., *J. Chem. Phys.*, **51**, 4025 (1969).
86. Kröger, F. A., Stieltjes, F., and Vink, H. J., *Philips Res. Rep.*, **14**, 557 (1959).
87. Schottky, W., in *Halbleiter probleme* (W. Schottky, Ed.), Vieweg, Braunschweig, (1959), Vol. 4, p. 235.
88. Albers, W., in *Physical Chemistry of 2-6 Compounds* (M. Aven and J. S. Prener, Eds.), North-Holland Publ. Co. (1967), Chapter 4.
89. Ong, J. W., Jr., *J. Electrochem. Soc.*, **114**, 1080 (1967).
90. Rees, A.L.G., *Chemistry of the Defect Solid State*, Wiley, New York (1954).
91. Schottky, W., and Stockmann, F., *in Halbleiter probleme* (W. Schottky, Ed.) Vieweg, Braunschweig (1954), Vol. 1, pp. 80, 135.
92. Besson, J., *J. Chim. Phys.*, **62**, 937 (1965).
93. Lander, J. J., in *Semiconductors* (B. Hannay, Ed.) Reinhold Publ. Co., New York (1959), Chapter 2.
94. Thomas, D. G., in *Semiconductors* (B. Hannay, Ed.) Reinhold Publ. Co., New York (1959), Chapter 7.
95. Grimley, T. B., and Mott, N. F., *Discussions Faraday Soc.*, **1**, 3 (1947); Grimley, T. B., *Proc. Roy. Soc. (London)*, **A201**, 40 (1950).
96. Lehovec, K., *J. Chem. Phys.*, **21**, 1123 (1953).
97. Eshelby, J. D., Newey, C.W.A., Pratt, P. L., and Lidiard, A. B., *Phil. Mag.*, **3**, 77 (1958).
98. Bassani, F., and Thompsen, R., *Phys. Rev.*, **102**, 1264 (1956).
99. Engell, H. J., in *Halbleiter probleme* (W. Schottky, Ed.), Vieweg, Braunschweig (1954), Vol. 1, p. 249; Mott, N. F., *J. Chim. Phys.*, **44**, 172 (1947), *Research*, **2**, 162 (1949); Cabrera, N., and Mott, N. F., *Rep. Progr. Phys.*, **2**, 163 (1949).
100. Schottky, W., *Naturwissenschaften*, **26**, 843 (1938).
101. Honig, J. M., *Trans. Faraday Soc.*, **65**, 2248 (1969); *Colloid Interface Science*, **31**, 545 (1969).

102. Caffyn, J. E., and Goodfellow, T. L., *Proc. Phys. Soc. (London)*, **79**, 1285 (1962).
103. Rueda, F., and Dekeyser, W., *Acta Met.*, **11**, 35 (1963).
104. Davidge, R. W., *Phil. Mag.*, **8**, 1369 (1963); *Phys. Status Solidi*, **3**, 2745 (1963).
105. Sonoike, S., and Wakabayashi, H. W., *J. Phys. Soc. Japan*, **14**, 1664 (1959); Wakabayashi, H. W., *ibid*, **15**, 2000 (1960).
106. Wagner, C., *J. Phys. Chem. Solids*, **33**, 1051 (1972).
107. Bacon, D. J., *Phys. Status Solidi*, **50b**, 607 (1972).
108. Magnéli, A., *Arkiv. Kemi.*, **1**, 513 (1950); *Acta Cryst.*, **6**, 495 (1953).
109. Wadsley, A. D., *Rev. Pure Appl. Chem.*, **5**, 165 (1955).
110. Gado, P., and Magnéli, A., *Mater. Res. Bull.*, **1**, 33 (1966).
111. Gado, P., *Acta Cryst.*, **16**, A182 (1963).
112. Bursill, L. A., and Hyde, B. G., *Progress in Solid State Chemistry*, H. Reiss and Y. O. McCaldin, Eds.), Pergamon Press, Oxford, New York (1972), Vol. 7, p. 177.

DISCUSSION on Paper by F. A. Kröger

HEUER: With regard to Wagner's suggestion that a precipitate – matrix interface can be a place where a space charge can build up, doesn't this depend on the nature of the interface? Shouldn't coherent precipitates act differently from incoherent precipitates in this regard, in that they cannot act as a source or sink for defects?

KRÖGER: All interfaces between different materials should, in general, be expected to accumulate charge, with either equal space charges of opposite sign at the two sides or a planar charge of one sign compensated by space charges of the opposite sign in the adjacent subinterface regions.

VINK: Regarding precipitates and atomic defects, the question arises as to how many atomic defects have to segregate before one can talk about precipitates. If a precipitate is to be considered as a phase in the true sense of phase-diagram thermodynamics such that it has to be included in the phase rule, then one, two, three, or say 20 atomic defects sitting together cannot be considered to form a separate phase. In my opinion the question is one of numbers. One has to talk about a different phase if the number of "defects" forming the precipitate is sufficiently high that statistical thermodynamics can be applied to that entity. This transition from clusters of defects to distinct phases is an intriguing question. Possibly the same consideration applies in the case of "blocks". One can apply the usual defect theories to blocks only if they are large or if they are all considered together.

KRÖGER: This seems to be a reasonable position. Comparing a set of N clusters of size n (with $n << N$) with a set of n particles of size N in the same volume, one might consider the latter to form a phase if the contribution of the configurational entropy $kT\ln n$ due to the distribution of the n particles to the chemical potential of the atoms in the particles is negligible. This will be the case if the number of atoms per particle is larger than the number of particles: $N > n$. This will also make $kT\ln n < g_{surface}$, the surface energy per atom, which is the driving force for the formation of a smaller number n' of larger particles, N' (Ostwald ripening).

THE REAL STRUCTURE OF DEFECT SOLIDS

J. S. Anderson

Inorganic Chemistry Laboratory
University of Oxford
Oxford, England

Any theoretical analysis of thermodynamic equilibrium, transport and diffusion, electronic properties, and other properties of solids needs to be based on an understanding of the role played by defects from perfect crystalline order. When one turns from crystals that deviate from ideal composition to a degree that is barely detectable chemically to solid-solution systems, one finds a situation that differs in a vital respect from that discussed by point-defect theory, viz., in the concentration of lattice sites that appear to be "wrongly" occupied. The validity of basic point-defect theory is not open to question; it is firmly rooted in the statistical thermodynamics of crystals. The equilibria between defects, and between electrons and positive holes, are defined by the energetic cost of creating defects and by the band structure of the crystal. There is a wide range of experimental evidence, drawn from nuclear magnetic resonance, electron-spin resonance, and other techniques, that, at low concentrations, the defects have the site symmetry and properties associated with a point-defect description in terms of vacant sites or interstitials. The theory leads to recognition that a crystalline compound is not strictly a stoichiometrically determinate structure: its composition varies with the chemical potential of its components, through the consequential displacement of the defect equilibria. Calculations of the energy of defect creation lead, in general – and particularly for the class of substances with which this Colloquium is largely concerned – to the view that native defect concentrations are small. With that consideration in mind, it is important that for solids with a wide band gap (and hence $K_i < K_f$ or K_s in Kroger's notation), there is a wide range of chemical potential over which the maximum deviation from ideal composition is less than the already small concentration of native defects, $(K_f,K_s)^{1/2}$. Only for solids with a narrow band gap and relatively high native defect concentration ($K_i > K_f$ or K_s; K_f,K_s not too small) can the composition vary significantly across the whole range of chemical potentials.

The immediate concern, however, is with substances for which there is an apparent wide discrepancy between the occupancy of cation sites and anion sites, as compared with the assigned crystal structure, implying – if a point-defect description is applicable – that vacancies or interstitials are associated with at least 1 to 20 percent of the lattice sites. Two sorts of systems display these chemically significant deviations from ideal stoichiometry: (a) solid solutions involving cations of different charge from those of the host, together with compensating defects (e.g., stabilized zirconia solid solutions such as CaO/ZrO_2); and (b) nonstoichiometric compounds proper, such as $Fe_{1-x}O$, in which cations of the same element are present in two charge states. These two classes do not differ fundamentally, except that the location of anomalous cationic charges in the solid solutions is fixed by some random distribution of cations and can be changed only by atomic diffusion. Insofar as the compensating defects and the charge anomalies are associated by coulomb interactions, they are therefore anchored in a fortuitous configuration. In mixed-valence nonstoichiometric phases, inequalities of charge can be redistributed by electron hopping from cation to cation, and configurational rearrangements – e.g., ordering processes – are kinetically more facile.

The apparent point-defect concentrations permitted in these systems are so high – e.g., an apparent cation vacancy concentration of 15 percent in the nominally rock-salt structure of $Fe_{0.85}O$, the oxygen-rich limit of the ferrous oxide phase – that defect structure and defect distribution must evidently be dominated by interaction effects. These interactions are not necessarily describable in terms of coulomb forces, association and clustering, Debye-Hückel effects, etc., as in the dilute systems. To understand the structure, thermodynamics, and solid-state chemistry of highly nonstoichiometric crystals, information is needed at three levels of definition.

(1) The *average structure of the crystal,* as determined by standard crystallographic methods. If the structural singularities or defect centers – whether they are point defects or more complex – are randomly distributed, the average unit of translational symmetry remains that of the parent structure. It can then be determined whether the compensating defect centers arise – in Wadsley's nomenclature – from subtraction or interpolation effects; these terms are preferable to the terms "vacancy" or "interstitial" in that they do not carry the structural or thermodynamic connotation of the point-defect concept.

(2) The *average structure of the defect centers.* Diffraction methods, and especially neutron diffraction, have been used increasingly in recent years, and in two distinct ways, to afford direct evidence of defect structure. Sufficiently precise measurements of the intensity of Bragg reflections can yield occupancy fractions of lattice sites and of interstitial sites, thereby providing a direct measurement of the notional point defects for comparison with the expectation from a point-defect model or alternative defect-complex models. A second kind of information can be derived from the diffuse, incoherent scattering and, for this, neutron scattering is especially suited. Randomized point defects give rise to an isotropic background scattering. If, however, pairs or more complex groups of defects are correlated in position, but otherwise randomized, each correlated group acts in neutron scattering as if it were a randomly oriented molecule embedded in the matrix of the crystal. One or more broad peaks indicative of the mean correlation distance between defects are then superimposed on the background of the Bragg reflections. If these correlated groups themselves cluster, in incipient long-range ordering, the background is increasingly modulated around the positions of the superstructure peaks which ultimately appear, as sharp Bragg reflections, when long-range ordering approaches perfection over domains of significant extent. These methods, based on those developed in X-ray diffraction for the study of ordering processes in alloy systems, have been applied to nonstoichiometric compounds and solid solutions by several workers, and particularly by Fender and his co-workers at Oxford. Some results of this direct investigation of defect structure will be considered below.

(3) The third level of investigation is that of *local structure in defective crystals.* If, as proves frequently to be the case, the microstructure of a nonstoichiometric crystal corresponds neither to a uniform, random distribution of defects nor to the development of extended regions of long-range order, the averaged defect structure gives an incomplete picture and may mask significant local fluctuations in structure and composition. These can be detected and identified only by methods that reveal the local structure and degree of order within small regions of crystal. No general solution to this problem is within sight, but for one

particular group of compounds – that for which fluctuations of order and structure can be treated as one-dimensional or two-dimensional problems – very direct evidence has come from the application of high-resolution electron microscopy. With the instrumental possibility, now available, of routinely operating at a resolving power of about 3 Å – i.e., a resolution comparable with the spacing of heavy scattering cations in oxide structures – it is now possible to observe and identify structure at the unit cell level. This topic will no doubt be discussed more fully in Professor Cowley's paper; the technique, largely pioneered by Allpress in Melbourne, has been actively exploited also by Hyde and Bursill in Australia, by Cowley, and by my own co-workers, and Tilley in Britain.

It is now clear that, in nearly all the cases hitherto investigated, the real defect structure is not adequately described by the point-defect model. Two distinct effects are discernible. The first is a tendency for *ordering* of defects, whether these be untenanted lattice sites, interpolated atoms, or more complex structural units. Carried to completion, such ordering generates a superlattice for particular compositions of the material and – particularly in the nonstoichiometric binary oxides and similar compounds, where diffusion of defect centers is directly coupled with migration of the compensating cationic charge – the low-temperature equilibrium state corresponds, in many systems, to the formation of a succession of phases of definite composition, all structurally derived from the parent compound. Vacant lattice sites, etc., are no longer defects, but are inherent elements of the superstructural pattern; true point-defect disorder is superimposed on this situation. At high temperatures there may be an order $\rightleftharpoons$ disorder transition (see the praseodymium oxide system, Fig. 1, with defined intermediate phases at low temperatures, and the disordered oxygen-deficient fluorite α-phase, of variable composition, at high temperatures); loss of long-range order relaxes the constraint of constant composition imposed by translational symmetry, and the high-temperature state in such systems may have a broad range of existence. In solid solutions, a random distribution of two species of relatively immobile cations may completely inhibit long-range ordering, and result in broad stoichiometric variability under all conditions. There is some direct, and a good deal of indirect, evidence that short-range order is not completely lost, and it has been maintained (e.g., by Wadsley, Ariya, and others[1]) that the type of extended ordering process observed in intermediate or superstructure phases persists within clusters or microdomains in the derived nonstoichiometric phase.

The second effect found is the *modification or reorganization of the host crystal structure* around the notional point-defect vacancy or interstitial, to form a new structural entity which may itself be arranged either in ordered or disordered fashion. This is more fully exemplified below; in effect, one or more point defects are replaced by a larger defect complex, and the displacements of atoms are not those that would arise from simple relaxation under the influence of central force fields. Point defects in the strict sense become minority species only.

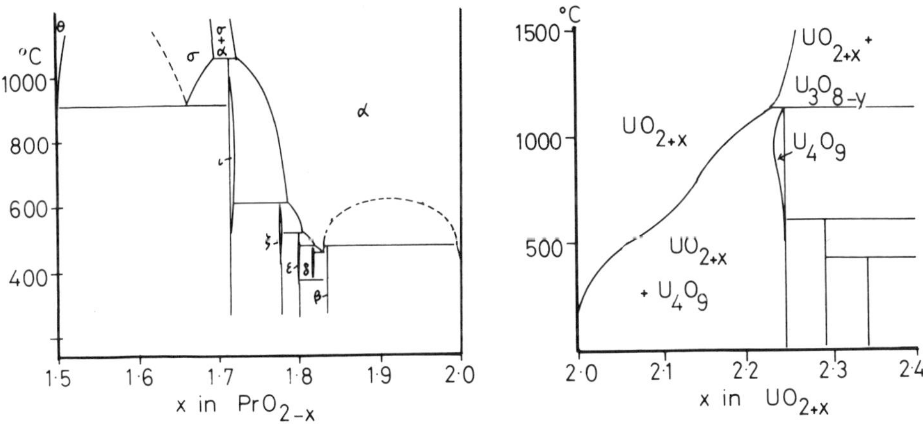

Fig. 1. Equilibria in nonstoichiometric fluorite systems: line phases of definite composition at low temperatures, and phases with broad composition range at high temperatures.

A distinction can be drawn between crystals that must be interpreted in terms of metallic bonding, with itinerant electrons and a Fermi surface, and those wide-band-gap insulators or semiconductors in which a non-integral total electron/atom ratio is accommodated by localized states – ions of discrete, different charge. In the former, there are no localized charges introduced by stoichiometric deviations; Coulomb interactions between defects are screened. The effect of changes in composition is to raise or lower the Fermi level, and the consequences of such changes must depend upon the density of states. Denker[2] has argued that in the transition metal carbides, nitrides, and oxides with the NaCl structure, the density of states in the upper, antibonding part of the d-band is such as to favor energetically the formation of defects that lower the ratio of valence electrons to lattice sites, by creation of vacant sites. The compounds referred to all have an extensive composition range, on the nonmetal-deficient side of the ideal formula; the nonstoichiometric, high-temperature modifications of TiO and VO not only span a wide stoichiometric range on both sides, but the stoichiometric material is highly defective, with vacant sites on both sublattices (Fig. 2). At first sight, these materials appear to conform to a classical vacancy description, as having unbalanced Schottky type disorder, but the total defect concentration is inexplicably high and there is evidence that the vacancies are distributed neither randomly nor so as to optimize vacancy-vacancy repulsions. An explicit structural study has been made only for the analogous carbide system NbC_{1-x}, by Fender and Henfrey, using diffuse neutron scattering. It showed that carbon vacancies have a strong third-neighbor site preference (Table I), bringing them into the same disposition as is required for the Nb_6C_5 superstructure that is fully developed at temperatures below 1025°C.[3] Direct evidence for TiO and VO is lacking, but the immiscibility of these "defect NaCl" structures and their quite different variation of cell dimensions with composition (Fig. 2) suggest that they differ structurally in significant degree. Below 900°C, stoichiometric TiO forms an ordered vacancy structure $Ti_{4/5}O_{4/5}$; from oxygen-rich TiO, the ordered phase $Ti_{4/5}O$ (Fig. 3) is formed. In both of these, all occupied lattice sites can be referred to sites of the NaCl parent structure and they can intergrow coherently with a common, defect-free interface.[4] This aspect of alternative superstructures on a common theme is taken up again later; it at once suggests a model for a nonstoichiometric phase with small, uncorrelated domains approximating to two distinct ordering patterns and local compositions. It should be noted that metallic properties and a Fermi surface do not suffice to cause stoichiometric variability. NbO, which could also be regarded as an ordered, defect rocksalt phase, has only a narrow range of composition: $0.97 \leqslant x \leqslant 1.03$ in NbO_x.

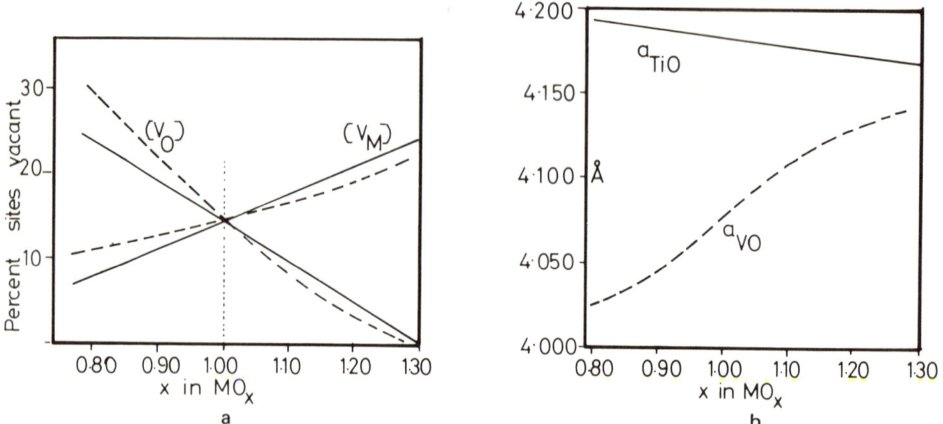

Fig. 2. TiO_x and VO_x $(0.8 \leqslant x \leqslant 1.3)$: (a) concentrations of metal vacancies and oxygen vacancies; (b) variation of cell dimensions. Solid lines TiO_x; dashed lines VO_x.

In more nearly ionic crystals, defects of site occupancy are, in general, compensated by excess ionic charges, localized on cations in a different valence state. The exceptions to this generalization are the substances (such as ReO_3 and the related nonstoichiometric tungsten bronzes M_xWO_3) in which electrons are present in a narrow, partially filled, nonbonding band, rather than in localized states. In the oxides, to a

Table I. Distribution of Vacancies in Nonstoichiometric Rock-Salt Structure NbC_{1-x} (Percentage of vacant sites in i^{th} neighbor anion position.)

Quench Temperature:	1850°C				1200°C				
Neighbor Anion Site:	1st	2nd	3rd	4th	1st	2nd	3rd	4th	Random Distribution
$NbC_{0.90}$	2	5	15	10	2	5	15	10	10
$NbC_{0.86}$	4	8	21	14	4	8	21	14	14
$NbC_{0.79}$	10	15	28	21	5	18	40	0	21
$NbC_{0.73}$	19	24	32	27	16	20	42	12	27

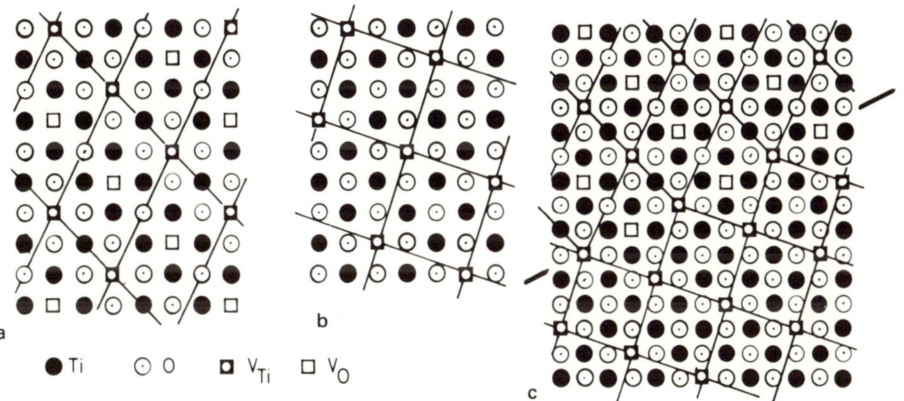

Fig. 3. Structures of (a) ordered $Ti_{4/5}O_{4/5}$, (b) ordered $Ti_{4/5}O$, (c) coherent intergrowth along (120) of rock-salt structure subcell. Note elements of overlapping structure at the interface.

far greater extent than in the isostructural fluorides, account must be taken of the d-orbital overlap between cations and anions which manifests itself in magnetic exchange interactions, and this covalent contribution to the bonding is particularly important in compounds based on cations of high formal charge (i.e., the oxides of the elements of Groups IV through VII) in the 3d and 4d transition series. Coulomb interactions between the excess charge of an altervalent ion and the charge associated with a vacant site or interstitial would lead naturally to association, as has been confirmed experimentally for defects at low concentration, in doped crystals.

By no means all structural types give rise to nonstoichiometric phases and solid solutions with broad composition ranges, and the average defect structures formed have been investigated, as yet, for rather few examples.

However, at the defect concentrations that are of interest in the present context, every case that has hitherto been studied has revealed that point defects are replaced by larger complexes, with which the compensating altervalent cations are associated. The host crystal structure around a notional point defect undergoes changes in the occupancy of its lattice sites, so that the original vacancy or interstitial is subsumed within an element of rebuilt crystal structure. Although the evidence is incomplete, it appears that the type of defect complex formed often depends upon the host crystal structure rather than on the particular chemical system concerned – e.g., a common structural principle runs through the anion-excess fluorite structure, both oxides and fluorides. It is not possible, at present, to predict the defect structure that will result from interactions and relaxations in a particular case; development of statistical thermodynamic or quasi-chemical treatments of equilibria must be based upon evidence of real structure, not upon generalized models. Some general principles can now be recognized, however, and a few experimental facts are next briefly reviewed.

Nonstoichiometric oxide phases with discrete defects are formed chiefly by two structural types – the rock-salt structure and the fluorite structure. In the former, the progressive change in properties along the 3d transition series, from the metallic TiO and VO already discussed, to the essentially ionic oxides, MnO and NiO, shows itself also in the extent and the structure of the nonstoichiometric phases. Common to these monoxides is variable composition on the oxygen-rich side of the ideal composition – significant for MnO at high temperatures, very wide for FeO which does not exist at all as a stoichiometric compound, and narrowing to small composition ranges in CoO and NiO. In solid solutions and doped oxides, there is one crystal structure – that of $MnMg_6O_8$ – in which vacant cation sites and altervalent ions are ordered in a superstructure, and one might infer that the dominant effect would probably be short-range ordering rather than local reconstruction. However, in the nonstoichiometric $Fe_{1-x}O$ and, almost certainly, in $Mn_{1-x}O$, the interaction between defects and host structure is more far reaching.

Ferrous oxide thermodynamically stable only above 570°C, has long been recognized as a cation-deficient compound, existing over the composition range $Fe_{0.95}O$ to about $Fe_{0.85}O$. If, as is usual, it is regarded as a defective rock-salt structure, with the deviation from stoichiometry produced by the simple point-defect reaction

$$2Fe^{2+} + 1/2O_2 \rightleftharpoons O^{2-} + V_{Fe} + 2Fe^{3+} \quad ,$$

all cations would reside on octahedral sites and the vacancy concentration would be equated directly to the stoichiometric defect (i.e., 5 to 15 percent cation vacancies). Roth's finding, from magnetic neutron scattering, that a substantial fraction of Fe^{3+} cations occupied tetrahedral sites implied at least the further structural rearrangement

$$Fe^{3+}{}_{Oct} \rightleftharpoons Fe^{3+}{}_{Tet} + V_{Fe}$$

and the formation of small defect complexes $(Fe_{Tet})(V_{Fe})_2$, leading to a maximum value of 2 for the ratio (octahedral cation vacancies)/(tetrahedral cations). From their X-ray crystallographic work on quenched, partially superstructure-ordered $Fe_{0.9}O$, however, Koch and Cohen[5] inferred that the defect complex was larger than this, with four tetrahedral iron atoms surrounded by 13 vacant sites (Fig. 4). This inference was definitely supported by the diffuse-neutron-scattering work of Fender and Childs and agrees well with conclusions drawn from Mossbauer work carried out by Greenwood and Howe[6], at high temperatures and under equilibrium conditions. It can be taken as well substantiated, and it is important that in Bragg neutron-diffraction studies by Fender and Cheetham[7], again carried out at high temperatures under chemical equilibrium conditions, the ratio (octahedral vacancies)/(tetrahedral cations) had, within error limits, the required value 3.25 over the composition range $Fe_{0.947}O$ to $Fe_{0.906}O$ at 800°C, and over the temperature range 800°C to 1200°C for one composition ($Fe_{0.906}O$) (Fig. 5).

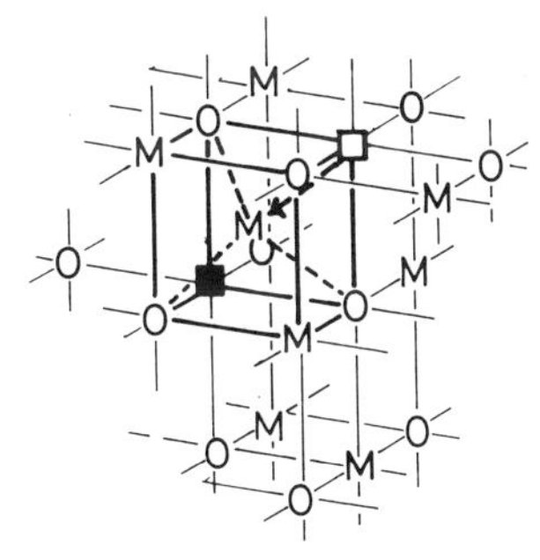

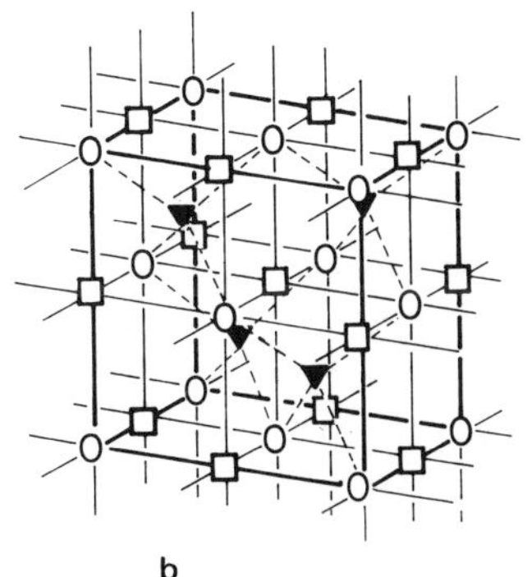

Fig. 4. $Fe_{1-x}O$. (a) Formation of tetrahedrally coordinated (interstitial) Fe^{3+}, the Roth model; (b) Koch-Cohen cluster. Squares indicate empty cation sites.

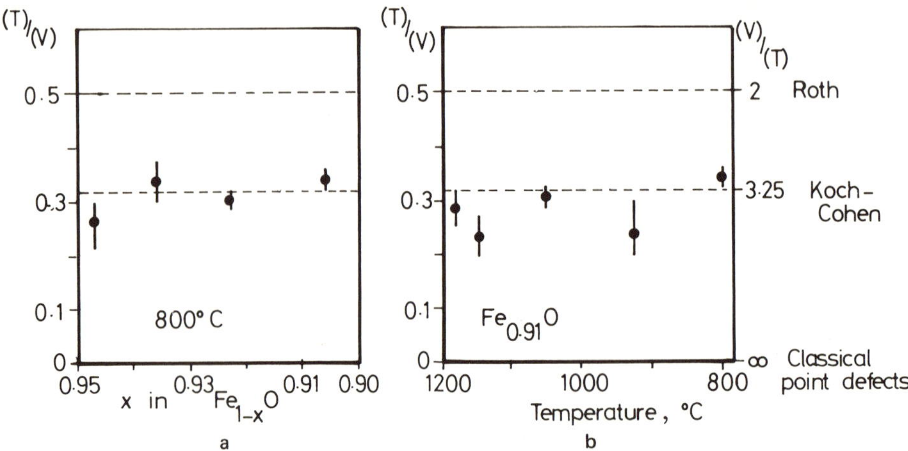

Fig. 5. $Fe_{1-x}O$. *Ratio of concentrations, (cations in tetrahedral sites):(vacant octahedral sites), (a) as f(x), at 800°C; (b) as f(T), for* $Fe_{0.91}O$.

What this means is that the defect center is much more complex than a simple cation vacancy, and is of a structure that cannot be formed by simple aggregation of vacancies. It arises from both an aggregation process and a local alteration of the rock-salt matrix, and represents, indeed, the complete transformation of the contents of one unit cell of NaCl structure into a new structure – virtually an element of zinc blende structure. Only the cation sites are thereby affected. The oxide ions in this unit appear to be but little shifted from sites proper to the original NaCl type, and the unit as a whole is dimensionally compatible with the matrix. Since it contains only 4/9 of the Fe^{3+} ions, and bears a net anionic charge, there must be a surrounding and compensating "atmosphere" of Fe^{3+} ions, preferentially occupying octahedral sites that can be assigned to the rock-salt matrix structure. The same complex is found irrespective of the oxygen excess, so that a change of composition implied only a change in the mean spacing between defect complexes within the host structure. Development of a superstructure, in the metastability region, indicates regular ordering of such complexes, and could, in principle, generate a family of discrete phases. The empty octahedral sites in the complex are not vacancies; those properties of the oxide that depend upon point-defect vacancies (e.g., diffusion) involve a dissociation of the complex, as has been discussed by Greenwood and Howe[8]. True cation vacancies are present only at low concentration.

In the fluorite structure, it is recognized that the native disorder arises from anionic vacancies and interstitial anions; the extensive series of nonstoichiometric oxides and solid solutions derived from it are properly described as either anion deficient, MO_{2-x}, or anion excess, MO_{2+x}. Each of these gives rise to its own type of defect complex, which does not depend upon the particular chemical system involved.

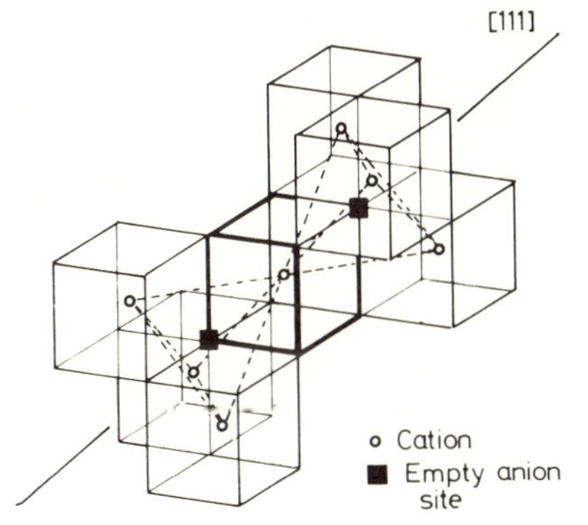

Fig. 6. Associated pairs of vacant anion sites in substoichiometric fluorite structures. Central cation 6-coordinate, surrounded by six 7-coordinate cations. All atoms at sites proper to the fluorite structure.

In the anion-deficient oxides, the complex can be regarded as arising from the tight association of pairs of vacant oxygen sites on diagonally opposite corners of the cubic coordination group around the cations of lower charge (Fig. 6).[9] Linear ordering of such clusters along a [111] direction of the basic fluorite structure is found in the ordered M_7O_{12} phases, and is probable in the other intermediate phases, as it is also in the related intermediate phases in the Y_2O_3 – YOF system. All the sites in the cluster are sites of a fluorite lattice; the cation positions are unchanged and shifts in anion positions are minimal. The cluster could therefore be "plugged in" without distortions, in four orientations, into the fluorite MO_2 structure, and it is a reasonable inference that the broad-range α phase, MO_{2-x}, is based on a random distribution of such cluster units, rather than on randomized, true vacancies. Confirmation of this view comes from neutron-scattering experiments by Fender and Greaves on Y_2O_3-stabilized ZrO_2: the diffuse scattering is consistent with the pairing of vacant anion sites as in the proposed cluster, and modulation of the background to the Bragg scattering affords direct evidence of short-range ordering of the clusters. That the α phase is not merely a random vacancy structure is supported by further evidence. There is a second "defect fluorite" phase, the σ phase MO_x with a broad composition range, $1.5 \leqslant x \leqslant 1.67$ (Fig. 1). The miscibility gap between the σ and α phases persists up to the highest temperatures (~2000°C). This σ phase is formed by inserting oxygen atoms into the cubic (Type C) L_2O_3 structure (L = Ce, Pr, Tb, etc.), which is itself derived from the fluorite structure by the ordered omission of one quarter of the anions; the empty fluorite sites are arranged in nonintersecting strings along the four <111> directions. Hence there is a structural difference, sufficient to cause immiscibility, between $LO_{1.71}$, with 14.5 percent of the anions removed from the fluorite structure, and $LO_{1.67}$, with 16.7 percent of potential fluorite sites in the related Type C structure remaining unoccupied. The cumulative evidence is strong that the defect fluorite structure is not a simple vacancy structure.

In anion-excess fluorites, there is incontrovertible evidence that excess anions at low concentrations – e.g., in rare-earth-doped CaF_2 – occupy the high-symmetry interstitial positions 1/2 1/2 1/2, etc. At higher concentrations, however, this is not the case. As was first established, by Willis, from neutron-diffraction work on U_4O_9 (with one additional anion per unit cell) and for $UO_{2.12}$, and as has recently been found by Fender, Cheetham, et al., for the analogous fluoride solid solutions[10] ($M^{3+}{}_x$, $M^{2+}{}_{1-x})F_{2+x}\cdot(M^{2+}$ = Ca, Sr, Ba; M^{3+} = a rare earth), each interstitial anion is replaced by a group involving substantial modification within one unit cell. The high-symmetry interstitial site is vacant; anions are displaced from the sites proper to the fluorite structure, creating anion vacancies and two kinds of interstitial anions (Fig. 7) which are considerably displaced from the high-symmetry positions. The basic unit has been described as a 2:2:2 cluster (i.e., 2 vacancies and two interstitials in each of two types of sites), but the diffuse-neutron-scattering results indicate that the clusters may enlarge, extending through more than one fluorite unit cell in such a way as to give variable ratios of vacant anion sites: type I interstitials:type II interstitials. In the fluorite solid solutions, these complexes remain disordered, being anchored by the randomly distributed, immobile compensating cations. They are randomized in the UO_{2+x} phase, and stable at high temperatures, but order completely at lower temperatures to form the superstructure of U_4O_9, which nucleates and grows as a separate phase.

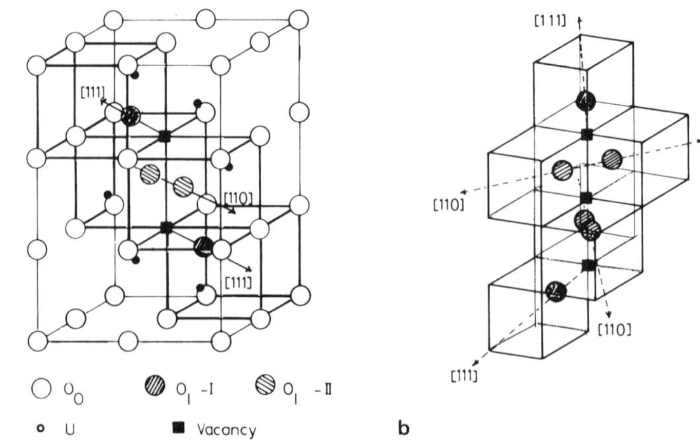

Fig. 7. "Interstitial" clusters in hyperstoichiometric fluorite phases: (a) the Willis 2:2:2 cluster; (b) a more extended 3:4:2 cluster.

The real defect structure of nonstoichiometric fluorite phases, on both sides of the ideal composition, is thus related to the native point-defect disorder but transformed into defect complexes. On the anion-deficient side, there is a strong energetic advantage in a sterically specific association of vacant sites in pairs; true, mobile anion vacancies arise only from the dissociation of such complexes. In the anion-excess structures, atoms are displaced from the ideal fluorite positions in such a way that excessively close anion-anion distances are avoided. These displacements are not random, but correlated, and it is of interest that recent calculations by Catlow[11] have shown that there is a substantial energetic advantage in forming the 2:2:2 cluster, as compared with a cluster in which fluorite anions are not displaced from their original sites.

The facts cited suffice to suggest a generalization of some importance. In crystals with an irrational anion:cation ratio, the crystal structure can tolerate without change only a low concentration of classical, randomized point defects. Above that limit, some measure of rearrangement sets in. As a minimum, this involves strong correlation between defects and the inception of superstructure ordering; more usually, in ionic structures, it may impose a new local structure. The constraint is that the energetically preferred structure should be capable of *coherent intergrowth with the parent structure.* The elements of new structure may be present as isolated units – extended defects or defect complexes – or may cluster in regular extended arrays of finite extent, which may (but do not necessarily) correspond to the ordering pattern that is observed in thermodynamically stable phases formed under different conditions (lower temperatures). Thus, the anion-excess complex of UO_{2+x} is the basis of the superstructure of U_4O_9, and the "Bevan cluster" (Fig. 6) in the α phase of PrO_{2-x} provides the structure of Pr_7O_{12}, but there is no known stable or metastable phase in the iron oxide system corresponding to completely ordered Koch-Cohen clusters. Coherent intergrowth requires (a) continuity of one sublattice through parent structure and extended defect (e.g., the cation sublattice in UO_{2+x} and in the αPrO_{2-x} and related phases; the anion sublattice in $Fe_{1-x}O$; both sublattices in nonstoichiometric TiO) and (b) minimal dimensional changes and shifts of atomic positions. These conditions minimize the strain or interfacial energy of the transition between parent structure and extended defect. The new coherently intergrown structure has its own repeating unit which will, in general, differ in size and symmetry from that of the parent structure, but as long as the fluctuations in structure and composition are not correlated over domains of significant size, the sole remaining translational symmetry remains that of the parent crystal.

In what has been considered thus far, the reconstructed units have been such as to intergrow coherently in three dimensions. The statistical thermodynamic problem is then that of inserting extended defects, unit cell by unit cell, into the host crystal. There is thus a configurational entropy, but not that which would be calculated from a point-defect description. The enthalpy of solution of defect structure in parent structure, involving the interfacial energy and a site-preference energy, is positive; both terms depend upon the dispersion of the defect-structure units. If the total enthalpy of solution (for either a completely dispersed or a polydisperse system) is small, the minimum free energy at high temperatures can correspond to formation of an equilibrium nonstoichiometric phase with a wide range of total composition. If it is not small, a succession of discrete, ordered phases is necessarily formed, even if the topological conditions for coherent intergrowth are fulfilled. These considerations apply to the binary nonstoichiometric compounds. In the case of the ternary solid solutions, the entropy of randomization of the cations favors formation of the disordered solid-solution structure, and ordering of cations is, in most cases, kinetically inhibited by the "freezing" of diffusion processes. For statistical thermodynamic purposes, a partition function would have to be based upon the real structure and energetics of the extended defects in each system. No generalized treatment is possible.

There is a second group of materials in which irrational anion:cation ratios are accommodated by locally ordered fluctuations of structure and composition which display full coherence with the parent structure in two dimensions only. The entire deviation from ideal composition is thereby taken up in the form of *planar "defects" or structural singularities* which can, in general, be formally described in terms of one or the other of two virtual processes:

(a) In anion-excess structures, by aggregation of interstitial atoms into an interstitial disk lying on some defined crystallographic orientation. To minimize the edge energy, its boundary climbs outwards, draining further interstitials from the structure, until the singularity extends across the full cross section of a crystal.

(b) In anion-deficient structures, by aggregating anion vacancies into a vacancy disk on an appropriate orientation. This undergoes a combined collapse and shear displacement and, again, drains vacancies from the structure until it extends across the crystal.

These formal descriptions relate the derived structures to the basic modes of native disorder, but do not necessarily – and probably do not at all – represent the way in which the planar singularities are formed. Several general points may be noted. The coordination environments of the atoms are significantly changed only in the planar singularity itself: in type (b), the cation coordination is invariably preserved and the anion coordination increased, while cation-cation distances are locally decreased through closer linkage of cation coordination polyhedra – changing from apex sharing to edge sharing, or from edge sharing to face sharing between octahedral groups. Secondly, and very important, a single planar singularity absorbs a very large number of notional defects. For significantly large deviations from stoichiometry, the concentration of planar singularities in a crystal is indeed high, but they constitute only a one-dimensional order-disorder problem and, if they are disordered, the configurational entropy remains small compared with that for a random point-defect structure of the same composition. The free energy of the system is therefore enthalpy determined, and – subject always to possible kinetic inhibition of diffusion controlled rearrangements – is minimized if the planar singularities recur at regular spacings. Such systems therefore tend to produce a multiplicity of ordered phases, with definite composition and homologous structures, rather than nonstoichiometric or solid-solution phases of variable composition. Across each planar singularity, slices of the parent structure lie in proper register (type a) or in an antiphase relation with respect to the cation sublattice (type b), and the total composition of a crystal, for a recurrent structure, depends upon the number of unit cells of parent structure enclosed as a slab between pairs of planar singularities. The structural rearrangement at the planar singularity imposes some distortions in the parent structure; these attenuate progressively, and minimization of the strain energy of the distortions is undoubtedly one energetic factor that favors equilibrium structures in which the planar singularities recur regularly – i.e., the formation of crystallographically defined compounds.

The first kind of structural readjustment is found in anion-excess fluorite structures, as an alternative to the localized Willis defect complex discussed earlier, in the supposed "cubic solid solutions" of YF_3 in YOF[12] and, over the same generalized composition range, in oxide structures, Nb_2O_5-stabilized ZrO_2[13]. Bevan and Mann found in the YF_3-YOF system between the compositions $YX_{2.130}$ and $YX_{2.220}$ a set of superstructure phases based on the homologous series of compounds $Y_nO_{n-1}F_{n+2}$. The structural principle is that all the additional anions segregate into every n^{th} layer of fluorite-type unit cells, along one of the original cubic axial directions. With some shifts of anions from the fluorite cell sites, the cation environment in this layer is related to that in YF_3, and the whole structure can be described as layers of modified YF_3 coherently intergrown with n-1 layers of fluorite structure YOF, slightly distorted but devoid of interstitial anions (Fig. 8). Regularly ordered basic structural units with $n = 4(MX_{2.250})$, $5(MX_{2.200})$, $6(MX_{2.167})$, $7(MX_{2.143})$, and $8(MX_{2.125})$ were identified, but the structural principle is yet more interesting than these basic units imply. For every composition examined, the diffraction patterns showed only a single superstructure; intermediate compositions were not biphasic mixtures of pairs of com-

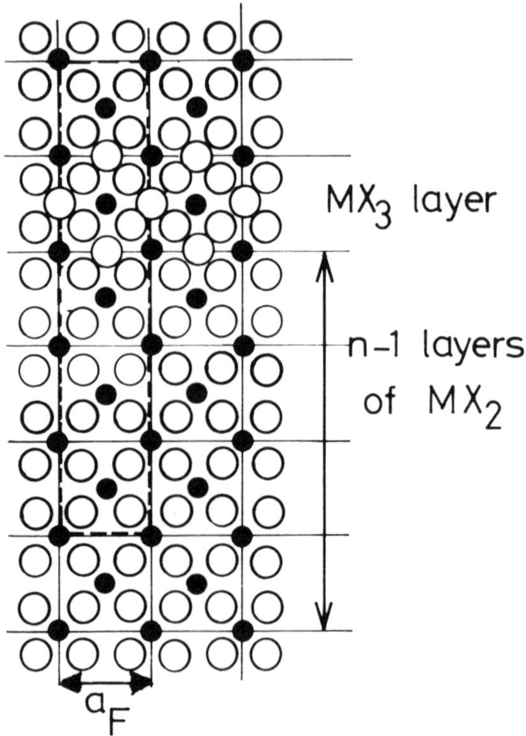

Fig. 8. Recurrent "interstitial layer" principle in fluorite-derived hyperstoichiometric phases. Structure of the $Y_5O_4F_7$ unit.

pounds. The superstructure reflections shifted continuously with composition and, for compositions close to those of the Y_nX_{2n+1} series, the main superstructure reflections broke up into spot groups indicative of a very large repeating unit. The meaning of this is that, even for an arbitrary composition within the indicated range, the interstitial anions adopt some ordered arrangement which may have a very long unit cell. It would appear that this anion-mobile system can be optimized for any composition by spacing the interstitial layers at one or the other of two intervals ($n = n_1$ or n_2) and, further, by regular ordering of the resulting slabs of structure to give a repeating unit with the total superstructure multiplicity $n = (a.n_1 + b.n_2)$. The total anion:cation ratio of such a super-superstructure would be $2 + \frac{a+b}{an_1 + bn_2}$, and if a crystal is indeed capable of attaining long-range order with very high values for the total multiplicity n, the most subtle gradations of composition could be accommodated within completely ordered structures. The interpretation assigned to the range of compositions actually examined is shown in Table II.

Table II. Ordered-Stacking Super-Superstructures in YOF-YF_3 System

Experimental Composition, x in $Y(O,F)_x$	Super-structure Multiplicity, n	Ordered Intergrowth	Calculated Composition	c axis Repeat Length, Å
2.136	23	$(8F)_2 + (7F)_1$	2.130	126.7
2.115	7	$(7F)_1$	2.143	38.6
2.148	19	$(7F)_1 + (6F)_2$	2.153	104.9
2.149	45	$(7F)_3 + (6F)_4$	2.155	248.2
2.165	6	$(6F)_1$	2.167	33.1
2.171	17	$(6F)_2 + (5F)_1$	2.176	94.0
2.183	47	$(6F)_2 + (5F)_7$	2.191	260.0
2.187	57	$(6F)_2 + (5F)_9$	2.193	315.4
2.220	28	$(5F)_4 + (4F)_2$	2.214	155.2
2.259	41	$(5F)_5 + (4F)_4$	2.220	227.4

The second kind of planar singularity is exhibited by the substoichiometric oxides of Ti^{4+}, V^{4+}, V^{5+}, Nb^{5+}, Mo^{6+}, and W^{6+}, cations that are octahedrally coordinated in oxide structures. Formation of an anion vacancy by removal of an oxygen atom from a shared octahedral apex or edge would leave the high formal charge on adjacent cations unscreened; there is also an important degree of cation-anion orbital overlap in these oxides. These may be contributory reasons why they accommodate stoichiometric changes by a structural adjustment *(crystallographic shear)*[14] that preserves the cation coordination but virtually eliminates from the crystal a complete sheet of anion sites, lying on a particular orientation, the *crystallographic shear plane* (CS plane). For a given orientation of the CS plane (hkl), the regular elimination of every nth (hkl) sheet that contains only anion sites generates a set of crystallographically defined structures. In phases derived from the ReO_3 structure, these conform to the common formula type MnO_{3n-m}, where m depends upon the CS plane orientation (Table III, Fig. 9). In the rutile and αPbO_2 structures, the principal intermediate phases are of the type M_nO_{2n-1}.

Table III. Crystallographic Shear in the ReO_3 and TiO_2 Structures

Parent Structure	Orientation of CS Plane	Series Formula	n	Examples
ReO_3	(100)	M_nO_{3n-1}	2	$R\text{-}Nb_2O_5$
			3	Nb_3O_7F
	(130)	M_nO_{3n-2}	20, etc.	$W_{20}O_{58}$, etc.
	(120)	M_nO_{3n-1}	$8 \leqslant n \leqslant 12$	Mo_8O_{23}
Rutile	(121)	M_nO_{2n-1}	$4 \leqslant n \leqslant 9$	$Ti_4O_7 - Ti_9O_{17}$
				$V_4O_7 - V_9O_{17}$
	(132)	M_nO_{2n-1}	$16 \leqslant n \leqslant 36$	$Ti_{16}O_{31} - Ti_{36}O_{71}$

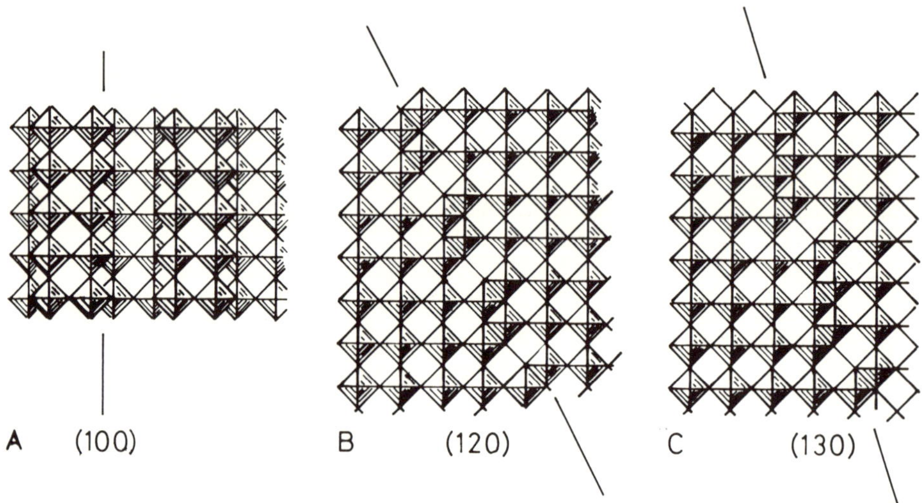

Fig. 9. Crystallographic shear in the ReO_3 structure: (A) $R\text{-}Nb_2O_5$; (B) CS plane in Mo_8O_{23}; (C) CS plane in $W_{20}O_{58}$.

A closer look at the nature of a CS plane is afforded by Fig. 10, which represents the trace of a (121) CS plane intersecting the (100) plane of the rutile structure. The structure has been slightly idealized, to bring the oxygen atoms into true hexagonal close packing, and this packing is continuous throughout the crystal. Across the CS plane, the rows of cations at any level, x = 0 or x = 1/2, are in antiphase relation, and at each junction there is one additional cation. Either of the cations X or Y could be regarded as being in an interstitial site with respect to the arrangement in one slab of rutile structure. Alternatively, the structure can be described in terms of the removal of all oxygen atoms from one (121) layer, followed by a displacement of the two slabs of good crystal. This shear displacement has a component normal to the CS plane (Fig. 10B). This is the structure found in the oxide series Ti_4O_7 to Ti_9O_{17}, originally identified by Magneli and his co-workers. The more compactly grouped cations in the CS plane build up columns of face-sharing octahedra (as in Ti_2O_3) extending along the [111] direction. It was later found (1970) that, in less-reduced rutile, CS planes lay on the different orientation (132), with the structure shown in Fig. 10C: columns of face-sharing octahedra along [111] alternate with cation rows in pure antiphase relationship.

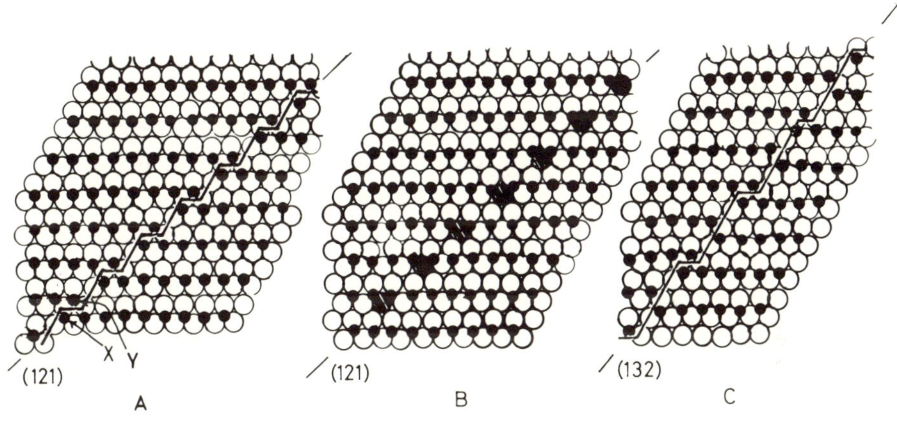

Fig. 10. Crystallographic shear in rutile: (100) plane of slightly idealized ball model showing (A) trace of (121) CS plane; (B) removal of shaded oxygen atoms lying on a (121) sheet, followed by displacement $\frac{a}{2}$ [011] across this sheet gives the structure (A); (C) structure of (132) CS plane.

Since these planar singularities give rise to fringe contrast in electron microscopy, it is possible by that technique to attack very directly the problems of the formation and ordering of planar extended defects. The first question is the relative importance of point defects and extended defects. Bursill and Hyde[15] found that rutile, reduced coulometrically to the composition $TiO_{1.9986}$, already contained numerous isolated CS planes on all the {132} orientations. In addition, there were packets of parallel, regularly spaced (132) CS planes, defining domains of the highest intermediate CS phase. Most of the oxygen deficiency had therefore already been taken up by crystallographic shear; from the average density of CS planes it could be inferred that the point-defect concentration (whether titanium interstitials or oxygen vacancies) in the rutile domains could not exceed 10^{-4}. Analogous results have been found for slightly reduced WO_3[16], in which {120} planar defects were numerous in oxide of composition $WO_{2.994}$ and sparsely present in $WO_{2.9993}$. Except at extremely low defect concentrations, a vacancy or interstitial description is not appropriate for such materials.

The next problem is that of ordering. That CS planes align themselves in parallel groups indicates a rather long-range interaction, not repulsive but leading to an equilibrium spacing. This apparently involves cooperative effects since, at least in the rutile CS structures, it is not independent of the size of the group; only in large enough packets is the spacing of the interior CS planes quite uniform and indicative of a composition about $Ti_{36}O_{71}$ for the least-reduced CS superstructure phase. More highly reduced oxides consist entirely of CS structure, but electron microscopy and electron diffraction have shown that, at least between the compositions $Ti_{16}O_{31}$ and $Ti_{36}O_{71}$ ($TiO_{1.937}$ to $TiO_{1.972}$), complete ordering and regular recurrence of CS planes is rarely attained. Only the (121) CS phases, M_4O_7-M_9O_{17} (M = Ti,V), build completely well-ordered crystals. Similar observations of variable spacing between CS planes – equivalent to the intergrowth of unit cell-thick lamellae of various members of the homologous series – have been made for the ReO_3-based oxide $W_{20}O_{58}$, an oxide considered, on the evidence of a single-crystal X-ray structure, to have a definite composition.

Finally, changes of composition can be attained in two distinct ways: by changes in the width of the slab of parent structure between CS planes, generating a homologous series of compounds corresponding to any particular orientation of the CS planes; and through changes in the orientation of the CS planes themselves, since the number of (hkl) anion sheets per unit cell itself depends upon (hkl). It may be seen from Fig. 10 that the CS planes in rutile are actually segmented. In (121) CS planes, every step eliminates an anion site. For (132) CS planes, the (121) steps alternate with antiphase steps parallel to (011) which involve no composition change. In principle, any combination $\underline{a}(121) + \underline{b}(011)$ would give a possible CS plane orientation in the same $[1\bar{1}1]$ zone and with the same displacement vector, and the CS planes could hypothetically pivot around absolutely continuously, from (121) to (132). For any orientation, one must expect the formation of a homologous series, in which slabs of variable width are enclosed between CS planes. This hypothetical structural response to changes of composition has indeed been found to occur, easily and reversibly, in the oxidation-reduction of rutile between $TiO_{1.900}$ and $TiO_{1.937}$.[17] In the Cr_2O_3-TiO_2 system, the lateral ordering of CS planes is good, but their orientation, for total composition between about $MO_{1.909}$ and $MO_{1.967}$, is variable from sample to sample, invariably lying between (121) and (132).[18]

Changes in composition that alter the anion:cation ratio, because of either solid solution or changes in oxidation state can thus be achieved in two quite distinct ways: (a) the introduction or elimination of CS planes from the structure and their lateral reshuffle to attain or approach uniformity of spacing, and (b) the rotation of the CS planes. The former process appears to be kinetically sluggish, while the latter, although restricted to certain ranges of composition, is unexpectedly facile. It involves a succession of unit diffusion jumps from original cation site to adjacent octahedral interstice, each jump requiring a cooperative displacement of a whole line of cations lying along the zone direction of the family of CS plane orientations. Although kinked and curved CS planes have been observed in the electron microscope (in $TiO_{1.90} - TiO_{1.93}$, for example), the evidence is that this configuration is metastable. It is inferred that the interfacial energies associated with CS structures are minimized by keeping the CS interface planar. This process enables an initially well-ordered crystal to respond to changes in the chemical potential of its constituents in a truly bivariant, hysteresis-free manner, as a classical nonstoichiometric phase, yet without adopting a randomized, defective structure at any stage.

Crystallographic shear evidently restricts the role of point defects, and indeed eliminates them, down to a low residual concentration. It further gives rise, both in the simple oxide systems and in solid solutions, to successions of phases with recurrent CS planes which are stable up to the highest temperatures and may, indeed, be stable in some systems only at high temperatures: CS planes are not the products of low-temperature ordering phenomena. In rutile, at least, the equilibrium solubility of cations of lower valency is very small at low temperatures, but phase-equilibrium studies indicate that Fe^{3+}, Cr^{3+}, etc., enter to the extent of several atom percent at high temperatures (e.g., above 1400°C). It is pertinent to inquire whether, under these conditions, the solid solutions represent a true point-defect disorder. Direct evidence is lacking, and X-ray diffraction evidence does not readily distinguish the highest members of an intermediate-phase series from rutile itself. There is, however, increasing electron-microscopic evidence that the so-called "true rutile solid solutions" are, in fact, CS phases which display genuine thermodynamically bivariant behavior in virtue of the facile rotation of CS planes considered above.

An extension of the foregoing discussion of crystallographic shear applies to the defect chemistry of niobium and tantalum oxides. In many of their derived phases, the basic structure – conditioned possibly through the covalent component of bonding – is essentially of the ReO_3 type, but the total anion-cation site ratio (2.4 to 2.6) is adjusted by the operation of two sets of intersecting CS planes. These divide the whole structure up into rectangular columns of ReO_3 structure. From extensive electron microscopy in several centers[19], and from recent precise X-ray and neutron-diffraction work at Oxford there is much evidence that these columns are essentially defect-free.[20] In oxidation-reduction reactions, and in solid solutions with oxides of different types, these materials simulate nonstoichiometric phases, but it is now clear that changes of composition are taken up without creation of compensating point defects, by means of localized change in the spacing of CS planes or by rotation of CS planes. The former inserts one or more lamellae of a different composition and column structure, exemplified by the insertion of isolated strips of $Nb_{28}O_{70}$(H-Nb_2O_5) structure in the $Nb_{25}O_{62}$($NbO_{2.480}$) structure, Fig. 11. The latter, for certain discrete orientations, generates a set of new structures with their own defined compositions. For intermediate orientations, as long as the mean CS surfaces remain planar, the transition from one composition to another proceeds through a series of coherent intergrowth phases based on the orderly recurrence of two structure types (Fig. 12).

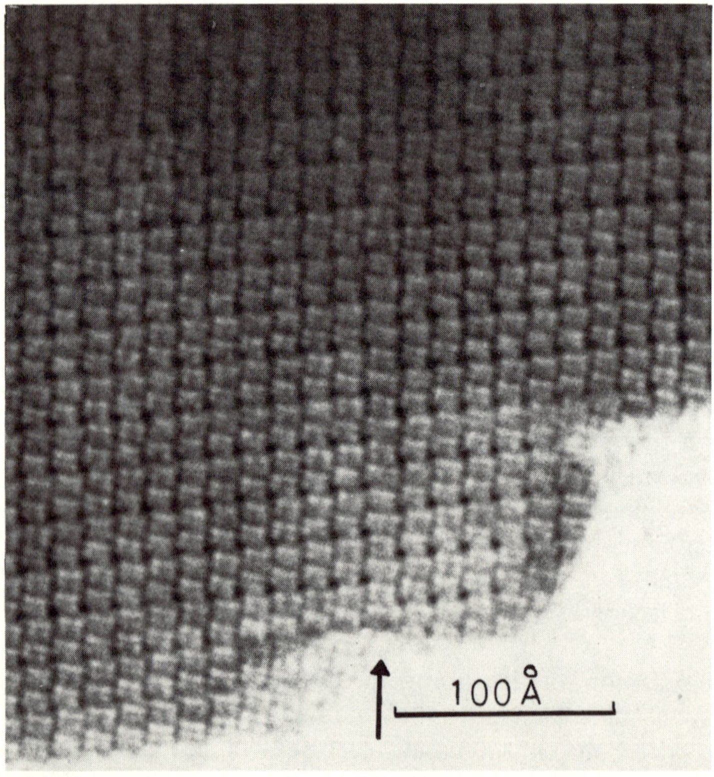

Fig. 11. Small change in stoichiometry in the $Nb_{25}O_{62}$ structure accommodated by coherently intergrowing a single lamella of $Nb_{28}O_{70}$ structure.

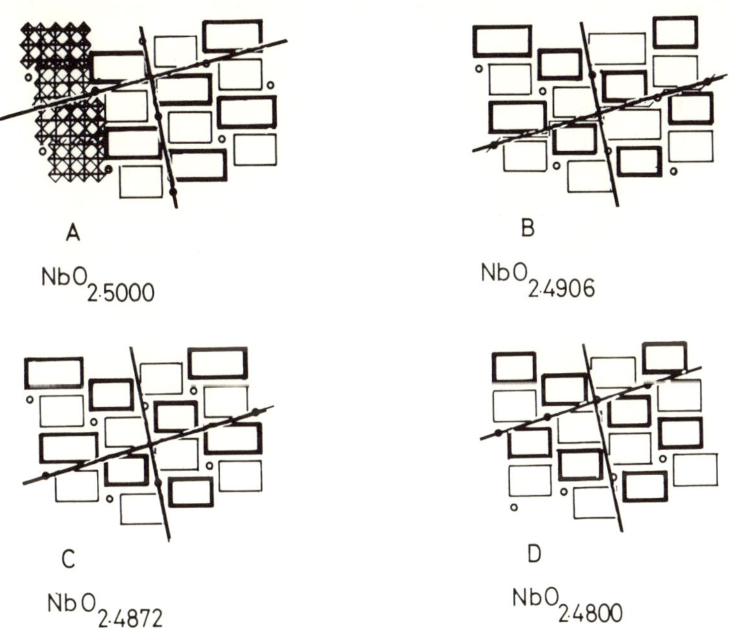

Fig. 12. Change of stoichiometry through rotation of CS planes in niobium oxide block structures. With one CS orientation in common, $(501)_R$, ordered structures of defined composition are formed with the second CS plane on (A) $Nb_{28}O_{70}$, $(10\bar{4})_R$; (B) $Nb_{53}O_{132}$, $(40\bar{1}\bar{5})_R$; (C) $Nb_{39}O_{97}$, $(50\bar{1}\bar{9})_R$; (D) $Nb_{25}O_{62}$, $(10\bar{5})_R$. Subscript R relates CS plane orientation to the ReO_3-type subcell.

Because these oxides involve two intersecting sets of CS planes with different displacement vectors, they possess great flexibility for accommodating small composition changes by local rearrangement, not only by insertion of extended lamellae of new structure but also by localized faulting within finite, coherent domains. In all such faulting, the structural principle and short-range ordering is conserved; compensation for changes in cationic charge (or anionic charge, by substitution of F^- for O^{2-}) is provided entirely by the manner in which elements of perfect structure are coherently spliced together. This has been the subject of much investigation by lattice-imaging electron microscopy during the last 3 years, and will not be reviewed in this place.

What is emerging from the lines of work briefly reviewed is a much clearer view of how ionic structures, at least, respond to changes in the anion:cation ratio. Beyond some rather low limiting concentration, the uniformly distributed fluctuations due to point defects are rather abruptly replaced by localized, but stronger, fluctuations, which impose a structural reorganization rather than a simple relaxation. If that reorganization allows coherence with the parent structure to be preserved, the macroscopic and thermodynamic character of the monophasic material can be retained. The true heterogeneity of local structure can be revealed only by appropriate techniques; in particular, those that can distinguish between an averaged defect structure and the true local structure are important.

Some outstanding problems need study. The possibility of forming superstructures with large repeating units directs attention to long-range interactions in crystals, which are little understood. There is clearly a large measure of chemical specificity in the way that an excess of point defects relaxes into a reorganized structure, and this militates against formulating a generalized statistical thermodynamic approach. The band structure and crystal physics of phases with disordered or regularly recurrent extended defects need elaboration, especially since in some cases – for example, the niobium oxides – the change of semiconducting properties appears to be consistent with a point-defect description while the real structure is now known to undergo a succession of rearrangements.

REFERENCES

1. Anderson, J. S., *Problems of Nonstoichiometry,* North Holland Publ. Co., Amsterdam (1970), p. 1; *Rev. Pure and Applied Chem.,* **21,** 67 (1971); *N.B.S. Special Publ.* 364, *Solid State Chem.,* p. 295 (1972).
2. Bilz, H., *Z. Physik,* **153,** 338 (1958); Denker, S. P., *J. Phys. Chem. Solids,* **25**, 1397 (1964); *J. Less-Common Metals,* **14**, 1 (1968).
3. Venables, J. D., and Meyerhoff, M. H., *N.B.S. Special Publ.* 364, *Solid State Chemistry,* p. 583 (1972).
4. Watanabe, D., Terasaki, O., Jostsons, A., and Castles, J. R., *Extended Defects in Nonmetallic Solids,* North Holland Publ. Co., Amsterdam (1970), p. 488.
5. Koch, F., and Cohen, J. B., *Acta Cryst.,* **B25,** 275 (1969).
6. Greenwood, N. N., and Howe, A. T., *J. Chem. Soc., Dalton Trans.,* (1972), pp. 110, 116, 122.
7. Cheetham, A. K., Fender, B.E.F., and Taylor, R. I., *J. Phys.,* Ser. C, **4**, 2160 (1971).
8. Greenwood, N. N., and Howe, A. T., *7th International Symposium on Reactivity of Solids,* Chapman and Hall, London (1972), p. 240.
9. Thornber, M. R., and Bevan, D.J.M., *J. Solid State Chem.,* **1**, 536 (1970).
10. Cheetham, A. K., Fender, B.E.F., Steele, D., Taylor, R. I., and Willis, B.T.M., *J. Phys.,* **C, 4,** 3107 (1971); Steele, D., Childs, P. E., and Fender, B.E.F., *J. Phys.*, Ser. C, **5**, 2677 (1972).
11. Catlow, C.R.A., *J. Phys.,* Ser. C, **6**, L64 (1973).
12. Mann, A. W., and Bevan, D.J.M., *J. Solid State Chem.,* **5**, 410 (1972).
13. Roth, R. S., Waring, J. L., Brower, W. S., and Parker, H. S., *N.B.S. Special Publ.* 364, *Solid State Chemistry,* p. 183 (1972).
14. Bursill, L. A., and Hyde, B. G., *Progress in Solid State Chemistry,* 7, 178 (1972).
15. Bursill, L. A., and Hyde, B. G., *Phil. Mag.,* **23**, 3 (1971).
16. Allpress, J. G., Tilley, R.J.D., and Sienko, M. J., *J. Solid State Chem.,* **3**, 440 (1971).
17. Merritt, R. R., Hyde, B. G., and Bursill, L. A., *Phil. Trans. Roy. Soc.,* in course of publication.
18. Bursill, L. A., Hyde, B. G., and Philp, D. K., *Phil. Mag.,* **23**, 1501 (1971); Gibb, R. M., and Anderson, J. S., *J. Solid State Chem.,* **4**, 379 (1972).
19. Iijima, S., and Allpress, J. G., *J. Solid State Chem.,* 7, 94 (1973); Iijima, S., *Acta Cryst.,* **A29,** 18 (1973); Anderson, J. S., Browne, J. M., and Hutchison, J. L., *Nature,* **237**, 153 (1972); Allpress, J. G., *J. Solid State Chem.,* **1**, 28, 66 (1969).
20. Anderson, J. S., Browne, J. M., Cheetham, A. K., von Dreele, R., Hutchison, J. L., Lincoln, F. J., Bevan, D.J.M., and Straehle, J., *Nature,* **243,** 81 (1973).

DISCUSSION on Paper by J. S. Anderson

EYRING: Professor Anderson alluded to the complex defect occurring in oxygen-deficient fluorite structures as a unit consisting of an oxygen double vacancy associated with a metal atom having threefold symmetry. I would like to comment that current work on the structures of the binary rare-earth oxides belonging to this type indicates that the extended defect to consist of planes, for example, in the odd n members of the R_nO_{2n-2} homologous series; the dumbbells are ordered in two dimensions. (This work has been done by P. Kunzmann in our laboratories.) There is a periodicity characteristic of all members of the series (n both odd and even) which is one dimensional and consists of strings of these divacancies where the dumbbells are oblique to the string. In short, it may be that the extended defect in the binary-fluorite-related series with oxygen deficiency may be more complex than heretofore thought.

ANDERSON: This is a very interesting finding. Except for the suggestion, based on the M_7O_{12} structures, that the Bevan clusters might line up along their threefold axes, there has hitherto been no firm information about how they might generate short-range or long-range order. It is therefore important to have definite evidence and to find that these substoichiometric fluorites provide another example of planar ordering. It may be worth noting that, unlike crystallographic shear planes which normally extend right across a crystal because they are bounded by a dislocation loop if they are finite in area, planar arrays of Bevan clusters could form finite disks without obvious edge strain. Thus there is the possibility not only of ordered or disordered sheets that define a homologous series of intermediate phases, but also of small domains. They provide a structural basis for the nonstoichiometric α phase in which the empty fluorite sites are segregated into organized regions to a considerable extent.

NOWICK: In answer to an earlier question of evidence for the mechanism of diffusion in oxides of the fluorite structure, I would like to point out the usefulness of considering dielectric loss measurements in slightly doped oxides of the form MO_2:CaO. In these cases, a dielectric loss peak is found whose relaxation rate is clearly related to the jump rate of an oxygen vacancy adjacent to a Ca^{++} ion. For CeO_2:CaO, the activation energy obtained is 0.86 eV [K. Lay and D. H. Whitmore, *Physica Stat. Solidi,* **43**, (6), 175 (1971)]; for ThO_2:CaO and ZrO_2:CaO, slightly higher values are obtained [J. B. Wachtman, Jr., *Phys. Rev.,* **131**, 517 (1963)]. These values generally agree quite well with oxygen self-diffusion [B. Steele and Floyd, *Proc. Brit. Ceram. Soc.*, **19**, 55 (1971)], both in nonstoichiometric and in more heavily doped oxides. It therefore seems reasonable that diffusion is controlled by the migration of single oxygen-ion vacancies.

ANDERSON: There is some apparent conflict, either in the evidence or in the interpretation. The existence of so many ordered phases in oxide and oxyfluoride structures, based on the pairing of empty anion sites across the diagonal of a cubic coordination polyhedron, suggests strongly that this is an energetically favored configuration. If the energy of dissociation is not large, it is true that the resulting complexes may be largely dissociated into mobile vacancies at high temperatures. However, Eyring's work on the α phase of the Pr-O system suggests that a good deal of structure is retained, and the character of the transition between the ordered low-temperature phases in the Pr-O, Ce-O, and Tb-O systems looks more like an order $\rightleftharpoons$ disorder change than a quasi-chemical dissociation. At low temperatures, the formation of the paired complexes is well established by structural methods. Notwithstanding any conclusions drawn from dielectric loss measurements, it seems definitely established by neutron scattering that in the ZrO_2-Y_2O_3 system most of the empty anion sites are paired.

STUBICAN: There are findings that in many nonstoichiometric and solid-solution systems, local reorganization of the host structure occurs. In some of these structures, only isolated units or domains are built in the parent phase. In other systems, coherent sheets of modified structure are built in the parent phase. How do we define a phase in the thermoanalytical sense (phase rule) in these cases?

ANDERSON: It is not an altogether simple matter to frame clear definitions, and the problems have already been discussed in the meetings at Scottsdale in 1969 [Extended Defects in Nonmetallic Solids. (Edited L. Eyring and M. O'Keeffe). North Holland Publ. Co. (1970)] and Washington in 1971 [*N.B.S. Special Publication* 364, *Solid State Chemistry* (1972)]. From some confusions that have arisen in this Colloquium, it is evident that we need to consider carefully the concepts of a *phase,* of a *compound,* and of *defects.*

It is best to define *phases* strictly in the original sense of thermodynamics and equilibria. The essential point is that we can recognize in a system discrete volumes or particles which are completely specified in their thermodynamic properties by the magnitudes of the intensive variables, including the appropriate number of chemical potentials. Each such element of volume is homogeneous, and all the volume elements that are assigned to the same phase have the same intensive thermodynamic properties. Structure, as such, does not enter into a thermodynamic definition but, in practical phase-equilibrium work, the phases present in a system are commonly identified by diffraction methods and described primarily in structural terms – the dimensions, symmetry, and contents of the statistically averaged unit cell. This is a macroscopic definition.

A *compound* in solid state chemistry can be defined only in structural terms. If there is some recognizable unit of translational symmetry, that unit, however large and complex, defines the composition and structure of a unique and distinguishable compound. What has come from the newer structural techniques is the recognition that the topology of structures may allow two distinct structural patterns, corresponding to two different compounds, to intergrow perfectly coherently within a single crystalline particle. If, as is usually the case, the two structures are based upon the smaller subcell of some parent structure, the diffraction pattern of the composite crystal may closely resemble that of either structural component, or that of the parent structure. If the intergrowth is regularly periodic, it furnishes a new, larger repeating unit, which must be accepted as a new compound. This is usually, but not necessarily, a matter of stacking lamellae of the two structural types – a one-dimensional ordering problem. If the stacking sequence is regular, there is no ambiguity about the status and composition of the resulting intergrowth compound. If we are to investigate the thermodynamic and other properties of such compounds, we need to prepare materials in which the regularity of intergrowth extends throughout each crystalline particle; the resulting homogeneous particles will then constitute a single phase in the strict sense. Structures intergrow; phases do not intergrow in this sense.

Our concept of defects has to be widened. Point defects and localized defect complexes raise no problems of definition. They are thermodynamic – their concentration is specified by the intensive variables – and as long as they are randomly distributed, the only translational symmetry of a macroscopic particle is still that of the idealized parent structure. Some precision of thought is needed, however, when planar defects are considered. For example, when TiO_2, WO_3, etc., are slightly reduced, or when cations of lower charge are introduced at low concentrations (e.g., Nb^{5+} in WO_3), sparsely distributed crystallographic shear planes are created, lying on all the equivalent hkl orientations of the CS operation. These are defects. At a higher oxygen deficit, the CS planes line up on one (hkl) orientation, and at true equilibrium form a perfectly regularly spaced array. They thereby generate a new set of compounds, with a unit cell given by the orientation of the CS planes and the spacing between them. The CS planes are then not defects in any sense of the word; they are an integral part of the structural arrangement. If, however, there are irregular spacings between CS planes, we have a (metastable) defective crystal. The defect is not the crystallographic shear plane; it is the imperfection of order. This can be described in different, but equivalent, terms as the interpolation of one or more layers with a structure that corresponds to a different unit cell – a coherent intergrowth of a lamella of another compound, belonging to the same homologous series. In practice, kinetic factors may make it difficult to obtain crystals that are homogeneous on the microscopic scale; coherent intergrowth may result in crystals that are highly irregular in their spacings of CS planes or other planar singularities of structure. The relations between composition, structure and thermodynamic properties are no longer unambiguously defined, and different crystalline particles, apparently coexisting in a reaction product, could then differ in microstructure even though they may be alike in composition. This disorder is, however, not a true equilibrium situation; there is no reason to suppose that the minimum of free energy does not correspond to the development of perfectly ordered structures.

There is one further consideration. If, starting from a perfectly ordered crystal of one CS compound, one effects a small change in composition, it is necessary in general either to eliminate CS planes from the structure or to introduce new ones; only in certain composition ranges can certain systems (TiO_{2-x}, TiO_2-Cr_2O_3) respond by pivoting the CS planes around. Thus, unit-cell thick lamellae of a second compound will be progressively formed and, unless disorder is frozen in, there are two possible consequences. The new lamellae might space themselves regularly, to form a super-superstructure which adapts itself to progressive changes in total composition. As far as is known, this is a rare phenomenon. Alternatively, and more frequently, the lamellae of each kind may group themselves into domains with the compositions and structures of the original compound and the product, respectively. Since they intergrow coherently, the total composition of a macroscopic particle can alter continuously as the relative proportions of the two kinds of domains alter; an example of this for a material with a total composition intermediate between Ti_7O_{13} and Ti_8O_{15} is given in J. S. Anderson and R.J.D. Tilley, *J. Solid State Chem.*, 2, 472 (1970). Macroscopically the material appears to be nonstoichiometric, but it is not microscopically homogeneous. If the variable-composition crystals are thought of as a "phase", the phase rule seems to be contravened and one needs an extra degree of freedom. The situation is, however, one that fits into T. L. Hill's thermodynamics of small systems, and the extra degree of freedom comes from the, probably small, energetic term that arises where two homologous, geometrically quite compatible crystal structures abut in coherent intergrowth – a particular form of interfacial energy. The free energy thereby depends upon the detailed microstructure – the size of domains, as influencing the area of the coherent interfaces etc. – and the behavior is not completely specified by the external variables.

COHEN: One wonders about the structure of TiO with its ordered sheets of vacancies. Many of the other materials with cation vacancies, and even anion vacancies, have defect arrays that are tetrahedra or near tetrahedra. Are there detailed studies of intensities for TiO? Or are the structures based mainly on positions of extra peaks in electron diffraction? If the latter is true, more work would be helpful.

ANDERSON: The structures are based largely on electron diffraction (Watanabe, Jostson and Castles), but there is X-ray diffraction evidence to support them. I agree that a neutron diffraction study would be desirable.

VINK: SiC shows a lot of simple "polytypes" that can be compared to the structures you have just mentioned. However, SiC does not have the "excuse" of a deviation from stoichiometry. Is there in your view any connection between these types of structures and the structures mentioned by you?

ANDERSON: Yes, in a formal sense at least. Polytypes are an example of material with a large – potentially an infinite – number of distinguishable structures corresponding to the same composition and built up by variations on a single structural theme: alternations of cubic and hexagonal stacking in differing proportions and different sequences. They raise the same problems as the compounds I referred to, as regards the mechanism by which crystals establish very large repeating units during growth or ordering reactions. Some discussion of these matters will be found in a paper by me in the *Dalton Transactions of the Chemical Society,* 1973 (March issue).

CARO: With respect to ordering phenomena, or polytypism, I would like to point out some of the results of the nonequilibrium thermodynamics recently developed by Glansdorf and Prigogine [reported in *Structures, Stabilities, and Fluctuations,* Wiley-Interscience, New York (1972)]. They show that when a system is far from equilibrium, the mathematical treatment brings in solutions periodic with respect to time or space for some parameters like chemical potentials. That is what they call "dissipative structures". Do you think that this thermodynamics of the "creation of structures" designed primarily for biological chemical systems may be somewhat relevant to the phenomenon of order in solids you described and to polytypism?

ANDERSON: I fully agree that attention should be paid to the irreversible thermodynamics of processes in crystals. The Glansdorf-Prigogine treatment is suggestive.

It may be worth noting that a second line of development would be an elaboration of the theory of fluctuations, in both structure and composition. There is already some literature on this, notably a book by Krivoglaz, but much of the existing treatment is of substitutional disorder in simple metallic crystals, and there is a considerable gulf between that and the problems of complex oxide structures.

ELECTRON TRANSPORT IN TRANSITION-METAL AND RARE-EARTH OXIDES*

John B. Goodenough

Lincoln Laboratory, Massachusetts Institute of Technology
Lexington, Massachusetts

ABSTRACT

Interpretation of transport properties begins with the construction of energy diagrams. Essential concepts required for this process are reviewed. In particular, it is indicated how electron correlations may induce spontaneous magnetism and localized atomic moments, suppression of superconductivity, mass or Stoner enhancement, and/or lattice instabilities. Transport of small carrier concentrations in nonmagnetic crystals is characterized by small-polaron formation, conventional itinerant behavior, or variable-range hopping, and in magnetic crystals by double or indirect exchange, band splitting below a Curie temperature, magnetic polarons, spin-disorder scattering, critical scattering, giant magnetoresistance, and/or strong mobility anisotropy. Compounds exhibiting these different processes are briefly discussed, and an indication of the systematics found in oxides is given.

*This work was sponsored by the Department of the Air Force.

1. MOBILITIES IN MAGNETIC SEMICONDUCTORS

1.1 Prologue

The electrical conductivity of a material is the sum of several individual contributions:

$$\sigma = \Sigma_i n_i q_i \mu_i \quad , \tag{1}$$

where n_i is the number of mobile charge carriers per unit volume of type i and $q_i\mu_i > 0$ is the charge-mobility product. This brief summary is restricted to consideration of the factors determining n_i and μ_i. Determination of the n_i is accomplished with conventional statistics once an energy diagram has been constructed. The next section is concerned with energy diagrams. This section summarizes some principal concepts that influence the carrier mobility.

At ordinary temperatures, any ionic contribution is generally ignored where there is good electronic conductivity. However, the case of tetragonal Na_xWO_3 and hexagonal K_xWO_3 as reversible electrodes for measurement of ionic conductivity in solid alkali-ion electrolytes[1] indicates that ionic conductivity may be important even in good electronic conductors. In oxides like $Na_2O \cdot 11Al_2O_3$[2] or $NaSbO_3$[3], which are electronic insulators and fast ion conductors, ion transport is the property of principal interest. This summary is restricted to electron transport, however tempting it is to comment on the conditions for fast ion transport.

In a semiconductor the concentration c of electrons or holes carrying a charge of magnitude e gives a contribution to Eq. (1):

$$\sigma_i = cN_i e\mu_i \quad , \tag{2}$$

where N_i is the number of atoms per unit volume available to the charge carriers. If N_i represents a periodic array of like atoms, three factors influence the carrier mobility: impurity scattering, electron-phonon interactions, and spin-spin interactions. The latter are important in magnetic conductors. If N_i represents a disordered array of like atoms, energy bands may develop localized states at the band edges and a mobility transition as the Fermi energy crosses a critical energy ϵ_μ. In this review, impurity scattering is not discussed.

1.2 Electron-Phonon Interactions

In semiconductors with $c \lesssim 0.1$, electron-phonon interactions separate electronic charge carriers into two groups: small polarons and itinerant carriers.

a. *Itinerant carriers:* Where a carrier moves between atomic positions too rapidly for the atomic constituents to respond to its presence, the electron-phonon interactions are weak and the carriers are only weakly perturbed from the conventional rigid-lattice solution:

$$\psi_{\underset{\sim}{k}} = \phi_{\underset{\sim}{k}}(\underset{\sim}{r})\, X = u_{\underset{\sim}{k}}(\underset{\sim}{r}) \exp (i\, \underset{\sim}{k} \cdot \underset{\sim}{r})\, X \quad . \tag{3}$$

In the effective-mass approximation (second order in the perturbation $2i\, \underset{\sim}{k} \cdot \underset{\sim}{\nabla}$), the energy of a carrier relative to the edge of the band of allowed energies is

$$\epsilon_{\underset{\sim}{k}}^{(0)} \approx \hbar^2 k^2/2m^* \quad ,$$

where X is the spin function, $\underset{\sim}{k}$ the carrier wave number, and m^* the effective mass. In this broad-band limit, the drift mobility is

$$\mu_d = e\tau_d/m^* \quad , \tag{4}$$

where τ_d is the mean time between scattering collisions. From the Uncertainty Principle, this description is valid only if the mean time between collisions is $\tau_d >> \hbar/w_b$, where $w_b \approx \hbar^2\pi^2/2m^*a_o^2$ is the bandwidth and a_o is a lattice parameter ($k \approx \pi/a_o$). This condition gives a lower limit to the bandwidth, and hence to μ_d, for weak electron-phonon coupling:

$$\mu_d >> e\, a_o^2/h \approx 0.4 \text{ cm}^2/\text{v-sec} \quad . \tag{5}$$

b. *Small polarons:* If the time it takes for a charge carrier to jump from one atomic position to another is long compared with the period ω_o^{-1} of an optical-mode vibration, the electron-phonon interactions are strong enough to trap the carrier in a local (though mobile) potential well. Such a "dressed" carrier is called a small polaron. In this limit, the electron-phonon coupling is introduced into the zero-order problem, and an isolated carrier has an energy that depends upon the configurational displacements x as:[4]

$$E(x) = \frac{1}{2} M\omega_o^2 x^2 - Ax - \frac{1}{2} w_b \exp\left\{- [x^2/(\hbar/M\omega_o)]\, [N_p + (1/2)]\right\}, \tag{6}$$

where M is an appropriate reduced mass, A is an electron-lattice coupling constant, and N_p is the phonon occupation number. Minimization with respect to x for $w_b = 0$ gives the maximum small-polaron stabilization energy

$$\epsilon_{am} = -A^2/2M\omega_o^2 \text{ at } x_m = A/M\omega_o^2 \quad . \tag{7}$$

The small-polaron bandwidth is reduced by an exponential factor:

$$w_\ell = w_b \exp(-S_o)$$
$$S_o(T) = (2N_p + 1)\epsilon_{am}/\hbar\omega_o \quad , \tag{8}$$

where the phonon occupation number N_p increases with temperature.

The small-polaron mobility may be described by diffusion theory as

$$\mu_\ell = eD/kT; \; D = \gamma a_o^2 \tau_\ell^{-1} = \gamma a_o^2 \; (\tau_h^{-1} + \tau_b^{-1}) \quad , \tag{9}$$

where $\gamma \sim 1$ is a geometrical factor. Since small polarons are formed only if the transition probability is $\tau_\ell^{-1} << z(\omega_o/2\pi) \approx 2z \times 10^{12}$ sec^{-1}, it follows that the upper mobility limit for small polarons is

$$\mu_\ell << z(ea_o^2/h)\,(\hbar\omega_o/kT) \approx 0.1 \text{ cm}^2/\text{v-sec at } 300 \text{ K} \quad , \tag{10}$$

where z is the number of like nearest neighbors.

Emin[5] has investigated theoretically a single carrier moving in a simple-cubic array subject to short-range electron-lattice interactions. With this model, he found that a carrier may exist either as a small polaron or as an itinerant carrier. In a narrow intermediate range of mobilities, corresponding to about $0.1 < \mu < 1$ cm^2/v-sec, the two types of carriers coexist, the domain of coexistence increasing with temperature. This opens up the possibility of observing abrupt mobility changes with temperature. The small polaron appears, or disappears, abruptly because of a regenerative feedback mechanism: w_ℓ/w_b decreases sensitively with increasing ϵ_{am}, and a decrease in w_ℓ increases A and hence ϵ_{am}.

For larger concentrations c, polaron-polaron interactions introduce many-body coulomb forces and reduce the energy ϵ_{am} to be achieved without polaron ordering and a cooperative lattice distortion. A cooperative lattice distortion produces inequivalent lattice sites, sharply reducing the concentration of carriers on any given subarray.

The transition probability τ_ℓ^{-1} of Eq. (9) has two components, corresponding to two competing transfer processes.[6] The first requires a thermal energy, equal to the change in Gibbs free energy due to small-polaron formation, $\Delta G = \epsilon_a - T\Delta S$, to make the energy level at an occupied site equal to that at an unoccupied, nearest-neighbor site. First-order tunneling occurs between states of equal energy, and

$$\tau_h^{-1} = z(1-c-c_i)\,(b/\hbar)\exp(\Delta S/k)\exp(-\epsilon_a/kT) \quad , \tag{11}$$

where c_i is the concentration of impurity atoms unavailable to the carrier and b is the nearest-neighbor transfer-energy integral

$$b_{ij} = (\psi_i, \mathcal{H}'\,\psi_j) \quad . \tag{12}$$

The energy $\mathcal{H}'$ is the perturbation of the atomic potential at position $\underset{\sim}{R}_j$ by a like atom at $\underset{\sim}{R}_i$, and the ψ_i, ψ_j are one-electron wave functions for the carrier localized at $\underset{\sim}{R}_i$ and $\underset{\sim}{R}_j$, respectively. Note that in tight-binding band theory, first-order perturbation theory gives

$$w_b \approx 2zb \quad . \tag{13}$$

The second process is carrier transfer from a polaron-potential well of depth ϵ_a to a nearest-neighbor site. Such a transfer requires second-order perturbation theory, including the band-narrowing factor:

$$\tau_h^{-1} = z(1-c-c_i)\,(b^2/\epsilon_a\hbar)\,\exp\,(-2S_o) \quad . \tag{14}$$

1.3 Spin-Spin Interactions

If the atoms of the periodic array carry a spontaneous magnetic moment, as commonly occurs in transition-metal and rare-earth oxides, consideration must also be given to the interaction between the atomic spin $\underset{\sim}{S}_n$ and the carrier spin $\underset{\sim}{s}$.

a. *Small-polaron limit.* Within a small polaron, $\underset{\sim}{s}$ and $\underset{\sim}{S}_n$ are strongly coupled: $\underset{\sim}{s}$ parallel to $\underset{\sim}{S}_n$ where allowed by the Pauli Exclusion Principle, $\underset{\sim}{s}$ anti-parallel to $\underset{\sim}{S}_n$ otherwise. Magnetic coupling between the polaron and the $\underset{\sim}{S}_m$ at a nearest-neighbor site is achieved via a small-polaron transfer. Since rotation of a spin state α,β through an angle θ to the state α',β' is given by

$$\begin{aligned} \alpha &= \alpha' \cos(\theta/2) + \beta' \sin(\theta/2) \\ \beta &= -\alpha' \sin(\theta/2) + \beta' \cos(\theta/2) \quad , \end{aligned} \tag{15}$$

conservation of spin angular momentum requires that the transfer integral of Eq. (12) be modified to

$$t_{ij} = b_{ij} \cos(\theta/2) \quad , \tag{16}$$

where θ is the angle between vectors $\underset{\sim}{S}_n$ and $\underset{\sim}{S}_m$. Substitution of t for b in Eqs. (9), (11), and (14) gives a mobility that, in an antiferromagnetic or ferrimagnetic material, may be strongly anisotropic below the magnetic-ordering temperature.

In addition, Eq. (16) introduces a ferromagnetic coupling, called *double exchange,* between $\underset{\sim}{S}_n$ and $\underset{\sim}{S}_m$ provided the atomic spin-orientation relaxation time is $\tau_R >> \tau_\ell$. If $\tau_R << \tau_\ell$, as tends to be the case for small polarons, the mobile spin $\underset{\sim}{s}$ has time to relax between hops, and it induces little magnetic coupling between localized spins. However, in the limit $\epsilon_a \to 0$, where $w_\ell \to w_b$, $\tau_R >> \tau_\ell$, and the band-forming stabilization energy is $\frac{1}{2} w_b \approx zt$, Eq. (16) substituting for b in Eq. (13). This leads to a double-exchange energy:

$$\mathcal{H}^D_{ex} = zc\,(1-c-c_i)b \cos(\theta/2) \approx zcb \cos(\theta/2) \quad . \tag{17}$$

b. *Itinerant-carrier limit:* The interaction between an itinerant carrier and the localized atomic spins is generally expressed as an isotropic exchange energy:

$$\mathcal{H}_{int} = -\Sigma_n J\,(\underset{\sim}{r} - \underset{\sim}{R}_n)\,\underset{\sim}{s} \cdot \underset{\sim}{S}_n \quad . \tag{18}$$

Scattering between itinerant states $\underset{\sim}{k}$ and $\underset{\sim}{k}'$ of Eq. (3) is described in second-order perturbation theory via the matrix elements

$$J_{\underset{\sim}{k}\underset{\sim}{k}'} = N_i\,(\phi_{\underset{\sim}{k}'}, J(\underset{\sim}{r} - \underset{\sim}{R}_n)\,\phi_{\underset{\sim}{k}}) \quad , \tag{19}$$

which leads to the Rudermann-Kittel-Kasuya-Yosida (RKKY) *indirect-exchange* energy

$$\mathcal{H}^I_{ex} = -2J\,(R)\,\underset{\sim}{S}_n \cdot \underset{\sim}{S}_m$$

$$J(R) \sim c(J^2/\epsilon_F)\,(y \cos y - \sin y)/y^4 \quad , \tag{20}$$

where $y = 2k_F R$, $\underset{\sim}{R} = \underset{\sim}{R}_n - \underset{\sim}{R}_m$, and $\epsilon_F = k_F{}^2\hbar^2/2m^*$ is the Fermi energy. The $\underset{\sim}{k}$ dependence of the intra-atomic interaction between localized spin and itinerant electron is neglected, making $J_{\underset{\sim}{k}\underset{\sim}{k}'} = J$ in Eq. (20). In the limit $k_F \to 0$, the RKKY formula reduces to Eq. (17).

Ferromagnetic ordering of the localized spins produces *spin splitting* of a conduction band and any donor impurity states, as illustrated in Fig. 1. This splitting can be calculated from first-order theory as[7]

$$\epsilon_{\underset{\sim}{k}}{}^{(1)}\,(\pm) = -J_{\underset{\sim}{k}\underset{\sim}{k}}\,\langle \underset{\sim}{s} \rangle \cdot \langle \underset{\sim}{S}_n \rangle = \mp \frac{1}{2} J_{\underset{\sim}{k}\underset{\sim}{k}} S(M/M_o) \quad , \tag{21}$$

where $(\pm)$ refers to the two possible states $\pm\frac{1}{2}$ of $\underset{\sim}{s}$ relative to the direction of the magnetization $\underset{\sim}{M}$, which has a saturation value at T = 0°K of $\underset{\sim}{M}_o = -N_i g S \mu_B$. In zero external magnetic field $\underset{\sim}{H}$, $M = M_s$ decreases with temperature, vanishing for $T > T_c$. However, an $\underset{\sim}{H}$ may induce an additional magnetization $\Delta M = \chi H$, where χ is the susceptibility. Since χ has a sharp maximum at the Curie temperature, ΔM has a corresponding maximum at T_c. In the absence of an $\underset{\sim}{H}$, short-range correlations produce a band splitting just above T_c in second-order theory.[8]

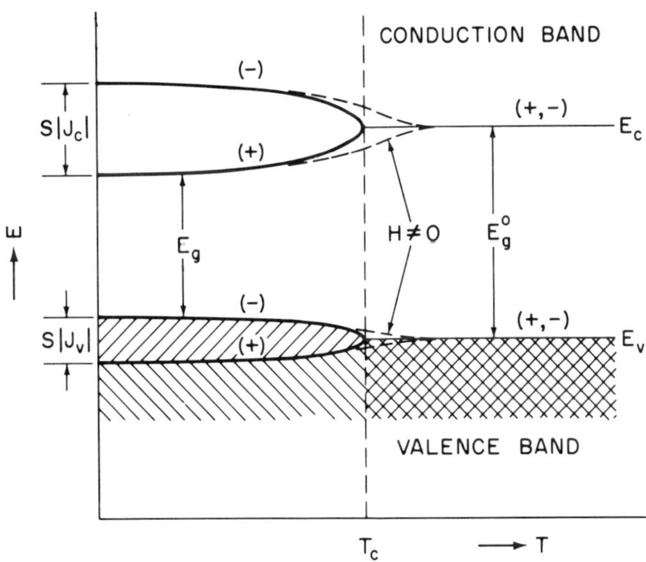

Fig. 1. First-order spin splitting, with and without an applied magnetic field, of the conduction and valence bands of a ferromagnetic semiconductor. E_g is the energy gap. A donor level would also be spin-split, a spin-up level tracking E_c^+ and a spin-down level tracking E_c^-. Therefore, a doubly occupied donor level could have an occupied spin-down level cross the E_c^+ band edge below T_c, giving rise to a semiconductor-to-metal transition with decreasing T.

Figure 1 illustrates the first-order band splitting in zero field and for a small c. At small c, the $J_{\underset{\sim}{k}\underset{\sim}{k}}$ is essentially a constant. The principal contribution to this energy is from intra-atomic exchange. Therefore, the interaction energy J_c for the conduction band should be stronger than the J_v for the valence band, since the conduction electrons, like the $\underset{\sim}{S}_n$, are primarily located on the cation array, and the holes on the anion array.

In the temperature interval of short-range magnetic order just above T_c, an isolated carrier becomes trapped in a spatial region of ferromagnetic order. An electron localized to a small volume of ferromagnetic order is called a *magnetic polaron*.[9] The mobility of a magnetic polaron is reduced to that of the region of short-range order.

Spin-disorder scattering adds an additional term to τ_d^{-1}, where τ_d is defined by Eq. (4). Scattering from state $\underset{\sim}{k}$ to state $\underset{\sim}{k}'$ can occur with or without spin flip. From second-order theory,[7] the transition probabilities for $s = \pm 1/2$ are, for small c where $J_{\underset{\sim}{k}\underset{\sim}{k}'} \to J$,

$$(\tau_m^{\pm})^{-1} = \Lambda J^2/(\hbar N_i g^2 \mu_B^2)$$

$$\Lambda^{-1} = \frac{8(2\pi)^{1/2}}{3} (N_i a_o^3) \left(\frac{\hbar^2}{m^* a_o^2 kT} \right)^{3/2} \int_0^\infty \frac{t \exp(-t)\, dt}{\chi_{||} + 2f^{\pm} \chi_{\perp} [1 \mp (\delta/t)]^{1/2}} , \tag{22}$$

where $\chi_{||}$ and $\chi_{\perp}$ are generalized susceptibilities proportional, respectively, to the two scattering processes. The parameter

$$\delta = (\epsilon_o^- - \epsilon_o^+)/kT = (SJM/M_o kT) \tag{23}$$

is a measure of the spin splitting of the band, and the multiplier $f^- = 1$, or $f^+ = 1$ for $t > \delta$ and 0 for $t < \delta$, takes into account the fact that below a definite energy the spin-flip processes are forbidden by energy conservation.

Because of band splitting, the number of charge carriers in the spin-split bands are unequal. From Eq. (23),

$$n^+/n^- = \exp \delta \quad ,$$

where $n^+ + n^- = cN_i$, and from Eqs. (2) through (4).

$$\sigma = cN_i e \left\{ \frac{\mu^+ \exp(\delta/2) + \mu^- \exp(-\delta/2)}{\exp(\delta/2) + \exp(-\delta/2)} \right\} \quad . \tag{24}$$

In metals, spin-disorder scattering causes the electrical resistivity to increase with temperature below T_c, particularly as $T \to T_c$ where $\chi_{||}$ and $\chi_{\perp}$ are large, but above T_c this contribution is temperature independent. In ferromagnetic semiconductors, critical scattering and the formation of magnetic polarons due to short-range magnetic order make an additional contribution in the vicinity of T_c, so that for small c the resistivity exhibits a sharp maximum at T_c. An applied $\underset{\sim}{H}$ changes both the mobility and the carrier concentrations n^+ and n^-, giving a *magnetoresistance*

$$\Delta\rho/\rho_o = [\rho(H) - \rho_o]/\rho_o \tag{25}$$

that, for small c, can have giant values in the neighborhood of T_c.

A nondegenerate, antiferromagnetic semiconductor shows spin-disorder scattering with

$$\Lambda^{-1} = \frac{8(2\pi)^{1/2}}{3} (N_i a_o^3) \left(\frac{\hbar^2}{m^* a_o^2 kT} \right)^{3/2} / (\chi_{||} + 2\chi_{\perp}) \quad , \tag{26}$$

but there is no critical scattering or magnetic-polaron formation to give a sharp maximum in the resistivity at T_N.

1.4 Disordered Systems

In disordered systems, the potential energies at like atoms may have large energy fluctuations. Anderson[10] first pointed out that, if the energy fluctuations are much larger than the bandwidth, electrons become trapped at discrete sites. Because of this *Anderson localization,* states at the edges of a band may become localized and form a decaying density of states rather than a sharply defined band edge. As the Fermi energy ϵ_F is raised across a critical energy ϵ_μ within the band, the mobility of the carriers changes abruptly.[11] If ϵ_F lies in a tail of localized states, *variable-range hopping*[12] gives an exponential temperature dependence that varies as $T^{-1/4}$ instead of as T^{-1}, the dependence for nearest-neighbor small-polaron hopping according to Eq. (11). This is because local-state energy fluctuations may make tunneling to a distant site more probable than that to a nearest-neighbor site. Mott argued that hydrogen-like carrier wave functions

$$\psi = \text{const. } \exp(-\alpha r) \tag{27}$$

would have an energy transfer integral, Eq. (12),

$$b \approx b_o \exp(-2\alpha R) \quad , \tag{28}$$

where R is the hopping distance. For an energy-density tail of localized states, the activation energy for hopping is proportional to $[(4\pi R^3/3) N(\epsilon_F)]^{-1}$, where $N(\epsilon_F)$ is the density of states at the Fermi energy. Therefore, from Eq. (11),

$$\tau_h^{-1} \sim \exp\left\{-2\alpha R - \epsilon_a (a_o/R)^3/kT\right\} \quad , \tag{29}$$

where ϵ_a is the activation energy for hopping to nearest-neighbor sites a distance a_o away. Minimization of the exponent with respect to R gives an exponent in Eq. (29) that varies as $T^{-1/4}$.

2 LOCALIZED VERSUS ITINERANT ELECTRONS

The number n_i of charge carriers is obtained from an energy profile of the density $N(\epsilon)$ of one-electron states and a knowledge of the number of electrons per molecule, which locates ϵ_F. Construction of an $N(\epsilon)$ profile requires a knowledge not only of the crystal structure, but also of whether the electrons have strong or weak correlations. Strongly correlated electrons are adequately described by crystal-field and superexchange theories, and weakly correlated electrons by conventional band theory.

Crystal-field theory describes a multielectron manifold localized to a

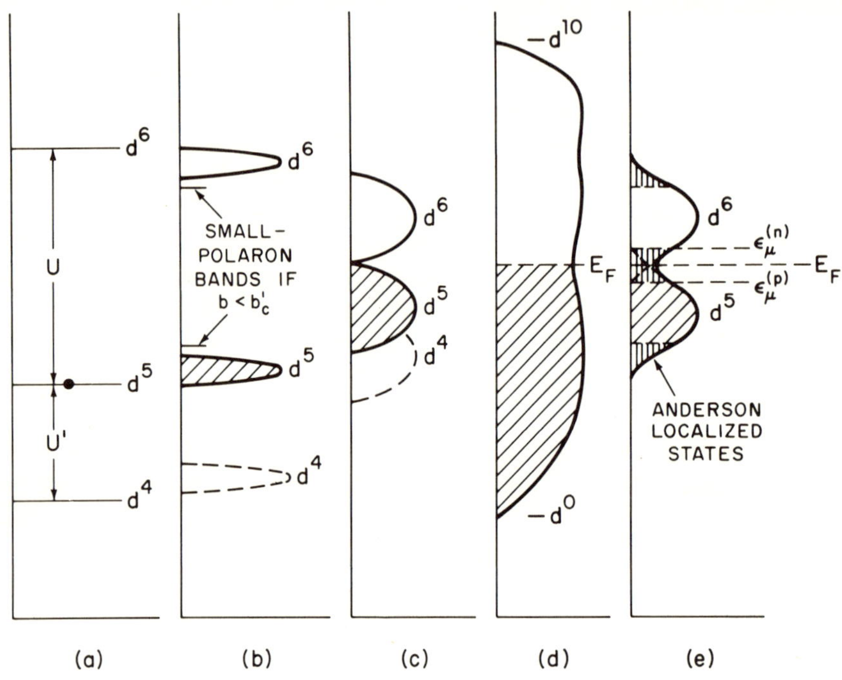

Fig. 2. Change of d^5 manifold with increasing transfer-energy integral b: (a) isolated atom (b = 0); (b) strongly correlated ($b < b_c$) crystalline states; (c) weakly correlated ($b \approx b_g$) crystalline states; (d) itinerant ($b > b_{cs}$) crystalline states; (e) weakly correlated ($b_m < bs$, H b_{cs}) states with Anderson localization at band edges.

discrete atomic position. As on a free atom, the localized-electron states may impart a net atomic spin, and the ground-state energies for successive numbers of electrons in the manifold are separated by discrete energies: U if the added electron enters a half-filled orbital, $U' < U$ if it enters an empty orbital. These energies are shown schematically in Fig. 2. They are intra-atomic, electron-electron coulomb energies that, in the free-atom limit, can be obtained from successive ionization potentials or spectroscopic data. They vary sensitively with the radial extension of the localized wave function, and Mott[13] gives, for a ψ described by Eq. (27),

$$U = \frac{5}{8}(e^2/\kappa)\alpha \quad , \tag{30}$$

where κ is the dielectric constant. Therefore, $U \approx 2$ to 5 ev for metal-atom outer s and p electrons, which have a relatively small α, whereas outer 4f electrons at rare-earth atoms have a $U > 20$ ev. At transition-metal atoms, the d-state manifolds have a $U \approx 10$ ev. In oxides these numbers are reduced by covalent mixing. In particular, *U decreases with increasing b.*

Interactions between localized electrons on neighboring atoms are described by admixing electron-transfer excited states. For an integral number of electrons per like atom on equivalent lattice sites, electron transfer is to a higher energy state, and the energy gained is given by second-order perturbation theory:

$$\Delta\epsilon = -t^2/U \quad . \tag{31}$$

From the Pauli Exclusion Principle, electron transfer between half-filled orbitals can conserve the spin angular momentum only if the spins of the electrons on the two atoms are antiparallel. Therefore, from Eq. (15) t = b sin (θ/2) and

$$\Delta\epsilon = -(b^2/U)\sin^2(\theta/2) = \text{const} + (2b^2/4S^2U)\,\underset{\sim}{S}_n \cdot \underset{\sim}{S}_m \quad . \tag{32}$$

Comparison of the spin-dependent part with the Heisenberg isotropic-exchange operator $\mathcal{H}_{ex} = -\sum_{nm} J_{nm} \underset{\sim}{S}_n \cdot \underset{\sim}{S}_m$ gives a superexchange contribution to the exchange parameter:

$$J^S_{nm}\,(1\text{-}1) = -(2b^2/4S^2U) \quad . \tag{33}$$

Transfer from a half-filled orbital to an empty orbital is not restricted by the Pauli Exclusion Principle, and a spin-dependent term arises only in third-order theory. Differentiation of Eq. (33) with respect to U gives

$$J^S_{nm}\,(1\text{-}0) = +(2b^2 J^{intra}/4S^2U'^2) \quad , \tag{34}$$

where J^{intra} is the intra-atomic-exchange energy and U′ has replaced U. The first interaction is antiferromagnetic, the second ferromagnetic. From molecular-field theory, the magnetic-ordering temperatures vary as

$$T_N \sim |J_{nm}|\,S\,(S+1) \sim 2b^2/U;\; T_c \sim 2b^2 J^{intra}/U'^2 \tag{35}$$

so long as the perturbation expansion remains valid.

In conventional band theory, U = 0 and the atoms carry no net atomic moment. From Eqs. (12) and (13), the tight-binding limit for cationic d orbitals gives

$$w_b \approx 2z\epsilon_{ij}\,(\psi_i, \psi_j) \quad , \tag{36}$$

where ϵ_{ij} is a one-electron energy and the localized wave functions include covalent mixing from the anions. At octahedral-site cations, the crystalline fields split the d manifold into three t_{2g} orbitals that π bond with the anions and two e_g orbitals that σ bond:

$$\begin{aligned} \psi_e &= N_\sigma\,(f_e - \lambda_s\phi_s - \lambda_\sigma\phi_\sigma) \\ \psi_t &= N_\pi(f_t - \lambda_\pi\phi_\pi) \quad , \end{aligned} \tag{37}$$

where f_e, f_t are atomic orbitals, the ϕ's are symmetrized anion s, p_σ, and p_π orbitals of e and t symmetry, and the λ's are covalent-mixing parameters.

Conventional band theory is applicable where $w_b \gg U$. In the intermediate region $w_b \approx U$, the energy U must be introduced explicitly in a perturbation of the one-electron states described by Eq. (3). One model of this type has been explored intensively by Hubbard[14] and many others:

$$\mathcal{H} = \mathcal{H}_o + U \sum_i n_{i\uparrow} n_{i\downarrow} - (n^2U/2N) \quad , \tag{38}$$

where $\mathcal{H}_o$ leads to the solutions of Eq. (3), and $n_{i\uparrow}$ and $n_{i\downarrow}$ are the occupation numbers of the states at $\underset{\sim}{R}_i$ with spin up and spin down. A principal result of this Hamiltonian, which is applicable to a half-filled band, is a splitting of the band in two for $w_b \ll U$, as anticipated in Fig. 2. This splitting occurs at a critical $b = b_g$, which separates semiconductor behavior from metallic behavior. It is a consequence of two identifiable spin states and the Pauli Exclusion Principle, which excludes polar states having electrons of parallel spin. Thus electrons of up spin are shared equally by all the atoms of up spin, those of down spin by the atoms of down spin, and conventional band theory is applicable to each subarray separately. However, transfer of an electron from one sublattice to the other requires the interaction $2d^n \rightarrow d^{n+1} + d^{n-1}$ and therefore costs an energy U. It follows that half of the states are raised by this amount of energy, as indicated in Fig. 2. Where $w_b < U$, there is a discrete energy gap ($b < b_g$) and each atom carries an identifiable atomic moment.

If intra-atomic exchange is much stronger than the interatomic interactions ($w_b \ll U$), then the superexchange theory is valid and contact is made with the localized-electron limit. If the bandwidth exceeds a critical value, corresponding to $b = b_c$, the interatomic interactions are strong enough relative to the intra-atomic exchange that the superexchange perturbation expansion breaks down. In the interval $b_c < b < b_m$, where b_m is the critical value of b for spontaneous magnetism, the magnetic-ordering temperature varies as a U_{eff}, increasing with decreasing b, whereas for $b < b_c$ it obeys Eq. (35) and increases with b.

These features are illustrated in Fig. 3 by the conceptual phase diagram for the case of a half-filled band. This and similar diagrams are discussed in more detail elsewhere.[15] In Fig. 3, b_c' demarks the transition from small-polaron to itinerant behavior for a hole or an electron introduced into the split bands, and b_{cs} is the lower bound for superconductivity at lowest temperatures. In the region $b_m < b < b_{cs}$, the electron

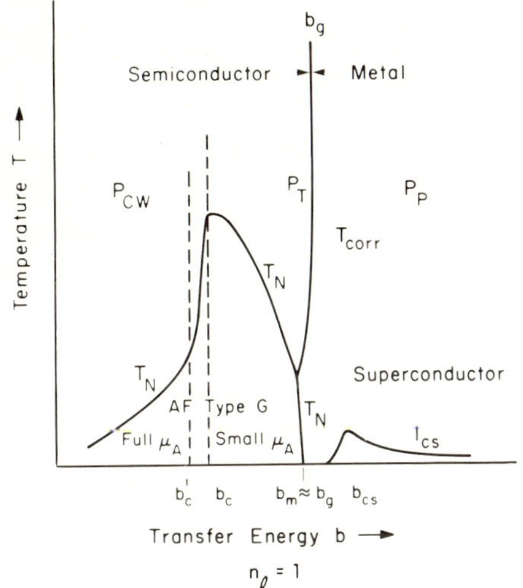

Fig. 3. Conceptual T-b phase diagram for a half-filled manifold.

correlations, although not strong enough to split the bands, create two overlapping bands. Some peculiar physical properties result. These include: (1) semimetallic behavior that may become semiconducting if sufficient disorder is present to produce Anderson localization of the states with overlapping energies[13]; (2) enhancement of the effective mass (by as much as 20 percent) due to a renormalization to satisfy the Luttinger requirement that the Fermi surface enclose a fixed number of states[16]; (3) enhancement of the paramagnetic susceptibility that reflects mass enhancement in the antiferromagnetic case, Stoner enhancement in the ferromagnetic case[17]; (4) suppression of the superconducting state[18]; and (5) lattice instabilities that may induce semiconductor-to-metal transitions[15,19].

3 ILLUSTRATIVE EXAMPLES

3.1 Itinerant s and p Electrons

Outer s and p electrons active in chemical bonding generally satisfy the criterion $w_b >> U$, and the broad s and p bands may be constructed by conventional techniques. For oxides, the primarily O^{2-}:2p bands are stable and filled. Introduction of holes by oxidation generally produces molecular anions $(O_2)^{2-}$, which raises substantially the hole energies. Although cations with filled 4d and 5d cores may have occupied 5s or 6s states, the transition-metal and rare-earth s and p bands are empty. Therefore, the number of d or f electrons in these oxides is known from the formal valence states of the cations.

3.2 Itinerant d Electrons

The outer d electrons are of intermediate character ($w_b \approx U$), which is why the transition-metal oxides exhibit so many interesting properties. Some d electrons are itinerant, without spontaneous magnetism, and many oxides are metallic for this reason. The condition $w_b > U$ may be due to either of two mechanisms: large direct overlap of orbitals on nearest-neighbor cations or large overlap over an intervening anion because of strong covalent mixing. Oxides with partially filled bands of the first type are distinguished as *Class 1 conductors,* and those of the second type as *Class 2 conductors.* In either case, a smaller U is conducive to a $w_b > U$.

From Eq. (30), U and U′ are sensitively smaller the larger the radial extension of the localized d wave functions of, for example, Eq. (37). Three factors influence the radial extension: (1) the effective nuclear charge, which increases with atomic number and with the effective atomic charge as the formal charge is increased or the screening due to covalent mixing of cation s and p orbitals is reduced; (2) the covalent mixing of cation d orbitals, which extends the orbitals out over the anions with increasing λ's in Eq. (37), and (3) intra-atomic exchange, which decreases screening from the nuclear charge by other electrons of the d manifold. Moreover, if the number of d electrons at octahedral-site cations in $n_d <$ 3, then $U' < U$ is the significant parameter.

Class 1 conductors are, from the above considerations, more probable for lighter elements in low valence states. Moreover, in order to enhance the cation-cation overlap, the cation-anion bond should be relatively short, which dictates that the σ-bond orbitals be empty. For octahedral-site cations, these are the e_g orbitals. In general, Class 1 conductors are found with octahedral-site cations and $n_d \leqslant 3$.

References to the experimental literature on transition-metal oxides may be found elsewhere.[15] Class 1 conductors include the superconductor TiO, the semimetal VO, metallic NbO, and the trioxides Ti_2O_3 and V_2O_3. In the spinels M^{2+} $[V_2^{3+}]O_4$, on the other hand, the 3d electrons are strongly correlated. However, with the introduction of holes in the d^2 manifold by Li^+-ion substitution, $Mg_{1-x}Li_x[V_2]O_4$ becomes metallic without spontaneous magnetism.

Class 2 conductors, unlike Class 1 conductors, are more probable with higher atomic number and formal valence state at the cation, where covalent mixing is greater. Superposed on this generalization is the influence of intra-atomic exchange, which would stabilize localized atomic moments and is stronger the larger the localized spin S to be stabilized. Therefore, for a given formal valence state and principal quantum number, localized atomic moments are most probable for half-filled manifolds.

Experimental confirmation of this qualitative reasoning is found in

Fig. 4, which provides a table of the ABO_3 "cubic" perovskites in which b_π and b_σ increase on descending any column, and contours map out domains between the critical parameters defined in Fig. 3.[20] In these oxides, the transition-metal cations form a simple-cubic array (with small distortions sometimes superposed) and the O^{2-} ions are located at or near the midpoint of the cube edges. The t_{2g} orbitals π bond with the oxygens, the e_g orbitals σ bond – see Eq. (37). Since $b \sim \lambda^2$ for cation-anion-cation interactions, it follows from geometry that $b_\pi < b_\sigma$. Interestingly, itinerant σ^* electrons (occupying band states from e_g orbitals) are associated with low-spin states.

Reduced $SrTiO_3$ is the only one of these compounds known to become superconducting.[21] Although it is generally accepted that the π^* electrons (occupying band states from t_{2g} orbitals) in n-type $SrTiO_3$ are itinerant, considerable controversy continues[22] over the nature of the π^* electrons in n-type $BaTiO_3$. From Fig. 4, itinerant electrons are anticipated; but $\mu_d \approx 0.5$ cm^2/v-sec at 300°K is at the lower limit for itinerant carriers. Not shown in Fig. 4 is the large Goldschmidt tolerance factor – $(r_{Ba} + r_O)/\sqrt{2}\,(r_{Ti} + r_O) \approx 1.04$, where the r's refer to ionic radii – in $BaTiO_3$, which means that the large Ba^{2+} ion has stretched the Ti-O bond length beyond its normal equilibrium value. This bond stretching reduces b_π, and that reduction is not accounted for in the placement of $BaTiO_3$ within the table.

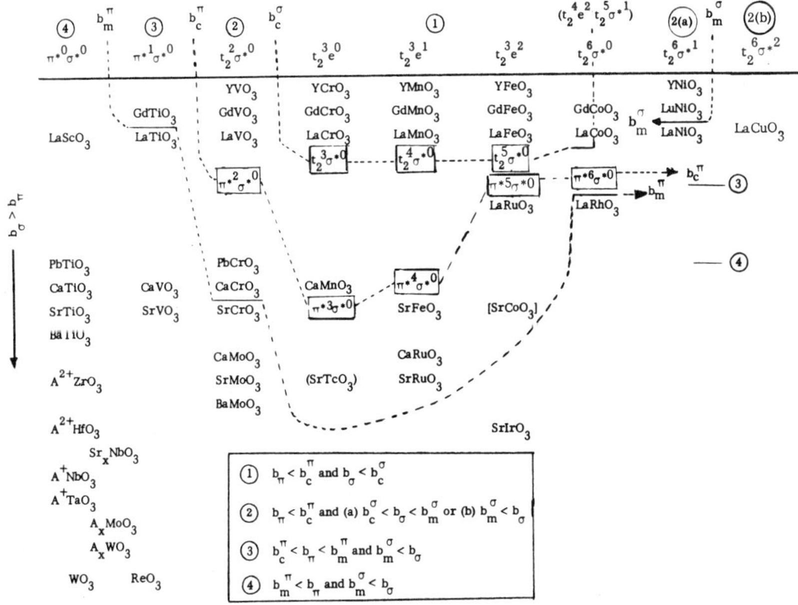

Fig. 4. Outer-d-electron configurations of "cubic" ABO_3 perovskites as a function of $b_\pi \sim N_\pi^2 \lambda_\pi^2$ or $b_\sigma \sim N_\sigma^2 (\lambda_s^2 + \lambda_\sigma^2)$ and the d-electron/atom ratio.

3.3 Localized 4f Electrons

In the rare-earth oxides, the 5d orbitals form bands of itinerant states that overlap the 6s band. In EuO, which has the rock-salt structure, the t_{2g} orbitals form a Class 1 band. The bottom of this band falls below the bottom of the 6s band, which is raised relative to the t_{2g} energies by stronger σ-bond covalent mixing. States at the bottom of this 5d conduction band may be occupied by electrons, since such electrons would contribute to cation-cation bonding. However, in stoichiometric EuO, the 5d bands are empty, as shown in Fig. 5.

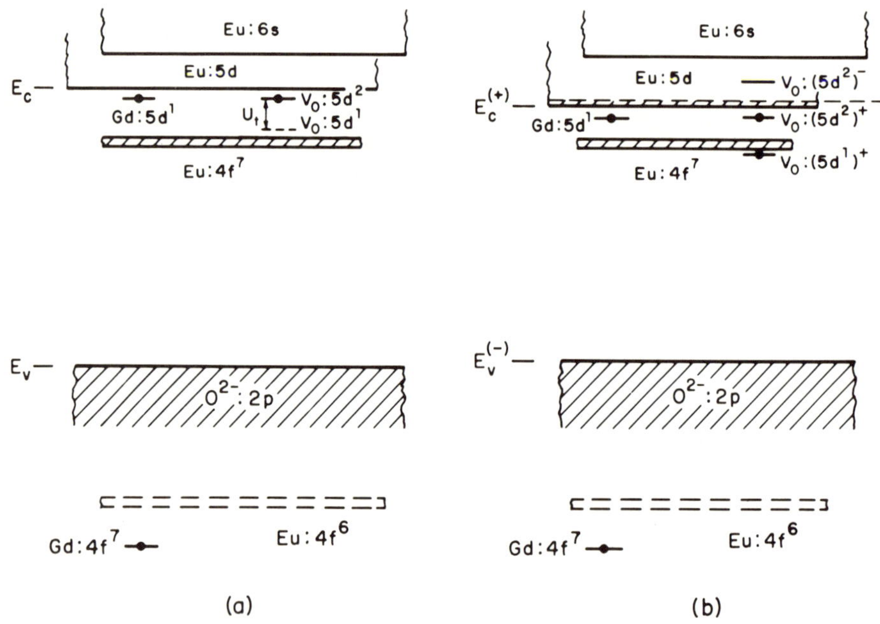

Fig. 5. Schematic band structure of (a) paramagnetic and (b) ferromagnetic EuO. Also shown are occupied gadolium and oxygen-vacancy donor levels. The down-spin oxygen-vacancy level donates its electrons to 5d band on crossing E_c^+, which makes donor level simply occupied V_o $(5d^1)^+$.

Outer 5f electrons satisfy the localized-electron criterion $w_b << U$. The energies U and U′ for the 4f manifold are greater than the energy gap between E_v, the top of the O^{2-}:2p bands, and E_c, the bottom of the 5d bands. Therefore, it is possible to have no $4f^n$-manifold energy in the gap. In this situation, only one 4f-core occupation is possible, since removal of an electron places a hole in the O^{2-}:2p band and an added electron enters the 5d bands. Thus gadolinium, for example, is found only in oxides with a $4f^7$ core. However, some rare-earth cations have $4f^n$-manifold energies falling in the energy gap, and two 4f-core occupancies are found at these ions. The 4f^n-manifold energies falling in the gap include $Ce^{3+}(4f^1)$, $Pr^{3+}(4f^2)$, $Sm^{2+}(4f^6)$, $Eu^{2+}(4f^7)$, $Tm^{2+}(4f^{13})$, and $Yb^{2+}(4f^{14})$. In EuO, the $Eu^{2+}(4f^7)$ ground state lies about 1.1 ev below E_c. Stoichiometric EuO is an insulator that orders ferromagnetically below a $T_c = 69°K$. The interaction is ferromagnetic because, in this case, the charge transfer that dominates the superexchange interaction is from a half-filled 4f to an empty 5d orbital.

Since the gadolinium atom can have only a $4f^7$ core, gadolinium substitutions in $Eu_{1-x}Gd_xO$ introduce 5d electrons. The larger effective nuclear charge at a gadolinium ion stabilizes a shallow $Gd^{2+}(4f^7 5d^1)$ donor state below E_c. However, this donor state differs from that of a typical shallow donor and the transport properties of $Eu_{1-x}Gd_xO$ differ from those of a typical semiconductor because the periodic cation array carries large, localized atomic moments ($7\mu_B$/atom).[23] For example, indirect exchange causes the trapped electrons to couple the moments of the nearest-neighbor Eu^{2+} ions parallel to that at the Gd^{2+} donor center, but the radial extension of the donor state depends on the range of the short-range magnetic order about the center. Free carriers, excited from the donor centers to the 5d band states, introduce a ferromagnetic indirect exchange that raises T_c sharply with x from 69°K to nearly 135°K. The band electrons exhibit spin-disorder scattering, the resistivity rising to a relatively large, constant value for $T \geqslant T_c$ in samples with larger x, but exhibiting a pronounced maximum and giant magnetoresistance at T_c for small x, typical of critical scattering and magnetic-polaron formation in the region of short-range order just above T_c.

By varying the annealing atmosphere, it is possible to prepare $Eu_{1-\delta}O$ or $EuO_{1-\delta}$. The transport properties of these two defect oxides are strikingly different[24] (see Fig. 6). In $Eu_{1-\delta}O$, holes are introduced into the $4f^7$ level, and small-polaron hopping is anticipated. As shown in Fig. 6, there is no anomaly in the resistivity at T_c. Moreover, T_c is not noticeably shifted. These results are compatible with small-polaron hopping and a $\tau_\ell > \tau_R$. In $EuO_{1-\delta}$, on the other hand, electrons are introduced into the 5d bands. From Fig. 6, a sharp metal-to-semiconductor transition is observed at $50°K < T_c$, and a pronounced maximum in the resistivity occurs just above T_c. The maximum shows a giant magnetoresistance and appears well described by the critical-scattering theory of Fisher and Langer.[25] The transition, on the other hand, is associated with an abrupt change in the number of charge carriers. This change is brought about by a splitting of the conduction band below T_c, as indicated in Fig. 1. The stabilized band edge crosses the donor level associated with the anion vacancies, so that at low temperatures the donor levels empty their electrons into the 5d band. To see why this might happen in $EuO_{1-\delta}$, but not in $Eu_{1-x}Gd_xO$, it is instructive to compare the nature of the donor levels in the two cases.

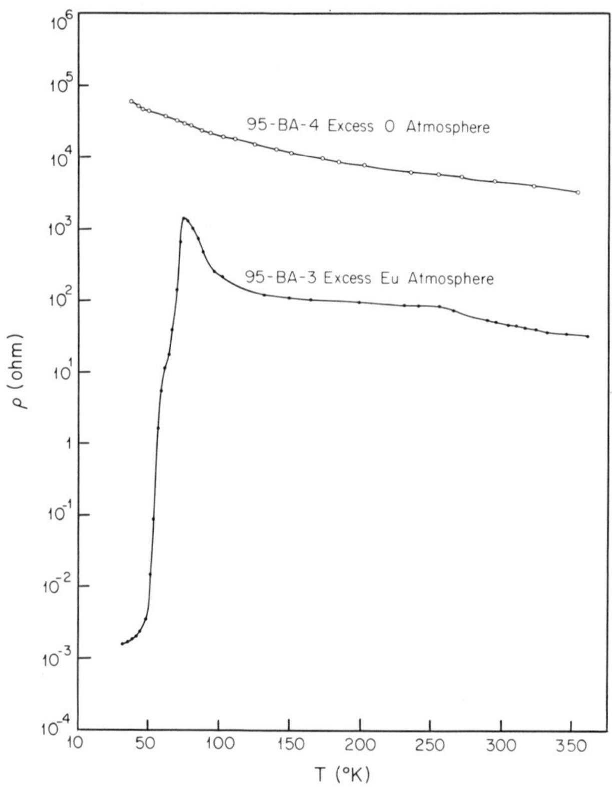

Fig. 6. Resistivity versus reciprocal temperature for EuO annealed in a europium-rich atmosphere ($EuO_{1-\delta}$) and an oxygen-rich atmosphere ($Eu_{1-\delta}O$), after M. R. Oliver.[24]

The $Gd^{2+}(4f^7 5d^1)$ donor level contains a single electron having its spin strongly coupled to the magnetic order about the gadolinium site. Therefore, this level is split, the occupied level moving with the parallel-spin band edge as spin splitting of the 5d bands occurs below T_c. In this case, crossing of the band edge and donor level are improbable. An oxygen vacancy, on the other hand, traps two spin-paired electrons. Because the trap states are localized, the double-occupancy donor level is separated from the single-occupancy trap state by an electrostatic energy U_t. Below T_c, the double-occupancy donor level is also spin split, but in this case the upper donor level is occupied and donates its down-spin electron to the up-spin band as crossing of the level and the band edge induces a spin flip.

3.4 Strongly Correlated 3d Electrons

a. *Small polarons versus large polarons in the monoxides.* The monoxides MnO, CoO, and NiO crystallize in the rock-salt structure, and their antiferromagnetic order at low temperatures shows that cation-anion-cation interactions dominate the cation-cation interactions. The cations have the high-spin configurations $t_{2g}^3 e_g^2$, $t_{2g}^5 e_g^2$, and $t_{2g}^6 e_g^2$. Thus each cation has half-filled, σ-bonding e_g orbitals, and, from Fig. 3, the e_g electrons are strongly correlated. The fact that T_N increases progressively with atomic number of the cation indicates a $b < b_c$, approaching $b \approx b_c$ in NiO, where $T_N = 520°K$ has an exceptionally high value. From the magnetic order, the t_{2g} electrons must be even more strongly correlated (more localized). At Ni^{2+} ions, filled t_{2g} orbitals act like closed-shell cores, although they may participate in 90-degree superexchange. At Co^{2+} ions, the t_{2g} electrons induce a Jahn-Teller distortion below T_N, a characteristic of localized electrons.[26] The t_{2g} electrons at neighboring Mn^{2+} ions interact strongly enough to produce a pronounced exchange striction below T_N, but not strongly enough to change the magnetic order.

Optical data place the $Mn^{2+}(d^5)$, $Co^{2+}(d^7)$ and $Ni^{2+}(d^8)$ ground levels about 3.5, 3.0, and 4.0 ev, respectively, below E_c, the bottom of the 4s band. Inability to stabilize Mn^+, Co^+, or Ni^+ ions therefore makes $U > 4$ ev for half-filled e_g orbitals at divalent ions, consistent with strong correlations. The optical data also show sharp-line spectra characteristic of localized, crystal-field excited states. However, there remains to ask whether the introduction of a hole into the d-state manifold will yield a small polaron or a broad-band (or large-polaron) charge carrier.

Although all three monoxides tend to have cation vacancies, Li^+-ion substitutions suppress these vacancies while introducing an acceptor state that is less complex. From simple electrostatic considerations, a Li^+ ion lifts a symmetrized orbital from the d manifold to a level about 0.4 ev about it. The transport properties of these oxides have been measured extensively.[27] It was found that in all three oxides, carriers trapped at the Li^+-ion centers act as small polarons, but the free holes act as small polarons only in MnO. In $Li_\delta Mn_{1-\delta}O$, $\epsilon_a \approx 0.3$ ev for both bound and free cariers. In $Li_\delta Co_{1-\delta}O$, a bound hole has $\epsilon_a \approx 0.2$ ev, whereas a free hole has an unactivated $\mu_d \approx 0.7$ cm^2/v-sec at room temperature. In $Li_\delta Ni_{1-\delta}O$, the bound hold has $\epsilon_a \approx 0.1$ ev, the free hole a $\mu_d \approx 5$ cm^2/v-sec at room temperature. These findings are consistent with an abrupt change from small-polaron to itinerant carriers, and hence with an increase in b for the e_g electrons from a $b < b_c'$ in MnO to a $b_c' < b \lesssim b_c$ in NiO (see Fig. 3) as a result of the increasing covalency with increasing atomic number. Schematic energy diagrams for $Li_\delta Mn_{1-\delta}O$ and $Li_\delta Ni_{1-\delta}O$ are shown in Fig. 7. APW calculations have been made for the stoichiometric oxides.[28]

b. *Observations on Fe^{2+} and Fe^{3+} ions.* The d^5 level at an Fe^{3+} ion should lie well below the d^5 level at a Mn^{2+} ion, because the Fe^{3+} ion

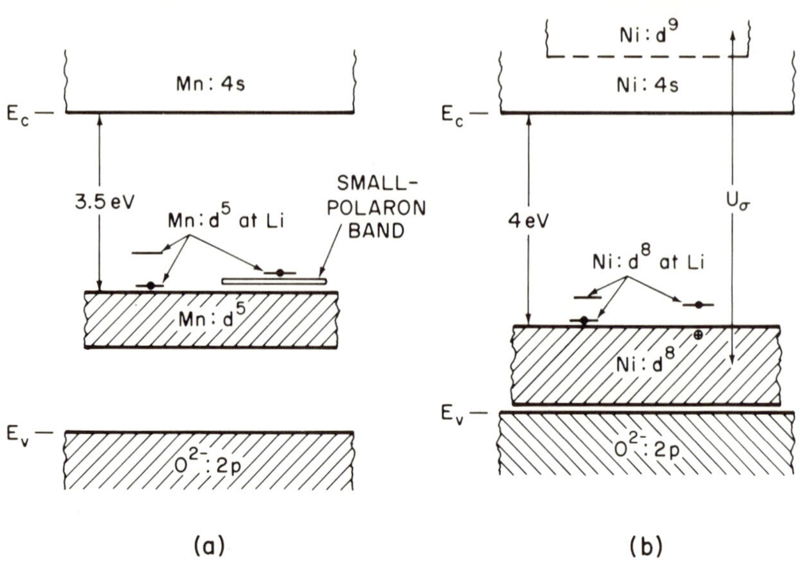

Fig. 7. Schematic energy diagrams for (a) $Li_{1-\delta}Mn_\delta O$ and (b) $Li_\delta Ni_{1-\delta}O$. Note that a Li^+-ion center is capable of trapping two holes. Polaron stabilization at Li^+-ion centers and in bulk are indicated.

has both a larger formal valence and a larger atomic number. The fact that an Fe^{4+} ion is only stabilized under oxygen pressure places the Fe^{3+}-ion level close to E_v. As is clear from the perovskite table of Fig. 4, the 3d electrons at high-spin Fe^{3+} ions are strongly correlated. The large Fe^{3+}-O^{2-}-Fe^{3+} interactions in oxides ($T_N = 750°K$ in $LaFeO_3$) suggest a $b \lesssim b_c$, as in NiO. The $Fe^{2+}(3d^6)$ level is located an energy U above the d^5 level. Since the Fe^{2+} ion is stable in oxides, the d^6 level must fall in the energy gap between E_c and E_v, which makes $U \lesssim 5$ ev for an octahedral-site Fe^{3+} ion. The d bands in the transition metals have a $w_b \approx 5$ ev, and narrower d bands may be anticipated for the oxides. Therefore, these estimates for U are consistent with strong correlations, but with $b \approx b_c$.

The sixth 3d electron at an Fe^{2+} ion enters a t_{2g} orbital with its spin opposite to that of the other five. Since the transfer-energy integrals for t_{2g} and e_g orbitals are $b_t < b_e$, the fact that b_e approaches b_c is consistent with a $b_t < b_c'$. Nevertheless, the radial extension of the down-spin t_{2g} electron must be greater than that of the up-spin electrons because it is not stabilized by intra-atomic exchange. Therefore, it is an open question whether in oxides an electron added to the d^6 level of an array of $Fe^{3+}(d^5)$ ions will have the properties of a small polaron or an itinerant carrier. For small carrier concentrations in ferrospinels and ferrogarnets, experiment unequivocally indicates the existence of small polarons. In Fe_3O_4, on the other hand, the situation remains ambiguous.

The spinel and garnet structures contain four inequivalent, edge-sharing octahedral sites, each with a trigonal crystal-field axis along a different ⟨111⟩ axis. The trigonal component to the crystal field splits the threefold-degenerate t_{2g} orbitals into twofold-degenerate $e_{g\pi}$ and non-degenerate a_{1g} orbitals. Stabilization of the $e_{g\pi}$ relative to the a_{1g} orbitals allows a first-order spin-orbit coupling within a small polaron. This coupling produces a large magnetostriction and a local anisotropy field favoring spin alignment along the unique axis. Therefore, if the atomic spin is oriented along the locally unique axis, the small polaron is more stable. At low temperatures, the polarons are not mobile, but mobility can be induced by infrared radiation. If a sample is cooled to 4.2°K in a demagnetized state, the polarons are randomly distributed. Subsequent magnetization along a particular ⟨111⟩ axis creates a bias stabilizing one of the four inequivalent sites relative to the other three. Irradiation with a pulse of infrared radiation excites the electrons, which subsequently assume a distribution reflecting the bias due to the magnetization direction. The change in the polaron distribution is reflected in changes of the bulk uniaxial magnetic anisotropy, strain, and linear dichroism. Elegant experiments of this type have been performed[29] on the ferrimagnetic garnets $Y_3Fe_{5-x}Si_xO_{12}$. Clearly $b_t < b_c'$ in these crystals.

In the spinel system $Fe^{3+}[Ni^{2+}_{1-x}Fe^{2+}_xFe^{3+}]O_4$, the activation energy for conduction is essentially the ϵ_a for a polaron hop, since all the electrons associated with the formal valence Fe^{2+} should be mobile. This activation energy decreases rapidly with x, from 0.5 ev for small x to less than 0.1 ev for $x > 0.2$. Moreover, the small-polaron formula for the Seebeck coefficient becomes inapplicable for $x > 0.2$.[30] These data indicate that at higher concentrations the carriers are being converted from small polarons to either itinerant electrons or "intermediate" polarons consisting of carriers trapped in discrete iron-pair molecular orbitals.[31] In the end member Fe_3O_4, there are no foreign cations to perturb the t_{2g} orbitals in a disordered way, which eliminates the possibility of Anderson localization, and the concentration of Fe^{2+} ions is too large for small-polaron formation. However, measurements[32] of the Hall mobility in Fe_3O_4 give $\mu_H \approx 0.2$ cm^2/v-sec at 300°K, at or below the lower limit for itinerant-carrier conduction. Moreover, a cooperative crystallographic distortion establishes, below 119°K, distinguishable octahedral sites for Fe^{2+} and Fe^{3+} ions. The number of mobile carriers changes by nearly three orders of magnitude across this transition, and a small $\epsilon_a \approx 0.05$ ev vanishes without any significant change in the magnitude of the Hall mobility.[32] The optical reflectivity in the cubic phase[33] has no sharply defined plasma edge because of the very low carrier mobility. This ambiguity of the experimental data for Fe_3O_4 has left theoretical interpretation a matter of dispute between small-polaron and itinerant-carrier protagonists.[34] It appears that, if the number of charge carriers is too large, small-polaron formation cannot be sustained by random, local lattice distortions, but only by a cooperative distortion.

In hexagonal, ferrimagnetic $BaFe_{12}O_{19}$, the magnetic order below T_c is such that polaron transport on the octahedral-site Fe^{3+} array encounters ferromagnetically coupled ions within basal planes, but antiferromagnetically coupled ions along the c-axis. According to Eqs. (9), (11), (14), and (16), this should introduce an anisotropy in the conductivity ($\sigma_{||} < \sigma_{\perp}$) that is not due to an anisotropy in ϵ_a, but to an anisotropy in the spin-dependent transfer-energy integral t. Bunget[35] has found just such an effect in n-type $BaFe_{12}O_{19}$. He reported a ratio $t_{||}/t_{\perp} \approx 10^{-2}$ at room temperature.

c. *Competition between Jahn-Teller stabilization and ferromagnetic ordering.* The $Mn^{3+}(d^4)$ ion in high-temperature $LaMnO_3$ occupies an octahedral site and has the high-spin configuration $t_{2g}^3e_g^1$ (see Fig. 4). A large intra-atomic-exchange energy removes any spin degeneracy and, from Eq. (16), the transfer-energy integral for the e_g electrons is $t_\sigma = b_\sigma \cos(\theta/2)$. The t_{2g} electrons are strongly correlated by the energy $U >> w_\pi$ separating the t_{2g}^3 and t_{2g}^4 configurations. However, the energy $U' < U$ separating the e_g' and e_g^2 configurations may approach the bandwidth $w_\sigma > w_\pi$ of the σ-bonding orbitals. Therefore, at lower temperatures the occupied σ^*-band states may be stabilized in first-order theory by either of two competitive mechanisms: a cooperative Jahn-Teller distortion below a T_t or ferromagnetic ordering below a T_c. The former enhances the potential energy at the expense of the band energy (w_σ/U' decreases); the latter enhances the band energy, splitting the spin states below T_c (see Fig. 1) and giving to itinerant electrons the same intra-atomic-exchange stabilization experienced by small magnetic polarons.

Low-temperature $LaMnO_3$ is an antiferromagnetic insulator with a structure and magnetic order that reflect a cooperative Jahn-Teller distortion. In the systems $La_{1-x}A_xMnO_3$, where A = Ca, Sr, Ba, or Pb, x holes per molecule are introduced into the d^4 manifolds. These mobile holes lower T_t and, from Eq. (17), introduce a ferromagnetic double-exchange energy $6xb_\sigma \cos(\theta/2)$. Since the superexchange coupling of Eq. (32) varies as $\cos\theta$, double exchange cants the manganese spins to introduce into the low-temperature magnetic order a ferromagnetic component. The Curie temperature T_c increases with x until $T_c \to T_t$ at a critical $x = x_c$. For $x > x_c$, there is no antiferromagnetic component, and band broadening below T_c quenches any cooperative Jahn-Teller distortion. The system $La_{1-x}Ca_xMnO_3$, for example,[20] is a ferromagnetic metal without any Jahn-Teller distortion in the interval $0.2 < x < 0.5$. However, in the interval $0.5 < x < 1.0$, polaron ordering creates a series of discrete antiferromagnetic phases and stabilizes cooperative Jahn-Teller distortions having a $T_t > T_N$. Moreover, the disordered Ca^{2+} ions create Anderson localized states at the top of the d^4 band, so that the initial holes have a low mobility and introduce little double-exchange coupling. A long-range ferromagnetic component (excluding the weak magnetization due to antisymmetric exchange) appears only for $x > 0.1$, which suggests that the Fermi energy crosses a mobility edge at the top of the d^4 band at $x \approx 0.1$. As x increases, the interval $(T_t\text{-}T_c)$ decreases, T_t disappearing abruptly at $x \approx 0.2$ as it enters the temperature interval of short-range order above T_c.

3.5 The Intermediate Case

The intermediate regime $b_c < b < b_{cs}$ is characterized by a variety of unusual physical properties. Oxides typical of this situation include the following:

a. *Weak correlations with Anderson localization.* TiO_x and VO_x exhibit a wide phase field and the presence of about 15 percent vacancies in the stoichiometric compounds. More dense TiO, prepared under high pressure, has a larger lattice parameter. Since TiO and VO are Class 1 conductors, it appears that the spontaneous introduction of vacancies may be caused by the enhanced metal-metal bonding it brings with it.

The system VO_x shows a change with increasing x from n-type to p-type conduction at $x \approx 1.0$, which indicates correlation splitting of the half-filled t_{2g} bands. Lack of any spontaneous antiferromagnetism requires a $b \gtrsim b_m$. Nevertheless, samples with $x > 1$ have a negative temperature coefficient of the resistivity, characteristic of a semiconductor. This paradox, according to Fig. 3, is resolved by the observation of a $T^{-1/4}$ dependence of $\ell n\sigma$ at low temperatures. The cation vacancies apparently create sufficient disorder to induce Anderson localization of the states in the overlapping edges of the correlation-split bands [see Fig. 2(e)]. However, there is an asymmetry with respect to $x = 1$ because the trap orbitals are different at anion and cation vacancies.[15]

$LaVO_3$ is an antiferromagnetic semiconductor ($b < b_m$) and $SrVO_3$ is a classic metal (see Fig. 4). In the system $La_{1-x}Sr_xVO_3$, the activation energy for p-type conductivity, which must come from the mobility, decreases with increasing x from 0.09 ev for small x to zero for $x > 0.15$.[36] Since a $b < b_c'$ is not anticipated in this system, the activated mobility would not be due to small-polaron formation, and Mott[13] has suggested it is due to Anderson localization. The Sr^{2+} ions are randomly distributed. They compete less strongly than the La^{3+} ions for the oxygen p_π orbitals, so that stronger covalent mixing with vanadium t_{2g} orbitals lifts the $V^{3+}(d^2)$ levels near neighbor to a Sr^{2+} ion above those near neighbor to only La^{3+} ions. Therefore, the holes tend to be trapped at the randomly distributed Sr^{2+} ions. On an energy diagram, the $V^{3+}(d^2)$ band develops localized states at its edges, and conduction for small x is via variable-range hopping. Although attractive, Mott's discussion fails to include the disappearance of spontaneous magnetism in the metallic range of compositions or why double exchange does not introduce spontaneous ferromagnetism. It would apppear that the vanadium ions near-neighbor to a Sr^{2+} ion not only trap holes, but also lose their spontaneous atomic moments. Thus strongly correlated and weakly correlated electronic states appear to coexist.

The spinel system $Mg_{1-x}Li_x[V_2]O_4$, which is a Class 1 conductor, shows a similar evolution with x of the conductivity and spontaneous magnetism, $Li[V_2]O_4$ being a classic metal. From the Seebeck data it is possible to infer the formation of itinerant electrons within the cluster of vanadium ions near neighbor to a Li^+ ion.[31]

b. *Mass and susceptibility enhancement.* $LaCuO_3$ has a half-filled σ^* band, whereas $LaNiO_3$ has a quarter-filled σ^* band (see Fig. 4). Both compounds are metallic without spontaneous magnetism, but each exhibits enhancement of the susceptibility. In $LaCuO_3$ the correlations are antiferromagnetic, which should mean a corresponding mass enhancement. In $LaNiO_3$ the correlations are ferromagnetic, which should mean a Stoner enhancement of the susceptibility without a corresponding mass enhancement.[17] Although differences in the susceptibility enhancements are compatible with this reasoning, electronic specific heats have not been measured.

V_2O_3 exhibits a first-order transition from an antiferromagnetic semiconductor to a Class 1 metal, indicating a $b \approx b_m$. The transition may be suppressed by 20-kbar hydrostatic pressure or by titanium substitutions, and measurements of the ratio of electronic specific heat to paramagnetic susceptibility in the metallic phase are consistent with an enhancement of the effective mass.[37]

c. *Lattice instabilities.* Structural instabilities in oxides are due to a variety of mechanisms. The structural changes across the first-order transition in V_2O_3 as well as the magnetic order in the low-temperature phase indicate removal of a twofold band degeneracy that stabilizes a $b < b_m$ in one band and a $b > b_m$ in the other. At higher temperatures, V_2O_3 and Ti_2O_3 exhibit electronic transitions owing to shifting of the relative energies of overlapping bands. The principal driving force for these band shifts appears to come from the interband and intraband correlation energies U_{ii} and U_{ij}, corresponding to U and U'.[38]

VO_2 shows a first-order semiconductor-to-metal transition that does not involve spontaneous magnetism. Antiferroelectric displacements of the cations lift one of two overlapping bands above the Fermi energy, and a cooperative pairing of cations splits a nondegenerate band in two. Such band splittings by lattice instabilities are characteristic of the intermediate region.[15,19,26] They always remove an orbital degeneracy (or quasi-degeneracy) so as to stabilize occupied states at the expense of unoccupied states. Electron-correlation splitting of the type illustrated in Fig. 2 does not induce a lattice instability, and where it competes successfully ($b < b_m$) any lattice instabilities are due to a Jahn-Teller removal of remaining localized-orbital degeneracies.

The system $V_{1-x}Cr_xO_2$ has a complex change with x of the low-temperature phase. This change can be interpreted in terms of a change with x in the relative occupancies of two overlapping, partially filled bands in metallic VO_2 and competitive instabilities having state energies sensitive to these occupancies as well as to the presence of foreign ions.[39]

Metallic Ti_4O_7 contains Magneli planes across which electrostatic forces favor an antiferroelectric displacement of the cations. This stabilizes Ti^{4+} ions at the boundaries. Below room temperature, a first-order semiconductor-to-metal transition is correlated[40] with the formation of Ti-Ti pairs, which trap the carriers in molecular orbitals, within the rutile slabs between Magnéli planes.

d. *High-spin versus low-spin states.* $LaCoO_3$ contains only diamagnetic, low-spin Co^{III} ($t_{2g}^6\sigma^{*0}$) ions at lowest temperatures, but the paramagnetic high-spin Co^{3+} ($t_{2g}^4e_g^2$) state is less stable by only about 0.05 ev. Above room temperature, half of the ions have a localized, high-spin configuration, and ordering of the two types of trivalent ions occurs. At 1100°K there is a first-order electronic transition owing to delocalization of the σ-bonding e_g electrons.[41]

REFERENCES

1. Whittingham, M. S., and Huggins, R. A., *Proc. 5th Materials Research Symposium,* R. S. Roth and S. J. Schneider, Jr. (Eds.), *N.B.S. Spec. Publ.* **364,** pp. 51, 139 (July, 1972).
2. Yao, Y. F., and Kummer, J. T., *J. Inorg. Nucl. Chem.,* **29,** 2453 (1967); Radzilowski, R. H., Yao, Y. F., and Kummer, J. T., *J. Appl. Phys.,* **40,** 4716 (1969).
3. Goodenough, J. B., and Kafalas, J. A., *J. Solid State Chem.,* **6,** 493 (1973); Hong, H. Y-P., Kafalas, J. A., and Goodenough, J. B., *ibid. (in press).*
4. Holstein, T., *Ann. Phys. (N.Y.),* **8,** 343 (1959).
5. Emin, D., *Adv. Phys.,* **22,** 57 (1973).
6. Friedman, L., and Holstein, T., *Ann. Phys. (N.Y.),* **21,** 494 (1963).
7. Haas, C., *Phys. Rev.,* **168,** 531 (1968).
8. Rys, F., Helman, J. S., and Baltensperger, W., *Phys. Kondens. Materi,* **6,** 105 (1967).
9. Methfessel, S., and Mattis, D., *Handbuch der Physik XVIII/1,* S. Flugge (Ed.), Springer Verlag, Berlin, 1968, p. 389.
10. Anderson, P. W., *Phys. Rev.,* **109,** 1492 (1958).
11. Cohen, M. H., Fritzsche, H., and Ovshinsky, S. R., *Phys. Rev. Letters,* **22,** 1065 (1968).

12. Mott, N. F., *J. Non-Cryst. Solids,* **1,** 1 (1969).
13. Mott, N. F., *Adv. Physics,* **21,** 785 (1972).
14. Hubbard, J., *Proc. Roy. Soc. (London),* **A276,** 238 (1963); **A277,** 237 (1964); **A281,** 401 (1964);**A285,** 542 (1965); **A296,** 82, 100 (1966).
15. Goodenough, J. B., *Prog. Solid State Chemistry,* **5,** 145 (1972). See also "New Developments in Semiconductors", P. R. Wallace, R. Harris, and M. J. Zuckermann (Eds.), Noordhoff Int. Publ., Leyden (1973), p. 105; *J. Solid State Chem.,* **3,** 26 (1971); *ibid.,* **7,** 428 (1973); *Phys. Rev. Ser. B,* **5,** 2764 (1972).
16. Brinkman, W. F., and Rice, T. M., *Phys. Rev. Ser. B.,* **2,** 4302 (1970).
17. Goodenough, J. B., Mott, N. F., Pouchard, M., Demazeau, G., and Hagenmuller, P., *Mater. Res. Bull.,* **8,** 647 (1973).
18. Berk, N. F., and Schrieffer, J. R., *Phys. Rev. Letters,* **17,** 433 (1966).
19. Hallers, J. J., *Theories on the Metal-Nonmetal Transitions,* Thesis, Univ. of Groningen, 1972.
20. Goodenough, J. B., and Longo, J. M., *"Landolt-Börnstein Tabellen Neue Serie III/4a",* Springer Verlag, Berlin, 1970, p. 126.
21. Frederikse, H.P.R., Thurber, W. R., and Hosler, W. R., *Phys. Rev.,* **134A,** 442 (1964).
22. Berglund, C. N., and Baer, W. S., *Phys. Rev.,* **157,** 358 (1967); Berglund, C. N., and Braun, H. J., *ibid.,* **164,** 790 (1967); Reik, H. G., and Heese, D., *Phys. Stat. Sol.,* **24,** 281 (1967); Gerthsen, P., Grath, R., and Härdtl, K. H., *ibid.,* **11,** 303 (1965); Bursian, E. V., Girshberg, Ya. G., and Starov, E. N., *ibid., (b)* **46,** 529 (1971).
23. Shafter, M. W., and McGuire, T. R., *J. Appl. Phys.,* **39,** 588 (1968).
24. Oliver, M. R., Thesis, Massachusetts Institute of Technology (1970); Oliver, M. R., Dimmock, J. O., McWhorter, A. L., and Reed, T. B., *Phys. Rev. Ser. B.,* **5,** 1078 (1972); Penney, T., Shafer, M. W., and Torrance, J. B., *Phys. Rev. Ser. B.,* **5,** 3669 (1972).
25. Fisher, M., and Langer, J., *Phys. Rev. Letters,* **20,** 665 (1968).
26. Goodenough, J. B., *Mater. Res. Bull.,* **2,** 37, 165 (1967).
27. Bosman, A. J., and Van Daal, H. J., *Adv. Physics,* **19,** 1 (1970) and references therein.
28. Mattheiss, L. F., *Phys. Rev.,, B.,* **5,** 290, 306 (1972).
29. Gyorgy, E. M., Dillon, J. F., Jr., and Remeika, J. P., *IBM J. Res. Develop.,* **14,** 321 (1970).
30. Nicolau, P., Bunget, I., Rosenberg, M., and Belciu, I., *IBM J. Res. Develop.,* **14,** 248 (1970).
31. Goodenough, J. B., *Mater. Res. Bull.,* **5,** 621 (1970).
32. Siemons, W. J., *IBM J. Res. Develop.,* **14,** 245 (1970).
33. Balberg, I., and Pankove, J. I., *Phys. Rev. Letters,* **27,** 596 (1971).
34. Cullen, J. R., and Callen, E., *Phys. Rev. Letters,* **26,** 236 (1971); Mott, N. F., *Adv. Physics,* **16,** 49 (1967); Adler, D., *Revs. Mod. Phys.,* **40,** 714 (1968); Camphausen, D. L., *Solid State Commun.,* **11,** 99 (1972).
35. Bunget, I., Thesis, Univ. Bucharest, 1968.
36. Dougier, P., and Casalot, A., *J. Solid State Chem.,* **2,** 396 (1970).
37. McWhan, D. B., Remeika, J. P., Maita, J. P., Okinhka, H., Kosuge, K., and Kachi, S., *Phys. Rev. Ser. B.,* **7,** 326 (1973).
38. Zeiger, H. J., *Mater. Res. Bull.* (in press).
39. Goodenough, J. B., and Hong, H. Y-P., *Phys. Rev. Ser. B.,* **8,** 1323 (1973).
40. Marezio, M., McWhan, D. B., Dernier, P. D., and Remeika, J. P., *Phys. Rev. Letters,* **28,** 1390 (1972).
41. Raccah, P. M., and Goodenough, J. B., *Phys. Rev.,* **155,** 932 (1967); Bhide, V. G., Rajoria, D. S., Rama Rao, G., and Rao, C.N.R., *Phys. Rev.,* **6B,** 1021 (1972).

DISCUSSION on Paper by J. B. Goodenough

HONIG: I wish to raise two minor points:

(1) The distinction between Class I and Class II compounds is certainly very useful, but it appears that at best the cation-anion, <100> direction interactions are very competitive with cation-cation, <110>

direction interactions. Hence it appears difficult to talk about purely cation-cation overlap for Class I materials.

(2) For pure NiO the latest experiments (Keem, Sinha, Honig, unpublished) indicate the existence of an activation energy of conductivity of $\approx$ 0.6 ev below the Néel temperature. One might consider this to indicate the presence of an activated conduction process because of our recent observation of a large, temperature-independent Seebeck coefficient.

GOODENOUGH: Answer to Point (1): Of course the d orbitals in question are not simple atomic orbitals, but include covalent admixtures of properly symmetrized anionic orbitals, as indicated in Eq. (37) of my paper. Nevertheless, the distinction between Class 1 and Class 2 conductors is useful not only for predictions of the condition for itinerant versus strongly correlated electrons, but also for understanding the type of lattice instabilities to be anticipated in the intermediate case, where $b_m \lesssim b < b_{cs}$. For octahedral-site cations, Class 1 conductors occur only where the σ-bonding e_g orbitals are empty, so that even if the cation-anion-cation overlap integrals are competitive, these orbitals are essentially unoccupied. In Class 1 conductors, there is a critical cation-cation separation for the transition from strongly correlated to itinerant electrons. This critical separation has been fairly well established for V^{3+} and V^{4+} ions in oxides. To my knowledge, oxides containing V^{3+} ions are metallic only if Class 1 conduction occurs, and the conductivity of VO versus the semiconduction of $LaVO_3$ and the spinels $M^{2+}[V_2^{3+}]O_4$ cannot be rationalized on the basis of cation-anion-cation interactions.

Answer to Point (2): Whether holes in the d^8 energy band form large or small polarons is not, of itself, critical to the general scheme I have outlined. However, on proceeding from holes in the d^5 band of MnO to holes in the d^7 and d^8 bands of CoO and NiO, there should be a systematic progression from a greater to a lesser tendency to form small polarons since b_σ increases on going to the heavier, more acidic cations. The data from the Bosman-Van Daal paper I have quoted does satisfy this anticipation.

VINK: In one of your slides, NiO, CoO, and MnO were shown to be large-polaron semiconductors. This is true for p-type NiO and p-type CoO, but titanium-doped n-type MnO was experimentally shown (by Crèvecoeur) to be of the large-polaron semiconductor type, whereas p-type MnO proved to be of the small-polaron hopping type.

GOODENOUGH: I believe you misunderstood me. I indicated that p-type

MnO is a small polaron conductor, and that the holes in p-type CoO and NiO are small polarons if trapped at a Li^+-ion impurity center, but large polarons if freely moving in the bulk. According to my Fig. 7, electrons would enter the 4s band. Electrons in these bands should always be itinerant in agreement with the results of Crèvecoeur that you quote. There is no disagreement with what I said. However, I assume that the titanium concentration in MnO is very small and that the titanium atoms form deep-donor levels. It should be difficult to dope these oxides n-type except via deep-donor levels.

TANNHAUSER: One probably must not compare electrons, which move in a 4s band, with holes, which move in a 3d or 2p band. In view of the different types of bands, it is not surprising that electrons have a free-charge character in MnO, but holes have a small-polaron character.

GOODENOUGH: That is the point I just made. However, I have not mentioned the O^{2-}:2p band. If you introduce carriers with a high-energy electron beam, as you do in your experiments, you may get holes in both the d^8 and the 2p bands, see my Fig. 7. In such a case, the holes in the 2p bands would be itinerant and have the higher mobility, giving a higher effective hole mobility than you would get from only d^8-band holes. I am assuming that the Bosman-Van Daal data provide a mobility characteristic of only d^8 band holes.

SOLID-STATE REACTIONS BETWEEN OXIDES

H. Schmalzried

Institute for Theoretical Metallurgy and
Applied Physical Chemistry, SFB 126
Technical University Clausthal, West Germany

1 HISTORICAL INTRODUCTION

The study of solid-solid reactions is relatively young, although the art of fabricating ceramics, making use of reactions in the solid state, is as old as the existence of fire and man. Faraday observed diffusion effects in metals[1], and Roberts-Austen did so more systematically[2]. Hedvall in 1912 was one of the first chemists[3] who found solid state reactions between inorganic nonmetallic phases in the sense of this review. A decade later Tammann, Jander, and Fischbeck started to study the kinetics of solid-state reactions.[4-6] But not before the advent of point-defect thermodynamics, connected with the names of Frenkel, Wagner and Schottky, and Jost, was a correct understanding of the atomistic basis of solid-solid reactions possible.[7-9]

In connection with the explanation of the parabolic rate law for metal oxidation (tarnishing), Wagner in 1936 proposed[10] that the experiments of Jander on the formation of silicates from SiO_2 and binary oxides and also the spinel reactions could be understood if counterdiffusion of cations in the reaction product was assumed to be operative. Thus, the explanation of the parabolic rate law of solid-solid reactions, with local thermodynamic equilibrium anticipated, could be given. But the correct calculation of the parabolic reaction-rate constant in terms of elementary kinetic and thermodynamic parameters was not possible until about 30 years later, when the statistical thermodynamics of point defects in higher than binary ionic crystals and especially the knowledge about the dependence of point-defect concentrations in higher ionic crystals upon all the relevant thermodynamic variables became available.[11]

2 SOME STATEMENTS ON POINT-DEFECT EQUILIBRATION

In order to survey the efforts in the field of solid-state reactions between oxides, rather than following the course of history, a number of relevant situations are discussed in some detail, pointing out achievements, problems, and future developments. Since the basic physicochemical aspects are best seen if the number of variables is kept as low as possible, the discussion is essentially restricted to reactions between compact materials, preferably single crystals, starting with plane interfaces.

Two classical situations are encountered and these are illustrated in Fig. 1. In the first case, the reactants form a complete series of solid solutions. In the second case, a heterogeneous reaction occurs, a new phase is formed, starting with nucleation, and then transport across phase boundaries and diffusion in the reaction product are operative.

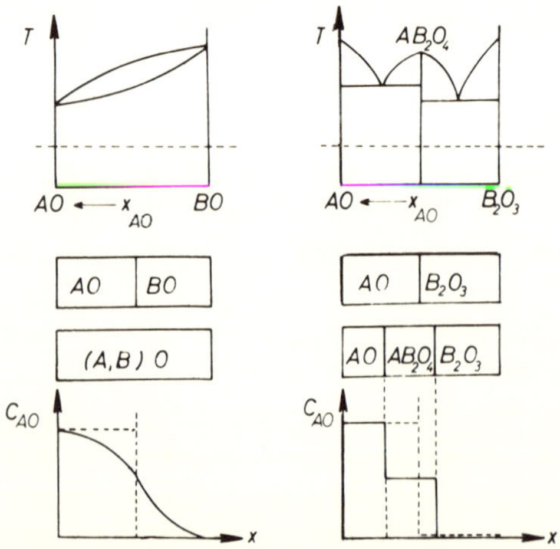

Fig. 1. Two limiting cases of solid state reactions between oxides.

If one wishes to make use of the knowledge of the Gibbs free energies in the system under consideration in order to explain the kinetics, local thermodynamic equilibrium is assumed to prevail throughout the course of the reaction. However, since the composition is necessarily locally changing with time, and point-defect concentrations responsible for the reactivity are composition dependent, one of the basic questions arising at the very beginning of a discussion on solid-state reactions is that concerning the relaxation of point-defect equilibria at phase boundaries and in the interior of the reaction product. This is illustrated in Fig. 2 for the special case of interdiffusion in the system MgO-CoO.

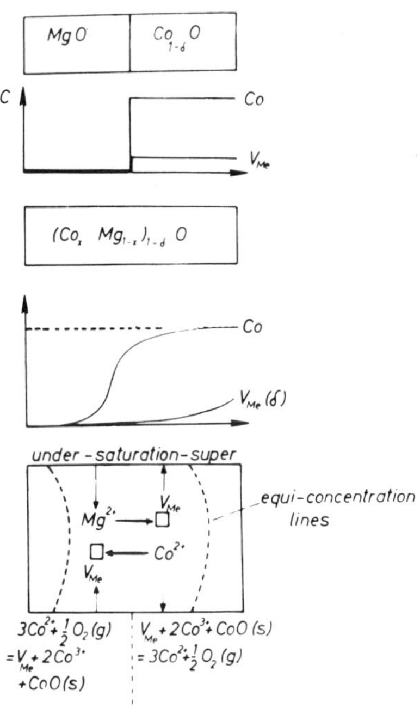

Fig. 2. Concentration of cations and point defects during interdiffusion between $Co_{1-\delta}O$ and MgO.

The interdiffusion in this system occurs in the cation sublattice via cation vacancies, the concentration (δ) of which depends upon p_{O_2} and the composition (x) of the solid solution $(Co_x Mg_{1-x})_{1-\delta}O$, as shown in Fig. 3.[12] Since δ is composition dependent and the only sources and sinks for cation vacancies are at the surface of the compact material, provided that dislocation pipes are not extremely fast diffusion paths for oxygen ions or oxygen atoms, the defect relaxation time is given by the vacancy diffusion coefficient (which may also depend on composition if the elementary jump frequencies of Mg^{2+} and Co^{2+} ions are different and/or composition dependent) and by the characteristic length L of the sample.

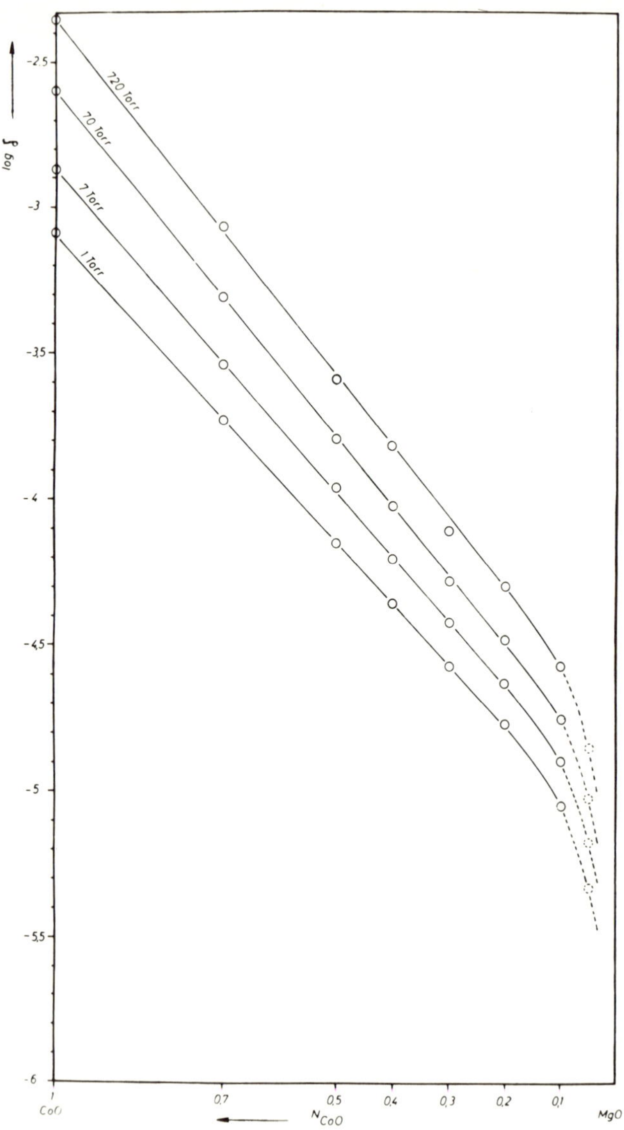

Fig. 3. Point defect concentrations δ in $(Co_xMg_{1-x})_{1-\delta}O$ at 1000°C and different oxygen partial pressures.

Since D_V is measured to be of the order of 10^{-6} cm^2/sec at 1300°C, a sample with a diameter of 0.3 cm has a relaxation time of several hours. As a result, one would expect equiconcentration lines after the first couple of hours of a solid-state reaction, as shown in Fig. 2. Some indications for this kind of effect have been obtained[13] through time-dependent interdiffusion coefficients.

In case of Schottky-type defects, the relaxation process for the defect pair may occur at grain boundaries or dislocations. The defect relaxation process is again a heterogeneous reaction as before, but the characteristic length L between sinks or sources for defects is given by the grain size or the dislocation density, as shown in Fig. 4.

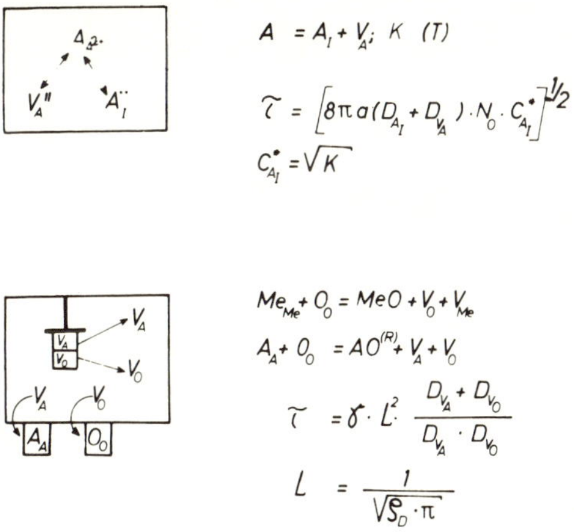

$$A = A_I + V_A;\ K\ (T)$$

$$\tau = \left[8\pi a(D_{A_I} + D_{V_A}) \cdot N_0 \cdot C^{\bullet}_{A_I}\right]^{-1/2}$$

$$C^{\bullet}_{A_I} = \sqrt{K}$$

$$Me_{Me} + O_O = MeO + V_O + V_{Me}$$

$$A_A + O_O = AO^{(R)} + V_A + V_O$$

$$\tau = \gamma \cdot L^2 \cdot \frac{D_{V_A} + D_{V_O}}{D_{V_A} \cdot D_{V_O}}$$

$$L = \frac{1}{\sqrt{\varrho_D \cdot \pi}}$$

Fig. 4. Point defect relaxation processes for Frenkel defects and for Schottky defects.

In case of Frenkel-type point defects, the defect relaxation process occurs as a homogeneous reaction, the theory of which is worked out in detail.[14] The relaxation times here are relatively short, compared with the foregoing cases. For the purpose of an estimate, the formula presented in Fig. 4 may be used.

Finally it should be mentioned that ionic fluxes in heterogeneous solid-state reactions are a result of corresponding defect fluxes which have to be annihilated at the phase boundaries if local thermodynamic equilibrium is assumed to prevail. It is not known in detail how the solid-solid interfaces are working to relax these defect fluxes, but it is well known that in the solid-solid reactions, the phase boundaries are often porous, which is probably the result of poorly achieved defect relaxation.

3 REACTIONS WITHOUT PHASE BOUNDARIES

As indicated in Fig. 2 and shown explicitly for the case of the reaction between MgO and NiO in Fig. 5, interdiffusion between isostructural oxides forming complete solid solutions results generally in a rather unsymmetrical concentration profile, the form of which is dictated by the concentration dependence of the interdiffusion coefficient. Two different cases are to be distinguished.

(a) The oxide solid solution behaves like a metal or a semiconductor: the transference number for electronic charge transfer is close to unity. Starting with the ionic fluxes in the lattice frame

$$j_i = -\frac{c_i D_i}{RT} \text{ grad } \eta_i ; \qquad \eta_i = \mu_i + z_i F \varphi \ , \tag{1}$$

where η_i is the electrochemical potential, the resultant form of the interdiffusion coefficient, assuming local equilibrium, reads

$$\tilde{D} = (1\text{-}x) \cdot D_{Ni^{2+}} + x \, D_{Mg^{2+}} \tag{2}$$

for $(Ni_x Mg_{1-x})_{1-\delta}O$. It is seen that $\tilde{D}$ has the form of the Darken equation for interdiffusion in binary metal systems. Since it is known from Fig. 3 that the defect concentration δ in these solid solutions depends approximately upon composition as

$$\delta = \delta^\circ \cdot \exp - \beta \cdot (1\text{-}x) \ , \tag{3}$$

one would also expect $\tilde{D}$ (x) to be exponentially dependent upon the composition variable x provided the jump frequencies of the individual cations are not too different. This can be confirmed by comparing calculated and measured concentration profiles (see Fig. 5) and has been found to be the case for similar systems as well.[16,17]

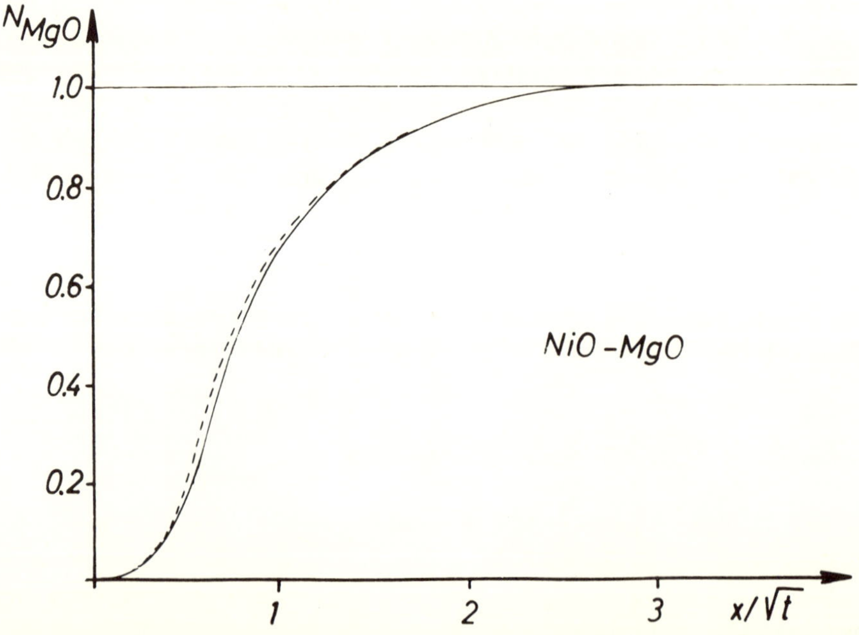

Fig. 5. Concentration profile of an interdiffusion experiment in NiO-MgO[15]. Abscissa in arbitrary units. T = 1370°C in air. - - - - - - - - calculated, ———— measured.

(b) The oxide solid solutions are conducting only by ions, the electronic transference number is close to zero. As a result, diffusion potentials occur which couple the ionic fluxes. The interdiffusion coefficient can again be calculated starting with the flux equations, eliminating the diffusion potential. The general result now reads for a crystal $(A_xB_{1-x})O$

$$\widetilde{D} = xD_B + (1-x)\,D_A - x\,(1-x) \cdot \frac{(D_A - D_B)^2}{x \cdot D_A + (1-x)\,D_B + D_O} \quad . \tag{4}$$

From Eq. (4) it is seen that Darken's equation results if $D_O >> D_A$ and D_B, whereas Nernst's equation results if $D_O << D_A$ and D_B. That is

$$\widetilde{D} = x\,D_B + (1-x)\,D_A \quad \text{for } D_O >> D_A \text{ and } D_B$$

$$\widetilde{D} = \frac{D_A\,D_B}{x\,D_A + (1-x)\,D_B} \quad \text{for } D_O << D_A \text{ and } D_B \quad . \tag{5}$$

In the first case, the faster cation is essentially rate determining. In the second case it is the slower one. By and large one notes that the interpretation of concentration profiles occurring in solid-state reactions between oxides resulting in a quasibinary single-phase inhomogeneous reaction product is quite complicated. This is especially noticed if one remembers that in Eq. (4) D_A and D_B are true self-diffusion coefficients which are usually set equal to tracer diffusion coefficients. The correlation effects, however, may become substantial in solid oxide solutions if the individual jump frequencies of ions of one sublattice are rather different.[18]

4 HETEROGENEOUS REACTIONS

An example of the classical heterogeneous solid-solid reaction is

$$AO + B_2O_3 - AB_2O_4 \quad ,$$

where two reactants form a single-phase reaction product. For the reaction product, the number of independent thermodynamic variables is four. Therefore, it is customary to fix the oxygen partial pressure p_{O_2} beside P and T, since oxygen is the component common to both reactants and the reaction product. At the phase boundaries, the system then is nonvariant, and therefore the kinetic boundary conditions are also determined unambiguously.

In order to calculate reaction rates, several assumptions must be made: (1) reactants and reaction product are compact with theoretical density, and the contact between the reacting phases is ideal; (2) the reaction product is one phase with a narrow range of homogeneity; (3) the reaction is isothermal; and (4) local thermodynamic equilibrium prevails in the bulk and at the interfaces. If these assumptions hold, the reaction rate can be calculated, provided the Gibbs free energy of formation of AB_2O_4 and the self-diffusion coefficients of all ions in AB_2O_4 are known as a function of the component activities.

Possible reaction mechanisms are shown in Fig. 6. Beside those characterized by ionic transport, mechanisms including gas transport in connection with electronic semiconduction in the product phase have also been found. A listing of the known ionic mobilities in ternary crystals reveals that diffusion coefficients of the different kinds of ions are, in general, quite different. In spinel systems, for example, it is true that the oxygen-ion mobility is usually relatively low. Thus, if ideal contact between reactants and reaction product is achieved, only the counterdiffusion of cations in the reaction product is left as the predominant reaction mechanism. This can be seen from Fig. 6.

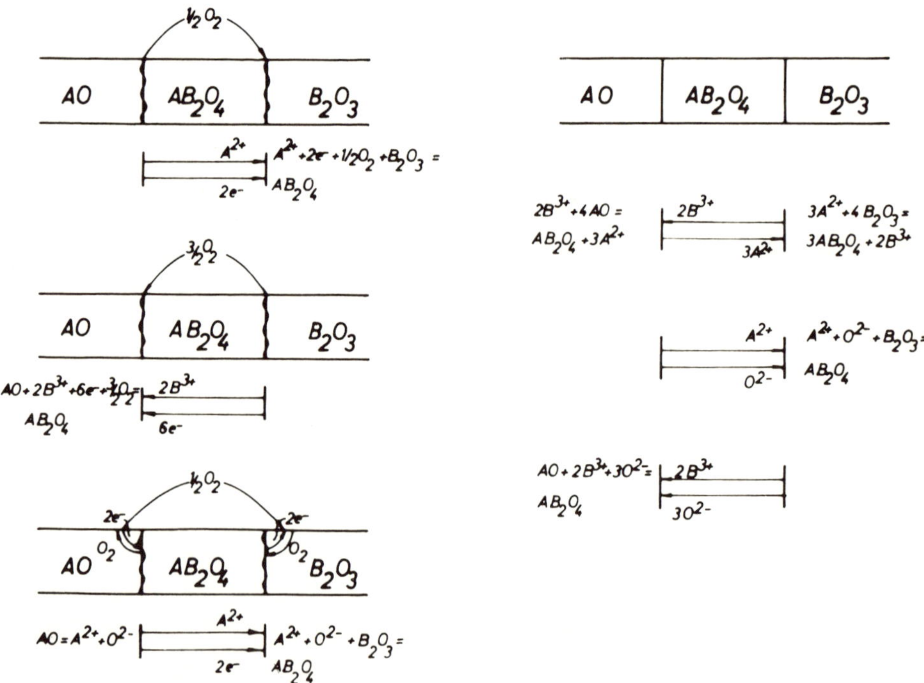

Fig. 6. Spinel reaction mechanisms and corresponding phase boundary reactions.

For the reaction-layer thicknesses (Δx) of more than 10^{-4} cm (1 μm), the parabolic rate law is usually found. An example is given in Fig. 7. The practical reaction-rate constant $\bar{k}$ ($\Delta x^2 = 2\,\bar{k}\,t$) is obtained by integrating the flux equations [Eq. (1)] over the reaction-layer thickness, taking into account the condition of electrical neutrality and of local thermodynamic equilibrium. The result reads:

$$\bar{k} = \frac{\gamma \cdot z_i\, c_i\, D_i^\circ}{m \cdot \Delta G^\circ_{AB_2O_4}} \left[1 - \exp \frac{m \cdot \Delta G^\circ_{AB_2O_4}}{RT} \right] \cdot v \quad , \tag{6}$$

where D_i° is the self-diffusion coefficient of the rate-determining ion when $a_{AO} = 1$; γ is a numerical factor of the order of one, m is a coefficient reflecting the disorder type of the reaction product, $\Delta G^\circ_{AB_2O_4}$ is its Gibbs energy of formation, and v is the increase of the product volume resulting from the transport of one equivalent of ions through the reaction layer. If the reaction mechanism is established, then v is known. A comparison between calculated and measured $\bar{k}$-values is presented in Fig. 7.

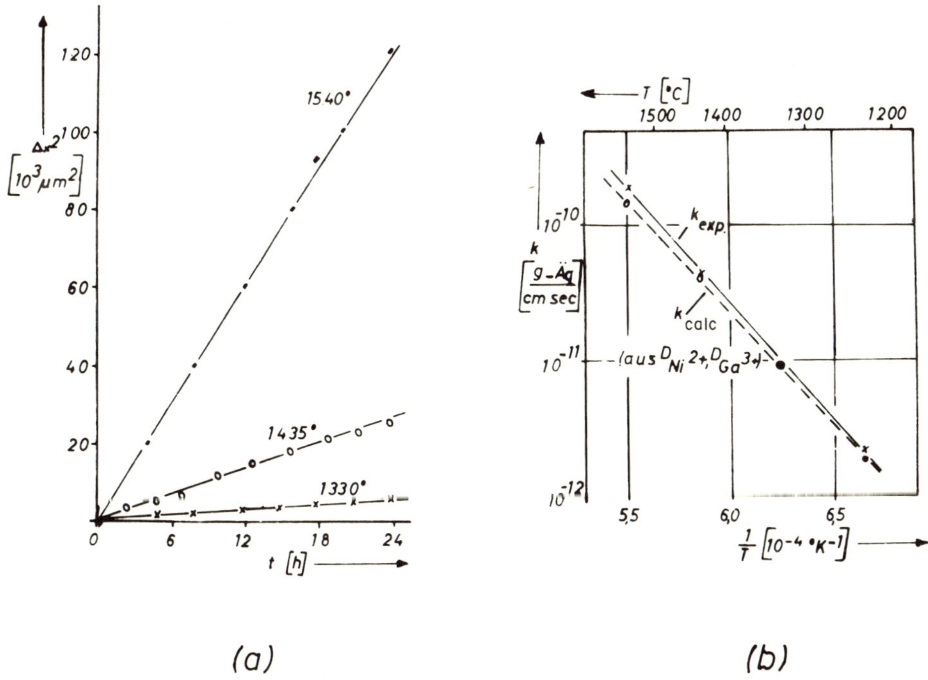

Fig. 7. The parabolic rate law (a) and the comparison between experimental and calculated rate constants $k = \bar{k}/v$ (b) for the reaction $NiO + Ga_2O_3 = NiGa_2O_4$[19].

The kinetic problem of a heterogeneous reaction, for example $AO + B_2O_3 = AB_2O_4$, could be solved since at the phase boundaries all the independent variables, defect concentrations, etc., were unambiguously determined. This is also true if multiphase reaction products form.[21] Since $\Delta x^{(k)} = (2\,\bar{k}^{(k)} \cdot t)^{1/2}$, $\Sigma \Delta x^{(k)} = t^{1/2} \,.\, \Sigma\,(2\,\bar{k}^{(k)})^{1/2}$. This means that under the given experimental conditions a multiphase reaction product forms according to a parabolic rate law too. If the elementary kinetic and thermodynamic parameters are given for every phase (k), then it has been shown that the overall growth rate can be calculated. Even in a system with a two-phase reaction product, the exact solution is rather tedious, leading to a quadratic equation for the two $\bar{k}$-values.[20] In case that $\bar{k}^{(1)} >> \text{or} << \bar{k}^{(2)}$, an approximate solution in terms of the ratio of the reaction layer thicknesses $\Delta x^{(1)}/\Delta x^{(2)}$ is

$$\frac{\Delta x^{(1)}}{\Delta x^{(2)}} \approx \frac{D^{(1)}}{D^{(2)}} \cdot \frac{\Delta\mu_{AO}^{(1)}}{\Delta\mu_{AO}^{(2)}} \,, \tag{7}$$

where $\Delta\mu_{AO}$ is the change in the chemical potential of the reactant AO over the reaction layer.

Either artificial markers such as thin platinum wires or natural markers such as pores, grain boundaries, etc., have often been used to mark the original interface between the reactants in order to obtain information about the atomistic reaction mechanism (in analogy to the Kirkendall marker experiment). Sometimes the conclusions about the reaction mechanism derived from other independent measurements (e.g., diffusion coefficients) could be confirmed. At other times, however, the location of the platinum wires or other inert markers after the reaction seemed to be in contradiction to the assumed reaction mechanism.[19] For example, even where counterdiffusion of cations in the reaction product was ascertained, markers were found close to the phase boundaries or even in the bulk phase of a reactant. An example of such behavior is given in Fig. 8.

The explanation of these effects needs the definition of reactions of first and second kind, and the corresponding treatment is given in the following section. It will be shown that without a knowledge of the mutual solubility of the reactants and their diffusivity, nothing can be concluded from the location of the markers in so-called marker experiments with respect to a reaction mechanism.

5 REACTIONS OF FIRST AND SECOND KIND

In context with quasibinary multiphase reaction layers and especially

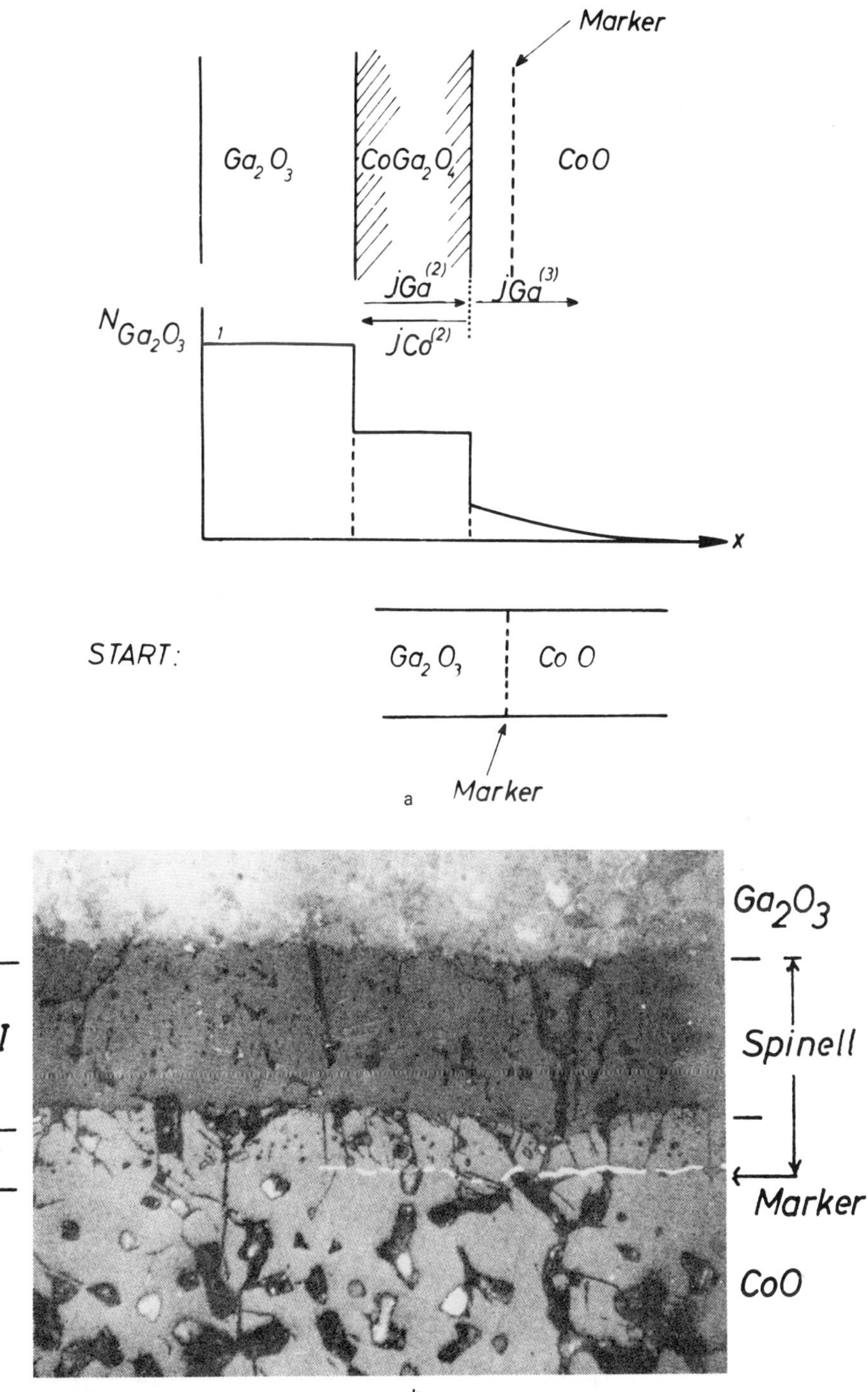

Fig. 8. Concentration profile and marker location in the case of the reaction $CoO + Ga_2O_3 = CoGa_2O_4$: (a) schematic; (b) sample photographed after reaction at 1340°C in air.[19]

with regard to reactions between solids with extended ranges of homogeneity, it is necessary to define reaction-rate constants ($\bar{k}$) of first and of second kind:

$$\bar{k}_I = \frac{\Delta x_I^2}{2\,t} \; ; \quad \bar{k}_{II} = \frac{\Delta x_{II}^2}{2\,t} \; ; \tag{8}$$

Δx_I is the reaction layer thickness where the layer is formed in a reaction starting with pure reactants (for example AO and B_2O_3). Δx_{II} is the reaction-layer thickness where the reaction layer is formed from the saturated adjacent phases. Examples for a one-phase reaction product and for a multiphase reaction product are schematically illustrated in Fig. 9. In order to obtain a general relation between the two reaction-rate constants, the following deduction is given:

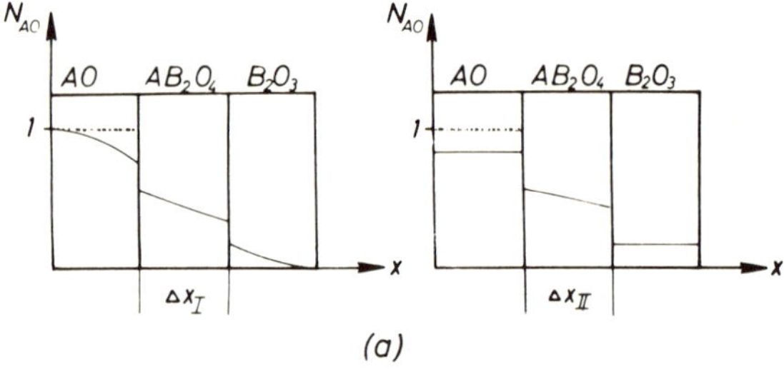

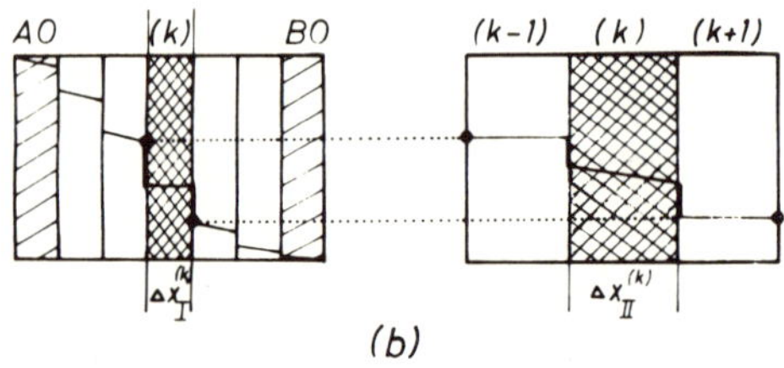

Fig. 9. Reactions of first and second kind and the corresponding concentration profiles in case of a monolayer reaction product (a) and of a multilayer reaction product (b).

It is well known that one can derive the interdiffusion coefficient $\tilde{D}$ from the concentration profile in a one-dimensional diffusion couple with the help of the Boltzmann-Matano method and, because of its very general assumptions, this is true even if the binary or quasibinary system involves a variable molar volume $V_m\,[N_2]$ and several different phases. The equation for $\tilde{D}$ then reads

$$\tilde{D}\,[N_2] = \frac{(N_2^+ - N_2^-)\cdot V_m\,[N_2]}{2t\,(\partial N_2/\partial x)_x}\left[(1-Y)\int_{-\infty}^{x}\frac{Y}{V_m}\,dx + Y\int_{x}^{\infty}\frac{1-Y}{V_m}\,dx\right], \tag{9}$$

where Y is equal to $(N_2-N_2^-)/(N_2^+-N_2)$. Equation (9) has been deduced by Sauer and Freise.[22] Accordingly, the diffusion coefficient $\widetilde{D}\left[N_2^{(k)}\right]$ in the kth layer of a multiphase reaction product can be written as

$$\widetilde{D}\left[N_2^{(k)}\right] = \frac{1}{2t\left(\dfrac{\partial N_2^{(k)}}{\partial x}\right)}\left[\frac{N_2^+-N_2^{(k)}}{N_2^+-N_2^-}\int_{-\infty}^{x(k-1,k)}\frac{V_m^{(k)}}{V_m}\cdot(N_2-N_2^-)\cdot dx + \frac{(N_2^{(k)}-N_2^-)/(N_2^+-N_2^{(k)})}{N_2^+-N_2^-}\Delta x^{(k)} + \frac{N_2^{(k)}-N_2^-}{N_2^+-N_2^-}\int_{x(k,k+1)}^{\infty}\frac{V_m^{(k)}}{V_m}(N_2^+-N_2)\,dx\right]. \tag{10}$$

From the definition of an average interdiffusion coefficient $(\overline{D})$ one obtains, using Eq. (10) and starting with pure reactants ($N_2^+ = 1$, $N_2^- = 0$),

$$\overline{D}^{(k)} = \frac{\Delta x^{(k)}}{2t\cdot\Delta N_2^{(k)}}\left[(1-N_2^{(k)})\int_{-\infty}^{x(k-1,k)}\frac{V_m^{(k)}}{V_m}N_2\,dx + N_2^{(k)}(1-N_2^{(k)})\cdot\Delta x^{(k)} + N_2^{(k)}\int_{x(k,k+1)}^{\infty}\frac{V_m^{(k)}}{V_m}(1-N_2)\cdot dx\right]. \tag{11}$$

In case that the different phases (k) in the reaction product have negligible ranges of homogeneity, the integrals in Eq. (11) degenerate into a sum of the form

$$\sum_{\nu}\frac{V_m^{(k)}}{V_m^{(\nu)}}\cdot N_2^{(\nu)}\cdot\Delta x^{(\nu)}\;.$$

Since the average interdiffusion coefficient $(\overline{D}^{(k)})$ of the kth reaction layer is an intrinsic property of that layer and does not depend upon the starting material in a reaction of the first or of the second kind, one may return to Eq. (11) in order to derive the relation between reaction-rate constants of first and second kind. By equating $\overline{D}^{(k)}$ in Eq. (11) for a reaction of second kind, taking into account the definitions of Eq. (8), it is found that

$$\bar{k}_{II}^{(k)} = \bar{k}_I^{(k)} \cdot \frac{N_2^{(k+1)} - N_2^{(k-1)}}{(N_2^{(k)}-N_2^{(k-1)})\,(N_2^{(k+1)}-N_2^{(k)})} \cdot \Bigg[(1-N_2^{(k)}) \cdot N_2^{(k)} + \frac{(1-N_2^{(k)})}{\Delta x^{(k)}} \int_{-\infty}^{x(k-1,k)} \frac{V_m^{(k)}}{V_m} \cdot N_2\, dx + \frac{N_2^{(k)}}{\Delta x^{(k)}} \int_{x(k,k+1)}^{+\infty} \frac{V_m^{(k)}}{V_m} (1-N_2) \cdot dx \Bigg], \tag{12}$$

which, for a reaction product consisting of phases with very narrow homogeneity ranges, obviously becomes

$$\bar{k}_{II}^{(k)} = \bar{k}_I^{(k)} \cdot \frac{N_2^{(k+1)} - N_2^{(k-1)}}{(N_2^{(k)}-N_2^{(k-1)})\,(N_2^{(k+1)}-N_2^{(k)})} \cdot \Bigg[(1-N_2^{(k)}) \cdot N_2^{(k)} + \frac{(1-N_2^{(k)})}{\Delta x^{(k)}} \sum_{\nu=2}^{k-1} \frac{V_m^{(k)}}{V_m^{(\nu)}} \cdot N_2^{(\nu)} \cdot \Delta x^{(\nu)} + \frac{N_2^{(k)}}{\Delta x^{(k)}} \sum_{\nu=k+1}^{n-1} \frac{V_m^{(k)}}{V_m^{(\nu)}} \cdot (1-N_2^{(\nu)}) \cdot \Delta x^{(\nu)} \Bigg] . \tag{13}$$

Equations (12) and (13) give the desired relation between the rate constants of the first and of the second kind if the diffusion profile is a function of the distance coordinate or, more specifically in case of Eq. (13), if the thicknesses and composition of the product phases are known. Similar equations have been first derived by Wagner.[23]

The most important conclusion, for our purposes, from Eq. (12) for a one-phase reaction product is deduced if phase (1) is the one reactant, phase (3) is the other reactant, and the index s marks the saturation concentrations of phase (1) in phase (3) and vice-versa (see Fig. 10).

$$\left(\frac{\bar{k}_{II}}{\bar{k}_I}\right) \cong \frac{(1-N_2)\cdot N_2 \cdot (N_{2,s}^{(3)} - N_{2,s}^{(1)})}{(N_2-N_{2,s}^{(1)})\,(N_{2,s}^{(3)}-N_2)} \cdot \Bigg[1 + \frac{V}{\Delta x} \Bigg\{ \frac{1}{V^{(1)} \cdot N_2} \cdot \int_{\text{phase (1)}} N_2^{(1)}\, dx + \frac{1}{V^{(3)}\,(1-N_2)} \cdot \int_{\text{phase (3)}} (1-N_2^{(3)})\, dx \Bigg\} \Bigg] , \tag{14}$$

where terms without superscript refer to the reaction product and the integrals go over the respective phases. For negligible solubility in the reactants, the integrals vanish and $\bar{k}_{II} = \bar{k}_I$. In any case, $\bar{k}_{II} \geqslant \bar{k}_I$.

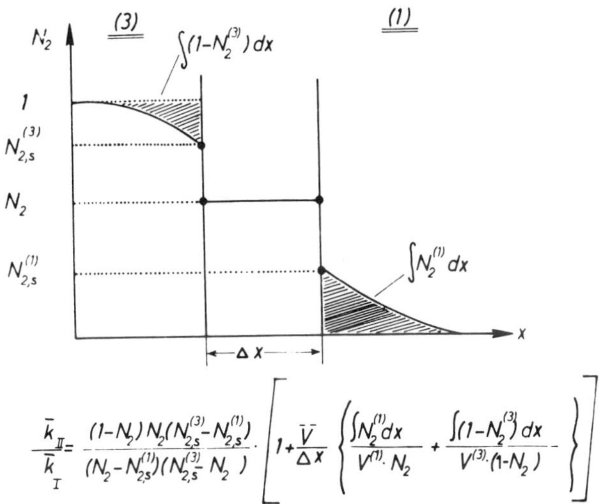

Fig. 10. Explanation of the symbols in the derivation of the ratio $\bar{k}_{II}/\bar{k}_I$ for a monolayer reaction product [see Eq. (14)].

Equation (14) is useful if, for example the diffusion profile and Δx have been measured by electron microprobe analysis after a solid-state reaction of the first kind in order to obtain the more fundamental reaction-rate constant of the second kind, from which the transport coefficients in the reaction product can immediately be calculated if the thermodynamics of the A-B-system is known [see Eq. (6) and Fig. 10)].

If, however, the diffusion coefficient in the reaction product and in the reactants and the thermodynamics of the system are known, and the task is now to calculate reaction-rate constants of the first and second kind, or, if marker experiments have been performed and the marker location has to be explained in a reaction of the first kind (for example if markers are found after the reaction embedded inside the reactant at a distance x from the phase boundary as is the case in the system $NiGa_2O_4$/NiO[19] or $NiCr_2O_4$/NiO, see Fig. 8), then it is necessary to calculate explicitly the different reaction rate constants as a function of the elementary parameters of the system. This has been done in a recent publication for a system with a single-phase reaction product.[24] In the explicit calculation, the analytical form of the interdiffusion coefficient as a function of composition has to be known. The calculations were done (1) for a constant $\tilde{D}$ and (2) for $\tilde{D}$ varying directly with concentration ($\tilde{D} = D^\circ \cdot c/c^\circ$). The latter dependence is important since in heterovalent ionic systems the dissolved species (Ga_2O_3 in CoO or NiO) determine the extrinsic point-defect concentrations which in turn determine the transport coefficients.[25]

Whether or not the markers are found inside the reactant after the reaction (see Fig. 8) depends only on the sum of the fluxes that arrive at the phase boundary through the reaction product and that leave the phase boundary into the adjacent reactant because of the solubility of the reaction product in the reactant.

It is of some interest to note that the explicit calculation yields $\bar{k}_I/\bar{k}_{II}$ to depend only upon a single parameter called r which is a function of $N_{2,s}^{(1)}/N_2$, $1\text{-}N_{2,s}^{(3)}/1\text{-}N_2$, $D^{\circ(1)}/\bar{D}$; $D^{\circ(3)}/\bar{D}$, and $\Delta G^{\circ}_{AB_2O_4}$. (The indices are explained in Fig. 10). To illustrate the theory, it has been calculated[24] that during the formation of $NiCr_2O_4$ from NiO and Cr_2O_3, the solubility of Cr_2O_3 in NiO at 1580°C (~15 percent) decreases the reaction rate constant $\bar{k}_I$ by about one-fourth compared with $\bar{k}_{II}$.

6 TOPOCHEMICAL REACTIONS AND THE STABILITY OF MOVING BOUNDARIES

In the field of solid-solid reactions, starting with plane interfaces, one often encounters roughening of the moving phase boundaries and/or morphologically interesting situations after some time of reaction. Examples of this behavior are shown in Figs. 11 through 13. The first of

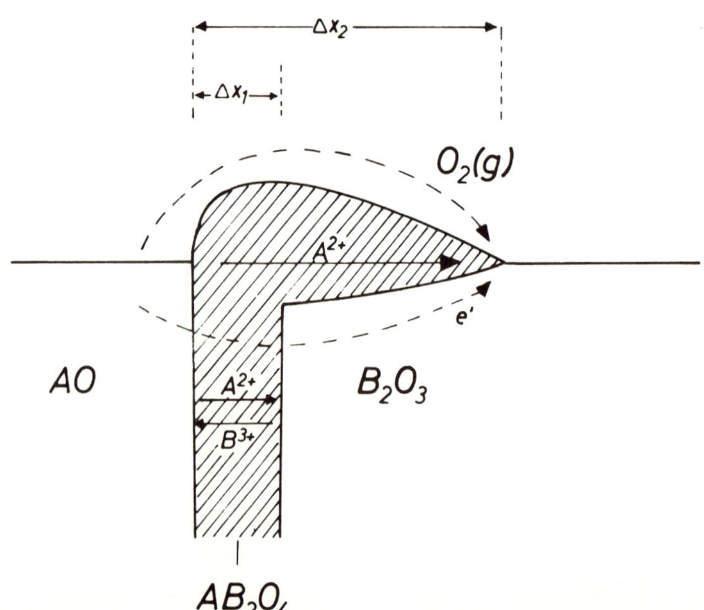

Fig. 11. Growth morphology for spinel formation $AO + B_2O_3 = AB_2O_4$ near the surface with adjacent gas phase.

these figures gives an example of a solid-state reaction morphology if a distinct and rapid diffusion path is available. The classical experiment for this type of kinetic behavior is shown in Fig. 14. Generalizing, one may say that the geometry of the reacting system conforms to a maximum overall rate, thus taking advantage of rapid diffusion paths such as adjacent gas or liquid phases, surfaces, grain boundaries, or dislocation pipes. Applying this principle, it may even occur that the reaction product is not forming between the reactants, but nucleates and grows such that a maximum reaction rate is achieved. In this case (see Fig. 15), and in analogy to the well-known principle of a local separation of the different independent reaction steps of an overall reaction in the electrochemistry of aqueous media, the phase-boundary reaction between two reactants and the production of the reactant occurs at two different loci.

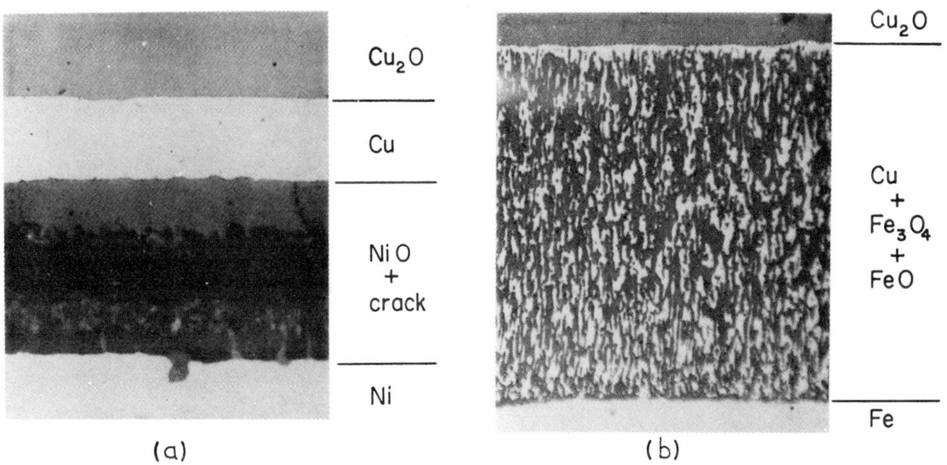

Fig. 12. Growth morphologies developing during displacement reactions: (a) $Cu_2O + Ni = NiO + Cu$; (b) $Cu_2O + Fe = FeO\ (+Fe_3O_4) + Cu$, according to ref. 28.

Another type of instability of moving interfaces has been shown to occur with displacement reactions. Double reactions of the form AO + BX = AX + BO and displacement reactions AO + B = BO + A have been discussed by Wagner and Jost[26,27] with regard to their morphologies. It was shown that flat and rugged interfaces may occur, leading to a monophase layer structure (layered morphology) for plane boundaries and a two-phase layer structure (aggregate arrangement) in case that the growing reaction product develops rugged interfaces. Again applying the principle of a maximum reaction rate, Rapp and co-workers[28] were able to quantitatively describe the kinetics of displacement reactions for flat phase boundaries as well as for nonstable flat phase boundaries when rugged interfaces and aggregate two-phase layers of reaction products are formed.

The underlying physical reason for the instability of flat moving interfaces is exactly the same as is operative in dendritic growth, in dry alloy oxidation, or in wet alloy corrosion, and is explained in Fig. 16. This simple picture will be complicated by mutual solubilities, by finite reaction resistances of the interfaces, by the influence of surface tension, and by fast transport along interfaces. Neglecting for the moment these complications, one would predict that flat interfaces and a layered structure are stable for the reaction between Cu_2O and Co (Ni), since the transport coefficient (D_ic_i) is larger in the reaction product copper than in the product CoO; the opposite is true for the reaction between Cu_2O and iron, where one would expect flat interfaces to be unstable and aggregate reaction-product structures to occur. Both cases have been verified recently. The corresponding morphologies are shown in Fig. 12.

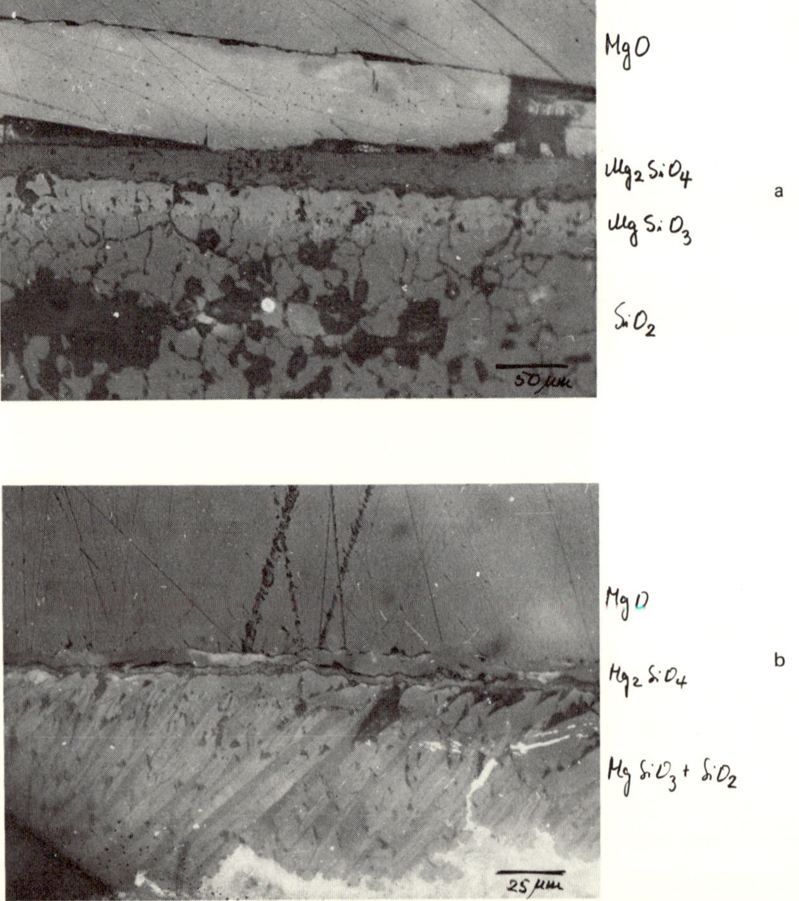

Fig. 13. The development of the morphology during the reaction between MgO and SiO_2: (a) 280 h, 1450°C, air; (b) 102 h, 1514°C, air.[29]

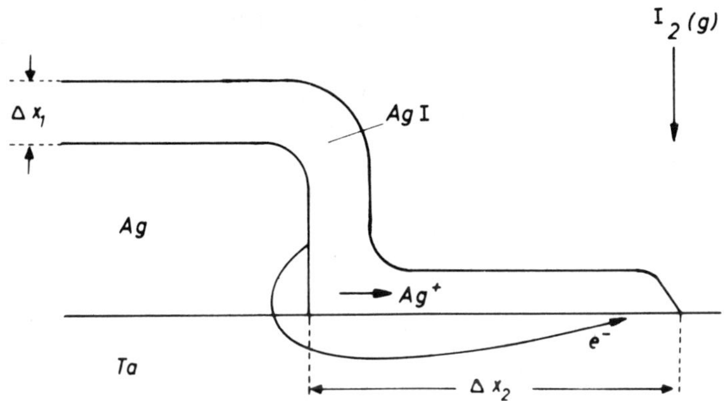

Fig. 14. Morphology of the reaction $Ag + \frac{1}{2} I_2\ (g) = AgI$. A tantalum foil is supporting the reaction product.[30]

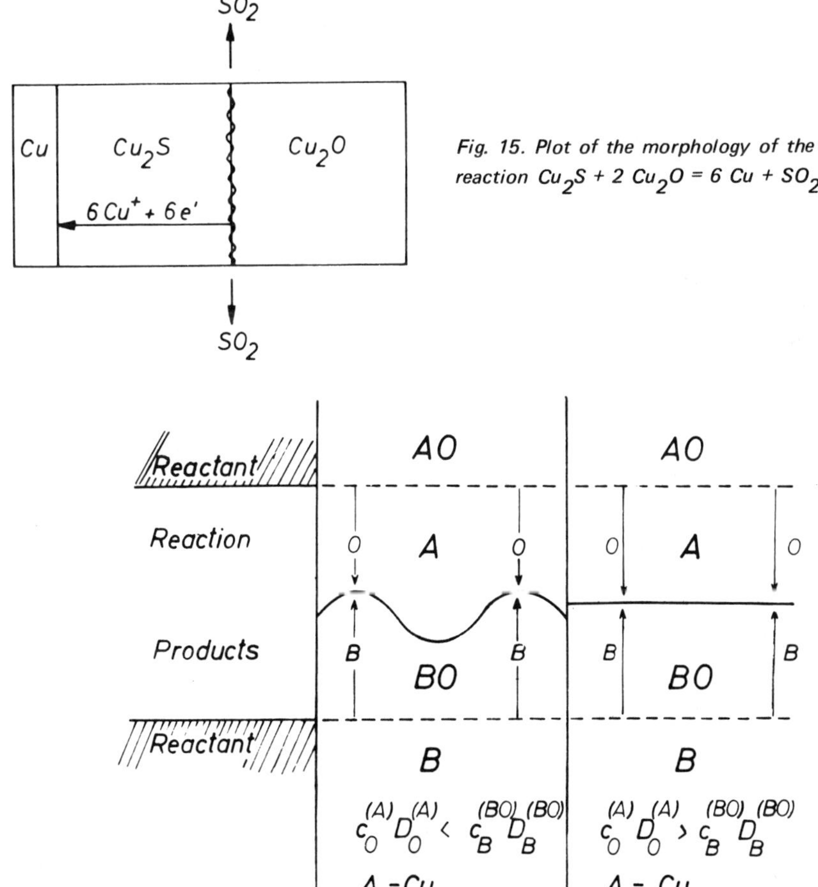

Fig. 15. Plot of the morphology of the double reaction $Cu_2S + 2\ Cu_2O = 6\ Cu + SO_2\ (g)$.[31]

Fig. 16. Plot of displacement reaction morphology explaining interface stability (see also ref. 28).

A third type of interface instability during solid-solid reactions has been observed in the system MgO-SiO_2. Since neither reactant nor the reaction products Mg_2SiO_4 and $MgSiO_3$ are known to have an appreciable range of homogeneity, the phase boundaries ought to be flat in a one-dimensional experiment. Nevertheless, as Fig. 13 shows, the interface $SiO_2/MgSiO_3$ starts to become uneven right from the beginning of the reaction, and after some time, an aggregate two-phase structure with a pronounced growth texture is found.

The explanation is probably that (1) SiO_2 as a reactant is not completely homogeneous in structure so that in the very beginning of the reaction a microscopically uneven interface develops; (2) the volume change between reactant and reaction product is appreciable so that a growth form is preferred that minimizes the overall strain; and (3) as it is well known, a marked increase in diffusivity is found along the surface or interface of silicates, as compared to the bulk diffusivities. The combination of these three effects could lead to a high growth rate of the platelike $MgSiO_3$ reaction product with a marked growth texture as observed. An estimate of interface diffusivities and activity changes due to strain is in the right order of magnitude.

7 PHASE-BOUNDARY REACTIONS

Any heterogeneous reaction starts with nucleation; thereafter, transport across phase boundaries and through the bulk of the newly formed phases is operative. The discussion up to here has been concerned with the diffusional transport in the bulk phases.

Strictly speaking, nothing is really known about phase-boundary reactions and their reaction mechanism on an atomistic scale in the solid systems of interest here. A pertinent analysis is still not available, although, in some cases, linear rate laws indicating phase-boundary control have been found when (see Fig. 17) the reaction-layer thickness was smaller than about 10^{-4} cm or 1 μm.

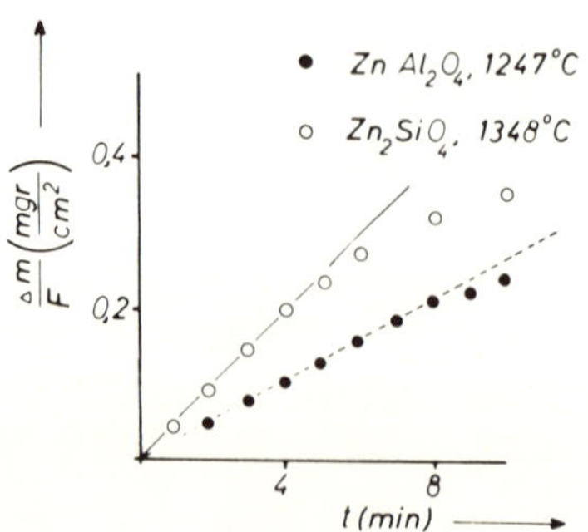

Fig. 17. Linear reaction rates for the reaction AO (g, a=1) + B_2O_3 (s) = AB_2O_4 (s) according to ref. 35.

Three different kinds of phase boundaries may be distinguished for the sake of a theoretical discussion:

(1) The close-packed anion sublattice is retained at the solid-solid phase boundary during reaction. This is, for example, the case at the boundary AO/AB_2O_4 if AO crystallizes in the NaCl structure. Thus the atomistic-reaction steps consist only of a rearrangement of the different cations on the octahedral and tetrahedral interstices. Accordingly, it has been observed that the reaction $MgO + Cr_2O_3 = MgCr_2O_4$ between single crystalline MgO and polycrystalline Cr_2O_3 results in a reaction product which has the same oxygen sublattice and is single crystalline so far as it is grown at the $MgO/MgCr_2O_4$ interface.[32]

(2) The close-packed anion sublattice is retained at the solid-solid phase boundary during reaction, but the stacking of this sublattice is changed. This is the case if at the phase boundary AB_2O_4/B_2O_3, B_2O_3 crystallizes in the α-corundum structure with hexagonal packing of oxygen ions (the spinel structure shows a nearly cubic closest packing). Here the atomistic phase-boundary reaction not only consists of a cation rearrangement as before, but also of a change in the oxygen packing sequence. It has been suggested that this change in the packing sequence is brought about by a correlated effect rather than by diffusional steps of oxygen ions. Since the (111) planes in a spinel and the (0001) planes in $\alpha\text{-}Al_2O_3$ are glide planes in a plastic-deformation process, one may anticipate that the simplest correlated phase-boundary reaction steps consist of the movement of bundles of partial dislocations under the action of chemical stress.* Accordingly, it is found that the spinel $NiAl_2O_4$ is single crystalline so far as it grows in the reaction $NiO + Al_2O_3 = NiAl_2O_4$ at the phase boundary $NiAl_2O_4/Al_2O_3$ from single crystalline Al_2O_3.[33]

$$\frac{dG}{d\xi} = \frac{1}{d\xi}\left[\frac{F \cdot d\xi}{V_{AB_2O_4}} \cdot \Delta\mu_{B_2O_3}(\xi)\right] = \frac{1}{d\xi}\left[P_{Ch} \cdot F \cdot d\xi\right] \text{ or } P_{Ch} = \frac{\Delta\mu_{B_2O_3}(\xi)}{V_{AB_2O_4}}.$$

(3) There is no crystallographic relation between reactant and reaction product. This is true for all sublattices. Accordingly, no epitaxial relations are found between reactants and reaction product. The solid-solid phase boundary consists of a disordered layer, which serves as a reservoir for ions from which the reaction product is built up.

In order to carry on an analysis of the phase-boundary reaction rate in terms of elementary thermodynamic and kinetic parameters, the different cases of phase boundaries should be treated separately. This simplest case is certainly found under (1), above, where only cation rearrangement will take place at the phase boundary. As can be seen from Fig. 18, in case of phase-boundary rate control, one has to expect a zone

*The term chemical stress (P_{Ch}) is used since (see Fig. 18)

in AO which is supersaturated with regard to B_2O_3. The question is how fast a cation rearrangement takes place in this zone in order that AB_2O_4 is formed. The relaxation time τ for the cation rearrangement in spinel phases can, in principle, be measured independently and has been measured for the case of $NiAl_2O_4$, disturbing the equilibrium cation distribution by a temperature jump, and following the cation relaxation by optical spectroscopy or other means.[33] A plot of the relaxation time versus temperature is presented in Fig. 19.

In the framework of this concept, the advancement of the phase boundary is a function of the mobility of B–ions in AO and the relaxation time τ for cation rearrangement

$$\frac{d\xi}{dt} = f\,(D_B^{(AO)}, \tau) \quad .$$

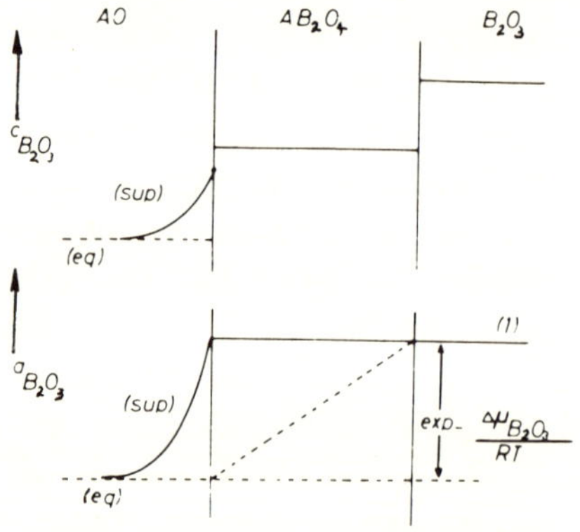

Fig. 18. Plots of concentration and activity (chemical potential) near the phase boundary AO/AB_2O_4 in the case of phase-boundary rate control.

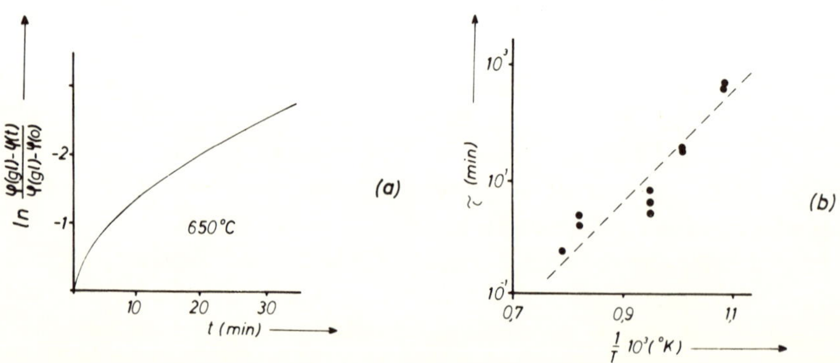

Fig. 19. (a) Relative change in lattice constant during cation distribution relaxation in $NiAl_2O_4$[33]; (b) temperature dependence of the average relaxation time τ for the cation redistribution in $NiAl_2O_4$[33].

For the purpose of an estimate, one may, in a zeroth approximation, set $d\xi/dt = \frac{\gamma \cdot \Delta\xi}{\tau}$, where $\Delta\xi = \left(2\, D_B^{(AO)} \cdot \tau\right)^{1/2}$ and γ is a numerical factor indicating the fraction of which is transformed into spinel in time τ (see Fig. 18). At 1250°C, $D_B^{(AO)}$ is of the order of 10^{-12} cm^2/sec[34] and τ is of the order of 1 sec[33]; thus $d\xi/dt$ is of the order of 10^{-6} to 10^{-7} cm/sec. This estimated value is of the same order of magnitude as that which has been measured for phase-boundary controlled reaction rates, as can be seen from Fig. 17. However, while measuring the rates plotted in Fig. 17, the reaction product AB_2O_4 was not in contact with solid AO, but with a gas phase the activity (a_{AO}) of which was kept constant and close to unity.[35] Therefore, different elementary reaction steps may have occurred at the phase boundary in comparison with those at a solid-solid phase boundary, and the rates may not be comparable.

There is no analysis available that would yield numerical values for phase-boundary reaction rates in case of the boundaries mentioned above under (2) and (3). This is a field of solid-state chemistry which seems to be a challenge for unemployed scientists.

ACKNOWLEDGMENT

I am indebted to Dr. W. Laqua and Dr. G. F. Yurek for permission to quote their experimental results.

REFERENCES

1. Faraday, M., *Quart. J. Sci.*, **9**, 319 (1820).
2. Roberts-Austen, W. C., *Proc. Roy. Soc.*, **59**, 283 (1896); **67**, 101 (1900).
3. Hedvall, J. A., *Einführung in die Festkörperchemie,* Vieweg, Braunschweig (1952), p. 3.
4. Tammann, G., *Lehrbuch der Metallkunde,* 4th edition, Voss-Verlag, Leipzig (1932).
5. Jander, W., *Z. anorg. Chemie,* **163,** 1 (1927).
6. Fischbeck, K., and Jellinghaus, W., *Z. anorg. Chemie,* **165,** 46 (1927).
7. Frenkel, J., *Z. Physik,* **35,** 652 (1926).
8. Wagner, C., and Schottky, W., *Z. phys. Chem.,* **B11,** 163 (1930).
9. Jost, W., *J. Chem. Phys.,* **1,** 466 (1933).
10. Wagner, C., *Z. Phys. Chem.,* **B34,** 309 (1936).
11. Schmalzried, H., in *Progress in Solid State Chemistry* (H. Reiss, Ed.), Pergamon Press, Oxford (1965), Vol. 2, p. 265 ff.
12. Schwier, G., Dieckmann, R., and Schmalzried, H., *Ber. Bunsenges. Physik. Chem.,* **77,** 402 (1973).
13. Chen, W. K., and Peterson, N. L., Argonne National Laboratory, Mat. Sci. Div., personal communication.
14. Waite, T. R., *J. Chem. Phys.,* **32,** 21 (1960).
15. Blank S. L., and Pask, J. A., *J. Am. Ceram. Soc.,* **52,** 669 (1969).
16. Schmalzried, H., and Holt, J. B., *Z. Phys. Chem.,* NF, **60,** 220 (1968).

17. Jones, J. T., and Cutler, I. B., *J. Am. Ceram. Soc.*, **54**, 335 (1971).
18. Wynnyckyj, J. R., Schmalzried, H., and Sockel, H.-G., *Z. Phys. Chem.*, NF, **68**, 30, 49 (1969).
19. Laqua, W., Doctoral Thesis, Technical University of Berlin, 1971.
20. Schmalzried, H., in *Reactivity of Solids,* (G. M. Schwab, Ed.), Elsevier Publishing Co., Amsterdam (1965).
21. Fink, G., *Z. Phys. Chem.*, NF, **44,** 122 (1965).
22. Sauer, F., and Freise, V., *Z. Elektrochem.*, **66,** 353 (1962).
23. Wagner, C., *Acta Met.*, **17,** 99 (1969).
24. Pelton, A. D., Schmalzried, H., and Greskovich, C. D., *Ber. Bunsenges. Physik. Chem.*, **76,** 543 (1972).
25. Greskovich, C., *J. Am. Ceram. Soc.*, **53,** 498 (1970).
26. Jost, W., *Diffusion und chemische Reaktion in festen Stoffen,* Steinkopff, Leipzig (1937).
27. Wagner, C., *Z. anorg. allgem. Chem.*, **236,** 320 (1938).
28. Rapp, R. A., Ezis, A., and Yurek, G. J., *Met. Trans.* **4,** 1283 (1973).
29. Daemgen, R. M., Diplom Thesis, Technical University Clausthal, 1972.
30. Ilschner-Gensch, C., and Wagner, C., *J. Electrochem. Soc.*, **105,** 198 (1958).
31. Wagner, C., *Trans. AIME,* **4,** 214 (1952).
32. Sockel, H.-G., Diplom Thesis, University Göttingen, 1965.
33. Rogalla, W., Diplom Thesis, University Göttingen, 1963.
34. Chen, W. K., Peterson, N. L., and Robinson, L. C., *J. Phys. Chem. Solids,* **34,** 705 (1973).
35. Duckwitz, C. A., and Schmalzried, H., *Z. Phys. Chem.*, NF, **76,** 173 (1971).

DISCUSSION on Paper by H. Schmalzried

STRINGER: Dr. Schmalzried has used pseudobinary diffusion theory throughout his presentation. Systems involving two oxides are generally three component systems, and although for much of his talk he restricted himself to cases where the oxygen was immobile, so that there are probably not too many differences, it is a little dangerous to use the same concepts in the interpretation of the development of two-phase structures during diffusion. It is not necessary to invoke short-circuit or other nonequilibrium processes to explain the appearance of two-phase regions in ternary diffusion systems; it is sufficient that the virtual diffusion path plotted on the ternary equilibrium diagram passes through a two-phase region. Of course, the processes mentioned by Dr. Schmalzried may be important in determining the exact *morphology* of the two phases.

SCHMALZRIED: It is not the mobility of ions, but the thermodynamic condition of quasi-binary-systems, that is important. And I do not know of any system, starting with binary oxides of narrow ranges of homogeneity (and SiO_2 as well as MgO certainly belong to that group), that could not be treated as quasibinary. Even the formation of ferrites, starting with "FeO" as one reactant, has never led to the formation of unusual growth morphologies. The reason for this can be found in any article on defects in ternary ionic crystals, *Progress in Solid State*

Chemistry (H. Reiss, Editor), Vol. 2, Pergamon Press, Oxford (1965). In conclusion, I think it is established without doubt that in solid-solid reactions of the quasibinary or binary type, there are mechanisms operative that are responsible for the instability of plane interfaces which are not related to diffusion theory in ternary systems. This is also independent of whether the oxygen is mobile or not.

SMELTZER: Is it possible that oxygen actually migrates when diffusion couples are formed with the metal-conducting vacancy-type divalent oxides, for example, MgO-NiO, and NiO-CoO? The few marker measurements that have been completed on these systems indicate that a Kirkendall effect does occur and, hence, that oxygen must migrate in the diffusion zones. This would possibly permit the metal cations to migrate under conditions approaching local equilibrium restraints during the complete period of diffusion.

SCHMALZRIED: The answer to your question depends certainly on the microstructure of the real materials used. For the case that the diffusion couple consists of rather perfect single crystals, one would assume that at least in the beginning of the interdiffusion experiment, when the gradients in composition are still steep, the surface of the material acts as an additional source or sink for defect equilibration. But all of these questions can and should be answered experimentally.

HEUER: Are isotopic tracers, e.g., ^{18}O for oxygen with subsequent detection by proton activation, not more reliable as markers than fine platinum wires or foils?

SCHMALZRIED: There is no general answer to the question. If you have enough information on the diffusional behavior of the materials in question, it may well be.

POINT DEFECTS IN OXIDES

A. B. Lidiard

Theoretical Physics Division
A.E.R.E., Harwell
Berkshire, United Kingdom

ABSTRACT

This article describes recent progress with the calculation of the structure, energy, and other characteristics of point defects and their simple aggregates in ionic crystals and oxides. Current approaches to the construction of physical models and to their mathematical evaluation are briefly reviewed. Recent results obtained at Harwell by means of Norgett's HADES program for ionic crystals are described, particular emphasis being given to the understanding of defect structure and other properties in lattices with the fluorite structure (alkaline-earth fluorides and UO_2) which these results provide.

1 INTRODUCTION

The theoretical descriptions of atomic transport used for the interpretation of experimental data (diffusion, ionic conductivity, solid-state reactions, etc.) are statistical theories.[1] These theories, for the most part, have been developed for assemblies of point defects of relatively simple structure; for example, vacancies are generally assumed to retain the point symmetry of the lattice site, while interstitials may be assumed to occupy the "obvious" interstitial sites, i.e., those that offer the most space to accommodate the extra atom or ion. These simple, largely geometrical, ideas extend also to the defect-jump directions under thermal activation, as well as to the structure of simple defect complexes, e.g., impurity-vacancy and impurity-interstitial pairs, and their thermally activated jumps. Accurate statistical theories of atomic transport processes and associated relaxation phenomena based on these simple assumptions have been developed and have been notably successful for the strongly ionic compounds and for metals. In these cases, it has often been possible, by careful experimental studies, to verify the underlying assumptions of defect structure, as for example when specific quantities such as the correlation factor, f, for diffusion can be isolated.

However, when approaching new and more complex systems, these assumptions may be inadequate and there may be considerable uncertainty over the appropriate model for statistical development. This is exemplified by the variety of statistical treatments[2] of nonstoichiometric UO_{2+x}. Even with a relatively simple oxide such as MgO, there are conflicts over the interpretation of diffusion data, associated with our ignorance of the formation energy of defects.[3] To remove these uncertainties, then, we need a theory of the properties of individual defects (particularly of their structure and characteristic energies) to guide us a priori towards the most appropriate model for statistical treatment. This might be deemed particularly important in a case like UO_{2+x} where no simple explanation of the unusual and complex interstitial defects inferred from neutron-diffraction experiments[4] has been forthcoming.

Early examples of such calculations are those of Mott and Littleton[5] for NaCl, KCl, and KBr and those of Huntington and Seitz[6] for copper. There have, of course, been many subsequent refinements and developments since that early work; most obviously, the availability of computers has allowed a wide range of models and "substances" to be studied with any individual computer program. Nevertheless, the development of mathematically accurate methods has been slow, and the assessment of the reliability of any given calculation was rather difficult until recently. In this paper, we want to draw attention specifically to recent substantial advances in the mathematical and numerical methods which now permit the precise evaluation of the defect properties of any specified physical model. We shall not go into these in detail since the essentials have already been published elsewhere.[7] We shall, however, illustrate the value of these calculations by describing some recent examples of their application to ionic crystals and oxides, especially MgO and UO_2. These illustrations concern the characteristic energies of the basic defects (Frenkel and Schottky), their structure, and the structure of their simpler aggregates; in particular, we shall show how recent work[8-10] reconciles the interstitial structures inferred from ESR and ENDOR studies of the alkaline-earth fluorides with the results of neutron-diffraction measurements on CaF_2 doped with YF_3 and on UO_{2+x}, as well as providing insight into a number of other aspects of defects in these materials.

2 PHYSICAL MODELS AND MATHEMATICAL METHODS[11]

These atomistic defect calculations for realistic models are almost always made for a static lattice. This simplification is an undoubted convenience, but justification for it can be found in the general quasi-harmonic approximation to lattice vibrations by which the vibrational thermal motions at any temperature T are analyzed about the mean lattice positions at this temperature, i.e. for a lattice with mean lattice spacing which is some given function of temperature, $r_o(T)$. With this approximation, it follows immediately that at high temperatures, where classical equipartition holds, the difference in internal energy of two different configurations of the lattice; Δu, e.g. with and without a defect, is

$$\Delta u = \Phi_2 - \Phi_1 \quad , \tag{1}$$

where Φ is the potential energy of the equivalent static lattice with parameter $r_o(T)$. Thus, internal energies of formation, activation, etc., of defects can be obtained immediately from Eq. (1) when the potential energy of the appropriately relaxed lattice configuration is known. This relaxed lattice configuration is generally obtained by minimizing the potential energy alone with respect to atomic displacements about the mean lattice configuration. Such a procedure does, of course, omit the dependence of the vibrational entropy upon the lattice relaxation, but such terms seem to be generally small compared with those arising from the potential energy and, indeed, also compared to the numerical uncertainty in many of the calculations made up to now. We shall, therefore, describe only work done in the static lattice approximation.

The principal questions of physics, therefore, center on the potential energy function, Φ. Here one uses models constructed on the basis of general theories and models of cohesion and interatomic forces.[12] The confidence that one may have in these is variable. For instance, in the simpler metals (e.g. the alkalis or aluminum), where pseudopotential theory holds, one has more securely founded models than in the case of those empirical models constructed for the transition metals.[13] Although models of ionic solids, including oxides, are generally elaborations of the classical Born model, they have now been extensively tested against a wide range of physical properties. Theories of mixed ionic and covalent bonding however, have not so far led to useful models of interionic forces; work on defects in oxides has therefore used ionic models constructed in the same way as for the more strongly ionic halides. This recent experience with ionic models points away from the frequently quoted Born-Mayer[14] and Fumi-Tosi[15] models in some respects:

1. It is important to incorporate as many empirical data as possible (e.g., dielectric constants, elastic constants, polarizabilities, Reststrahl frequency, etc.), as well as the general experience from lattice dynamics and the results of *ab initio* quantum mechanical calculations of interatomic and interionic interactions.
2. It is possible to incorporate such a large amount of empirical information by using shell models[16] (which contain one or more internal degrees of freedom). Polarizable-point-ion models, which have often been used for defect calculations in the past, by contrast can account only for a very limited number of crystal properties. As a result, so-called deformation–dipole extensions of these models have been made and used for the prediction of lattice dynamic and defect properties.[17]
3. Both polarizable-point-ion and so-called deformation–dipole models suffer from inherent instabilities[18,19] – "polarization catastrophes".
4. Careful examination of the nonelectrostatic terms in the potential energy shows that, in addition to the exponential Born-Mayer form of repulsion due to overlap of the outer electronic shells of two ions, there is also an attractive term between pairs of anions which, at the normal anion-anion separation, can be as much as 10 times larger than the usual van der Waals attraction. This conclusion seems to hold for alkali halides[19,20], alkaline-earth fluorides[9,10], MgO[21], and UO_2[10]. Its origin is not yet clear, although it could be a result of configuration interaction between the perfectly ionic ground state and excited charge-transfer states.

When we turn from the details of the model to the evaluation of the energy of a specified defect, then we can describe the general strategy of all calculations as dividing the lattice into two regions: an inner region (I) containing the defect with a certain number of neighbors and an outer region (II) representing the rest of the lattice. The total potential energy is then of the form

$$\Phi = \Phi_{I} + \Phi_{I,II} + \Phi_{II} \quad . \tag{2}$$

The object of the division is to draw the boundary between I and II so that in the outer region II, all displacements are small enough to employ a harmonic, or even a continuum, description of Φ_{II}, i.e. so that general methods of finding the equilibrium solution can be used there. The minimization of Φ with respect to the variables of the inner region I must then be done numerically since this region, by definition, is anharmonic. There exist three principal methods. The first two, that due to Kanzaki[22] and that developed mainly by Tewary[23], both deal with region II in the full harmonic approximation; the Kanzaki approach uses a Fourier transformation method, while Tewary takes over the real-space Green function technique developed originally for the study of vibrations in defect lattices[21]. In practice, in all applications of these methods made so far, region I contains only the defect itself, although such a restriction need not be imposed. These techniques are largely analytical; perhaps for this reason, no general purpose computer programs suitable for a wide range of defects and models have been developed from them, although it should be possible to do so.

The third method, a generalization of the basic principle behind the original Mott and Littleton calculation, assumes the limiting continuum solution for region II. As this limiting behavior is necessarily reached farther away from the defect than harmonic behavior, generalized Mott-Littleton methods must deal with much larger inner regions (I). Fortunately, numerical analysts have made substantial progress in the field of minimization[25] in the last 10 years or so, and it is now possible to deal efficiently and relatively speedily with functions of several hundred variables, although for the simpler, more symmetrical defects, it may be necessary to include only a few shells of neighbors to the defect.

These minimization methods fall into three classes: (1) those which iterate using only evaluations of the (energy) function itself (search methods), (2) those which also use derivative evaluations in the choice of iterative steps (e.g. conjugate gradient methods[26]), and (3) those which additionally introduce evaluations of the second derivatives or actually the inverse of the matrix of second derivatives – and, in practice, for large problems this means approximations to this inverse (modified Newton-Raphson methods). Speed of convergence is achieved by the incorporation, either implicitly or explicitly [as in (3)], of information on the second derivatives. Thus, a quasi-Newton-Raphson method may require a small number of iterations largely independent of the number of variables, n. The conjugate gradient method of Fletcher and Reeves[26] includes, implicitly, information related to second derivatives but does not evaluate them directly from the function: in principle it converges in a number of iterations proportional to n, although, in practice, the overall comparison with quasi-Newton-Raphson methods may be more favorable than this would suggest. (An example of the application of this method to a problem of dislocation structure is given in ref. 28.) By contrast, even good search methods which evaluate only the function require $O(n^2)$ iterations to converge. The development of a general-purpose program (HADES) based on the generalized Mott-Littleton method and a quasi Newton-Raphson minimization routine[27] thus represents a major step forward in our ability to evaluate accurately the energy and stable configuration of specified defects in ionic lattices.

With the insertion of appropriate constraints, the program is also capable of evaluating the energies of saddle-point configurations and, therefore, the activation energies. In practice, in complicated cases the precise "reaction coordinate" may not be obvious, i.e. it may not be obvious which combination of position variables is to be constrained, although a first approximation to it will be the straight line (in n-coordinate space) joining the initial and final stable defect positions. An automatic procedure for the location of saddle points has recently been described by Sinclair and Fletcher[29] and applied in another defect problem (crack propagation).

These and other examples show that rapid progress is now being made with methods and programs for the direct real-space simulation of defects in model crystal lattices. For ionic crystals, HADES is probably the most advanced of these currently available. It has been used so far for the modelling of point defects and impurities (especially rare-gas atoms) in the alkali halides (both NaCl and CsCl structures), in the alkaline-earth fluorides, and in MgO and UO_2. We comment below particularly on the recent oxide calculations.

3 APPLICATIONS, GENERAL

The previous evaluations of defect properties in the strongly ionic halides – or more strictly in ionic (shell) models constructed as described in the preceding section – have demonstrated not only that one can predict known defect energies with good numerical accuracy, but also that one may obtain new insights concerning complex and hitherto uncertain situations.

Thus, the calculated formation energies of the dominant intrinsic defects in the alkali halides[19,20,30] and in the alkaline-earth fluorides[9,10] generally agree quite well with values inferred from many careful experiments[31,32]. These same calculations on the alkaline-earth fluorides also confirm the nature of the dominant defects (anion Frenkel defects) by showing that the formation energies of the other possibilities (Schottky and cation Frenkel defects) are much higher. Shell-model calculations of

Frenkel defects in alkali halides have not been made, but evaluations of deformation dipole models by Schulze and Hardy[17] confirm that Schottky defects have the lowest formation energy and that anion Frenkel defects have the highest. Calculations such as these thus give confidence that reliable defect energies can be obtained by these methods. But the value of them to the understanding of experimental situations is better shown by the work on the energies of rare-gas atoms in ionic solids.[7,33] These calculations of the activation energies for interstitial gas-atom migration and of binding energies into vacancies and vacancy clusters have provided new insight concerning the behavior of gas atoms introduced into ionic solids by irradiation or other means. Another example of their value as an aid to understanding is provided by Catlow's recent work on interstitial clusters in the alkaline-earth fluorides.[8,10]

These successes, which we have already commented on in more detail elsewhere[7,34], encourage us then to approach the long-standing problem of modelling the oxides, in particular MgO and UO_2, which we expect to show strong analogies to the alkali halides with the NaCl structure and to the alkaline-earth fluorides, respectively. Indeed, just because this analogy has often been used in the interpretation of experiments, one object of the calculations must be to examine how far it is valid. Thus, in the next section, we shall summarize some of the recent results on UO_2 in parallel with those on the alkaline-earth fluorides. We shall not report much about MgO (since the work is not so advanced) except to say that the formation energy of Schottky defects is high (7 to 8 ev) as one would expect from general analogy to the alkali halides[31] and as is consistent with the recent magnesium self-diffusion experiments of Wuensch et al.[3] However, before turning to the detailed results on defects in UO_2, we must comment on one particular aspect of the ionic model which applies to oxides generally.

The cohesive energy of oxides such as the alkaline-earth oxides and UO_2 is generally in agreement with an ionic model based on O^{2-} ions. Nevertheless, the obvious difficulty with this model is that no stable O^{2-} ion exists in the free state and that an additional potential is needed to stabilize these ions in the crystal. For the normal lattice ions, this stabilization is provided by the Madelung potential. For O^{2-} ions in interstitial positions where this Madelung potential is less favorable, stabilization may still occur, however, through the field created as a result of the polarization of the crystal by the ion (i.e., self-trapping modified by the central O^- ion). Obviously, the electron distribution of the O^{2-} ion depends on this stabilizing field so that one might expect the Born-Mayer overlap repulsion potential to depend upon the particular environment. Attempts to evaluate theoretically the potential of interaction of two O^{2-} ions in a suitable point-ion potential (to simulate the Madelung field) by means of the IBMOL-V molecular orbital program, however, soon suggested that the two outermost electrons were in rather diffuse states which contributed little to the interaction energy. Catlow[10] therefore proposed that one could take the O^{2-} - O^{2-} overall interaction to be equal to that of a pair of O^- ions, independently of crystal environment. This actually leads to a potential decreasing much more rapidly with interionic separation than that for a pair of F^- ions.

4 DEFECTS IN URANIUM DIOXIDE

In trying to understand defects in nonstoichiometric UO_2, especially UO_{2+x}, we often, either implicitly or explicitly, rely on analogies with the alkaline-earth fluorides or other materials with the fluorite lattice. For example, excess anions may be introduced in CaF_2, etc., by doping with trivalent rare-earth or other trivalent cations, and the understanding of their behavior may be used to provide models of UO_{2+x}. One of the important tasks for the theory of defect properties must be to study how far this analogy holds and where it breaks down. Important basic questions are:

(1) Are the intrinsic defects of anion Frenkel type (as in CaF_2)?
(2) If so, what are their charge states? For example, since the O^{2-} ion is not stable in free space, is it sensible to consider O^{2-} interstitials, as often done by analogy to the fluorite compounds?
(3) What is the manner of incorporation of excess oxygen in UO_{2+x}?
(4) More particularly, how does one understand the complex interstitial defect structures first inferred from neutron-diffraction studies of UO_{2+x} by Willis[4] in 1964?

And so on. We shall here report some of the results of recent theoretical studies[9,10] using ionic shell models of UO_2 and the HADES program. Parallel studies of defects in the alkaline-earth fluorides were also made; direct comparison with these is therefore possible. Some of the important results are as follows.

4.1 Intrinsic Defects

Table I gives calculated energies of formation of Frenkel and Schottky defects in UO_2. From these values it will be seen that the majority defects in pure stoichiometric UO_2 should be doubly charged oxygen Frenkel defects. Although this is analogous to CaF_2, it is nevertheless somewhat surprising in view of the instability of the free O^{2-} ion and

Table I. Calculated Energy of Formation of Frenkel and Schottky Defects in UO_2

Defects	Energy, ev
Anion Frenkel – O_i'' and $O_v^{\cdot\cdot}$	5.3–5.5[a]
Anion Frenkel – O_i' and $O_v^{\cdot}$	8.9
Schottky – U_v'''' and $2O_v^{\cdot\cdot}$	10.3
Cation Frenkel – U_v'''' and $2U_i^{\cdot\cdot\cdot\cdot}$	18.5

(a) Two different models were evaluated in this case: the lower value is for the fully ionic model (U^{4+} and O^{2-} ions), the higher value for one with reduced ion charges; the fully ionic model predicts the cohesive energy better (to within 3 percent).

the unfavorable Madelung potential at the (¼,¼,¼) interstitial position*. In fact, in contrast to CaF_2, the stable position of the O^{2-} ion in this model is not exactly at (¼,¼,¼) but a small distance away in the <110> direction; however, the basic reason for the stability of the doubly charged oxygen Frenkel defects over the singly charged ones in this model is the much greater magnitude of the polarization energy – a negative quantity which the Jost continuum formula indicates should be proportional to q^2_{eff}. What this means in the present case is that the extra electron on the O^{2-} interstitial is self-trapped in the polarization field of the lattice (superimposed on that provided by the O^- ion itself).

It might well be questioned whether it is not going too far to use a classical ionic model to describe this situation – and the same model as one uses for the normal substitutional ion at that! While a more fundamental approach to this is undoubtedly desirable, it may nevertheless be noted that the ionic model appears internally consistent on this point. Thus it predicts that this extra electron is bound by 2.3 ev relative to a state of zero kinetic energy outside the UO_2 crystal. Therefore, unless the electron affinity of UO_2 is greater than +2.0 ev – which seems unlikely**

*The fluorite lattice is made up of a simple cubic lattice of anions with an interlocking fcc lattice of cations. Each cation is at the center of a cube of eight anions, while every other such cube (in the <100> directions) is empty. The centers of these empty anion cubes are the "obvious" interstitial positions; they are (¼,¼,¼) positions relative to the overall fluorite lattice cell.

**Indeed, the ionic model gives the affinity of UO_2 as -3.5 ev, indicating a surface barrier at the surface of opposite sign to that arising in the alkali halides, for example, which have affinities of ~½ ev.

– this indicates a state of small radius (~1 Å or less in the effective mass approximation); the ionic model thus appears internally consistent.

4.2 Nonstoichiometric UO_{2+x}

Having found that the intrinsic defects in UO_2 are O_i'' and $O_v^{\cdot\cdot}$ Frenkel pairs*, it is natural to assume that excess oxygen is incorporated by the addition of further O_i'' accompanied by electron holes in the form of U^{5+} ions in the cation sublattice ($U_s^{\cdot}$). However this does not follow, and it is necessary to consider the stability of the O_i'' and the $U_s^{\cdot}$ defects against several possible reactions, viz.,

$$2U_s^{\cdot} \rightarrow U_s^{\cdot\cdot} + U_s \ , \qquad (3)$$

$$U_s^{\cdot} + O_i'' \rightarrow O_i' + U_s \ , \qquad (4)$$

$$O_i'' + \tfrac{1}{2}O_{2(gas)} \rightarrow 2O_i' \ . \qquad (5)$$

The calculated energies of these reactions are given in Table II. It will be seen that they are all quite large and positive, which indicates that the predominant point defects in UO_{2+x} are O^{2-} interstitials and substitutional U^{5+} ions. There thus appears a close analogy between UO_{2+x} and CaF_2 doped with trivalent cations, one often assumed before but until now not verified by energy calculations.

Table II. Calculated Energies of Defect Reactions in UO_2

Reaction	Energy Required, ev
$2U_s^{\cdot} \rightarrow U_s^{\cdot\cdot} + U_s$	1.0
$U_s^{\cdot} + O_i'' \rightarrow O_i' + U_s$	3.8
$O_i'' + \frac{1}{2}O_{2(gas)} \rightarrow 2O_i'$	2.7

4.3 Oxygen Interstitial Structures

We have already mentioned the difference between the stable structure of the O_i'' defects in UO_2 and the F_i' defects in CaF_2 and the other alkaline-earth fluorides. This is a consequence of the shorter range of the

*We use here the chemical notation in which defect *effective* charges are denoted by a superscript $^{\cdot}$ (for +e) or $'$ (for −e), while interstitials and vacancies are denoted by subscripts i and v; substitutional ions are denoted by a subscript s. There being no possibility of exchange disorder in this material, we do not need a more elaborate notation.

calculated oxygen-oxygen potential (the Born-Mayer constant, ρ, being only one-half its value of approximately 0.3 Å for the $F^- - F^-$ interaction). However, the relatively small displacement of O_i'' from the (¼,¼,¼) interstitial site does not allow us to explain the complex structures proposed by Willis[4] following his neutron-diffraction observations* in UO_{2+x}.

Although these so-called 2:2:1 and 2:2:2 structures have been incorporated into statistical models of UO_{2+x}, their origin has remained unclear, thus throwing doubt upon some of the detailed assumptions made in these statistical calculations.[35]

Recently, analogous structures have been found[36] in Y^{3+}-doped CaF_2, while Catlow's calculations using the HADES program have shown very clearly that these arise as a result of defect clustering.[8] The simplest complexes are $Y_s^{\cdot} - F_i'$ pairs which in fact maintain the <100>-tetragonal symmetry one would expect from bringing a cube-center interstitial F_i' up as neighbor to a substitutional Y^{3+} ion. This symmetry and structure have been confirmed by many ESR and ENDOR experiments on alkaline-earth fluorides doped with trivalent rare earth and other magnetic ions. However, when two such pairs are brought together, head to tail in a (100) plane forming a square defect

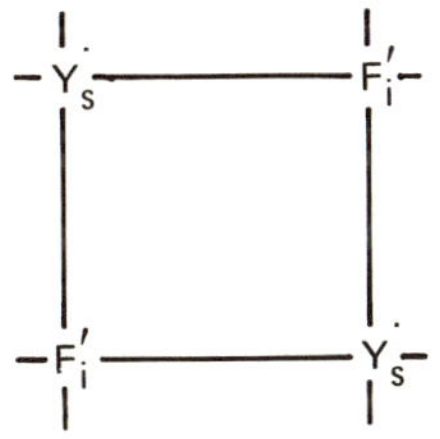

$$\begin{array}{ccc} Y_s^{\cdot} & — & F_i' \\ | & & | \\ F_i' & — & Y_s^{\cdot} \end{array}$$

then the situation changes quite remarkably. First, the two F_i' are pulled in towards the center of the square, i.e., along the <110> axis, essentially by the electrostatic attraction of each by the two $Y_s^{\cdot}$ ions. (The $Y_s^{\cdot}$ are, of course, firmly held in position, particularly by the Madelung field, while the F_i', as we have already noted, are in a repulsive Madelung field and so move off these sites rather readily). Second, this inward movement of the

---

*What Willis directly determined was the average structure and the numbers of oxygen or uranium atoms on the various sites. In addition to the normal sites of the fluorite structure, he found two kinds of oxygen interstitials, one corresponding to a <110> displacement from the (¼,¼,¼) position to (¼,0.19,0.19), the other corresponding to a <111> displacement from the normal anion lattice site to the position (0.2,0.2,0.2). The fractions of oxygen atoms on these sites were roughly x and 2x, while the deficiency of oxygen on the normal anion sublattice was also about 2x. No oxygen ions were found on the (¼,¼,¼) cube-center interstitial sites. No defects were detected in the uranium sublattice. These experiments could not give correlations in occupancy of these various sites, but Willis proposed clustered structures containing 2 <111> interstitials, 2 vacancies and either 2 or, most probably, 1 <110> interstitial, so called 2:2:2 or 2:2:1 structures.

interstitials brings into play strong repulsive interactions with the two F^- ions situated on the normal lattice sites above and beneath the midpoint of the square. Third, these ions relax off-site in directions which minimize these repulsive forces, which means not in the symmetry-preserving <100> direction, but towards the neighboring vacant cube-center interstitial sites, i.e., in <111> directions. The net result then is a Willis 2:2:2 cluster. The energy gain resulting from these symmetry-lowering relaxations accounts for most of the binding energy of these clusters relative to two separated $Y_s^{\cdot} - F_i'$ pairs (1 to 1½ ev in the three alkaline-earth fluorides). The calculated magnitudes of the F^--ion relaxations were in fairly good agreement with experiment. Qualitatively similar relaxations were also found in larger clusters. Thus there seems little doubt that Catlow's results enable us to understand the neutron-diffraction results in terms of defect aggregation, at the same time reconciling these with the simpler, more symmetrical defects seen in ESR and ENDOR experiments.

We should expect an analogous situation in UO_{2+x}. As we have already remarked, the basic defects here are predicted to be O^{2-} interstitials and U^{5+} ions. The simplest complex to show the same pattern of low-symmetry relaxations is the electrically neutral one formed from one O_i'' and two $U_s^{\cdot}$ in neighboring positions. The O_i'' moves towards the same position along the <110> direction and the two closest normal oxygen ions then move off-site in <111> directions, generating a Willis 2:2:1 structure. Thus Willis' neutron-diffraction results on UO_{2+x} would appear to be largely explicable in terms of the localization of two electron holes on two uranium neighbors to an oxygen interstitial. The binding energy of such a cluster is over 1 ev, and the clusters must therefore form in large proportions at the temperatures and degrees of nonstoichiometry of the specimens studied in these experiments. However, as comparable binding energies are also obtained for some other simple clusters, and as larger clusters are also stable, it is likely that the real situation in Willis' specimens was rather complicated. Experiments at low values of x (<0.05) would appear to be necessary to separate the effects of the different complexes.

4.4 Defect Mobility

Calculations of defect mobility should be equally capable of providing activation energies and detailed information about the most probable displacement paths. In general, there is less experience with such calculations though, as previously remarked, calculations of rare-gas mobility in a variety of ionic solids have been notably successful. The methods have therefore been applied to the calculation of defect mobility in the alkaline-earth fluorides and in UO_2. We shall limit ourselves to summarizing several general points.

(1) In UO_2, as in the alkaline-earth fluorides, the anion vacancy is more mobile than the anion interstitial. The calculated activation energies appear to be too low compared with experiment, although the experimental values are themselves rather uncertain both for the fluorides and for UO_2. In the case of UO_{2+x}, the measured oxygen diffusion may well refer to the motion of 2:2:1 or higher clusters, and although their activation energies have not yet been evaluated, they are almost certainly higher than those of isolated O_i'' defects in view of the large relaxations in these clusters.

However, the 2:2:1 and 2:2:2 clusters in UO_{2+x} will still be relatively mobile since the electron hole (U^{5+} state) moves easily, whereas the corresponding defects in CaF_2/YF_3 are firmly anchored by the substitutional Y^{3+} ions. We thus expect qualitative differences in the dependence of anion mobility upon defect concentration between UO_{2+x} and CaF_2/YF_3.

(2) The interstitial anion moves most frequently via an interstitialcy or replacement mechanism. In this and other modes of movement, the relaxation of saddle-point ions in <111> directions (i.e. toward vacant interstitial sites) proves to be important in lowering the activation energy. Indeed, one may couple this observation with the structure of the clustered defects and infer that the possibility of large <111> relaxations by normal anions close to defects must always be considered in the fluorite lattice.

5 CONCLUSION

These results show that for the first time we have a successful model of UO_2 with which to calculate its defect properties; indeed, this is the first time we have a successful model for any oxide. Previous calculations of defect energies in ionic models of the alkaline-earth oxides[37,38] and of UO_2[39] were plainly incapable of giving satisfactory results. The calculations of Dickens et al.[40] on the divalent 3d transition metal oxides were apparently more successful, but this is probably because the mathematical approximations made (the same as those of the original Mott-Littleton calculation) obscure the inadequacies of the physical model employed (a polarizable-point-ion model).

The important basic advance of the last few years is the development of accurate and efficient general-purpose defect programs, such as HADES, which are capable of handling a wide variety of different defects without program modification. This has enabled the rapid acquisition of experience with different ionic models, and this experience has been incorporated into the approach taken to the modelling of UO_2. It should also be noted that the models now being constructed are built up on the basis of empirical information obtained from the perfect crystals, supplemented with quantum mechanical calculations of the interionic forces; they do not contain ad hoc adjustments made to obtain agreement between calculated and experimental defect quantities. The results summarized here show that calculations made by these means are capable of offering useful physical insight concerning defect structure and properties. This understanding in turn allows the more confident selection of models for statistical mechanical theories of the thermodynamic and transport properties of ionic solids and oxides.

REFERENCES

1. For comprehensive accounts see, e.g., Kröger, F. A., *The Chemistry of Imperfect Crystals,* North Holland Publishing Co., Amsterdam (1964); Adda, Y., and Philibert, J., *La Diffusion dans les Solides,* Presses Universitaires de France, Paris (1966); Flynn, C. P., *Point Defects and Diffusion,* Clarendon Press, Oxford (1972).
2. See, e.g., Lidiard, A. B., *J. Nucl. Mat.,* **19**, 106 (1966); Kröger, F. A., *Z. Phys. Chem.,* **49**, 178 (1966); Thorn, R. J., and Winslow, G. H., *J. Chem. Phys.,* **44**, 2632 (1966); Atlas, L. M., *J. Phys. Chem. Solids,* **29**, 1349 (1968); Atlas, L. M., *The Chemistry of Extended Defects in Non-Metallic Solids,* L. R. Eyring and M. O'Keeffe (Eds.), North Holland Publishing Co., Amsterdam (1970), p. 425.
3. See Harding, B. C., and Price, D. M., *Phil. Mag.,* **26**, 253 (1972), and Wuensch, B. J., Steele, W. C., and Vasilos, Y., *J. Chem. Phys.,* **58**, 5258 (1973).
4. Willis, B.T.M., *Proc. Brit. Ceram. Soc.,* **1**, 9 (1964).
5. Mott, N. F., and Littleton, M. J., *Trans. Faraday Soc.,* **34**, 485 (1938).
6. Huntington, H. B., and Seitz, F., *Phys. Rev.,* **61**, 315 (1942); Huntington, H. B., *Phys. Rev.,* **61**, 325 (1942).
7. Lidiard, A. B., and Norgett, M. J., *Computational Solid State Physics,* F. Herman, N. W. Dalton, and T. R. Koehler (Eds.), Plenum Press, New York and London (1972), p. 385.
8. Catlow, C.R.A., *J. Phys.,* **C6**, L64 (1973).
9. Catlow, C.R.A., and Norgett, M. J., *J. Phys.,* **C6**, 1325 (1973); also *J. Phys. (Paris),* **34**, Colloque C-9, 45 (1973).
10. Catlow, C.R.A., D. Phil. thesis, University of Oxford (1973).
11. This section summarizes and brings up to date the fuller discussion contained in ref. 7. Additional background material will be found in the earlier reviews by Barr, L. W., and Lidiard, A. B., *Physical Chemistry – An Advanced Treatise,* W. Jost (Ed.), Academic Press, New York and London (1970), Vol. X, p. 152, and by Hardy, J. R., and Flocken, J. W., *C.R.C. Critical Reviews in Solid State Sciences,* **1**, 605 (1970), as well as in the recent article by Corish, J., and Jacobs, P.W.M., *Chemical Society Specialist Periodical Reports (Surface and Defect Properties of Solids)* (1973), Vol. 2, Chap. 7.
12. For a general discussion see Torrens, I. M., *Interatomic Potentials,* Academic Press, New York and London (1972).
13. See, e.g., the papers by Johnson, R. A., and Wilson, W. D., and by Perrin, R. C., Englert, A., and Bullough, R., in *Interatomic Potentials and Simulation of Lattice Defects,* P. C. Gehlen, J. R. Beeler, and R. I. Jaffee (Eds.), Plenum Press, New York and London (1972), pp. 301 and 509, respectively.
14. See, e.g., Seitz, F., *Modern Theory of Solids,* especially Chap. II, McGraw Hill, New York and London (1940).

15. Fumi, F. G., and Tosi, M. P., *J. Phys. Chem. Solids,* **25**, 31 (1964); Tosi, M. P., and Fumi, F. G., *ibid.,* p. 45.
16. See, e.g., the reviews by Bilz, H., *Computational Solid State Physics,* F. Herman, N. W. Dalton, and T. R. Koehler (Eds.), Plenum Press, New York and London (1972), p. 309, or by Cochran, W., *C.R.C. Reviews in Solid State Sciences,* **2**, 1 (1971).
17. See Karo, A. M., and Hardy, J. R., *Phys. Rev.,* **B3**, 3418 (1971); Schulze, P. D., and Hardy, J. R., *Phys. Rev.,* **B5**, 3270 (1972) and **B6**, 1580 (1972), and references cited there.
18. Faux, I. D., *J. Phys.,* **C4**, L211 (1971).
19. Faux, I. D., Ph.D. thesis, University of London (1971).
20. Müller, M., and Norgett, M. J., *J. Phys.,* **C5**, L256 (1972); *J. Phys. (Paris),* **34**, Colloque C-9, 159 (1973).
21. Catlow, C.R.A., Faux, I. D., and Norgett, M. J., unpublished work (1973).
22. Kanzaki, H., *J. Phys. Chem. Solids,* **2**, 24 and 37 (1957).
23. Tewary, V. K., AERE Report T.P. 388 (1969) and *Advances in Physics,* **22**, 757 (1973); see also Bullough, R., and Tewary, V. K., *Interatomic Potentials and Simulation of Lattice Defects,* P. C. Gehlen, J. R. Beeler, and R. I. Jaffee (Eds.), Plenum Press, New York and London (1972), p. 155.
24. See, e.g., Maradudin, A. A., *Rep. Prog. Phys.,* **28**, 331 (1965).
25. See, e.g., the reviews by Fletcher, R., *Comp. Phys. Commun.,* **3**, 159 (1972), and by Powell, M.J.D., *Mathematical Programming,* **1**, 26 (1971).
26. Fletcher, R., and Reeves, C. M., *Comput. J.,* **7**, 149 (1964).
27. Norgett, M. J., and Fletcher, R., *J. Phys.,* **C3**, L190 (1970).
28. Norgett, M. J., Perrin, R. C., and Savino, E. J., *J. Phys.,* **F2**, L73 (1972).
29. Sinclair, J. E., and Fletcher, R., AERE Report T.P.540; *J. Phys.,* **C**, in press.
30. Faux, I. D., and Lidiard, A. B., *Z. Naturforsch.,* **26a**, 62 (1971).
31. See, e.g., Barr, L. W., and Lidiard, A. B., ref. 11.
32. See, e.g., Hayes, W. (Ed.), *Crystals With the Fluorite Structure,* Clarendon Press, Oxford (1974).
33. Felix, F., *J. Phys. (Paris),* **34**, Colloque C-9, 149 (1973).
34. Lidiard, A. B., *Pure and Appl. Chem.,* in press.
35. See, e.g., the review of thermodynamic properties by Roberts, L.E.J., and Markin, T. C., *Proc. Brit. Ceram. Soc.,* **8**, 201 (1967); the analysis of anelastic relaxation modes by Socino, G., de Batist, R., and Gevers, R., *Proc. Brit. Ceram. Soc.,* **9**, 73 (1967); and the calculations of correlation factors for oxygen diffusion by de Bruin, H. J., and Murch, G. E., *Phil. Mag.,* **27**, 1475 (1973).
36. Cheetham, A. K., Fender, B.E.F., and Cooper, M. J., *J. Phys.,* **C4**, 3107 (1971); Steele, D., Childs, P. E., and Fender, B.E.F., *J. Phys.,* **C5**, 2677 (1972).
37. Yamashita, J., and Kurosawa, T., *J. Phys. Soc. Japan,* **9**, 944 (1954).
38. Boswarva, I. M., and Franklin, A. D., *Mass Transport in Oxides,* J. B. Wachtman and A. D. Franklin (Eds.), National Bureau of Standards, Washington (1968), p. 25.
39. Tharmalingam, K., *Phil. Mag.,* **23**, 181 (1971).
40. Dickens, P. G., Heckingbottom, R., and Linnett, J. W., *Trans. Faraday Soc.,* **64**, 1489 (1968).

DISCUSSION on Paper by A. B. Lidiard

SELTZER: As an explanation for low diffusion activation energies, Wuensch has pointed out that in all studies using the magnesium radiotracer to measure magnesium diffusion in MgO, extrinsic activation energies have been obtained because of the short half-life of this tracer. The silicon daughter product self-dopes the MgO. On the other hand, activation energies for creep in MgO have been reported to be as high as

4 to 7 ev, which suggests that the theoretical calculations for the Schottky formation energies in MgO are correct.

LIDIARD: The theoretical values of Schottky defect formation energy in MgO calculated by Faux are sufficiently high that for all the materials so far studied experimentally, the measured diffusion should have been in the extrinsic region (typical impurity contents $\sim 10^{-4}$ mole fraction). We thus agree with the interpretation offered by Wuensch and co-workers.

WELCH: How did you determine the interaction between fluorine ions and yttrium ions in the fluorite calculations? How sensitive are the results to the details of this interaction?

LIDIARD: Catlow followed Franklin [*J. Amer. Ceram. Soc.*, **50**, 648 (1967)] and assumed that Y^{3+} was equivalent to Ca^{2+}, apart from its additional charge, for the reason that their radii are closely equal. However, the calculations were repeated for ions of other sizes, e.g., La^{3+}. The qualitative results were unaltered, but the binding energies of a 2:2:2 cluster in La^{3+}-doped alkaline-earth fluorides (relative to two separated La^{3+}-F_i^- pairs) are only about half those for Y^{3+}-doped crystals.

DRAGOO: Was your empirical potential a central force potential? If so, which elastic constants did you use for the rock-salt-structure compounds?

LIDIARD: Yes, the potentials used were central-force potentials. For the alkali halides and MgO, C_{11} and C_{12} originally were fitted together with other empirical data (lattice parameter, dielectric constants, etc.). More recently for MgO C_{11}-C_{12} and C_{44} were fitted but the calculated results so far are not sensitive to this change. In the fluorite structure, C_{12} need not equal C_{44} and here all three elastic constants were fitted, again using a central-force potential. Actually, it turns out that all six third-order elastic constants, C_{ijk}, are well described by this model of the alkaline-earth fluorides [Catlow and Norgett, *J. Phys.*, **C6**, 1325 (1973)]. Obviously, though, for the rock-salt structure we must try to include three-body interactions and examine their effect in detail.

FRANKLIN: The *ab initio* calculations of the anion-anion interactions may be more appropriate for lattice anions than for anions in interstitial sites. This seems particularly true for the attractive part which seems to arise as an indirect result of the lattice-anion interaction with the lattice cations.

LIDIARD: Possibly, but until we are sure we understand the origin of these attractive terms, it is difficult to argue one way or the other. However, our view is that these calculated anion-anion interactions are good at close distances of approach and that it *is* valid to retain the attractive term down to the separations typical of interstitials and their neighbors. Certainly, if one drops the attractive term at these close separations while retaining it at "normal" separations, the calculated results are less satisfactory [Catlow and Norgett, *J. Phys.*, **C6**, 1325 (1973)].

TURCOTTE: Concerning the extension of these calculations from dilute point defects to concentrated defects, has the effect of concentration been considered? This would seem important with respect to the extent of relaxation.

LIDIARD: No, we have not in these calculations considered any statistical distribution of other defects in the relaxation region. This is really a matter of defect interactions and, except at close distances, these should be describable by electrostatics and elasticity theory. Insofar as the electrostatic interactions generally dominate when we have defects bearing a net change, the appropriate statistical description will be provided by Debye-Hückel theory as in the alkali and silver halides, etc.

TURCOTTE: With regard to the Willis defect models, the work by Catlow suggests that rather large cation displacements accompany the anion shifts. In fact, the cell symmetry is no longer cubic and should be observable experimentally (by diffraction methods). Has this distortion been observed for either the CaF_2-YF_3 or related UO_{2+x} systems?

LIDIARD: These various clusters do, of course, have a low symmetry; but, in general, unless they order in some way, the overall symmetry of the specimen will remain cubic. I do not know of any such ordering, although given that sufficient defect mobility is retained, one would expect it to occur at low enough temperatures.

FRANKLIN: In reply to the question of whether the complexes formed from two or more rare-earth ion – fluorine interstitial pairs in CaF_2 increase in number as the rare-earth concentrations increase, there is such evidence in the optical studies of Makovsky with W. Low. This work shows that for concentrations above about 1 mole percent, the rare-earth ions (particularly Gd^{3+}) are in either cubic sites or in unspecified clusters, and the cluster concentrations increase with rare-earth concentration.

SCIENTIFIC UNDERSTANDING OF THE MANUFACTURE AND PROPERTIES OF SOLID OXIDIC INDUSTRIAL MATERIALS

H. J. Vink

Philips Research Laboratories
Eindhoven, Holland

In this after dinner talk it seems to me a good thing not to dwell heavily upon purely scientific problems. Today we all have had the privilege to listen to five lectures by famous scientists. In order to prepare us for the coming days of contributions and discussions the speakers have given us excellent reviews of highly interesting fields in solid state science.

Tonight I would like to say something about the relations between solid state science and the industrial application of solid oxides. I am an industrial scientist and therefore somewhat more at home in this borderland. With the aid of a number of examples, I would like to elaborate a bit about how, on the one hand, the results and the way of thinking of solid state science have been useful and even indispensable to solve industrial problems, i.e., to make it possible for industry to manufacture its products in a reproducible way according to narrow specifications. On the other hand however, it will also become clear that industry, at the same time, creates problems for science, that because of their complexity or of their "strangeness" have so far seemed to resist a scientific understanding.

(1) We all know about the long period, starting in the mid-thirties and extending to about 1963, during which it was uncertain whether p-type conduction in NiO and CoO or n-type conduction in α-Fe_2O_3 was of the small-polaron hopping type or of the large-polaron semiconductor type. It was not before 1963 that it was concluded from measurement of homogeneous single crystals or of polycrystalline material prepared in a special way, that conduction in these oxides was of the large-polaron semiconductor type. The main cause of the confusion was the fact that during the preparation of the polycrystalline ceramic-type samples, high-ohmic grain-boundary layers were formed, the properties of which completely obscured the real properties of the material enclosed by these boundaries. These grain-boundary layers were formed during the cooling down to room temperature of samples fired at high temperatures. For a partial pressure of oxygen equal to that of air, the degree of oxidation of NiO or CoO or the degree of reduction of α-Fe_2O_3 decreases strongly with decreasing temperature. This, together with the fact that the diffusion velocity of oxygen in those oxides also diminishes strongly with decreasing temperature, leads to the formation of thin high-ohmic surface layers in p-type doped NiO and CoO and n-type doped α-Fe_2O_3.

Doing Seebeck and d-c conductivity measurements on those samples and not taking into account the presence of these high-ohmic barriers led to the erroneous conclusion of a mobility that could be thermally activated by a hopping mechanism. As, in the late fifties, the small-polaron hopping model of conductivity had been worked out, a theoretical backing was present to explain this erroneous experimental interpretation.[1]

The well-known NTC thermistors (NTC = negative temperature coefficient of the resistance) are an example of materials for which scientific considerations – such as those just mentioned – have been applied to guarantee reproducible manufacture of an industrial product within narrow specifications. By suitable doping, thermistors can be made out of p-type NiO or CoO and n-type α-Fe_2O_3, with a large range of conductivities and of temperature coefficients. Reproducible results can be obtained only by careful control of the outer layers of the grains. For n-type conductors, the adsorption of oxygen is harmful, while for p-type conductors, the depletion of oxygen is undesirable. Both situations give rise to high-ohmic boundary layers and, as a consequence thereof, to unwanted frequency and voltage dependence. To avoid this, the gas atmosphere during firing and cooling of those ceramic products must be carefully controlled, and, in addition, their microstructure should not show an open porosity, in order to minimize possible reactions leading to boundary layers.[2]

(2) The existence of layers so troublesome for NTC thermistors is of the utmost importance in a number of other industrial applications. Take, for example, the PTC effect of antimony-doped $BaTiO_3$ (PTC = positive temperature coefficient of the resistance). This effect could in principle be explained by the presence, due to surface states, of a potential barrier at the surface of the grains which is built up with increase in temperature because of the fact that the effective dielectric constant – above the ferroelectric Curie point – decreases with increasing temperature. However a great many phenomena remained unclear, and reproducible PTC resistors could be made only by adhering strictly to manufacturing procedures that were not really understood. Only recently, by rigorously applying the whole apparatus of defect chemistry to donor-doped $BaTiO_3$, has definite progress been made.

It now seems that the negative surface charge has to be replaced by a negative space charge. This negative space charge originates from the fact that the interior and the surface of the grain differ in oxygen content. Because of the doping with donors, the concentration of oxygen vacancies throughout the grain is very low; this impedes the oxygen diffusion and consequently leads to a surface layer having a different composition. On the basis of defect chemistry, it can be calculated that $BaTiO_3$ containing 0.3 at.% of a donor and fired in air at high (sintering) temperatures is n-type (with ρ = 100 ohm-cm), whereas the same material in equilibrium with air at room temperature would be p-type with an extremely high resistivity. Therefore, the electrical properties of the interior of the grains and of their surface may be easily quite different and consequently give rise to a depletion layer. In short, to get good PTC-$BaTiO_3$ resistors, one must aim for the presence of precisely that type of boundary layer that must be avoided to get good NTC-α-Fe_2O_3 resistors.

The replacement of the surface states by acceptors distributed in a thin surface layer does not change the essential features of the accepted explanation of the PTC effect, if and when the thickness of that layer is thin enough (e.g., 300 Å) to have its energy levels completely filled in the temperature range above the Curie point, where ϵ changes rapidly.[3]

(3) The newly discovered large-VDR-effect ZnO varistors are also interesting (VDR = voltage dependent resistance). After three to four decades of intensive research on ZnO, it suddenly and unexpectedly turned out that ceramics of ZnO containing about 0.5 mole % of Bi_2O_3 and other oxides showed Zener-type VDR characteristics. It seems that this "varistor" effect is directly related to the microstructure, which is that of conductive ZnO grains (of 30-μ diameter) separated and surrounded by an amorphous

phase of Bi_2O_3 containing some ZnO and having a thickness between 500 and 1000 Å. The nonlinear conduction seems to originate in this intergranular phase.

Although the simple addition of Bi_2O_3 gives a good varistor effect, the effect can be further enhanced by the addition of CoO, MnO, Cr_2O_3, and Sb_2O_3; especially if those oxides are all present simultaneously.

In principle, several mechanisms could explain such a nonlinearity. In order to be able to choose between possible explanations such as space-charge limited current or tunneling, more information is needed. In this respect, an important point is the investigation of the nature and properties of the interfacial layer. For instance, it should be established whether it is indeed an insulator or perhaps a semiconductor before a more definite conclusion about its conduction mechanism is possible.[4]

(4) We have talked somewhat about NTC resistors, PTC resistors and VDR resistors. Therefore, it will be proper to talk also about resistors that do not change at all. I have in mind the metal-glass thick-film systems. The metals (and possibly a metal-oxide) are formed by thermal decomposition of an organometallic solution: the resinates of gold and rhodium in an organic solvent. The glass component is a lead boroaluminosilicate. For the purely scientific mind, this system seems to be about the most horrible example of repulsive cooking chemistry. Nevertheless, it is possible with this system to get resistors with good mechanical and chemical stability and a wide range of square resistances from 10^3 to 10^5 ohms. They show (for applied voltages below 10^2 v/cm) a truly ohmic character and, in addition, a temperature coefficient that, between 30 and 700°K is less than 1×10^{-6} per degree! It appears that these layers – for a Rh/Au ratio lying between 13 and 30 percent – consist of three phases: glass, pure gold, and a phase containing rhodium. Most probably, the small gold particles of 200 Å diameter are covered by 5 to 10 Å-thick layers of possibly amorphous Rh_2O_3. Furthermore, these particles are embedded in the glass and in contact with each other through the thin amorphous layers. Conduction now takes place by means of a tunneling process through these layers. This process is essentially temperature independent and shows an ohmic character, because of the fact that as a consequence of the smallness of the gold particles, there are very many intermediate layers, each taking up only a very small part of the applied voltage. However, it must be emphasized that it has not been explained why it has to be rhodium and not any other (similar) metal to give the best results and also that the presence of an insulating layer of Rh_2O_3 has not really been proven.[5]

(5) As another example of the use of layers, mention is briefly made of the Mn-Zn ferrites used in high-frequency LC filters in telephony. In order to get a high initial permeability, ceramic Mn-Zn ferrites must be used, the grains of which do not contain pores. Further, in order to compensate for the temperature coefficient of the condenser, the Mn-Zn ferrites used should show a very low and constant positive temperature coefficient of their permeability. This can be obtained if these ferrites do contain $Ti^{4+}Fe^{2+}$-ion pairs and free Fe^{2+} ions on octahedral sites. Further, to get a low overall conductivity ($\sigma < 10^{-2}$ ohm-cm^{-1}) these grains must be smaller than 5 μ and surrounded each by a nonconductive layer of $CaO\text{-}SiO_2$ compounds or mixtures. To obtain all this, the manufacturing process (after the reaction phase) consists roughly of three consecutive steps. In the first one, the sintering takes place at high temperatures and high partial oxygen pressure. The concentration of the cation vacancies then is high, leading to a high mobility of the pores relative to the velocity of the grain growth. Therefore the pores, during this first sinter phase, diffuse out of the grains and coalesce between the grains. In the second step, the temperature is lower. As a consequence, the grain growth stops and the solubility of Ca^{2+} and Si^{4+} ions decreases. These can therefore diffuse out of the grains to form the insulating layer. During this second step also, the partial oxygen pressure is lowered after a while, leading to a low concentration of cation vacancies. As a consequence of this, the Ti^{4+} ions have to be compensated electrically by Fe^{2+} ions (which we need!) instead of by cation vacancies, as was the case in the first phase. When in such a way the precise chemical composition and the valencies have been established, the third step starts. This is a cooling stage during which, in order to keep this composition fixed, the oxygen pressure must be changed continuously, following a well defined pattern as a function of temperature and time.[6]

(6) To end this part of my talk, I want to draw the oxide-coated cathodes to your attention. For quite a number of years it was thought generally that the conduction mechanism of oxide-coated cathodes, which runs parallel to the pore conduction and predominates at low temperatures, was n-type conduction through the bulk of the grains. Only a few reliable experiments on the conductivity of p- and n-type single crystals of BaO as a function of the equilibrium oxygen pressure have been published by different authors. Nevertheless, it has proved to be possible to embed those results nicely in a general picture by a straightforward application to BaO of the well-known arguments of defect chemistry. From this picture one has to conclude that for the pressures of barium around a working oxide-coated cathode, the concentration of native donors in BaO is much too small to explain the so-called bulk conductivity at lower temperatures.

As a matter of fact, one should expect the BaO to be virtually intrinsic or, if SrO is present, to be actually p-type.

Now, it can be shown theoretically, that – a bit contrary to expectation – the low work function, characteristic for oxide-coated cathodes, can be explained only by the absorption of barium ions and barium atoms, not on n-type BaO or SrO, but on intrinsic or, even better, p-type BaO or SrO. Therefore, intrinsic or p-type BaO or SrO that are to be expected from the arguments of defect chemistry are just what is needed to explain the low work function.

Further, as a consequence of this, one has reason to assume that the relatively high conductivity predominating at low temperatures most probably has to be ascribed to surface conduction, owing to the absorbed barium ions and barium atoms. The surface donor levels associated with these absorbed barium atoms probably will lie close to the conduction band. This then explains, at the same time, why the low-temperature conduction of oxide-coated cathodes exhibits activation energies smaller than those of the bulk conduction in single crystals.[7]

(7) In the foregoing, the importance of the presence or the nonpresence of layers in polycrystalline, often ceramic-type, materials has been stressed. In some cases, those layers were present in the form of distinct phases. The study of phase diagrams of course is of prime importance for the full understanding of the systems under investigation. As an example, where the precise knowledge of phase diagrams of pseudo binary or ternary systems has helped greatly to make better industrial products, the lead-titanium zirconates are briefly dealt with. The mixed crystals of these compounds form, at not too high temperatures, a tetragonal phase for Ti/Zr ratios below about 50 percent and a rhombohedral phase above that ratio. At higher temperatures, both those structures transform into a cubic phase. The figure of merit for these piezoelectric materials is the so-called coupling factor k, the square of which gives the efficiency of the transformation of electrical energy into mechanical energy and vice versa. This coupling factor is proportional to the remanent dielectric polarization and to the square root of the dielectric constant. Generally, these two properties counteract each other. Now it appears that the remanent polarization changes strongly at the tetragonal-rhombohedral phase transition, reaching a maximum at the rhombohedral side, mainly because the maximum degree of orientation is naturally higher in a rhombohedral phase than in a tetragonal phase. The dielectric constant has a maximum for compositions at the phase transition, as can be made understandable from general thermodynamic arguments.

The addition of 2 to 5 mole % of lanthanum on lead sites, under certain conditions of partial pressures of PbO and O_2, causes the formation of half that concentration of lead vacancies. In those cases, the addition of lanthanum causes the tetragonal/rhombohedral phase transition to occur at higher zirconium concentrations and, further, brings down the Curie temperature considerably. The latter effect is probably connected to the fact that the transition temperature to the cubic phase is considerably lowered by the addition of lanthanum.

Now, by carefully studying the tetragonal/rhombohedral phase transition as a function of the Zr/Ti and La/Pb ratios, compositions could be found with about 5 mole % lanthanum and 56 mole % zirconium lying in the phase-transition region (and probably containing both phases) which showed coupling factors as high as 0.7. For these compositions, the dielectric constant is much higher than that for the lanthanum-free lead, titanium-zirconates at the phase-transition boundary and is only partially counteracted by a decrease of the remanent polarization.[8]

(8) In the context of this conference, a word or two about non-stoichiometry and existence regions cannot be omitted. Of course, this has already been dealt with implicitly in a number of the foregoing examples. Now I want to talk somewhat more explicitly about nonstoichiometry, taking $LiNbO_3$ as an example. Although the existence region of many oxidic substances may be very small, the chemical and physical properties may change considerably over this small region, as will be immediately clear if one looks at the equilibrium pressure of one of the components at the limiting concentrations of the existence region at a given temperature. $LiNbO_3$ has a rather broad existence region lying between 50 and 46 mole % of Li_2O at about 1100°C, and melts congruently at a composition of 48.6 percent Li_2O. This means that if solid $LiNbO_3$ is prepared from a liquid containing, say, equimolecular amounts of Li_2O and Nb_2O_5 – for instance by the Czochralski method – the crystals will show changes of composition (increasing Nb_2O_5 content) along the axis of growth. Now the Curie temperature and the extraordinary index of refraction, n_e, among other physical properties, vary quite a lot with composition. This change of n_e along the axis of growth makes it then impossible to get crystals with a constant phase-match temperature for second harmonic generation over the length of the crystal. Only by starting from a liquid with the congruential composition can one avoid these composition changes, and thus obtain much better phase-match crystals. It remains to be investigated of course what kind of native disorder goes along with those deviations from stoichiometry and what their relation is to the value of n_e.[9]

(9) Interesting also is the case of photon-induced changes in permeability in single-crystal YIG. It is known that these effects are caused by photon-induced valency exchanges between Fe^{2+} ions lying next to a compensating Si^{4+} ion, and Fe^{3+} ions having no Si^{4+} ions in their immediate surroundings. Starting from very pure base materials, sintered YIG garnets could be prepared that were still very pure and then were heated to equilibrium in an oxygen atmosphere at 1200°C for more than 100 hours. These preparations (not containing Si^{4+} ions in any appreciable concentration), also showed photon-induced changes in susceptibility, their values depending on the partial oxygen pressures according to about $p_{O_2}^{-1/12}$. A possible explanation of this could be that in those preparations, the Fe^{2+} ions needed for the effect are created as a charge compensation for the oxygen-ion vacancies introduced by the equilibrizing at 1200°C and not – as in the case for the single crystals – to compensate for the extra charge of the impurity Si^{4+} ion.[10]

(10) The possibility of having oxygen-vacancies present in garnets could be of importance in explaining the growth-induced uniaxial anisotropy in magnetic garnet layers grown on nonmagnetic garnet substrates, which is essential for the formation of bubble domains. The phenomenon of bubble formation in these cubic materials is not well understood. It seems certain that strains caused by lattice mismatch cannot be the direct cause. A possibility might be a growth-induced pairing or a site preference of the rare-earth ions in iron garnets containing two or more rare-earth ions. However, because of the fact that a substantial growth-induced anisotropy has also been reported for garnets that have no magnetic ions on the rare-earth sites or, like YIG, contain only iron as the magnetic ion, other mechanisms apparently also have to be considered. Instead of a pair interaction, a single-ion description of the induced anisotropy may be appropriate, in which the crystal field varies locally owing to a growth-induced ionic ordering process; a growth-induced ordering of oxygen vacancies has been proposed as an example of such a mechanism. However, if the known data for the diffusion of oxygen in YIG are applied, it would appear that the ordering should have been annealed out during preparation. The whole picture is therefore rather in the dark, and other factors, for example the influence of the ions incorporated from the flux, apparently have to be taken into account.[11]

(11) Properties of materials often result from an interaction of various native and/or foreign defects. In the foregoing, implicit examples of this have been given. It is worthwhile to give one example where this phenomenon is dealt with more explicitly. For this I have chosen the phenomenon of magnetic relaxation. There are quite a number of different

relaxation mechanisms. I shall restrict myself to two examples of orientation relaxation in ferrites. One of them is connected with the movement of vacancies and the other with the movement of electrons.

When Co^{2+} ions are present, the amount of magnetic energy after a rotation of the applied magnetic field can decrease further if Co^{2+} ions can move to more favorable octahedral sites. Now, in the case of mixed crystals of $Co^{2+}Ni^{2+}$ ferrites and γ-Fe_2O_3, cation vacancies will be present. These cation vacancies make possible the transport of the Co^{2+} ions to the more favorable sites. Therefore, it is understandable that the amount of the change in anisotropy is proportional to $[Co^{2+}]$ and independent of $[V_{cat.}]$, but that the relaxation frequency is proportional to $[V_{cat.}]$.

Fe^{2+} ions can have an influence on relaxation phenomena in ferrites only if the extra electron of a Fe^{2+} ion is localized in its immediate surroundings; only in this case, can a change in the surroundings be the cause of a change in the anisotropy leading to relaxation.

For (inverse) nickel ferrites that have some Fe^{2+} ions on octahedral sites, the relaxation frequency is dependent on the temperature, with an activation energy decreasing from 0.3 ev to 0.05 ev with increasing Fe^{2+} concentration. These activation energies are identical with those of the electrical conductivity. This means that the electrical conductivity in those ferrites is of the hopping type, with a thermally activated mobility that is equal for electron movement in the immediate surroundings of a lattice defect or over larger distances.[12]

(12) In the course of this talk problems related to sintering have been mentioned a number of times. In the manufacturing of oxidic materials with their mostly very high melting points, sintering, of course, is often the only means to get a material with a high density. It cannot be the purpose of this talk to deal with the very complicated problems that are connected with all the aspects of sintering. It is a very fascinating subject and although considerable progress has been made many problems remain still to be solved by the application of solid state science in all its aspects.[13]

Also, so far in my lecture I have limited myself to oxidic materials that are in use nowadays in the electronic industries. Now, we all know that this field is a small one compared with the widespread use of oxidic materials everywhere in our surroundings. I have only to mention glass, pottery, and construction materials. These are old crafts, and one's understanding of these complicated systems has much to gain from the application of solid-state science in all its aspects. Further, the general tendency to go to higher and higher temperatures in the production or use of energy will lead to an increasing need of noncorrosive refractory materials, to which class many oxides certainly belong. In this field, much research is presently being done that, apart from the study of sintering proper, can be indicated by the words comminution, forming, shrinkage, strength, machinability, and resistance to wear. For quite a number of the products of the electronics industry, these topics are also rapidly becoming of importance.

I cannot completely neglect this vast area of present and future research and, therefore, I propose to end my talk by dealing briefly with Al_2O_3 as a lamp-envelope material. To get sodium-gas-discharge lamps with a much broader spectrum, a much higher temperature of the sodium vapor is needed. As a consequence, glass can no longer be used and a new type of lamp-envelope material had to be found that at the high temperature of 1300°C is both chemically inert, such as not to react with sodium vapor, and mechanically strong. In addition, it should be transparent or translucent. As an appropriate material, Al_2O_3, of course, jumps to the mind; it is very stable. Furthermore, just like ice – also of hexagonal crystal symmetry – it should be translucent if not transparent when densely packed. This last requirement proved to be a serious problem however, because of the occurrence of the phenomenon of exaggerated grain growth during the sintering of Al_2O_3. As a consequence of this, pores cannot move out of the grains, and this leads to a material with a relatively high concentration of small pores. These pores scatter the incident light rays, and thus cause the material to be opaque.

It was found that by adding 0.1 to 0.2 percent of MgO to Al_2O_3, a solid phase of $MgAl_2O_4$ will be formed at the grain boundaries during sintering at 1650°C. Exaggerated grain growth then does not occur, and therefore the pores will remain at the grain boundaries. Subsequent firing at 1900°C under very low oxygen pressure lets the MgO evaporate out of the material and, furthermore, causes the remaining pores to coalesce. The material, Lucalox, obtained in this way is translucent, chemically inert, and relatively strong and is used as a material for lamp envelopes. In recent research on this material, the theory of scattering of electromagnetic radiation developed to explain phenomena related to the existence of interstellar dust was applied. It was found in this way that the remaining scattering of this material depended not only on the concentration of the pores but also on their dimensions and their size distribution.

By applying the method of continuous hot pressing, a very dense material could be obtained using a sintering temperature as low as 1400°C, without the addition of MgO. This new type of sintered Al_2O_3 showed a homogeneous distribution of grain size of about 1 μ and a pore size of about 0.1 μ with a small distribution around that size. Notwithstanding the birefringence of Al_2O_3, this material, in agreement with theoretical expectation, proved indeed to be transparent.[14]

In the course of this talk, I think it has been demonstrated that solid-state science has been indispensable for the understanding and development of a number of industrial products. As an industrial scientist, I would like to point out to you quite clearly that, without the help of science, it would have been impossible for industry to develop and to manufacture those products. I can imagine that sometimes people engaged in scientific problems that seem to have no relation with the more apparent needs of society feel that they are wasting their time. Let me assure you that such is not the case. These scientists in doing their research create a background of facts and ideas without which industry could not have succeeded in developing the products that now do serve the needs of society.

On the other hand, it also has been demonstrated in this talk – although somewhat more implicitly – that industrial products will never be completely understood. Reality is complexity. This can even be said of the more simple experimental models used in scientific investigations. However, as we all know, by dealing with these complex problems with an open mind, without prejudice and always ready to change one's ideas in the face of new facts, one can go a long way in finding the right questions and the right answers.

REFERENCES

1. For a review see: Bosman, A. J., and Van Daal, H. J., *Advances in Physics,* **19** (77), 1 (January, 1970).
2. Jonker, G. H., and Stuijts, A. L., *Philips Tech. Rev.,* **32,** 791 (1971).
3. Heywang, W., *Solid State Electronics,* **3** (1), 51 (1961); Heywang, W., and Brauer, H., *Solid State Electronics,* **8,** 129 (1965); Brauer, H., *Z. Angew. Phys.,* **23,** 373 (1967); Brauer, H., *Z. Angew. Phys.,* **29,** 282 (1970); Seuter, A.M.J.H., Thesis, Leyden University, Holland (1973), to be published.
4. Matsuoka, M., *Japan. J. Appl. Phys.,* **10,** 736 (1971).
5. Biesterbos, J.W.M., *J. Appl. Phys.,* to be published.
6. Broese van Groenou, A., Bongers, P. F., and Stuijts, A. L., *Mater. Sci. Engng.,* **3,** 317 (1968/1969).
7. For a review see: Zalm, P., *Advances in Electronics and Electron Physics,* **25,** 211 (1968).
8. Carl, K., and Härdtl, K. H., *Ber. Deut. Keram. Ges.,* **47,** 687 (1970); Carl, K., Thesis, Universität Karlsruhe (1972).
9. Lerner, P., Legras, C., and Dumas, J. P., *J. Cryst. Growth,* **3,** 231 (1968); Byer, R. L., Young, J. F., and Feigelson, R. S., *J. Appl. Phys.,* **41,** 2320 (1970); Carruthers, J. R., Peterson, G. E., Grasso, M., and Bridenbaugh, P. M., *J. Appl. Phys.,* **42,** 1846 (1971).
10. Metselaar, R., and Huyberts, M.A.H., *J. Phys. Chem. Solids,* to be published.

11. Stacy, W. T., and Rooijmans, C.J.M., *Solid State Commun.,* **9,** 2005 (1971); Gyorgy, E. M., Sturge, M. D., Van Uitert, L. G., Heilner, E. J., and Grodkiewicz, W. H., *J. Appl. Phys.,* **44,** 438 (1973).
12. Broese van Groenou, A., *Appl. Phys.,* **2** (2), 47 (1973).
13. For a review see: "Sintering and Related Phenomena", *Proceedings of the 3rd Internat. Conf. Notre Dame,* G. C. Kuczynski (Ed.), *Materials Science Research,* Vol. 6, Plenum Press, New York, London (1973); Stuijts, A. L., "Synthesis of Materials From Powders by Sintering", to appear in *Ann. Rev. Mater. Sci.,* Vol. III.
14. Coble, R. L., and Burke, J. E., *Prog. Ceram. Sci.,* **3,** 197 (1963); Coble, R. L., and Paladino, A.E.P., *J. Am. Ceram. Soc.,* **46,** 133 (1963); Peelen, J.G.J., *Science of Ceramics,* **6,** XVII (1971); Peelen, J.G.J., and Metselaar, R., *J. Appl. Phys.,* to be published.

Part Two

THEORY AND CHARACTERIZATION

CALCULATION OF DEFECT-FORMATION ENERGIES IN ALKALINE-EARTH OXIDES: INTERIONIC FORCES

A. D. Franklin and A. L. Dragoo

National Bureau of Standards
Gaithersburg, Maryland

ABSTRACT

A model potential for the alkaline-earth oxides was developed and tested. The interionic potential contains the usual Coulombic and two-body short-range terms, such as those which have been used in previous alkali halide calculations, plus many-body terms which are incorporated by means of the Lundquist three-body approximation in which only the spherically symmetric term in the overlap-charge-density expansion is retained. The three-body interactions contribute to both the long-range and the short-range forces. Additional short-range central-force interactions between nearest-neighbor ions and next-nearest-neighbor ions are included; these are modeled by means of a Born-Mayer plus van der Waals potential. The electronic polarizabilities of the ions are described by a shell model. Parameters in the model potential and in the shell model are evaluated using the elastic constants and their pressure derivatives, the low- and high-frequency dielectric constants, the infrared dispersion frequency, and the equilibrium interionic separation. Three constraints on the shell model are required. The model does not give satisfactory results in that (1) the trend in an effective charge parameter, χ_o, through this family of oxides runs counter to that for the Phillips ionicities and (2) the Born-Mayer potential parameters appear to be highly anomalous compared with those for other ionic crystals. The principal weakness of the model apparently resides in its neglect of bond-bending (noncentral force) terms, which results in an overestimate of the overlap contribution to the bond-stretching terms, in excessive deviations in the effective charges from the classical valences, and in departures of the Born-Mayer potential parameters from the values found in alkali halide and alkaline-earth fluorides.

1 INTRODUCTION

Calculation of the formation energies of point defects in ionic crystals has had considerable success[1] in recent years, particularly as applied to alkali halides and alkaline-earth fluorides.[2,3] The most sophisticated calculations have been based upon the shell model with two-body, central-force interactions, with fits to experimental elastic and dielectric data being used to evaluate parameters in the Born-Mayer expressions for the short-range interactions between nearest-neighbor (NN) and between next-nearest-neighbor (NNN) ions (van der Waals forces also are usually included). The success of these calculations has depended heavily upon obtaining reasonable expressions for these short-range interactions, and this has not always been possible. For instance, in studying the lattice dynamics of the alkaline-earth fluorides, Axe[4] found that, while familiar repulsive parameters could be obtained for CaF_2, for BaF_2 the F^--F^- interactions were attractive rather than repulsive. This situation has also been found in simple oxides, such as MgO[5-7] (not, however, in the work reported in ref. 8) and UO_2[9], as shown in Table I. While attractive interactions could result from van der Waal's forces, which should be included along with the Born-Mayer repulsions, the magnitude of the attractive part seemed unreasonably large. If, on the other hand, the attractive term is due to a covalent contribution to the bonding, then use of the Born-Mayer function seems suspect.

Furthermore, there is the long-recognized fact that many-body, noncentral-force interactions are also present in these ionic crystals[10,11] as evidenced, for example, by the failure of the alkali halides and the alkaline-earth oxides to satisfy the Cauchy equality, $c_{12} = c_{44}$, of the elastic constants which is predicted by the central-force condition[12-15] (Table II).

Lundquist[11] has approximated the many-body contributions by the interactions of an overlap-charge distribution between a pair of NN-ions and all other ionic centers in the crystal. Verma and Singh[16,17] have included Lundquist's three-body potential in shell-model calculations of the phonon-dispersion curves of several alkali halides with good results. Dick and Overhauser[18,19], in their exchange charge model, have approximated the dependence of the overlap-charge magnitude on the NN-separation by an equation of the form of the Born-Mayer potential. Arguing from the Pauli exclusion principle, they have assumed that the overlap (exchange) charge is effectively positive.

In contrast to the positive exchange charge, Phillips[20] has developed a bond charge model of covalency on the basis of a simplified dielectric dispersion theory. Gillis[21] employed this model in the calculation of the phonon-dispersion curves for MgO, obtaining fits to the data which were comparable to those obtained from the shell model.[6]

In addition to providing a description of noncentral forces, Lundquist's three-body interactions, and such models as the exchange charge and bond charge, have been proposed as the corrections that might remove the apparent artifact of the attractive Born-Mayer potential between the anions. This proposition has as yet not been tested. Furthermore, Catlow and Hayns[22] have recently shown, in *ab initio* quantum mechanical calculations of the interactions between F^- ions, that the total energy could be represented by a sum of the usual repulsive term, closely approximated by the Born-Mayer form, plus a large, longer range attractive term. Hence, rather than being an artifact, the attractive short-range interaction noted for some fluorides and oxides may be real.

In this paper we report on the early stages of a study of the nature of the interionic forces in the alkaline-earth oxide crystals, performed by fitting a modified shell model to experimental data at zero wave number. In particular, we explore the role of three-body interactions of the Lundquist type by studying the apparent ionic charges required by the model and the short-range central-force overlap potentials, for both first-neighbor and second-neighbor interactions, expressed in the form of Born-Mayer repulsive plus van der Waals attractive potentials. By comparing the ionic charges with the Phillips ionicities[23] and the Born-Mayer potentials to those obtained with ionic crystals such as alkali halides and alkaline-earth fluorides, we conclude that the Lundquist model is probably inadequate for the alkaline-earth oxides.

2 MODEL

Our model is a shell model in which the two-body short-range interactions are assumed to operate only between shells, and are carried as far as second neighbors. In fitting the model to the data, the explicit dependence of these short-range interactions upon interionic separation does not appear. Rather, the first, second, and third derivatives of the interaction energy with respect to the separation are evaluated and then independently compared with possible functional forms. We have introduced three-body interactions using Lundquist's formulation as developed by Verma and Singh.[16] The equations we have used are those of Verma and Singh extended to include short-range interactions among second neighbors and developed for additional experimental quantities.

The model is based upon a potential energy per unit cell

$$\Phi(a) = e^2 \left[\frac{(-\epsilon^2 \alpha_M)}{r} \left\{ 1 + (12/\epsilon) f(r) \right\} + 6 \left\{ V_{11}(r) + V_{12}(r) + V_{22}(r) \right\} \right]_{r=a} , \qquad (1)$$

where e is the electronic charge, ϵ is the magnitude of the classical ionic charges (two for the alkaline-earth oxides), α_M is the Madelung constant (+1.7476), r is the interionic distance and a is its equilibrium value, $e^2 V_{ij}(r)$ is a potential of short-range interaction between ions of type i and j (12 denotes nearest neighbors, and 11 and 22 second-neighbor cations and anions, respectively), and f(r) is a nearest-neighbor function appearing in the three-body energies between the nearest-neighbor ions in question and the other ions in the lattice. As we have used it, the model contains 13 disposable parameters (the equations are given and discussed briefly in the appendix):

f(r) and its first and second derivatives with respect to r, evaluated at r = a

The first, second, and third derivatives of quantities closely related to $V_{12}(r)$ and $[V_{11}(r) + V_{22}(r)]$ with respect to r, evaluated at the equilibrium separations

The shell charges and shell-core spring constants for both ions.

We used ten pieces of experimental data (Table III):

Three elastic constants and their pressure derivatives

The dielectric constants at high and low frequency, with respect to the lattice-vibration frequencies

The infrared dispersion frequency

The equilibrium interionic separation.

Hence there are three more parameters than there are pieces of experimental data, and it is necessary to add three additional conditions in order to evaluate the parameters.

Table I. Short-Range Interactions

	Cation-Anion		Anion-Anion		
	ϕ'	ϕ''	ϕ'	ϕ''	Ref.
CaF_2	−1.58	5.91	−0.13	1.30	4
SrF_2	−1.42	5.10	−0.03	0.09	4
BaF_2	−1.30	4.63	0.04	−0.08	4
MgO	−3.46	12.50	0.34	−0.96	6
MgO	−1.49	6.84	−0.51	1.87	8
UO_2	−6.96	18.46	0.53	−1.62	9[a]

Note: ϕ' and ϕ'' are the first and second derivatives, respectively, of the potential of short-range interaction between the ions noted, with respect to interionic separation. The units of ϕ' are 10^8 ev/cm and those of ϕ'' are 10^{16} ev/cm^2.

(a) Model IV of Dolling et al. is quoted, but with their values divided by a factor of 2 to produce consistency with the experimental data (private communication, J. D. Axe).

Table II. Comparison of Elastic Constants – Deviation From the Cauchy Relation

Substance	$(C_{12}-C_{44})/C_{11}$	Ref.
LiF	−0.136	12
NaF	−0.038	12
KF	0.031	12
RbF	0.059	12
MgO	−0.230	13
CaO	−0.100	14
SrO	−0.072	14
BaO	−0.045	15

Table III. Properties of Alkaline-Earth Oxides

Property	MgO	CaO	SrO	BaO
Elastic Constant, Mb, c_{11}	3.06	2.4	1.8	1.1
Elastic Constant, Mb, c_{12}	0.942	0.59	0.45	0.34
Elastic Constant, Mb, c_{44}	1.646	0.83	0.58	0.39
Pressure Derivative, dimensionless $(\partial C_{11}/\partial P)_T$	9.975	10.5	11.3	–
Pressure Derivative, dimensionless $(\partial C_{12}/\partial P)_T$	2.166	3.73	3.33	–
Pressure Derivative, dimensionless $(\partial C_{44}/\partial P)_T$	1.431	0.6	–0.2	–
Cation-Anion Distance (a), Å	2.103	2.405	2.585	2.761
Low-Frequency Dielectric Constant (η_o), dimensionless	9.742	11.1	13.1	34
High-Frequency Dielectric Constant (η_∞), dimensionless	2.945	3.33	3.46	3.558
Lattice Vibration Frequency (ω_o), THz	76.85	55.57	42.76	21.7[a]

(a) ω_o for BaO was estimated from the first Szigeti[24] relation

$$e_s/e = \omega_o \left(\frac{\eta_o - \eta_\infty}{4\pi}\right)^{1/2} \left(\frac{3}{\eta_\infty + 2}\right) \frac{(\bar{M} 2a^3)^{1/2}}{\epsilon e},$$

where $\bar{M}$ is the reduced mass of the cation and anion. We have estimated e_s/e as 0.6 on the basis of Szigeti's[24] values for the other alkaline-earth oxides.

3 RESULTS

Only for MgO is there available the complete range of sufficiently precise experimental data (Table II), and the model in all its generality was therefore applied only to MgO. Ideally, all experimental quantities should refer to T = 0, but in this preliminary exploration, available experimental data at other temperatures have been used.

I. Description

A. MgO

The three additional conditions applied were:

Condition a – The shell-core spring constant for the Mg^{2+} ion was taken equal to the free-ion value, as derived from the Pauling polarizability and the lowest energy in the Mg^{2+} absorption spectrum (5.6542 J/cm^2).

Conditions b and c – The core charges for both ions were fixed at the free-ion value. For O^{2-}, which does not exist in the free state, this was done by following Dick and Overhauser[18] [6.8 elementary (electronic) charge units for Mg^{2+}, -2.8 for O^{2-}].

B. CaO, SrO, and BaO

Condition a – The spring constant of the oxygen ions was kept the same as found for MgO.

Conditions b and c – As for MgO, the core charges of both ions were fixed at the free-ion values.

II. Assessment

The behavior of a model of these compounds can be assessed on the basis of a number of points:

A. Ionicity

The model contains a parameter χ_o given by

$$\chi_o^2 = I + 12f(a)/\epsilon \quad . \tag{2}$$

This parameter occupies somewhat the same position in the equations for the elastic constants and cohesive energy that Z, the total ionic charge, does in the usual shell model. Hence χ_o is in a sense a measure of the ionicity, and can be compared to other measures of ionicity, such as suggested by Phillips.[20]

Table IV shows the values of χ_o for all four oxides compared with the ionicity values (f_i) on the Phillips model[23], in the first and second columns, respectively.

There appears to be no way to relate the absolute values of χ_o and f_i, but it is probably significant that while the spectroscopic ionicity (f_i) increases from MgO to BaO, the apparent ionicity represented by the value of χ_o decreases. The increase in f_i would appear to be at least consistent with the decrease in deviation from the Cauchy relation exhibited by the elastic-constant data in Table II.

B. Short-Range Interactions

The short-range interactions are contained in the potentials $e^2V_{ij}(r)$. For the cation-anion, nearest-neighbor interaction in our model, this potential contains the central-force overlap and van der Waals interactions, together denoted here by $\phi_{ij}(r)$, plus a purely formal term arising from the three-body forces:

$$e^2V_{12}(r) = \phi_{12}(r) + 2e^2\epsilon f(r)/r \quad . \tag{3}$$

In the present study, we have used the Born-Mayer form for the overlap term,

$$\phi_{12}(r) = A \exp(-r/\rho) - cr^{-6} \quad . \tag{4}$$

The van der Waals constants (c) were taken from the work of Boswarva[25], and A and ρ were obtained from our model fits. The results are shown in the first two columns of Table V.

The model produces parameters very unlike those found in alkali halides[26] and other ionic crystals[2,4], and must be considered unsatisfactory as a modification of the ionic model.

The problem appears to lie in three-body corrections that are too large. The seriousness of these terms is illustrated by columns 3 and 4 of Table V, which were derived setting f(r) and its derivatives equal to zero throughout the entire calculation. There is a large change in the parameters, back toward more reasonable values for CaO, SrO, and BaO. This procedure amounts to using a central-force model, which is clearly wrong for MgO with its large Cauchy discrepancy. However, it does illustrate the large effects the three-body terms exert on the short-range parameters of the model.

For the second-neighbor interactions, $V_{11}(r)$ and $V_{22}(r)$ always occur together as a sum in the equations for the rock-salt structure, and cannot be separated without some additional condition. This condition is outside of and does not affect the fit of the model to the experimental data, but is important in interpreting the model. Since in our model the three-body effects are considered to operate only among nearest neighbors, they make no direct contribution to the second-neighbor interactions, and these latter can in principle be handled in a straightforward way.

The fit of the model to the experimental data produces derivatives of $[V_{11}(r) + V_{22}(r)]$ with respect to the interionic separation, r_{ij}. These in turn can be fitted to trial potential functions $[\phi_{11}(r) + \phi_{22}(r)]$, and the results compared with those from other ionic compounds. At the present stage in this study, we have examined only the fit to a Born-Mayer function for the central-force overlap, plus an attractive term similar to a van der Waal's energy, as in Eq. (4).

In dealing with MgO, we neglect the Mg^{2+}-Mg^{2+} interaction altogether and use the second-neighbor contributions to evaluate the parameters in the oxygen-oxygen short-range interaction function. With this, by using the parameters from MgO, the oxygen-oxygen contributions in the other oxides are estimated and subtracted from the second-neighbor terms, leaving the cation-cation contribution as the difference. The results are shown in Table VI.

The Born-Mayer parameters, A and ρ, have considerably larger values than usually found for ionic crystals.[26] The coefficients c are orders of magnitude larger than the corresponding van der Waal's coefficients as calculated by Boswarva[25] from the ion crystal polarizabilities. We thus find once again that the short-range interactions produced by this model are very different from what we have come to expect in ionic crystals.

Table IV. Ionic Character in the Alkaline-Earth Oxides

	χ_o	f_i
MgO	0.949	0.841
CaO	0.765	0.913
SrO	0.717	0.926
BaO	0.658	

Table V. First-Neighbor Short-Range Interactions

	Three-Body Terms			
	Included		Not Included	
	A, ev	ρ, Å	A, ev	ρ, Å
MgO	46	0.974	4.70×10^8	0.093
CaO	−0.46	−1.03	636	0.349
SrO	−0.13	−0.796	193	0.499
BaO	−0.02	−0.582	90	0.654

Table VI. Second-Neighbor Short-Range Interactions

	O^{2-} -O^{2-}	Cation-Cation		
	MgO	CaO	SrO	BaO
A, 10^5 ev	0.428	1.199	1.554	1.522
ρ, Å	0.371	0.450	0.489	0.526
c, 10^5 ev-Å^6	0.142	1.264	2.655	4.091

4 DISCUSSION

The deviation of the elastic constants from the Cauchy condition, $\Delta c = c_{12}$-c_{44}, plays a critical role in determining how inclusion of three-body effects influences the parameters of a model fit to the data. As shown by Eqs. A-1b, A-1c, A-10b and A-10c in the appendix, Δc and its pressure derivative completely determine f_1 and f_2, essentially the first and second derivatives of the three-body function, f(r), appearing in Eq. (1).

The Lundquist model, used here, includes only bond-stretching effects in the three-body terms. That is, a change in separation between two ions causes a redistribution of electronic charge between them and therefore a change in their Coulombic interactions with other ions. The impact of the deviation of the elastic constants from the Cauchy relation is thus transmitted through the whole set of equations, for both the elastic and dielectric properties, through its effect upon the Coulombic interactions.

This is not the case for bond-bending effects, nor for the breathing-shell model[27], which like the Lundquist model introduces only bond-stretching terms, but acting through the shell-shell short-range forces rather than through the Coulombic interactions.

It appears from the behavior of our model, particularly the relatively drastic effects that introduction of the three-body interactions has upon the ionicities (Table IV) and the first-neighbor short-range interactions, that the model overcorrects for the Cauchy deviation. This tendency to overcorrect probably lies in the purely Coulombic nature assumed for the three-body effects. It thus seems necessary to incorporate some form of non-Coulombic three-body effects, perhaps in terms of bond-bending parameters derived from higher terms in the expansion of the overlap charges, as discussed by Lundquist (ref. 11, 1955). In the present model, only the first, spherically symmetric term in such an expansion is used.

ACKNOWLEDGMENT

We wish to thank Richard Spal for his invaluable assistance with the computations and, in particular, for his ingenious manipulation of the equations which enabled an exact solution to be obtained for each parameter.

APPENDIX: EQUATIONS

Elastic Constants

Expressions for elastic constants c_{11}, c_{12}, and c_{44} and their pressure derivatives in terms of atomic quantities resemble the familiar rigid-ion results. Shell-model parameters do not occur in those expressions. However, the three-body interactions introduce some changes in the Coulombic terms. These corrected terms have been worked out by Lundquist.[11]

The short-range terms remain formally the same as in the rigid-ion model. This is a consequence of restricting the overlap function, f(r), in the three-body potential to nearest neighbors and to a dependence upon the nearest-neighbor separation only. This three-body potential contributes a central, two-body term to the nearest-neighbor short-range potential.

It is convenient to express the elastic constants in a reduced form $\tilde{c} = (a^4/e^2\epsilon^2)c$, where $\tilde{c}$ is the reduced quantity and c is an elastic constant. The elastic-constant equations are:

$$\tilde{c}_{11} = -1.27802\ \chi_o^2 + 2.33008\ f_1 + 1/4\ (\psi_{+-}^{(2)} + \psi_{++}^{(1)} + \psi_{--}^{(1)} + \psi_{++}^{(2)} + \psi_{--}^{(2)}) \qquad \text{(A-1a)}$$

$$\tilde{c}_{12} = 0.34775\ \chi_o^2 + 2.33008\ f_1 + 1/8\ (\psi_{++}^{(2)} + \psi_{--}^{(2)} - \psi_{++}^{(1)} - \psi_{--}^{(1)}) \qquad \text{(A-1b)}$$

$$\tilde{c}_{44} = 0.34775\ \chi_o^2 + 1/8\ (\psi_{++}^{(2)} + \psi_{--}^{(2)} - \psi_{++}^{(1)} - \psi_{--}^{(1)})\ , \qquad \text{(A-1c)}$$

where $\chi_o^2 = \left\{1 + (12/\epsilon)\ f(a)\right\}$, $f_1 = (1/\epsilon)(rdf/dr)_o$.

The ψ's are defined in terms of the derivatives of the effective potential between a given pair of ions, the superscript indicating the order of that derivative. For nearest-neighbor ions (NN), the effective potential is

$$\tilde{V}_{NN} = V_{CF} + 2\epsilon\left(1/r - \alpha_M/a\right) f(r)\ , \qquad \text{(A-2)}$$

where

$$e^2 V_{CF} = \phi_{12}(r)\ ; \qquad \text{(A-3)}$$

ϕ_{12} is the customary central-force potential. For second-neighbor ions (NNN),

$$e^2 V_{NNN} = e^2 V_{CF} = \phi_{11}(r) + \phi_{22}(r)\ . \qquad \text{(A-4)}$$

For first-neighbor interactions,

$$\psi_{+-}^{(n)} \equiv \left(4a/\epsilon^2\right) \left(r^n d^n \tilde{V}_{NN}/dr^n\right)_{r = a}\ , \qquad \text{(A-5)}$$

and for second-neighbor interactions,

$$\psi_{\sigma\sigma}^{(n)} \equiv \left(2a/\epsilon^2\right) \left(r^n d^n \tilde{V}_{\sigma\sigma}/dr^n\right)_{r = a\sqrt{2}} \qquad \text{(A-6)}$$

where $\sigma\sigma$ = ++, – –.

The derivatives have the property that

$$(rd\psi^{(n)}/dr_\phi)_o = (n+1)\psi^{(n)} + \psi^{(n+1)}\ , \qquad \text{(A-7)}$$

i.e., a variation of the ionic separation acts both upon the derivative of potential and the volume factor multiplying it.

The reduced pressure derivatives

$$\tilde{c}^p = \left[a^4(c_{11} + 2c_{12})/e^2\epsilon^2\right] (\partial\tilde{c}/\partial p)_T \tag{A-8}$$

can be easily obtained from the elastic-constant expressions by use of the formula

$$\tilde{c}^p = 4\tilde{c} - (r\partial\tilde{c}/\partial r)_{r=a} \quad . \tag{A-9}$$

These are found to be

$$\begin{aligned}\tilde{c}^p_{11} = 4\tilde{c}_{11} &+ 13.00616\ f_1 - 2.33008\ f_2 \\ &-1/4\left(3\psi^{(2)}_{+-} + \psi^{(3)}_{+-} + 2\psi^{(1)}_{++} + 2\psi^{(1)}_{--}\right. \\ &\left.+4\psi^{(2)}_{++} + 4\psi^{(2)}_{--} + \psi^{(3)}_{++} + \psi^{(3)}_{--}\right) \quad ,\end{aligned} \tag{A-10a}$$

$$\begin{aligned}\tilde{c}^p_{12} = 4\tilde{c}_{12} &- 6.50308\ f_1 - 2.33008\ f_2 \\ &+1/8\left(2\psi^{(1)}_{++} + 2\psi^{(1)}_{--} - 2\psi^{(2)}_{++} - 2\psi^{(2)}_{--}\right. \\ &\left.-\psi^{(3)}_{++} - \psi^{(3)}_{--}\right) \quad ,\end{aligned} \tag{A-10b}$$

$$\begin{aligned}\tilde{c}^p_{44} = 4\tilde{c}_{44} &- 4.17300\ f_1 \\ &+1/8\left(2\psi^{(1)}_{++} + 2\psi^{(1)}_{--} - 2\psi^{(2)}_{++} - 2\psi^{(2)}_{--}\right. \\ &\left.-\psi^{(3)}_{++} - \psi^{(3)}_{--}\right) \quad ,\end{aligned} \tag{A-10c}$$

where $f_2 = (1/\epsilon)\,(r^2 d^2 f/dr^2)_{r=a}$.

Dielectric and Infrared Properties

The low-frequency dielectric constant η_0, the high-frequency dielectric constant η_∞ (obtained by extrapolating the refractive index at frequencies below that of the onset of optical absorption to infinite frequency [Boswarva[25]]), and the infrared dispersion frequency ω_0 were described by use of the shell model. The effective ionic charge is obtained from Lundquist's result (Eq. 22' of Lundquist's 1955 paper, ref. 11). In the Lundquist model, the overlap charge is partitioned between the pair of neighboring ions on an equal energy basis and is assumed to reside on the centers of the ions when the ions move as rigid particles. We have used this aspect in two ways.

First, the change in the ionic charge resulting from the partition of the overlap charge is assumed to alter only the shell charge. The core charge is taken as constant and equal to the free-ion value in the case of the Mg^{2+} ion.

Second, since the overlap charge is "redistributed" on the ions, the Lorentz condition for the inner field is assumed, i.e.,

$$E^{eff} = E^{appl} + (4\pi/3)P \quad . \tag{A-11}$$

Alternatively, one might leave the ionic charges and shell charges at their full values and adjust the inner field according to the method of Guertin and Stern.[28] In the latter case, the overlap charge in the Lundquist theory must be related to the field parameters in the Guertin-Stern theory. However, the former approach was simpler and was the one employed here.

Equations for the low- and high-frequency polarizabilities and the IR dispersion frequency were derived from the equations of motion and the polarization equation by solving the appropriate secular determinant according to the method of Cowley.[29]

Again it is convenient to define reduced quantities and we write

$$\begin{aligned}\tilde{\alpha}_o \equiv (\alpha_o/e^2\epsilon^2) &= \left(\frac{3v_a}{4\pi e^2\epsilon^2}\right)\left(\frac{\eta_o - 1}{\eta_o + 2}\right) \\ &= \left[\frac{\chi_1^2}{c_o} + \frac{(\chi_1 - \Delta_+)^2}{\kappa_+} + \frac{(\chi_1 + \Delta_-)^2}{\kappa_-}\right]\end{aligned} \tag{A-12a}$$

$$\begin{aligned}\tilde{\alpha}_\infty = (\alpha_\infty/e^2\epsilon^2) &= \left(\frac{3v_a}{4\pi e^2\epsilon^2}\right)\left(\frac{\eta_\infty - 1}{\eta_\infty + 2}\right) \\ &= \left[c_o(\Lambda_+ + \Lambda_-)^2 + \kappa_+\Lambda_-^2 + \kappa_-\Lambda_+^2\right] / \left[c_o(\kappa_+ + \kappa_-) \right. \\ &\quad \left. + \kappa_+\kappa_-\right] \quad ,\end{aligned} \tag{A-12b}$$

$$\tilde{\omega}_o^2 = \left(\frac{\eta_o + 2}{\eta_\infty + 2}\right) \overline{M}\,\omega_o^2 = c_o\kappa_+\kappa_-/[c_o\,(\kappa_+ + \kappa_-) + \kappa_+\kappa_-] \quad , \tag{A-12c}$$

where

$$c_o = \left(e^2\epsilon^2/v_a\right)\ (2\psi_{+-}^{(1)} + \psi_{+-}^{(2)}) \tag{A-13}$$

is the shell-shell force constant,

$$e\epsilon\chi_1 = e\epsilon[\chi_o^2 + 4f_1]^{1/2} \tag{A-14}$$

is the effective charge on the cation (anion charge = - cation charge), κ_+ is the cation-force constant, κ_- is the anion-force constant, $e\epsilon\Delta_+$ is the cation-shell charge and $e\epsilon\Delta_-$ is the anion-shell charge. The core charges on the cation and anion are $e\epsilon(\chi_1 - \Delta_+)$ and $-e\epsilon(\chi_1 + \Delta_-)$, respectively.

Reduced pressure derivatives of the polarizabilities,

$$\tilde{\alpha}^p \equiv \left(\frac{c_{11} + 2c_{12}}{e^2\epsilon^2}\right)\left(\frac{\partial\alpha}{\partial p}\right)_T \quad , \tag{A-15}$$

can be obtained using the formula

$$\tilde{\alpha}^p = -(r\partial\tilde{\alpha}/\partial r)_{r=a} \quad . \tag{A-16}$$

Assuming that κ_+, κ_- and the core charges are unaffected by the small isotropic strains, the pressure derivatives were found to be

$$\tilde{\alpha}_o^p = \left[\frac{\chi_1{}^2 c_o{}^{(1)}}{c_o{}^2} - \frac{(16f_1 + 4f_2)}{c_o}\right] \tag{A-17}$$

$$\begin{aligned}\tilde{\alpha}_\infty^p = &\left\{\left[-\tilde{\alpha}_\infty\kappa_1\kappa_2 + \kappa_1\Delta_2^2 + \kappa_2\Delta_1^2\right]\left(c_o^{(1)}/c_o\right)\right.\\ &\left.+ (\kappa_1\Delta_2 - \kappa_2\Delta_1)(16f_1 + 4f_2)/\chi_1\right\}\\ &\cdot\left[c_o(\kappa_1 + \kappa_2) + \kappa_1\kappa_2\right]^{-1} \quad ,\end{aligned} \tag{A-18}$$

where

$$c_o^{(1)} \equiv (rdc_o/dr)_{r=a}$$

$$= \left(\frac{e^2\epsilon^2}{v_a}\right)\left(\psi_{+-}^{(3)} + 2\psi_{+-}^{(2)} - 2\psi_{+-}^{(1)}\right) \quad . \tag{A-19}$$

REFERENCES

1. For recent reviews, see Barr, L. W., and Lidiard, A. B., in *Physical Chemistry, An Advanced Treatise, Vol. 10, Solid State,* W. Jost (Ed.), Academic Press, New York (1970), and Franklin, A. D., in *Point Defects in Crystals,* Vol. I, J. H. Crawford, Jr., and L. M. Slifkin (Eds.), Plenum Press, New York (1972).

2. Franklin, A. D., *J. Phys. Chem. Solids,* **29**, 823 (1968).
3. Norgett, M. J., *J. Phys.,* **C4**, 298 (1971).
4. Axe, J. D., *Phys. Rev.,* **139**, A1215 (1965).
5. Yamashita, J., and Kurosawa, T., *J. Phys. Soc. Japan,* **9**, 944 (1954).
6. Peckham, G., *Proc. Phys. Soc.,* **90**, 657 (1967).
7. Boswarva, I. M., and Franklin, A. D., in *Mass Transport in Oxides,* J. B. Wachtman, Jr., and A. D. Franklin (Eds.), National Bureau of Standards Special Publication 296 (1968).
8. Verma, M. P., and Dayal, B., *Phys. Stat. Solidi,* **19**, 751 (1967).
9. Dolling, G., Cowley, R., and Woods, A.D.B., *Can. J. Phys.,* **43**, 1387 (1965).
10. Löwdin, P. O., in *Advances in Physics,* Vol. 5, N. F. Mott (Ed.), Taylor and Francis, Ltd., London (1956).
11. Lundquist, S. O., *Ark. Fys.,* **6**, 25 (1953); **9**, 435 (1955).
12. Spangenberg, K., and Haussühl, S., *Z. Krist.,* **109**, 4 (1957).
13. Anderson, O. L., and Andreatch, P., Jr., *J. Am. Ceram. Soc.,* **49**, 404 (1966).
14. Bartels, R. A., and Vetter, V. H., *J. Phys. Chem. Solids,* **33**, 1991 (1972).
15. Vetter, V. H., and Bartels, R. A., to be published in *J. Phys. Chem. Solids.*
16. Singh, R. K., and Verma, M. P., *Phys. Stat. Solidi,* **36**, 335 (1969); **38**, 851 (1970).
17. Verma, M. P., and Singh, R. K., *J. Phys.,* **C4**, 2749 (1971).
18. Dick, B. G., Jr., and Overhauser, A. W., *Phys. Rev.,* **112**, 90 (1958).
19. Dick, B. G., *Phys. Rev.,* **129**, 1583 (1963).
20. Phillips, J. C., *Phys. Rev.,* **166**, 832 (1968); **168**, 905 (1968); *ibid.,* pp. 912 and 917.
21. Gillis, N., *Phys. Rev.,* **B3**, 1482 (1971).
22. Catlow, C.R.A., and Hayns, M. R., *J. Phys.,* **C5**, L237 (1972).
23. Van Vechten, J. A., *Phys. Rev.,* **187**, 1007 (1969).
24. Szigeti, B., *Proc. Roy. Soc.,* **A204,** 51 (1950).
25. Boswarva, I. M., *Phys. Rev.,* **B1**, 1698 (1970).
26. Tosi, M. P., in *Solid State Physics,* Vol. 16, F. Seitz and D. Turnbull (Eds.), Academic Press, New York (1964), p. 1.
27. Schröder, U., *Solid State Commun.,* **4**, 347 (1966).
28. Guertin, R. F., and Stern, F., *Phys. Rev.,* **134**, A427 (1964).
29. Cowley, R. A., *Proc. Roy. Soc.,* **A268**, 121 (1962).

DISCUSSION on Paper by A. D. Franklin

HUGGINS: There has been a good measure of work done, especially in the chemistry of the transition elements, in which the electron distribution is typically described in terms of nonspherical arrangements in space, e.g., different configurations of lobes. Such models of the electron-density distribution have played an important role in furthering the understanding of the observed structural features of many materials.

On the other hand, the calculations discussed in this paper are based upon a central-force assumption.

Would the authors care to comment on the future convergence of these two approaches to the structure of solids?

FRANKLIN: The two approaches are indeed converging. People are now treating shell models (central force) with the addition of noncentral, short-range (shell-shell) interactions. The valence-force field model, which has been applied with great success to silicon and similar highly covalent materials, is basically a noncentral-force model, as is Gillis' bond-charge treatment of MgO. There is no problem in mixing in noncentral-force interactions. The problem seems to be to find the right ones to use.

WELCH: The work you have described here is all within the harmonic approximation; are there functional forms implicit in the Lundquist theory, analogous to the Born-Mayer potential, for example, which will permit the characterization of anharmonic effects which are very important in defect calculations?

It seems to me that the bond-charge model is a simple way to describe "bond-bending" potentials in covalent materials; however, it appears that the usefulness of this approach in defect calculations is very limited. Have you given this any thought?

FRANKLIN: The Lundquist model treats the overlap charges purely on the basis of an expansion in spherical harmonics about the center of the ion shell charge. We have retained only the first term, which implies an r^{-1} dependence. By including higher terms, we could include r^{-3}, r^{-5}, etc.

In answer to the second question, we share the doubts you express concerning the validity of using interaction potentials derived from lattice ions for interstitials. However, it does seem reasonable to try to see what happens if we do make this naive assumption and to try to see how far the model goes. Fortunately, it appears that in MgO, etc., it is vacancies that are of real interest, and here the model is on much firmer ground.

GOODENOUGH: As an outsider to the type of calculations you describe, I am impressed by the cations with $6s^2$ or $5s^2$ cores, such as Ti^+, Pb^{2+}, Bi^{3+}, or Sn^{2+}, and Sb^{3+}, and by the Ag^+ ion, which has a $4d^{10}$ core. These ions assume crystallographic positions or displacements from a center of symmetry that strongly suggest hybridization of cation orbitals that deform the shape of the core or spontaneously polarize the ion. I do not see how your formulation introduces the possibility, or the consequences, of cation-orbital hybridization in response to lattice vibrations, hybridizations that will either deform the core charge from spherical shape or introduce an anisotropic covalency in response to optical-mode lattice vibrations.

FRANKLIN: I believe the effects you are referring to would be included in what I have called bond-bending terms. These are not included in the Lundquist model as we have treated it. One can include such effects purely phenomenologically by introducing bond-bending terms with the short-range interaction coefficients, or perhaps by going beyond the first-order Lundquist 3-body term, which corresponds to the first term in the expansion of the overlap charge in spherical harmonics about each ion shell-charge center, and putting in the higher harmonics, guided by the symmetries expected on the basis of reasonable hybridizations.

DRAGOO: I would like to comment on David Welch's first comment.

We had hoped to avoid committing ourselves at the beginning of our computation to the sign of the overlap charge. We did not want to say that this was either a positive exchange charge or a negative bond charge. The Lundquist overlap function, f(a), seemed to offer a general way of going about this. By evaluating f(a) and its first and second derivatives, we tried to allow the empirical data to tell us the sign of the charge and its functional dependence on ionic separation.

We found, however, that the derivatives diverged. This might be taken as another sign that the Lundquist three-body formulation does not properly treat the overlap contribution.

ILSCHNER: Is it possible to reconcile the high "ionicity" given as, e.g., 0.949 for MgO in your Table IV with the importance claimed for the contribution due to (covalent) bond bending? Is it correct to say that the latter contribution shows up primarily in the binding-energy *derivatives*?

FRANKLIN: The short answer to the first question is, probably not. The deviation of the elastic constants from the Cauchy relation seems to require a strong three-body contribution of some sort, which, in turn, suggests the importance of covalency in the bonding. If this is true, then the high ionicity appears to be questionable. This, together with the trend of decreasing ionicity from MgO to BaO, in contrast to the opposite trend in the Phillips ionicity, makes us feel that the model is not very satisfactory in the optical properties as well as the elastic properties. In answer to the second question, the model deals only with the derivatives of the short-range energy terms, and this includes bond-bending as well as bond-stretching effects.

HEUER: Is it correct to assume from the trend in deviations from Cauchy behavior that the ionicity increases from MgO to BaO?

FRANKLIN: It is correct to say that this trend is consistent with an increase in ionicity from MgO to BaO, but I'm not sure that the trend necessarily implies an increase in ionicity.

HEUER: Why is there no need to introduce three-body forces in calculations of the alkali halides?

FRANKLIN: There is a need to introduce these three-body effects even in the alkali halides, but they are so small that good fits to most data can be obtained without them. Note, however, that even the alkali halides do not completely obey the Cauchy relation $c_{12} = c_{44}$, and this discrepancy requires the three-body terms.

CONCENTRATED DEFECTS IN PuO_2

R. P. Turcotte and T. D. Chikalla

Battelle
Pacific Northwest Laboratories
Richland, Washington

ABSTRACT

The accumulation of concentrated defects in $^{238}PuO_2$ by self-radiation damage is shown to proceed by a simple exponential function for samples which are initially defect free. Anomalous behavior for both ingrowth and damage recovery is related to helium inventory but not to uranium, which is the daughter product of the alpha decay. The annealing of samples having achieved a saturation of defects at 25°C proceeds through a two-stage mechanism; the first having an activation energy of 0.17 ev. Both defect production and annealing behavior are characterized by X-ray diffraction and electrical-conductivity measurements for $25°C < T < 1000°C$.

1 INTRODUCTION

The nature of defects in solid materials has been of fundamental and practical concern for many years. Dilute defects have been extensively studied, and because complex interactions do not occur at extreme dilution, a high level of success has been achieved. The concepts of chemical equilibria appear to apply well to the dilute defect state comprising vacancies, interstitials, and simple complexes as discussed by Kroger.[1]

At high defect levels, as in the present study, and especially for radiation-induced defects, the chemistry becomes complicated by the simultaneous existence of many types of defects and by defect clustering phenomena. In the radioactive actinide systems however, there are important simplifications which arise from the completely random decay process and the exponential time dependence. Thus, defects generated by self-damage are introduced homogeneously and at rates which, in principle at least, can be accurately known by correlation to the half-life of the isotope of interest. Moreover, the same sample can be conveniently studied as a function of defect concentration, eliminating sampling and/or impurity problems.

Historically, radiation-induced defects in the actinides or their compounds have been examined with a view to circumventing potential problems related to nuclear-energy applications. The first observation of self-irradiation damage in plutonium compounds was reported by Rand, et al.,[2] who followed the room-temperature lattice-constant changes as a function of time for plutonium-239 oxides, carbides, and nitrides. Since that time, there have been many similar studies extending to various plutonium isotopes (or mixtures thereof), as well as americium and curium; in general, most attention has been given to the oxides. At the start of this investigation, however, there had been no in-depth reports on the annealing kinetics of self-radiation damage in any actinide system, nor any quantitative analyses of line profile changes accompanying the self-damage process. Recently, however, a brief annealing study has been reported[3] and there has also been an extensive study of the line broadening accompanying self-radiation damage in PuO_2.[4] The low-temperature $[4 < T(°K) < 300]$ annealing behavior of PuO_2 has also been reported.[5] The purpose of this investigation was to study in some detail the damage ingrowth phenomena in $^{238}PuO_2$ and to define the annealing behavior of the same oxide after reaching the saturation damage level.

Plutonium-238 ($t_{1/2}$ = 89 years) decays to uranium-234 ($t_{1/2}$ = 2.5 x 10^5 yr) accompanied by a 5.49-Mev alpha particle. The light alpha particle creates displacements over a long range and the heavy recoiling daughter (in this case 93 kev) causes severe local damage. It has been estimated[2] that about 1500 Frenkel defects are produced per single alpha disintegration. In very short-lived alpha emitters, e.g., Po-210 ($t_{1/2}$ = 138 days), the crystal becomes amorphous after a self-irradiation of about 3 weeks. The self-damage process may also induce crystallographic alterations, as evidenced by the radiation-damage-induced transformation of cubic Cm_2O_3 to hexagonal Cm_2O_3 at room temperature.[6-8] Monoclinic Cm_2O_3, on the other hand, apparently resists radiation-damage attack and remains structurally stable after prolonged self-irradiation.[9] Such complications are nonexistent in the present study since stoichiometric PuO_2 exists only with the fluorite structure and apparently retains a high level of crystallinity throughout the self-damage process.

2 EXPERIMENTAL

The X-ray diffraction data were obtained using a GE-XRD-5 diffractometer employing copper radiation with $\lambda(K_{\alpha_1})$ = 1.54050 Å. For lattice-parameter measurements, an internal standard technique was employed in which the high-angle reflections of both the ThO_2 standard (a_o = 5.5971 Å) and the PuO_2 samples were simultaneously recorded as a continuous trace at 0.1 degrees θ/inch. Peak positions of α_1 at one-half peak height were used and could be read to ±0.001 degrees θ. The average accuracy in terms of lattice parameter was ±0.0005 Å, with those patterns being broadened by radiation damage approaching errors of ±0.001 Å. In most cases, a sufficient number of data were taken to self-define reproducibility.

The electrical-conductivity measurements were made using a three contact, guard-ring apparatus similar to that described by Bates and Schemmel.[10] Samples were sintered PuO_2 pellets, 1.13 cm in diameter and 0.31 cm thick with a 91 percent theoretical density. Platinum electrodes were applied by painting and firing coatings of platinum paste. Current and voltage leads were platinum and the sample holder was constructed entirely of high-purity alumina. The resistance measurements (1 to 10^8 ohms) were made using a constant dc power supply with applied voltages at 10 or 1 volt at high and low resistances, respectively. Currents were determined by measuring the voltage drop across calibrated resistors in series with the sample by means of a Keithley 160 digital voltmeter. Measurements were taken continuously as a function of temperature by recording sample temperature and the direct-current output of the voltmeter on a two-pen recorder. Range-change effects, which were at an error level of ~1 percent, were minimized by fitting the data to a 5th-order polynomial. Since the interest was primarily in resistance differences for the same sample, i.e., those due to production or annealing of defects, much of the system-induced error was automatically removed from consideration.

All temperatures were measured potentiometrically with precalibrated Pt/Pt-10% Rh thermocouples. Furnace temperatures were controlled to ±1°C using a Thermac proportional band controller. Furnace profiles indicated plateaus of ±1°C over a 2-inch length and samples were positioned systematically within 3/4 inch of the measuring thermocouple. An accuracy of T ±5°C for an annealing temperature, which is a worst case estimate, has a negligible effect on all results to be discussed. Also for the X-ray work, since ThO_2 and PuO_2 have nearly identical coefficients of thermal expansion, variation in room temperature in comparing different X-ray traces is automatically corrected to 25.0°C, on the basis of the ThO_2 lattice constant chosen.

As will be shown later, the helium inventory and, consequently, the thermal history of the samples studied play an important role in subsequent defect behavior. A detailed description in this respect is given elsewhere[11] for the X-ray samples. The resistivity samples studied were at similar levels of decay, with uranium ~1 percent and $0 < He < 1$ percent. A direct measurement of helium release was made on a resistivity sample containing a theoretical 0.5 at.% helium. Degassing did not begin until 1000°C and was not complete at the limiting high temperature of 1200°C. This result falls in the broad range suggested by published diffusion results.[12-14]

Both the X-ray and resistivity samples were prepared by oxidation of high-purity electrorefined metal. Actinide impurities, excluding uranium, totaled 0.4 percent. It should be pointed out that since the radii of actinides are all similar, it is virtually certain that they enter into complete solid solution at these levels. Nonactinide impurities totaled <0.2 mole percent.

3 RESULTS AND DISCUSSION

3.1 Damage Ingrowth – X-ray

Several $^{238}PuO_2$ samples differing in prior history were monitored by lattice-parameter measurements during the self-damaging process. Although samples annealed at $1000 < T < 1200$°C gave somewhat broadened patterns compared with those annealed at 1500°C, no significant difference in the cell parameter was observed, and the value for undamaged $^{238}PuO_2$ is a_o = 5.3954 Å at 25°C. The ingrowth of self-damage at room temperature is shown plotted in terms of fractional unit-cell increase versus time in Figs. 1 and 2. Figure 1 shows the result for sample E annealed initially at 1540°C and some additional data from

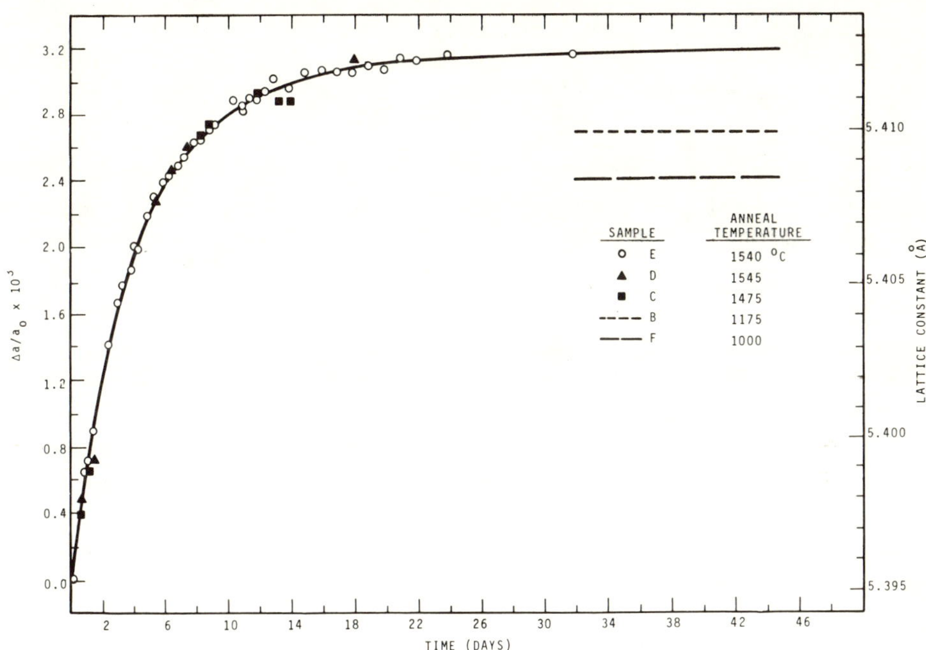

Fig. 1. Ingrowth of defects in high-temperature-annealed $^{238}PuO_2$ plotted as fractional unit cell increase versus time.

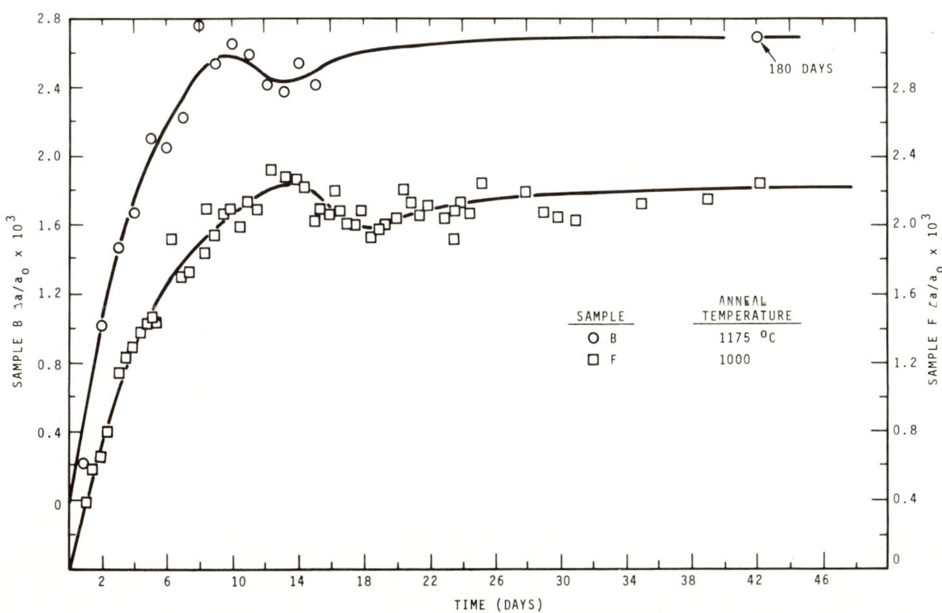

Fig. 2. Ingrowth of defects in low-temperature-annealed $^{238}PuO_2$ plotted as fractional unit cell increase versus time.

samples C and D annealed at 1475°C and 1545°C, respectively. The curve was obtained by the least-squares method giving (for t in days)

$$\Delta a/a_o = 3.20 \times 10^{-3} \, [1 - \exp(-0.24t)] \quad .$$

As pointed out by. Adam and Rogers[15], exponential behavior is to be expected if regions once damaged by a recoil atom undergo no further change as a result of passage of additional atoms. The lattice constant at the saturation level is 5.4126 Å on the basis of the above equation, while samples run at 32 and 52 days gave 5.4125 and 5.4127 Å, respectively. The agreement of data from samples C, D, and E indicates that variation in uranium concentrations from 0.7 to 1.3 percent has no significant effect on the rate or saturation level of self-damage.

Figure 2 shows the room temperature damage ingrowth for samples B and F, annealed initially at 1175°C and 1000°C, respectively. An apparent maximum exists in the damage-ingrowth curve for both of these samples, similar to that reported by Kapshukov, et al.[16,17] The saturation value indicated for curve B is based on three samplings taken after 4 months giving a = 5.4098 ± 0.0005 Å and five samplings taken after 6 months giving 5.4100 ± 0.0005 Å.

As suggested by the saturation values shown in Fig. 1, there is a regular dependence on prior annealing temperature and, therefore, on helium concentration. While we can assume with some confidence a zero helium concentration in samples annealed at approximately 1500°C, only partial release occurs at the lower temperatures, as mentioned earlier. Since the lattice-parameter increase after "saturation" as suggested by the data reported by Roof[4] does follow that expected for simple substitution of U^{+4} for Pu^{+4}, the simultaneous increase in helium concentration would appear to have a negligible effect on the "saturation" level. Electron-microscopy studies by Douglass and Bronisz[18] do show in PuO_2 and ThO_2 that "black death" defects of 70 Å diameter form and are stable to $T > 900°C$ when they cluster to form loops. In NpO_2, "black death" defects were replaced by "voids" at $T \sim 400°C$ and were stable to 1300°C. These regions are likely effective sinks for the defects produced by the self-damaging process at room temperature, and it is expected that the concentration of these and other sinks such as dislocation loops and helium bubbles is least in the case of the 1540°C anneal and greatest for the 1000°C preparation. That portion of the room-temperature self-annealing process whereby radiation-induced defects are annihilated at sinks is, therefore, greatest for the 1000°C preparation, which results in a lower saturation damage level. Similarly, the maximum point-defect concentration, and hence dilation, is highest for the most perfect crystal – the 1540°C preparation.

The Russian work[16] shows a large dip in the Δa/a versus time curve, and maxima in our own data are suggested for both the 1000°C and the 1175°C preparations. It had been suggested by Griffen, et al.,[19] that such a perturbation should not occur in the case of self-damage generated by alpha decay compared with neutron fission (the heavy-particle energies are ~80 kev and ~100 Mev, respectively). The Russian sample showing a large dip contained approximately 1 percent impurities, and the above discussion concerning sink concentrations may also be applied here. If there is a large concentration of sinks as induced by impurities and clustering of defects or by formation of helium bubbles as in our low-temperature preparations, a supersaturation of point defects relative to the sink concentration may occur and the mechanism suggested by Roberts, et al.,[20] for irradiated UO_2 may be applied. That is, the maximum is a result of migration of defects to the sink regions, which thereby reduces the supersaturation and the lattice dilation. In our own case as well as in the Russian work, the magnitude of the observed maximum in $\Delta a/a$ and the final saturation value do appear directly related to crystal perfection. In our 1175°C sample, the maximum occurs at a lower self-radiation dose than it does in the 1000°C sample. The Russian maximum likewise occurs at a lower dose for the more pure material.[16] By approaching saturation at different temperatures, Kapshukov et al. have similarly shown[17] that with increasing defect mobility (higher temperature), the dip occurs at lower values of $\Delta a/a$.

3.2 Annealing Behavior

a. X-Ray

Portions of the sample B were isothermally annealed at six temperatures. The data were taken over a 1-month period when the helium inventory was $0.4 < \text{He} < 0.8$ percent, depending on the degree of helium release at the initial heating at 1175°C. During the 1-month duration of the anneal experiments, the helium concentration increased by 0.05 percent. Figure 3 shows the short-time ($t < 150$ min) kinetic data plotted versus time. A stage I process, which we attribute to interstitial vacancy (I-V) annihilation, is virtually complete in less than 10 minutes. As pointed out previously, the concentration of defects at saturation is between 1/2 and 1 Frenkel pair per unit cell. It is anticipated that such close pair defects will be very easily annealed and because of mobility considerations, the definition of "close" will be temperature dependent. Thus, many more defects will be removed at 800°C than at 400°C, as shown by the observed increasing degree of recovery (plateau values) with increasing temperature (Fig. 3).

In Fig. 4, the longer time data are included and are plotted in terms

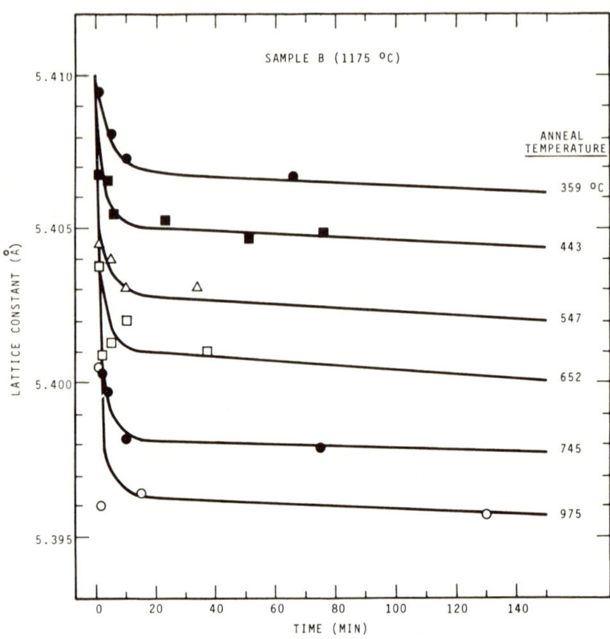

Fig. 3. Isothermal defect annealing plotted as lattice constant versus time.

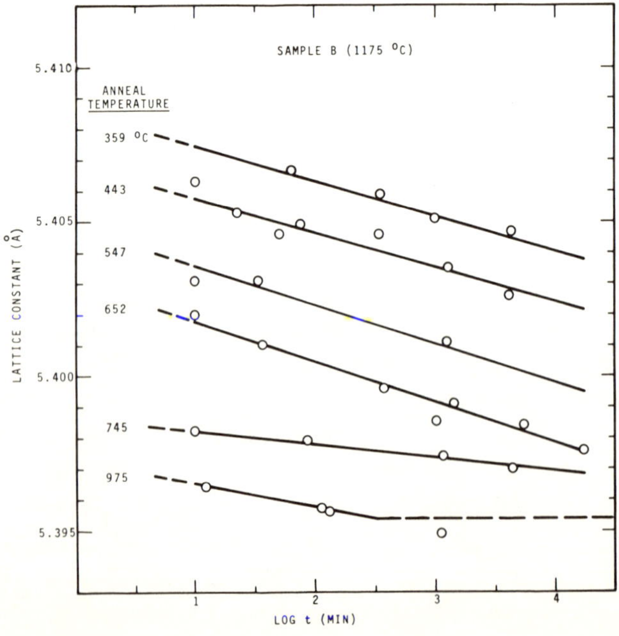

Fig. 4. Isothermal annealing plotted as lattice constant versus log t(minutes).

of lattice constant as a function of lot t (min), and it is apparent that annealing continues beyond the stage I "plateau" for at least 10^4 minutes. Although the log t dependence is drawn as linear, there is no theoretical justification for this dependence except that at long times an exponential approach to a steady-state lattice constant (due to annealing versus damage-ingrowth competition) is expected.

The stage I data do not offer sufficient detail to obtain more than a series of two-point lines by the well-known method of cross cuts. In this method, a cut through the fraction-recovery-time curves at a given recovery level yields points of equal defect concentration. For a series of such points, a plot of ln t_i versus 1/RT should be linear with slope –E, and if a single process is operative, cuts at different fraction recovery should yield a constant value for the activation energy. Since it is virtually impossible in so rapid a process to obtain reliable information by the X-ray method (except at lower temperature), we have simply drawn, on an expanded scale, straight lines connecting zero recovery at t = 0 and the plateau values at t = 10 minutes for each of the isotherms. Although this is equivalent to the assumption of a zero reaction order, it can be shown that higher-order reactions give a family of curves yielding similar activation energies.

The plots of ln t_i versus $1/RT_i$ so obtained for various fraction recoveries do yield excellent straight lines of constant slope with E = 0.14 ev, although the value is somewhat dependent on the choice of the time chosen as representing attainment of the plateau. If a time of 1 minute is used, for example, a value E = 0.20 ev is obtained.

We have also examined the isochronal annealing behavior of four samples of different thermal history: B_1, B_2, D, and E. The data obtained in the 18-hour anneals are plotted in Fig. 5 as lattice parameter versus temperature. A large shift is observed along the temperature axis, and the step which is quite definite for sample D becomes very small for sample B_2. Since the uranium concentration of samples D and E is the same while the helium concentration varies by more than a factor of three, it can be concluded that an increased helium content shifts the annealing curve to lower temperatures. Samples B_1 and B_2 suggest the same dependency. It is likewise clear that the samples initially heat treated at 1175°C are more readily annealed than those heat treated at 1545°C. (E and B_2 have comparable uranium and helium inventories.) According to earlier discussion, the high-temperature-annealed samples D and E may be considered as retaining helium predominantly in a random distribution. Samples B_1 and B_2, in addition, likely contain a large number of helium bubbles and/or other extended defects such as dislocation loops. Because of the higher sink concentration, annealing should occur at lower temperatures in the less perfect sample B, as observed. The stepwise annealing of self-damage

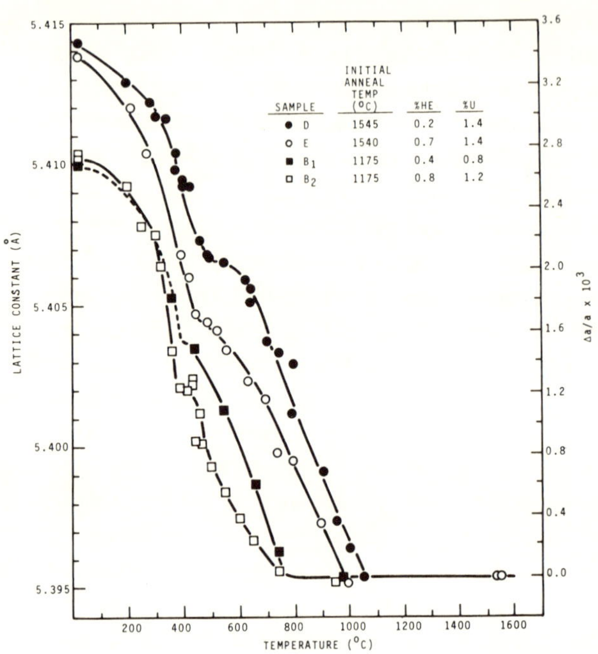

Fig. 5. Isochronal (18 hours) anneal curves for various $^{238}PuO_2$ samples.

in PuO_2 is evident for the samples annealed at 1540°C. The data suggest that the demarcation is less distinct with increasing helium content and for the 1175°C sample the step is little more than an inflection point. Although the choice of 18 hours as an annealing time is well suited for comparison with previous work, the curves are unfortunately the sum of the rapid stage I (I-V) annealing and a much slower process as defined by the kinetic results discussed earlier. Moreover, the (I-V) portion is an increasing fraction of the total recovery with increasing temperature, varying from ~52 percent at 350°C to 90 percent at 800°C.

In order to separate stage I from the long-time behavior, a series of 10-minute isochronal anneals was performed on one sample (D). Although less distinct, a midrange step was observed, indicating that even the very rapid initial annealing stage is more complex than simple annihilation. Although strictly conjecture at this point, as suggested previously[16], the low-temperature behavior is likely related to interstitial removal, while the less mobile vacancies anneal at the higher temperatures.

The X-ray diffraction patterns of the various samples differed in some respects. In the 1175°C annealed case, the high-angle reflections are quite broad, whereas the 1540°C preparation gave a pattern with well-resolved $\alpha_1 \alpha_2$ doublets, nearly equivalent to the ThO_2 standard. Although little variation could be measured in the already broadened 1175°C pattern at different levels of annealing recovery, the 1540°C annealed sample broadened appreciably as the annealing proceeded, becoming distinctly sharp only after the 18-hour anneal at 1050°C. Figure 6 shows the peak half-breadths for the (422) and (511) reflections after 18-hour anneals at the temperatures indicated. Though there is much scatter, the tendency to broadening at intermediate temperatures is apparent and is taken as an indication of nonhomogeneity. As the annealing proceeds, it must be concluded that strain fields or actual distortions arise owing to defect clustering. A more detailed analysis of line-broadening changes[11] suggests little variation in crystallite size and small strain values of $\epsilon \sim 1 \times 10^{-3}$, which are of the same order of magnitude as reported by Roof[4]. The fit of the data to classical theory as by Williamson and Hall[21] was, however, not entirely satisfactory with excessive curvature at the higher angles. Further, a systematic variation in half-widths of the low-angle reflections suggests a possible distortion from true cubic symmetry which, if assumed rhombehedral, requires $\theta \cong 89.94$ degrees.

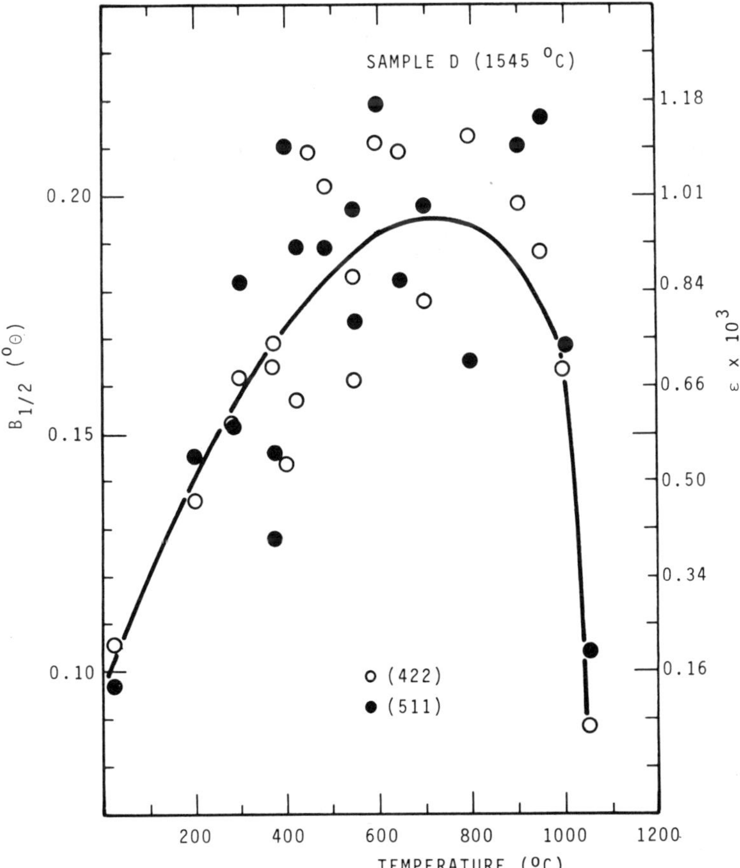

Fig. 6. Line breadth $\beta_{1/2}$, for the (422) and (511) reflections versus annealing temperature.

3.3 Annealing Behavior

b. Conductivity

It has been well established that electrical-conductivity changes can be a sensitive means of observing defects in solid materials. The application does suffer from a lack of knowledge as to the resistivity increment per unit concentration of defects and, in our case, with a multitude of defect types (and configurations); an extensive study will be required to interpret fully the observed changes. Coupled with information on other physical properties, such as the diffraction behavior described in the preceding section, a relatively detailed understanding of defect behavior will develop. The results that we present here already suggest a strong correlation of the dimensional changes and conductivity behavior.

In Fig. 7, we show a typical annealing cycle plotted as log $(1/\rho)$ versus 1/T(°K) for a sample allowed to radiation damage for 18 hours at ambient temperature and then heated and cooled at 2°C/min, during which the resistance was measured. The curves that should be linear for a singly activated process show substantial curvature at lower temperature and a change in slope at T ~ 500°C on the heating cycle. The cooling curve is linear down to T ~ 300°C where the curvature is due to the generation of new defects by decay and to other extrinsic conduction mechanisms. As indicated in the preceding section (Fig. 5), the maximum temperature of 950°C does cause complete recovery of the cell parameters, and a similar recovery of the conductivity is apparent by the overlap of data at the highest temperatures. The activation energies for conduction are reproducibly 1.02 ± 0.03 ev for the annealed samples of low helium inventory but, as shown in Fig. 8, shift to lower values with increasing helium concentrations.

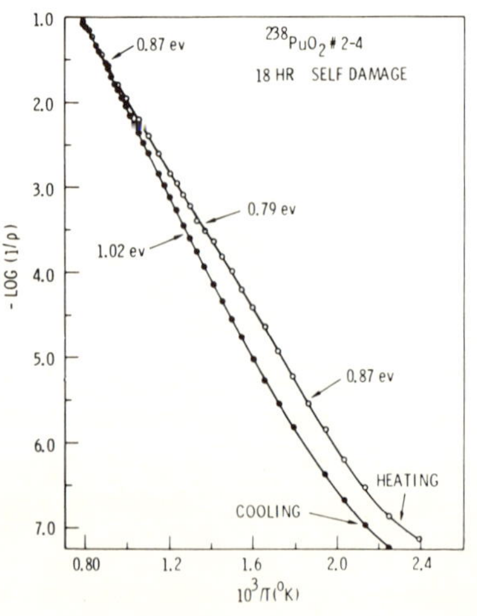

Fig. 7. Defect annealing after 18 hours of self-damage plotted as log $(1/\rho)$ versus 1/T(°K).

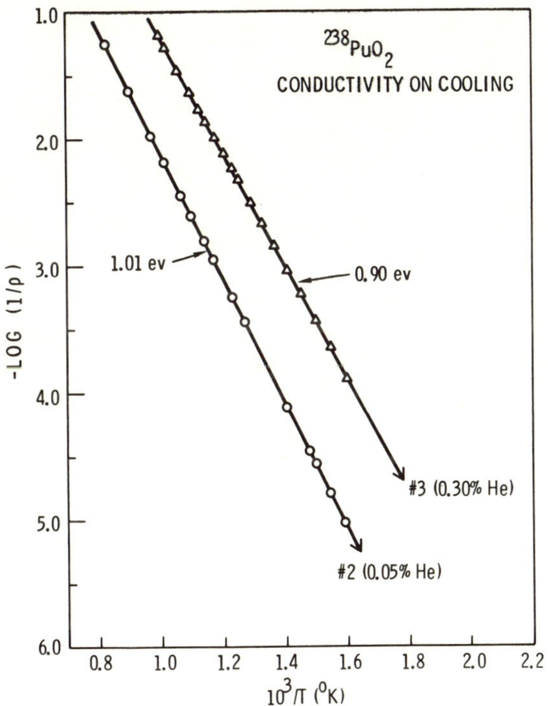

Fig. 8. Electrical conductivity plotted as log (1/ρ) versus 1/T(°K) for samples of low and high helium inventory.

In Fig. 9, we show the results of a series of runs for samples containing a theoretical helium inventory of ~0.6 percent but which were allowed to self-damage for periods of 1 to 100 days at ambient temperature. The curves are given as the difference in log $(1/\rho)$ between the heating and cooling cycles, plotted as a function of temperature. It is to be noted that induced damage causes a resistance change of more than an order of magnitude at ambient temperature in less than 1 week, and the sensitivity is consequently quite high even at low defect levels. The curvature to lower Δ log $(1/\rho)$ values at $T < 200°C$ again reflects the ingrowth of damage during the run. At higher temperatures, thermal annealing largely reduces the problem, but the curves are all shifted systematically to lower Δ log $(1/\rho)$, or to a greater extent the lower the temperature. Although several runs suggested a discontinuity near mid-range recovery, we show only the smoothed curves here until a more detailed examination is completed. By analogy with the X-ray results, some perturbation is expected in this region.

Previous studies of conductivity in long-lived $^{239}PuO_2$, also on polycrystalline samples, have been reported by McNeilly[22] and by Chereau and

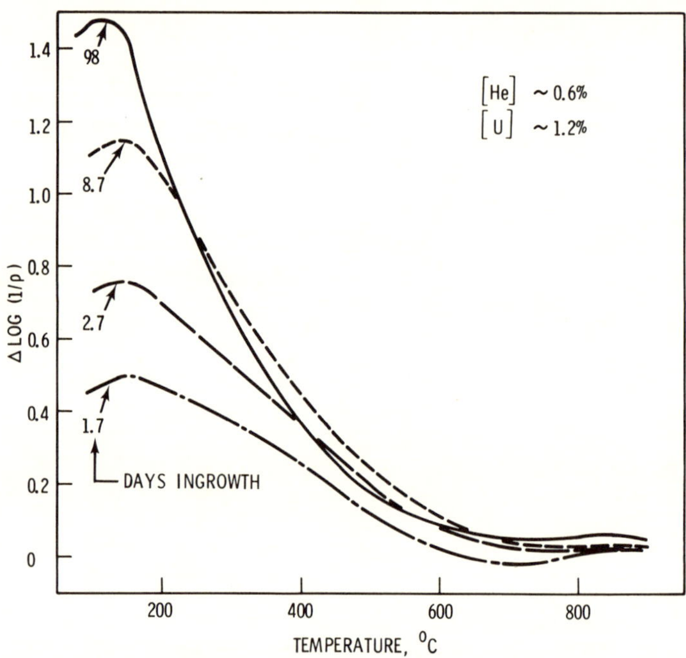

Fig. 9. Defect annealing for samples of various defect concentrations (days of self-damage) plotted as the difference in conductivity (Δ log 1/ρ) versus temperature (°C).

Wadier[23]. For $PuO_{2.00}$, both studies suggest an activation energy for conduction of 1.80 ev, considerably higher than found for the present $^{238}PuO_2$ samples on cooling. Both report a sharp stoichiometry dependence, with the most recent work giving E = 0.36 ev for O/Pu = 1.95, with a much slower drop with increasingly lower compositions. The present study was carried out under conditions of sufficiently high oxygen partial pressure (~10 ppm O_2 in N_2) so that there could be no reduction below O/Pu = 2.00. Thus, defect levels introduced either by non-stoichiometry in $^{239}PuO_2$ or by self-damage in stoichiometric $^{238}PuO_2$ result in reduced activation energies. The shift and similar energy reduction given in Fig. 8 reflect a similar effect owing to the presence of helium.

ACKNOWLEDGMENT

We would like to thank J. A. Leary and L. J. Mullins of the Los Alamos Scientific Laboratory for supplying the $^{238}PuO_2$ samples, H. E. Kissinger and J. L. Bates for helpful discussions, and E. H. Shade and R. W. Stephens for assistance in the experimental work.

REFERENCES

1. Kroger, F. A., *The Chemistry of Imperfect Crystals,* John Wiley and Sons, Inc., New York (1964).
2. Rand, M. H., Fox, A. C., and Street, R. S., *Nature,* **195**, 567 (1962).
3. Kapshukov, I. I., Sudakov, L. V., and Yakovlev, G. N., *Soviet Physics – Solid State,* **12**, 1718 (1971).
4. Roof, R. B., Jr., USAEC Report LA-4742 (1971).
5. Jacquemin, J., "Elongation, by Self-Irradiation at 4.2°K, of Plutonium and Some of Its Compounds", Thesis submitted to the University of Paris, 1970.
6. Wallmann, J. C., *J. Inorg. Nucl. Chem.,* **26**, 2053 (1964).
7. Kapshukov, I. I., Sudakov, L. V., Rykov, A. G., and Yakovlev, G. N., *Radiokhimiya,* **11**, 465 (1969).
8. Noe, M., Fuger, J., and Duyckaerts, G., *Inorg. Nucl. Chem. Letters,* **6**, 111 (1970).
9. Haug, H., *J. Inorg. Nucl. Chem.,* **29**, 2753 (1967).
10. Bates, J. L., and Schemmel, R. R., USAEC Report BNWL-1671 (1971).
11. Chikalla, T. D., and Turcotte, R. P., *Radiation Effects,* **19**, 93 (1973).
12. Zielinski, R. E., "Helium Release from $^{238}PuO_2$ Microspheres Test Program", presented at PuO_2-Helium Release Information Meeting, AEC Headquarters, Washington, D.C., August 7, 1969.
13. Stark, W. A., Sandia Laboratories Report, SC-DC-69-1892 (July, 1969).
14. Northrup, C.J.M., Sandia Laboratories Report, SC-DC-69-1949 (August, 1969).
15. Adam, J., and Rogers, M. D., *Reactor Sci. Tech.,* **14**, 51 (1961).
16. Kapshukov, I. I., Yakovlev, G. N., Skiba, O. V., Timofeev, G. A., and Sudakov, L. V., *Atomnaya Energiya,* **27**, 337 (1969).
17. Kapshukov, I. I., Sudakov, L. A., and Yakovlev, G. N., *Atomnaya Energiya,* **31**, 630 (1971).
18. Douglass, D. L., and Bronisz, S. E., in *Plutonium 1970,* W. N. Miner (Ed.), AIME, New York (1971), Vol. II.
19. Griffin, C. S., Mendelssohn, K., King, E., Lee, J. A., Rand, M. H., and Street, R. S., in *Plutonium 1965,* A. E. Kay and M. B. Waldran (Eds.), Chapman and Hall, London (1966).
20. Roberts, L.E.J., Brock, P., Findlay, J. R., Frost, B.R.T., Russell, L. E., Sayers, J. B., and Wait, E., *Proc. Third Int. Conf. on Atomic Energy, Geneva,* **11** 464 (1965).
21. Williamson, G. K., and Hall, W. H., *Act. Met.,* **1**, 22 (1953).
22. McNeilly, C. E., *J. Nucl. Mater.,* **11**, 53 (1964).
23. Chereau, P., and Wadier, J. F., *J. Nucl. Mater.,* **46**, 1 (1973).

DISCUSSION on Paper by R. P. Turcotte

SMELTZER: Do the annealing-recovery experiments give any distinction between plutonium, helium, or uranium interstitials?

TURCOTTE: In metallic systems, it has been possible to interpret differences in cell-parameter recovery and conductivity recovery to differences in interstitial and vacancy annealing, since changes in volume are much more influenced by interstitials than by vacancies. In an ionic compound, the interpretation will be much more difficult, especially with helium present, but this is a long-range objective of these studies. We are not aware of this having been achieved in any actinide oxide system.

KROGER: Helium production is accompanied by uranium formation. Is not the effect on resistivity you ascribe to helium really due to the uranium?

TURCOTTE: Yes, this is a possibility. We have no way of knowing the origin of the defect levels responsible for the observed changes.

WELCH: Do you know whether your observed conductivity is predominantly ionic or electronic?

TURCOTTE: Although we have not yet examined this question, it has been reported to be primarily electronic by McNeilly (ref. 22).

VINK: How large are the holes where the helium is fitting in? If they are very small, it would perhaps be interesting to do experiments at very low temperatures where helium gets into some very special states.

TURCOTTE: On the basis of literature reports, the bubbles grow from ~40 Å to ~150 Å at higher temperatures before migrating out of the crystal. We have not done any low-temperature work.

EYRING: Have metallographic and other high-magnification observations been made on these specimens?

TURCOTTE: We have not examined these particular samples by optical methods. The general behavior is, however, fairly well established by many studies of the heavier fission gases in UO_2 and also helium in $^{238}PuO_2$ using electron microscopy (see ref. 18).

WAGNER: To which process did you ascribe the activation energy of 0.15 ev?

TURCOTTE: The very rapid annealing observed isothermally must involve only very short-range migration to account for the low (0.15 ev) activation energy.

WAGNER: Did you state that the sample containing more helium exhibited a lower activation energy for electrical conduction (about 0.9 ev) than the samples containing little or no helium (about 1.2 ev)?

TURCOTTE: Yes, the sample with the lower activation energy contained a high helium concentration.

WAGNER: Does your sample gain or lose oxygen at the elevated temperature?

TURCOTTE: Under the conditions of these experiments (~10 ppm O_2 in N_2), PuO_2 is stable. The reduction of any Pu^{+4} to Pu^{+3} requires dry hydrogen and higher temperatures.

LIDIARD: Your interpretation of the annealing of radiation damage in PuO_2 follows the previous pattern developed for metals and other monatomic substances. It should be remembered, however, that in other compounds with the fluorite structure, the anion interstitial is less mobile than the anion vacancy which is in contrast to the situation in metals. I would thus like to know if you have estimated the relative numbers of anion and cation displacements in the damage process in these experiments. It would seem to me one should attempt to do this before setting out to interpret the annealing stages seen.

TURCOTTE: We have not attempted an estimation of the relative number of anion and cation displacements during the damage process. The samples we have examined were self-irradiated for ~3 to 4 months, a period of time sufficient that every ion has been displaced many times over, most of course being self-annealed. It is agreed that an interpretation of the annealing stages II and III is especially open to conjecture at this time.

THE DEFECT STRUCTURE OF TRANSITION-METAL MONOXIDES

*M. Hayakawa**, *M. Morinaga, and J. B. Cohen*

Department of Materials Science
The Technological Institute
Northwestern University, Evanston, Illinois 60201

ABSTRACT

Quantitative measurements of the diffuse scattering from wüstite (Fe_xO) at 840°C are reported. The clusters of vacancies and tetrahedral ions found previously in quenched specimens are also found at temperature in the stable one-phase field. These clusters are smaller and contain more octahedral iron than those found in a quenched specimen. As x increases the clusters become smaller, and the concentration of tetrahedral ions in them decreases, even more rapidly than the octahedral ion concentration.

Portions of the diffuse scattering from VO_x are shown to be related to the Fermi surface and hence to long-range electronic interactions. As x increases the electron concentration in the Brillouin zone decreases.

*Currently a Postdoctoral Fellow, Linköping University, Linköping, Sweden.

1 INTRODUCTION

The monoxides of the first transition series are generally considered to have a structure like that for NaCl (Strukturbericht Notation Bl). But this designation is far too general, for these oxides contain large numbers of defects and wide ranges of stoichiometry. Toward the end of the transition series, the defects are primarily cation vacancies and tetrahedral ions. The detailed evidence for such defects (in wüstite and NiO-CoO solid solutions) is based on measurements of densities, and/or lattice parameters, and/or integrated intensities of Bragg peaks with neutrons and X-rays, and in one case, of diffuse scattering.[1-5] In FeO, when about 10 percent of the total number of iron cations are missing, another 4 percent are in tetrahedral sites[2-4], so that ≈ 14 percent of the octahedral cation sites are vacant. For quenched $Fe_{0.902}O$, it appears that these vacancies and the tetrahedral ions are grouped in a very definite array – a central octahedral site and its twelve nearest-neighbor octahedral sites being vacant, the group "occluding" four tetrahedral ions.[4] Cations around such a cluster are displaced toward it and anions are displaced away from it. These groups are placed periodically in the host lattice. At low temperatures, FeO decomposes into iron and Fe_3O_4; this clustering is a metastable state. There are other metastable phases as well, at other compositions[6,7], but the detailed structures have not yet been derived.

At high temperatures in the stable one-phase field, there is no evidence (from the shape of X-ray peaks or lattice parameters) for any subphases, as had been suggested from thermodynamic studies[8], but there are some qualitative diffraction data that suggest that the clustering persists at high temperatures[4,9]. (See also ref. 51.)

While these clusters allow for at least a qualitative interpretation as to why the Néel temperature of Fe_xO is nearly independent of composition[10] and why diffusivity decreases with increasing deviation from stoichiometry[11,12], and the variation in semiconductivity with composition[13], no detailed quantitative analyses based on the defect arrangement determined experimentally have yet been reported.

In NiO-CoO solid solutions, measurements of integrated intensities at 1100 to 1500°C indicated 10 percent vacant cation sites and tetrahedral occupation factors of 2 to 5 percent.[5] It appears that measurements of departures from stoichiometry (of the order of 1 percent in this system) by such methods as thermogravimetry, measure only the difference in occupation of these two sites, *not* the cation vacancy concentration. Interestingly, the concentration of tetrahedral ions is similar to the concentration of trivalent ions reported for this material.

At the other end of the transition series, e.g., for TiO and VO the defects involve both cations and anions, with large concentrations of vacancies in both sublattices, even at stoichiometry (the order of 14 percent.[14] There is evidence for tetrahedral occupation for VO.[15-17] Quite complex diffuse scattering which is a function of heat treatment and composition has been reported in the literature for both TiO and VO[15,17,18] implying nonrandom defect arrays, and there are indications that long-range electronic interactions are important in determining the local defect order in TiO.[18]

There is evidence for ordered phases at low temperature in both compounds.[15,16,19-23] In VO, at least, there is a cluster of vacant octahedral sites and a tetrahedral vanadium ion, similar to FeO. These clusters are ordered, but above the ordering temperature the cluster seems to be similar. The defect arrangement in TiO is quite different. The conductivity of these two oxides is that of a semiconductor above the ordering temperature, but more like that of a metal below this point.[24] VO_x is an n-type semiconductor for $x < 1$, but is p-type for $x > 1$.[13] There are maxima in the Seebeck coefficient versus composition in both regions[14], which suggests some form of defect clustering or local order. Finally, there is an unresolved dispute over a transition in VO at 125°K, below which some authors report that the resistivity increases by 10^6; others could not reproduce these results.[25-29] Here too, there is a possibility of an effect of defect arrangement which could vary with heat treatment.

This vast array of defects, defect arrangements, and unusual properties in this group of oxides was virtually unknown only two decades ago. The phases without long-range order are not stable at room temperature in many of these oxides, for example, VO, TiO, and FeO. Yet much of the prior work on the disordered state has been on quenched specimens. In this paper, detailed studies are reported on the defect arrangement in Fe_xO *at* 840°C, in its one-phase field. Also, the relationship between the diffuse scattering from VO_x and its Fermi surface is presented.

2 Fe_xO

Absolute scattering intensities have been measured at 840°C; the theory for these measurements and the experimental procedures are presented in the Appendix. The focus here is on the results.

First some qualitative information is presented. With a crystal of $Fe_{0.902}O$ quenched from 1000°C, a quite complex pattern of supplementary peaks has been reported.[4] This is shown schematically in Fig. 1. The intensities of the spots are indicated by their size, and the asymmetry is quite clear. (Incidentally, this asymmetry is *not* noticeable on electron-diffraction patterns.) Furthermore, these extra peaks are stronger near 200 and 600 fundamental peaks than near the 400 peaks.[4] The positions of these peaks, at least for quenched specimens[4], is independent of composition. New results are given in Fig. 2 which show that this is so *at* 840°C in the one-phase field as well. While there are many physical reasons to

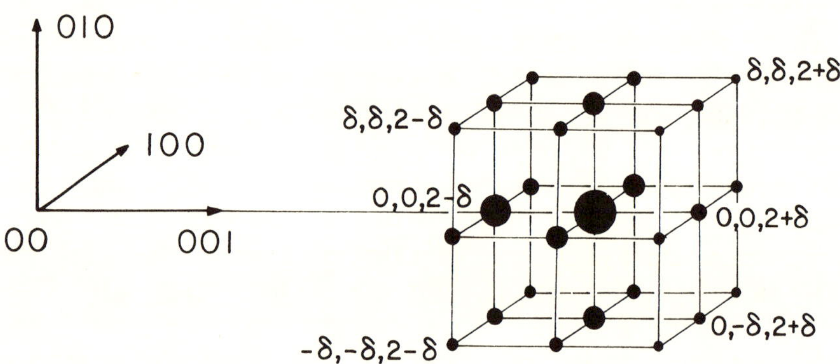

Fig. 1. Schematic representation of the supplementary peaks around the 002 peak from $Fe_{0.902}O$ quenched to room temperature from 1000°C.

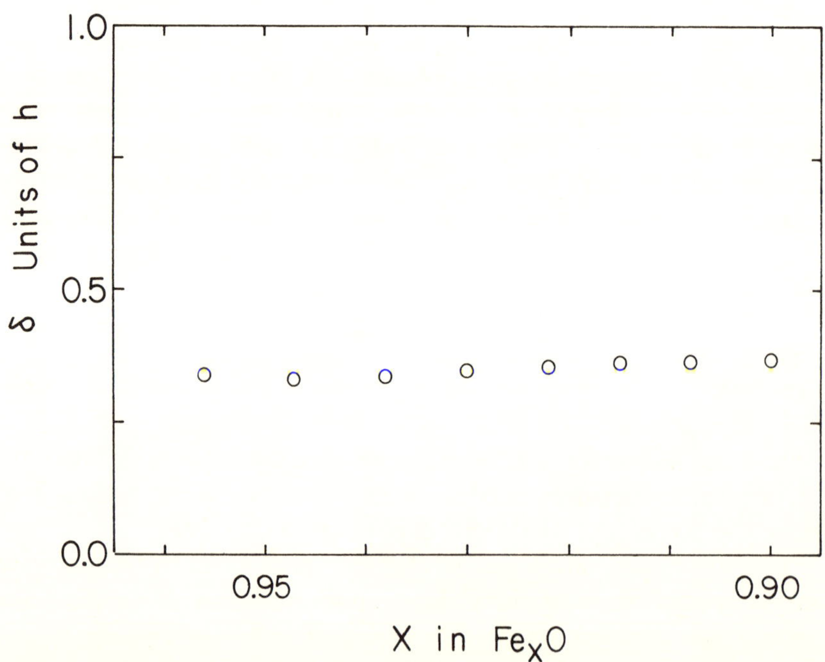

Fig. 2. The separation (δ) of the 2-δ,0,0 peak from the 200 versus composition at 840°C. (δ at temperature is similar to δ on quenched specimens (see ref. 4).

suspect cation-vacancy clustering in this system, rather than an ordered array, the fact that these extra peaks do not change position with composition clearly precludes any ordering scheme involving single vacancies. Koch and Cohen[4], from the intensities of many peaks in the pattern, were able to show that for x = 0.902:

1. There are vacancy clusters in quenched specimens involving a cation vacancy and 12 nearest-neighbor vacant cation sites. These clusters are arranged periodically on an approximately 3X cell.

2. The asymmetry of the supplementary peaks is due to displacements of cations toward a cluster, and anions away from it.

3. The stronger intensities of these "extra" peaks near the 200 and 600 peaks, compared to that ncar thc 400 peaks, is due to tetrahedrally coordinated iron cations in the vacancy cluster. The ratio of octahedral vacancies to tetrahedral iron is about three, *not* the value of two which would be expected if the local structure was like Fe_3O_4. This has been confirmed at high temperatures by Cheetham, Fender, and Taylor.[30]

4. At least some of the extra peaks remain at high temperatures.[4,31] This is clearly illustrated in Fig. 3. Note that the pattern of extra peaks becomes simpler as the temperature increases.

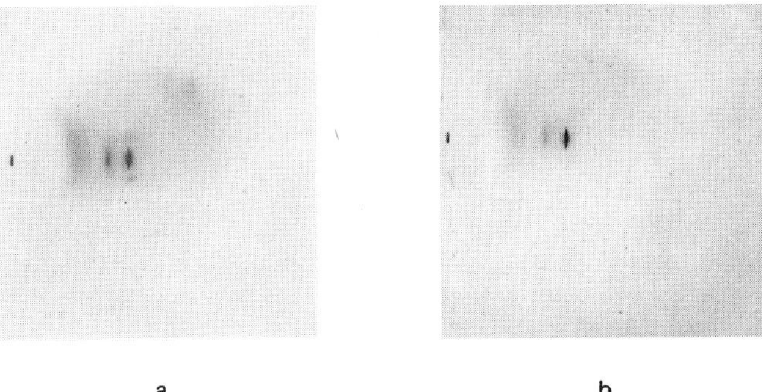

Fig. 3. Diffuse intensity around 200 diffraction peak, $Fe_{0.907}O$: (a) 575-578°C, (b) 699-702°C. (Films supplied by Dr. J. Manenc, IRSID, Saint-Germain-en-Laye, France.)

For a quenched specimen, the intensities of all the peaks were accounted for by the model, but it was not possible to fit the positions of the supplementary peaks; these actually correspond to a noninteger multiple of the NaCl cell even though they are sharp, which implies some regular mixing of more than one period. As this mixing can affect the intensities[31,32], the detailed model of vacancy clustering needs to be validated. More importantly, it is the local structure at high temperatures that is of most value in understanding the electrical properties and diffusivity, not the structure of a quenched specimen. It is for these reasons that the present study was undertaken at high temperatures.

The intensities of these extra peaks increase smoothly as x decreases at 840°C (Fig. 4); this implies that the density of vacancies in a cluster is decreasing but not the spacing, as stoichiometry is approached. (Recall that the positions of these peaks are unaffected by composition, Fig. 2.) As the widths of these peaks do not change with composition (Fig. 5), the range and regularity of the correlation between the vacancy clusters are not changing. It is quite interesting that the ratio of the 4-δ,0,0 intensity to the average of 2-δ,0,0 and 6-δ,0,0 approaches unity as x approaches one (Fig. 6). This implies that the tetrahedral-iron concentration in the clusters is decreasing to zero as stoichiometry is approached, and more rapidly than the octahedral-vacancy concentration is decreasing.

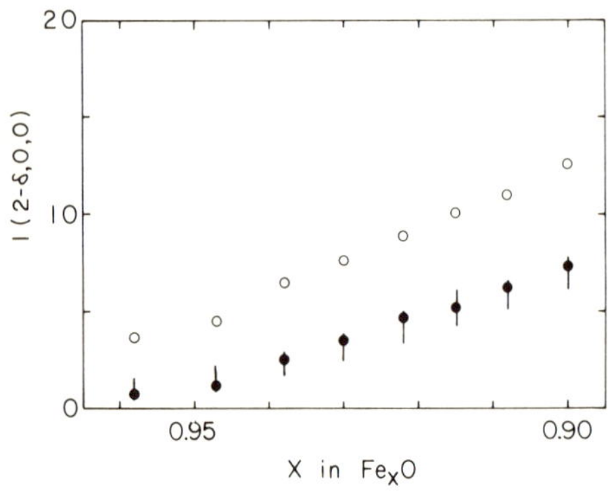

Fig. 4. Intensity of the 2-δ,0,0 supplementary peak versus x at 840°C: ●, intensity above background (error bars indicate range depending on extremes of possible background lines); ○, intensity over a fixed angular range, including background.

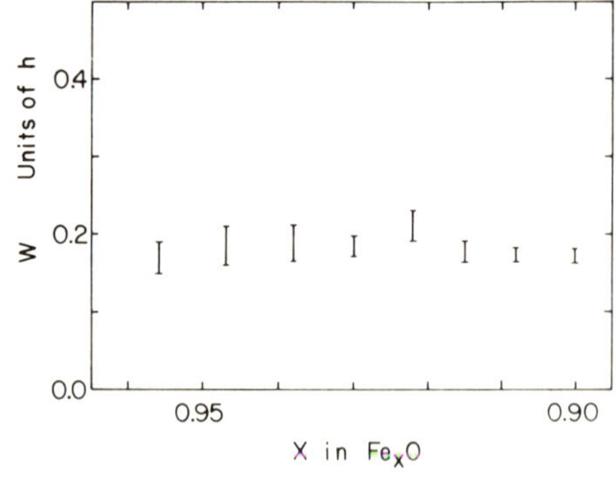

Fig. 5. Half-breadth (w) of the 2-δ,0,0 peak versus composition at 840°C.

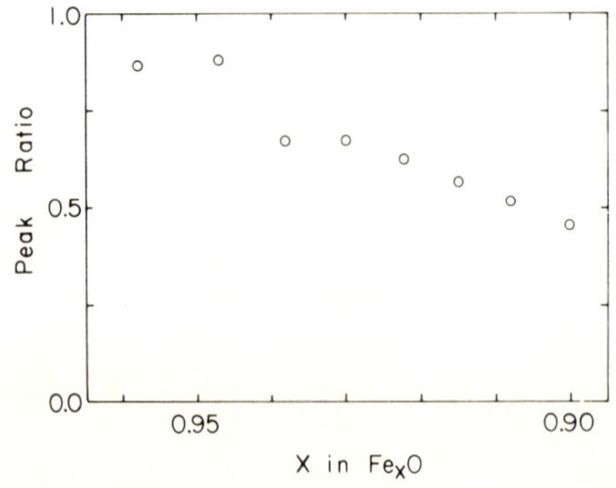

Fig. 6. Ratio of the 4-δ,0,0 intensity to average of 2-δ,0,0 and 6-δ,0,0 intensities versus composition at 840°C.

Quantitative studies of the diffuse scattering (the extra peaks and the background between them) were undertaken for x = 0.900 and x = 0.956 (see the Appendix). From these investigations, the local-order parameters α (ℓmn) were determined for interatomic vectors between octahedral sites and between octahedral and tetrahedral sites. (The values between tetrahedral sites were not determined for either composition.)

$$\alpha^{\mu\nu}_{ij}\,(\ell mn) = 1 - \frac{P^{ij}_{\mu\nu}}{x^{j}_{\nu}}\;, \qquad (1)$$

where ℓmn represents the coordinate of an interatomic vector in the form ¼[abc] for Fe_xO, where a,b,c are integers and x^j_ν is the fraction of species j on sublattice ν. The term $P^{ij}_{\mu\nu}$ is the conditional probability that if there is an atom of type i on the μth sublattice, there is one of type j on the νth sublattice. In this case, j is a vacancy, i is iron, and μ and ν represent the octahedral or tetrahedral sublattice. The α values are given in Table I. The negative values for the near-neighbor vectors between octahedral and tetrahedral sites implies that $P^{FeV}_{OT} > X^{V}_{T}$; that is, there is a tendency for

Table I. Local Order Parameters α(ℓmn) for Various Interatomic Vectors (ℓmn) in Fe_xO

	X = 0.900		X = 0.956	From Ref. 4 (X = 0.902),
	<ℓmn> · 4	α^{FeV} (ℓmn)	α^{FeV} (ℓmn)	α^{FeV} (ℓmn)
Vectors	220	0.199	0.178	0.35
between	400	0.059	0.067	0.15
octahedral	422	0.013	−0.004	0.02
sites	440	0.040	−0.011	−0.04
	620	0.028	−0.013	−0.25
α^{FeV}_{OO} (ℓmn)	444	−0.074	−0.104	−0.25
	642	−0.043	−0.004	−0.25
	800	0.067	+0.114	–
	12,0,0	0.036	–	–
	16,0,0	0.005	–	–
	20,0,0	0.005	–	–
Vectors	111	−0.0023		
between	311	−0.0012		
octahedral	331	0.0003		
(0) and	333	0.0005	Not determined	Not determined
tetrahedral (T)	511	−0.0005		
sites	531	0.0003		
	533	0.0009		
α^{FeV}_{OT} (ℓmn)	551	0.0006		
	553	0.0011		
	555	0.0012		

the octahedral cation to be surrounded by vacant tetrahedral sites, rather than by tetrahedral iron. The values for vectors between octahedral sites are about half those reported in ref. 4 for a quenched specimen, for the first few shells. These values point to a tendency for vacancy-vacancy pairs, not Fe-V pairs on the octahedral sublattice.

Employing these $\alpha(\ell mn)$, computer simulations[33,34] were carried out, so that a 10 x 10 x 10 lattice array of the octahedral sublattice (4,000 lattice sites) satisfied the first six $\alpha^{FeV}_{oo}(\ell mn)$. The resultant configurations were then examined in detail. The fractions of various near-neighbor triplet and quadruplet Fe-V arrays in the simulation are given in Table II.

Table II. First Nearest-Neighbor Triplets in Computer Simulations With Measured $\alpha(\ell mn)$

Type	Quenched Specimen[(a)]	840°C, X = 0.900	Random, X = 0.900	840°C, X = 0.956	Random, X = 0.956
FeFeFe	0.743	0.683	0.625	0.847	0.820
FeFeV	0.154	0.215	0.318	0.120	0.168
FeVV	0.070	0.084	0.054	0.029	0.012
VVV	0.028	0.017	0.003	0.005	0.000
FeFeFeFe	0.699	0.630	0.534	0.813	0.768
FeFeFeV	0.174	0.216	0.363	0.135	0.210
FeFeVV	0.058	0.107	0.092	0.031	0.021
FeVVV	0.055	0.041	0.010	0.014	0.001
VVVV	0.015	0.007	0.000	0.002	0.000

(a) Data from ref. 4 (X = 0.902).

Included are values from a simulation with the data on a quenched specimen from ref. 4 as well as calculated values for a random array of vacancies. There is clearly a tendency for vacancy clustering in Fe_xO, this tendency being larger for the quenched specimen, less for x = 0.900, and still less near stoichiometry for x = 0.956. This is brought out even more clearly in Fig. 7, where the fraction of vacancies with n vacancy near-neighbors is given for x = 0.900. It is interesting to note that the differences from a random vacancy array become small beyond the first shell for the specimen measured at high temperature, but in the third shell for the quenched specimen. These results imply a more dilute vacancy cluster at high temperatures, with a few cations replacing vacancies within a cluster. The simulations were also examined for the size, shape, and spacing of such regions. In the arrays for both the quenched specimen and for the two measurements at 840°C, vacancies are almost always linked by ¼ <220> and ¼ <400> vectors. The near-neighbor octahedral vacancy clusters in these simulations are three-dimensional in shape. The distribution of cluster sizes is shown in Fig. 8 for computer simulations from the measurements at 840°C, from the data on quenched $Fe_{0.902}O$, and for a near-random array of vacancies. The similiarities between the distribution for x = 0.9 at 840°C and after quenching are striking.

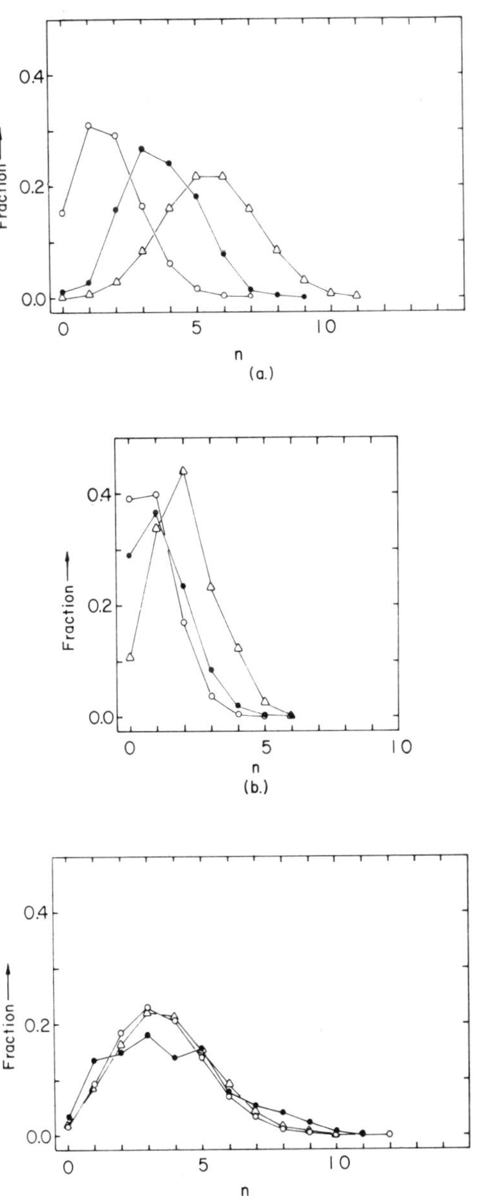

Fig. 7. Fraction of vacancies with n vacancies in the ith neighbor shell: ● simulated structure from measured α's for $Fe_{.900}O$, △ from αs in ref. 4, ○ random array of vacancies; (a) i = 1, (b) i = 2, and (c) i = 3.

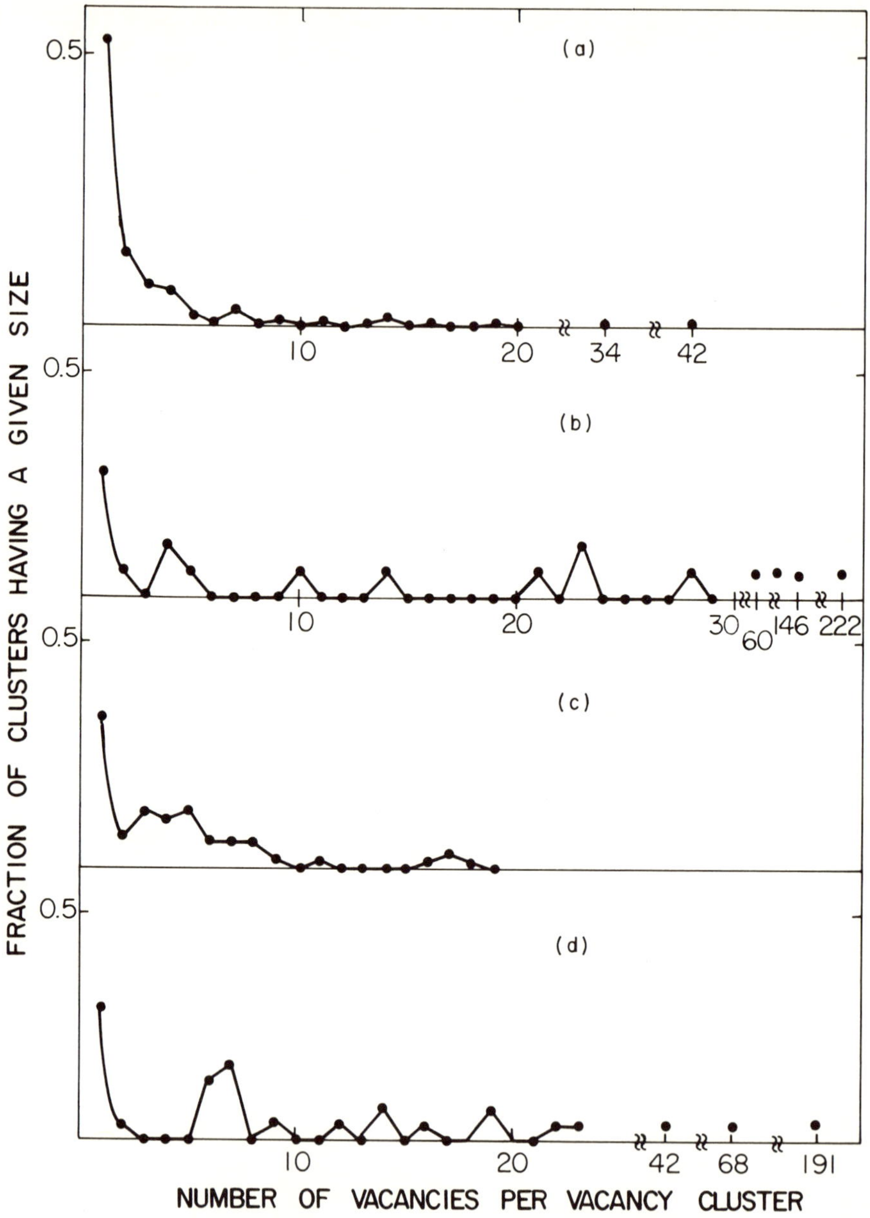

Fig. 8. Fractions of near-neighbor octahedral vacancy clusters in Fe_xO with a given number of vacancies, from computer simulations with αs: (a) x = 0.900, 1st 6 αs = 0.0; (b) x = 0.900, 1st 6 αs measured at 840°C; (c) x = 0.956, 1st 6 αs measured at 840 °C; (d) x = 0.902, quenched to room temperature from 1000 °C, 1st 3 αs taken from ref. 4. (In all cases, the octahedral vacancy concentration was adjusted to take into account that the ratio of octahedral vacancies to tetrahedral ions is 3.2.) There were 31 clusters for the quenched specimen, and 18 for the at-temperature simulations. Also, the spacing of clusters in the simulation was that measured from the positions of the supplementary peaks in the diffraction pattern.

3 VO_x

The diffuse scattering is inversely related to the Fourier transform of the pair potential[35], V(k). It has been shown that the portion of the conduction electron energy associated with order can be written in a pairwise form.[36] As a result, when the transform of the electron energy is small, the diffuse intensity is large. Singularities occur for V(k) for vectors in k space that span the extremities of the Fermi surface, 2 k_F. These singularities are especially marked for vectors spanning flat regions of the Fermi distribution[37], and hence the diffuse intensity will be large for such vectors. Thus, from diffraction patterns it is possible to obtain some information on the shape and dimensions of the surface in situations where other methods are difficult to use. This procedure has been employed successfully in alloys; for example, see ref. 38. Castles, et al.,[18] have been able to relate the diffuse intensity from quenched TiO specimens to the calculated Fermi surface; they showed how this surface differed from calculations and changed with composition. The supplementary peaks in Fe_xO may well have a similar origin. However, there are no theoretical calculations of the Fermi surface and, furthermore, the supplementary peaks do not vary in position with composition. But, as cation vacancies increase, the electron concentration should decrease.

The diffuse intensity from VO_x is quite complex and has not been thoroughly analyzed, but it is clear from the literature that there are marked changes in the pattern with composition. Accordingly, small single crystals of various compositions were quenched from 600 to 1000°C and the electron-diffraction patterns from small chips were examined in an Hitachi HU-200 F microscope operated at 200 kv. Sections of reciprocal space near the (hoo) and (hho) planes are shown in Fig. 9 for $VO_{1.02}$ and $VO_{1.17}$. Superimposed on these drawings are vectors representing the largest distance between flats on the calculated Fermi surface[39], shown in Fig. 10. These appear to be the same vector in both sections, a vector between the <103> and <113> directions. The length of this vector is considerably larger than that predicted from the calculated Fermi surface. (In Fig. 10, the solid line represents the calculated surface, and the segmented band represents the range of values observed.) This large discrepancy is not surprising in view of the fact that the calculations were made ignoring the vacancies and for monovalent ions, whereas the actual charges are much larger.[40] In fact, a recent calculation for divalent ions lies just *beyond* our measurements.[41] Also, clustering or local ordering can affect the Fermi surface[42], but were not included in the calculations. In Fig. 11 the measured 2 k_F is plotted versus x along with the vanadium vacancy concentration and lattice parameter. As the octahedral vanadium vacancy concentration increases, the concentration of electrons decreases

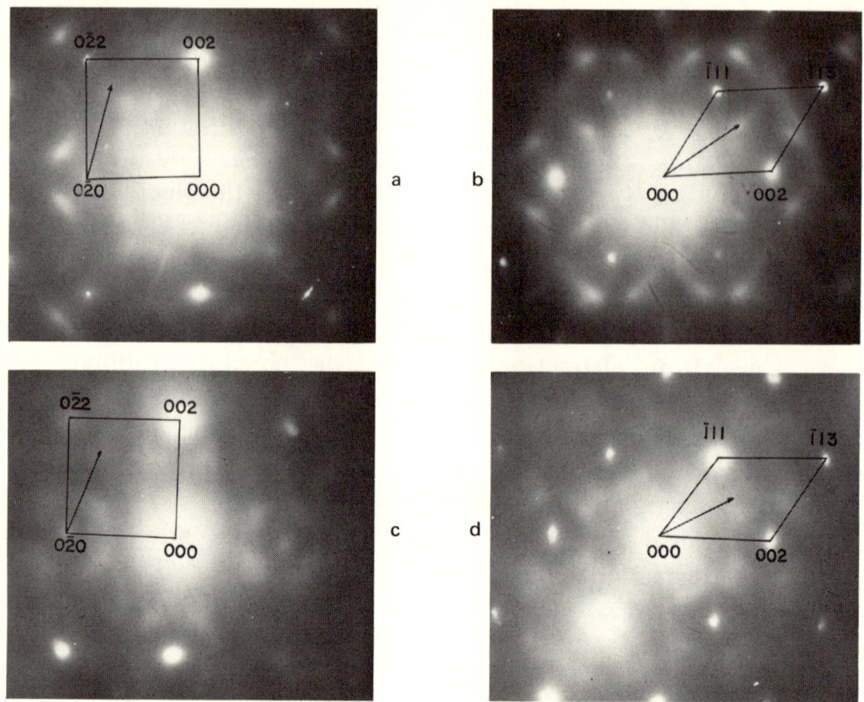

Fig. 9. Electron diffraction patterns, from VO_x, taken at 200 kv: (a) near (hoo), $VO_{1.02}$; (b) near (hho), $VO_{1.02}$; (c) near (hoo), $VO_{1.17}$; (d) near (hho), $VO_{1.17}$; vectors represent 2 k_F, as shown in Fig. 10.

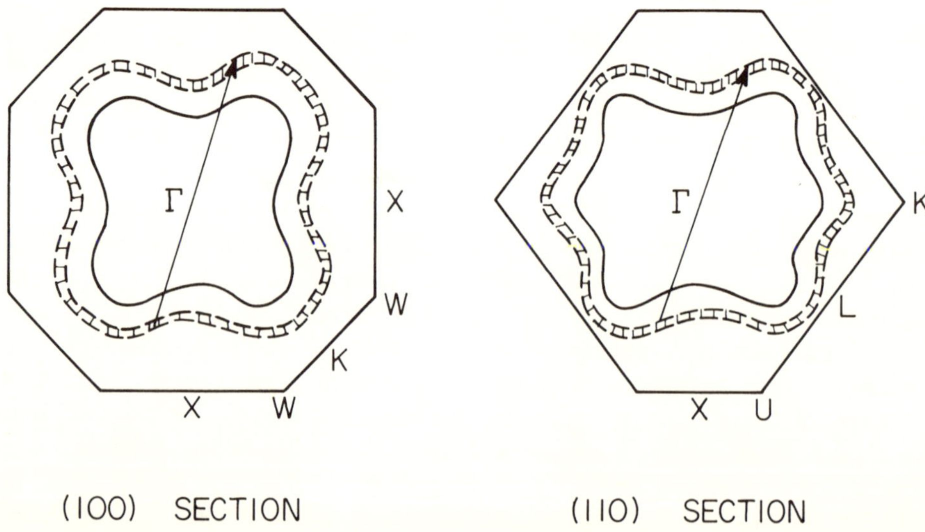

Fig. 10. Calculated Fermi surface from ref. 39 for VO (solid line) and measured values (segmented band) for the range of compositions $VO_{0.80}$ to $VO_{1.24}$. The vectors 2 k_F in Fig. 9 are shown.

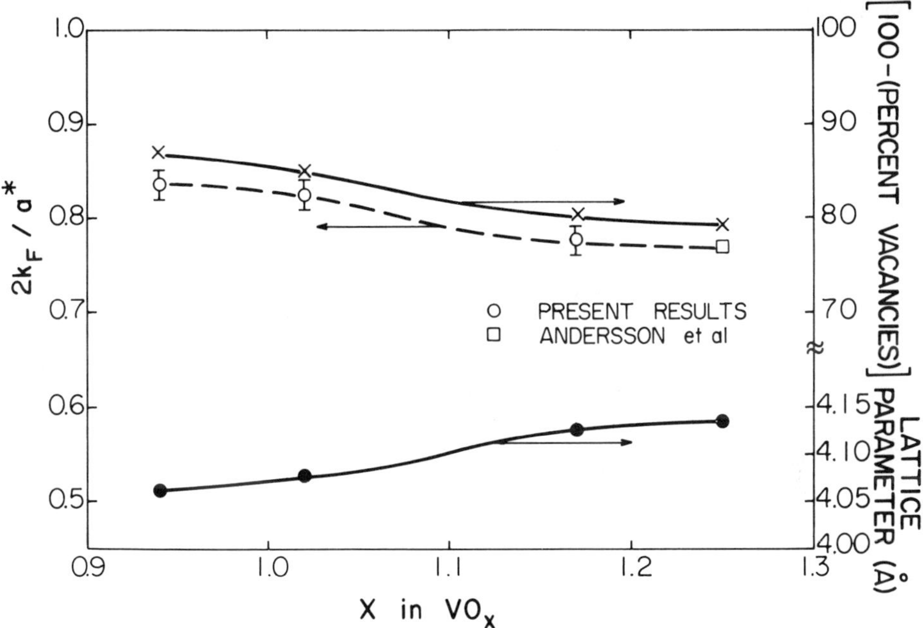

Fig. 11. Lattice parameter (●), octahedral vacancy concentration (from ref. 14) (x), and 2 k_F (0) versus x in VO_x.

and the Fermi surface shrinks. This is similar to what occurs in TiO. These kinds of measurements could be quite useful in comparing theory to the actual material, even for a semiconductor like VO.

In this paper we have tried to emphasize the vast amount of detail now available on the defect structure of the transition-metal monoxides as a result of scattering experiments. It is just these details that may be needed to allow us to understand in a quantitative fashion the properties and behavior of these oxides.

ACKNOWLEDGMENTS

We are especially indebted to Dr. T. B. Reed, Lincoln Laboratories, who kindly provided us with crystals of FeO and VO. Portions of the work on wüstite were presented in June, 1973, by M. Hayakawa in partial fulfillment of the requirements for the Ph.D. degree at Northwestern University. All of our efforts in this area were supported by the Air Force Office of Scientific Research under AF Contract 68-1457. We gratefully acknowledge their continued interest.

APPENDIX

The theory of quantitative measurements of diffuse scattering from oxides will be briefly described, as well as the data manipulations and experimental procedures specifically employed for Fe_xO.

Theory

Expressions for measurements of diffuse scattering due to local order and atomic displacements have been derived for alloys with one atom per lattice point[43,44], but for oxides, structures are more complex, often involving several sublattices. A more general theory has recently been developed[31]; its application to this study will be outlined here.

Along a line in reciprocal space, with coordinates h, I_D, the total coherent diffuse intensity in absolute units is divided by the Laue monotonic scattering for Fe_xO, $N(X_o^{Fe}X_o^V + 2X_T^{Fe}X_T^V)f_{Fe}^2$. The total number of lattice points in the scattering volume is N, X_μ^i is the sublattice fraction of component i on sublattice μ, "o" specifies an octahedral site, "T" specifies a tetrahedral site, V represents a vacancy, and f is the atomic scattering factor. The resultant I_D along a line in reciprocal space may be written as:

$$I_D\,(h) = I_{SRO}(h) + hQ(h) + h^2R(h)\,[\exp(b\,(h)h^2 - 1)] \quad . \qquad (A\text{-}1)$$

It is only I_{SRO} that concerns us here. It is a Fourier cosine series whose coefficients are the local-order parameters. The term Q(h) is a sine series whose coefficients involve the average static displacements of atoms from lattice sites. The term R(h) is a cosine series whose coefficients involve mean-square static and dynamic displacements and cross products of these displacements. The quantity in brackets is an approximate expression for higher-order displacements, assuming that such displacements are independent of atom type. As measurements on oxides often need to be made at high temperatures, such a term can be important.

The nature of each series (sine or cosine) and their coefficients lead to certain periodicities and symmetries, which permit the terms in Eq. (A-1) to be separated. The argument of all the trigonometric terms is $(2\pi[\ell h_1 + mh_2 + nh_3])$, where h_i are coordinates in reciprocal space and ℓmn are atomic coordinates. For wüstite, with octahedral and tetrahedral iron, these coordinates have the form 1/4, 1/4, 1/4; 2/4, 2/4, 0, etc. For, say, an hoo line, the argument is 2π n/4 and thus the periodicity is 4 along such a line, and $I_{SRO}(h) = I_{SRO}(h{+}4)$. The periodicity is 2 along [hho] and 4 along [hhh]. Each term also has certain symmetry. As I_{SRO} is a cosine series, $I_{SRO}(h) = I_{SRO}(-h)$. The term Q(h) is a sine series so that, Q(h) = -Q(-h). Adding the periodicity, Q(h) = -Q(4-h). Similarly, R(h)

= R(-h) and b(h) has the same symmetry as R(h). Because of these symmetries and periodicities, it is possible to separate the terms by taking differences in four different half-periods. Measurements from 100 to 900 are required for the separation for the hoo direction for Fe_xO, to recover I_{SRO} for $0 \leqslant h \leqslant 2$. For the rest of the period $2 \leqslant h \leqslant 4$, I_{SRO} can be obtained from the first region, as this term is symmetric across h = 2. This is a large region to measure, and in separating a few data points, it was clear that, for the weak diffuse intensity away from the Bragg and supplemental peaks (see main text), there was little difference if the data were treated by ignoring the tetrahedral sites, i.e., by treating the material as if it had an NaCl structure, involving only octahedral iron cations and vacancics and oxygen ions. The periodicity is half as great and only measurements from 100 to 500, 110 to 550, and 111 to 555 are needed. For such lines, the separation equations for $I_{SRO}(h)$, for $0 \leqslant h \leqslant 1$ (half the period, the other half being obtained by symmetry), are as follows:

$$I_D(2+h) - I_D(2-h) = 4Q(h) + R(h)\left(e^{b(h)(2+h)^2} - e^{b(h)(2-h)^2}\right) , \qquad \text{(A-2)}$$

$$I_D(4-h) - I_D(2+h) = -6Q(h) + R(h)\left(e^{b(h)(4-h)^2} - e^{b(h)(2+h)^2}\right) , \qquad \text{(A-3)}$$

$$I_D(4+h) - I_D(4-h) = 8Q(h) + R(h)\left(e^{b(h)(4+h)^2} - e^{b(h)(4-h)^2}\right) , \qquad \text{(A-4)}$$

$$2I_D(4-h) + I_D(2+h) - 3I_D(2-h) = R(h)\left(2e^{b(h)(4-h)^2} + e^{b(h)(2+h)^2} - 3e^{b(h)(2-h)^2}\right) , \qquad \text{(A-5)}$$

$$3I_D(4+h) + I_D(4-h) - 4I_D(2+h) = R(h)\left(3e^{b(h)(4+h)^2} + e^{b(h)(4-h)^2} - 4e^{b(h)(2+h)^2}\right) . \qquad \text{(A-6)}$$

And finally,

$$\frac{2I_D(4-h)+I_D(2+h)-3I_D(2-h)}{3I_D(4+h)+I_D(4-h)-4I_D(2+h)} = \frac{2e^{b(h)(4-h)^2} + e^{b(h)(2+h)^2} - 3e^{b(h)(2-h)^2}}{3e^{b(h)(4+h)^2} + e^{b(h)(4-h)^2} - 4e^{b(h)(2+h)^2}} . \qquad \text{(A-7)}$$

The last equation is solved numerically for b(h) first, and then the other separations are made.

The term I_{SRO} may then be inverted to obtain its Fourier coefficients A_n along each line. These A_n are linear combinations of $\overline{\alpha}_{\ell m n}$, the short-range order parameters. Ignoring interatomic vectors for tetrahedral sites, and assuming that the short-range order parameters vanish beyond $\alpha_{\ell m n} = \alpha_{\frac{1}{4}(800)}$, there are five A_n along [hoo], six along [hho] and four along [hhh]. These 15 A_n's are functions of nine α's, and a least-squares solution is possible. For example:

$$A_o^{hoo} = \bar{\alpha}_{000} + 4\bar{\alpha}_{\frac{1}{4}(220)} + 4\bar{\alpha}_{\frac{1}{4}(400)} + 4\bar{\alpha}_{\frac{1}{4}(440)} + 8\bar{\alpha}_{\frac{1}{4}(620)} + 4\bar{\alpha}_{\frac{1}{4}(800)}. \quad \text{(A-8)}$$

The higher A_n's, beyond the ones employed in this solution were the order of 10 to 15 percent of A_1, justifying the above assumptions as to the range of the $\bar{\alpha}$'s.

Each of the $\bar{\alpha}$'s has the form[31]:

$$\bar{\alpha}_{\ell mn} = \frac{X_o^{Fe}X_o^V\alpha_{oo}^{Fe\text{-}V}(\ell mn) + 2X_T^{Fe}X_T^V\alpha_{TT}^{Fe\text{-}V}(\ell mn)}{X_o^{Fe}X_o^V + 2X_T^{Fe}X_T^V},$$

$$\text{for} \begin{cases} \ell,m,n\ \frac{2p}{4} \\ \ell+m+n = 4q/4 \end{cases}, \quad \text{(A-9a)}$$

$$= \frac{2X_T^{Fe}X_T^V\alpha_{TT}^{Fe\text{-}V}(\ell mn)}{X_o^{Fe}X_o^V + 2X_T^{Fe}X_T^V}, \text{ for} \begin{cases} \ell,m,n\ 2p/4 \\ \ell+m+n = (4q+2)/4 \end{cases}, \quad \text{(A-9b)}$$

$$= \frac{2\,(X_o^{Fe}X_T^V\alpha_{OT}^{Fe\text{-}V}(\ell mn) + X_T^{Fe}X_o^V\alpha_{TO}^{Fe\text{-}O}(\ell mn)}{X_o^{Fe}X_o^V + 2X_T^{Fe}X_T^V}, \text{ for} \quad \text{(A-9c)}$$

ℓ for ℓ,m,n (2p+1)/4 .

$\alpha_{\ell mn} = 1 - \dfrac{P_{\mu\nu}^{ij}}{X_\nu^j}$, where $P_{\mu\nu}^{ij}$ is the probability of an atom of species j on sublattice ν when an i atom is on the μth sublattice.

The $\bar{\alpha}$'s for Eq. (9b) have not been obtained, as it has been assumed that the tetrahedral ions are not present except for α_{000}. Typically for wüstite, $X_o^{Fe} = 0.9$, $X_o^V = 0.1$, $X_T^{Fe} = 0.015$, $X_T^V = 0.985$. Then, since $0 \leqslant P_{\mu\nu}^{ij} \leqslant 1$:

$$0.09 \leqslant X_o^{Fe}X_o^V\alpha_{oo}^{FeV}(\ell mn) \leqslant -0.010 \;,$$

$$0.015 \leqslant X_T^{Fe}X_T^V\alpha_{TT}^{FeV}(\ell mn) < 0.0 \;,$$

$$0.887 \leqslant X_o^{Fe}X_T^V\alpha_{OT}^{FeV}(\ell mn) \leqslant -0.013 \;,$$

$$0.000 \leqslant X_T^{Fe}X_o^V\alpha_{TO}^{FeV}(\ell mn) < -0.001 \;.$$

Thus, in Eq. (A-9c), the last term may be ignored and it is unlikely that there is any appreciable contribution of the last term in Eq. (A-9a). The terms α_{oo}^{FeV} and α_{OT}^{FeV} may therefore be obtained in this trilinear analysis of the weak diffuse scattering. This procedure was followed for X = 0.900 and X = 0.956. However, this method is not adequate for the relatively sharp supplementary peaks found in this system for X = 0.900. In this case the weak scattering was extrapolated under these peaks for trilinear analysis and the peaks were then analyzed by inversion of volume data:

$$\bar{\alpha}_{\ell mn} = \frac{1}{v} \int_v I_{SRO}(h_1,h_2,h_3)\cos 2\pi(\ell h_1 + mh_2 + nh_3)dh_1\, dh_2\, dh_3\ , \quad \text{(A-10)}$$

where v is the volume of the unit cell in reciprocal space. For these peaks the symmetry was clearly not that of an NaCl structure; the 2-δ,0,0 and 6-δ,0,0 supplemental peaks are strong, while the 4-δ,0,0 is weak. This is illustrated in Fig. A-1(a). These were symmetrized as in Fig. A-1 (b); the symmetrized intensity then has the periodicity of an NaCl structure, with the contribution due to vectors involving tetrahedral positions removed, and the separation equations (Eqs. A-2 through A-7) can be applied. It is the atomic displacements that cause the supplemental peaks to occur on only one side of a Bragg peak. After separation, I_{SRO} appears as in Fig. A-1(c). The ratio of the asymmetric part of the intensity to the symmetric part is 1:3 in this schematic representation [compare (a) and (b) in Fig. A-1]. Thus I_{SRO} can be readjusted as in Fig. A-1(d); this is the pattern of I_{SRO} *with* tetrahedral occupation.

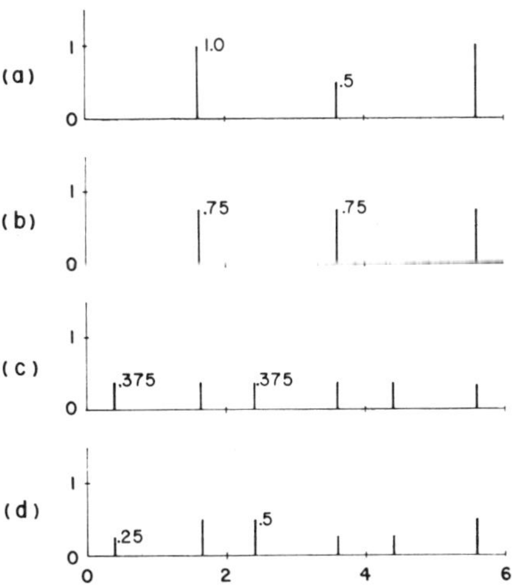

Fig. A-1. Schematic representation of the separation of the supplementary diffraction peaks from Fe_xO: (a) actual intensity; (b) symmetrized intensity, to simulate the intensity for an NaCl structure; (c) separated I_{SRO}; (d) I_{SRO} from (c) altered to represent the true I_{SRO} from Fe_xO.

The first octant of reciprocal space is shown in Fig. A-2(a). As I_{SRO}, being a cosine series, is symmetric across $h_i = 2$, this is all the volume that need be examined, but this region must be generated from the data taken only along lines. Consider the three groups of ℓ,m,n for Fe_xO. Each term in I_{SRO} is of the form $\overline{\alpha}_{\ell mn} \cos 2\pi(\ell h_1 + mh_2 + nh_3)$ which, due to symmetry, can be reduced to $\overline{\alpha}_{\ell mn} \cos 2\pi \ell h_1 \cos 2\pi m h_2 \cos 2\pi n h_3$. For ℓ,m,n = 2p/4, $\cos 4\pi\ell = 1$. Therefore:

$$\begin{aligned} \cos 2\pi h_1 \ell &= \cos 2\pi h_1 \ell \cos 4\pi\ell, \\ &= \tfrac{1}{2} [\cos 2\pi(h_1 + 2)\ell + \cos 2\pi(h_1 - 2)\ell], \\ &= \cos 2\pi(h_1 - 2)\ell = \cos 2\pi(2 - h_1)\ell. \end{aligned}$$

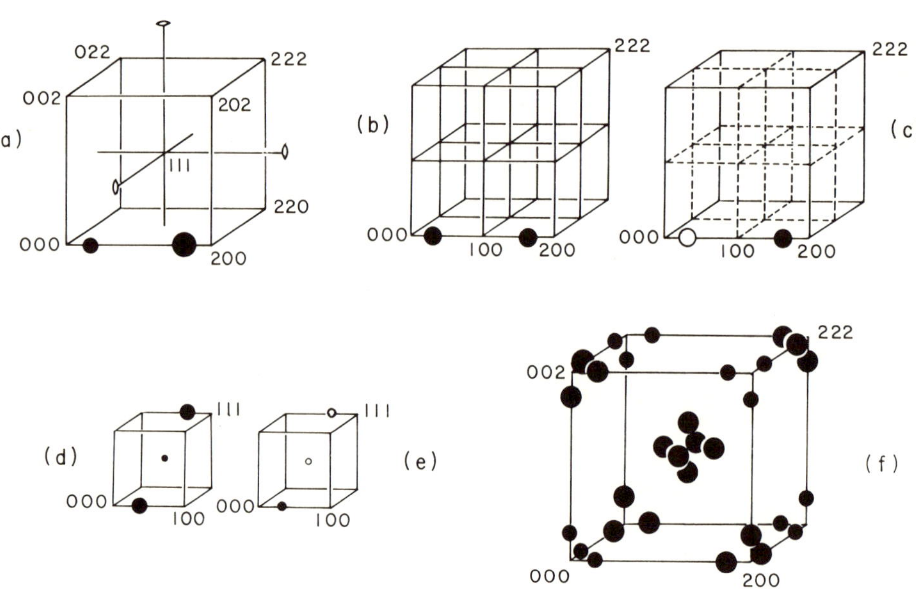

Fig. A-2. (a) Schematic of the supplementary diffraction peaks from Fe_xO in the first octant of reciprocal space. (b) Symmetric component of (a) across the planes $h_i = 1$, which contributes to $\overline{\alpha}_{\ell mn}$ for ℓ,m,n all 2p/4. (c) Antisymmetric component of (a) across the planes $h_i \geq 1$, which contributes to $\overline{\alpha}_{\ell mn}$ for ℓ,m,n all (2p+1)/4. (d) Symmetric component of (b) across 0.5,0.5,0.5, contributing to $\overline{\alpha}_{\ell mn}$ for ℓ,m,n all 2p/4 and ℓ+m+n = 4q/4. (e) Antisymmetric component of (b) across 0.5,0.5,0.5, contributing to $\overline{\alpha}_{\ell mn}$ for ℓ,m,n all 2p/4 and ℓ+m+n = (4q+2)/4. (f) Representation of the supplementary peaks in the first octant of the unit cell in reciprocal space, after applying all symmetries.

Thus $I_{SRO}(h_1,h_2,h_3) = I_{SRO}(2-h_1,h_2,h_3)$; I_{SRO} is symmetric across $h_i = 1$. Similarly, for ℓ,m,n = (2p+1)/4, $I_{SRO}(h_1,h_2,h_3) = -I_{SRO}(2-h_1,h_2,h_3)$, so that I_{SRO} for these terms is antisymmetric across $h_i = 1$. The intensity of these two portions of Fig. A-2(a) can be obtained as in

Fig. A-2(b) and Fig. A-2(c). For $\ell,m,n = 2p/4$ and $\ell+m+n = 4q/4$, I_{SRO} is symmetric across $h_1 = h_2 = h_3 = 0.5$, whereas for $\ell,m,n = 2p/4$ and $\ell+m+n = (4q+2)/4$, it is antisymmetric. These relationships are illustrated in Fig. A-2(d) and Fig. A-2(e). In order to carry out these last symmetry operations, the $1-\delta,1,1$ supplemental peak is needed. As it was not measured, the asymmetric term across 0.5,0.5,0.5 was neglected. It contributes only to Eq. (A-9b) involving tetrahedral ions, and as this term was not determined with the weak diffuse scattering there was no point in obtaining it here. From data on quenched specimens[4], the asymmetric term was only 5 percent of the total, so that the error is not large.

Application of all these symmetries results in the distribution of I_{SRO} from the supplemental peaks shown schematically in Fig. A-2(f). This pattern is quite similar to that shown in Fig. 3b. Although the intensity was actually measured only along lines, from the photographic study in Fig. 3, the supplemental peaks are essentially spherically symmetric. With this additional assumption, I_{SRO} was available for inversion in a volume in reciprocal space. This portion of the intensity contributes to all α,s, whereas the weak diffuse intensity analyzed with NaCl symmetry contributes only to the $\bar{\alpha}_{\ell m n}$ in Eq. (A-9a). Thus, the $\alpha^{FeV}_{oo}(\ell mn)$ were determined for both compositions, and for X = 0.900, the $\alpha^{FeV}_{OT}(\ell mn)$ were also determined.

Experimental Procedures

The preparation and purity of the single crystals employed in this study are described in ref. 8. Pieces 4 to 7 mm on a side and 1 mm thick were used, within 4 degrees of the (100) plane for measurements along the [hoo] direction, and within 1 degree of the (331) plane for measurements along [hho] and [hhh] directions. The high-temperature beryllium hemisphere attachment allowing for chi and phi tilts, is described in ref. 45. A platinum coil heater, nickel specimen stage, and heat shield were the only modifications. The temperature (840°C) was controlled with a Labac-Thermac controller to ±2°C. The actual surface temperature was determined with a dummy Fe_xO crystal, with a thermocouple embedded in the surface. All measurements were made in mixed CO/CO_2 gases. The gas train, gas purity, and source for data relating the composition to the mixture are also given in ref. 8. In addition to the three lines measured for both X = 0.900 and X = 0.956, other compositions were examined along the [hoo] line.

A modified G.E. XRD-5 diffractometer[46] was employed, with a perfect silicon crystal to monochromate the incident beam; with this crystal, the beam was narrow enough for the small specimens. The radiation was MoK_α, with the X-ray tube operated at 50 kv and 20 ma. The scintillation detector was set for 95 percent acceptance of the K_α radiation. Most of the data were taken with a zirconium filter in front of the detector to reduce iron fluorescence to negligible amounts, but even without this foil, this fluorescence was at most 7 percent of the weak scattering. The half-breadths of fundamental peaks were ≅ 0.5 degree, while that of the supplementary peaks was 2 degrees or more. Thus slit broadening was not important.

To carry out the separations of the coherent scattering for the two compositions for which all three lines in reciprocal space were explored, the data was first converted to absolute units to allow the subtraction of calculated Compton scattering. The power of the direct beam ($\approx 5 \times 10^6$ cps) was measured by integrating several peaks from an aluminum powder compact[47] placed in the beryllium hemisphere in the appropriate atmosphere. The scattering of the gas mixture was obtained from measurements at room temperature, in the gas and in vacuum, taking into account the absorption in the former case of the primary and incident beams. After subtracting this and Compton scattering, the intensity was divided *not* by the Laue monotonic, but by the average scattering per atom $(f_o + Xf_{Fe})^2/4$. Tests with simulated data indicated that a better separation is obtained in this way when thermal diffuse scattering is large.[48] After the separation, the data were divided by the ratio of the average value (along each line) of the Laue monotonic to the average scattering factor, to place the data in Laue units per atom prior to analysis.

Scattering factors and dispersion corrections for Fe^{+2} and $O^=$ were obtained from ref. 49; interpolated values were employed for $O^=$. The scattering factor for $O^=$ was taken from ref. 50 and no dispersion corrections were made. The ratio of octahedral to tetrahedral occupation of 3.25/1 was assumed for all compositions, as was found at similar temperatures in ref. 30. All calculations, separations, and inversions were carried out on a CDC 6400 computer.

The lowest data point was 500 counts (in 2 minutes), the highest 5800 counts. Data taken at 0.1-degree2θ intervals and smoothed by hand were extrapolated under Bragg peaks in all three directions to the same value, which was indicated by a "shoulder" on I_{SRO} in the [hhh] direction. Details of the extrapolation under the supplemental peaks for X = 0.900 are not important, as what was not in these peaks was included in the weak diffuse scattering and both were analyzed. Some typical data are presented in Fig. A-3.

The success of these procedures is indicated by the fact that $\bar{\alpha}_o$ = 1.28 for X = 0.900, and 0.77 for X = 0.956. These values are the total I_{SRO} in a unit cell in reciprocal space and theoretically should be unity; these differ from unity by an amount similar to values obtained with alloys at room temperature. Furthermore, the term $b(h)h^2 = 2B(h) \sin^2\theta/\lambda^2$, where $B(Å^2)$ is essentially a temperature depression ($h^2 = h_1^2 + h_2^2 + h_3^2$). The values of B were generally of the same order, 1.6 to 4 $Å^2$, as those measured from Bragg peaks in the same temperature region (2.5 $Å^2$).

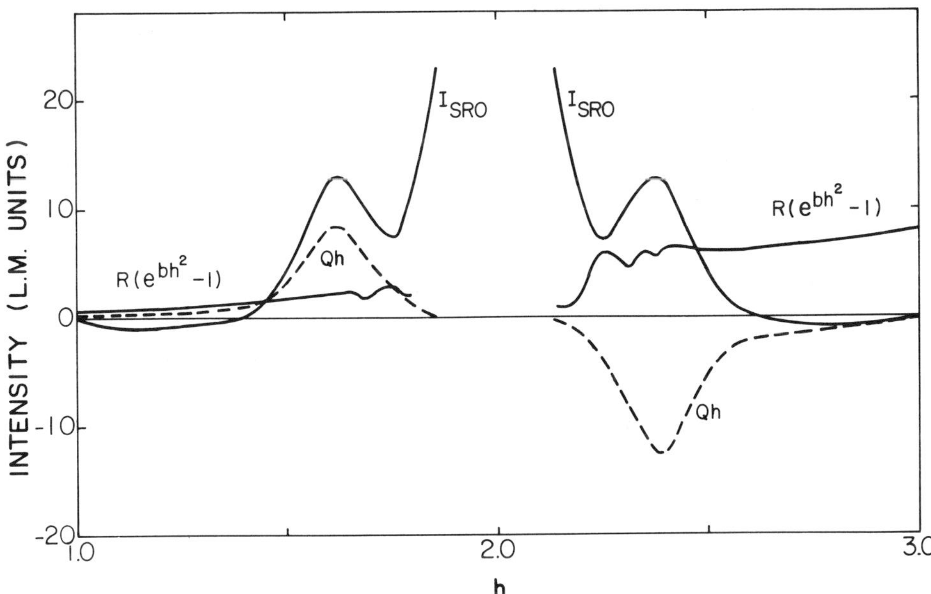

Fig. A-3. Diffuse intensity components in Laue monotonic units along the [hoo] direction, $Fe_{0.900}O$, 840°C.

REFERENCES

1. Jette, E. R., and Foote, F., *J. Chem. Phys.*, **1**, 29 (1933).
2. Roth, W. L., *Acta Cryst.*, **13**, 140 (1960).
3. Smuts, J., *J. Iron and Steel Inst.*, **204**, 237 (1966).
4. Koch, F. B., and Cohen, J. B., *Acta Cryst.*, **B25**, 275 (1969).
5. Stiglich, J. J., Jr., Whitmore, D. H., and Cohen, J. B., *J. Am. Ceram. Soc.*, **56**, 211 (1973).
6. Manenc, J., *J. Phys. Rad.*, **24**, 447 (1963); Manenc, J., Bourgeot, J., and Bénard, J., *Compt. Rend. Acad. Sci. Paris*, **256**, 931 (1963).
7. Herai, T., Thomas, B., Manenc, J., and Bénard, J., *Compt. Rend. Acad. Sci. Paris*, **258**, 4528 (1964).
8. Hayakawa, M., Cohen, J. B., and Reed, T. B , *J. Am. Ceram. Soc.*, **55**, 160 (1972).
9. Manenc, J., Mem. Scientifique Rev. Metallurg. LXIV. No. 718 (1967); Bénard, J., Herai, T., and Manenc, J., *Ann. Chim.*, **5**, 240 (1970).
10. Koch, F. B., and Fine, M. E., *J. Appl. Phys.*, **38**, 1470 (1967).
11. Anand, H. R., and Mullen, J. G., to be published.
12. Hembree, P. L., and Wagner, J. B., *Trans. Met. Soc. AIME*, **245**, 1547 (1969).
13. Geiger, G. H., Levin, R. L., and Wagner, J. B., *J. Phys. Chem. Solids*, **27**, 947 (1966).

14. Banus, M. D., and Reed, T. B., in *The Chemistry of Extended Defects in Non-Metallic Solids,* L. Eyring and M. O'Keeffe (Eds.), North-Holland Publishing Co., Amsterdam (1970), p. 488.
15. Andersson, B., and Gjϕnnes, J., *Acta Chem. Scand.*, **24**, 2250 (1970).
16. Bell, P. S., and Lewis, M. H., *Phys. Stat. Sol.*, 7, 431 (1971).
17. Andersson, B., and Tafto, J., *Proc. 5th European Conf. on Electron Microscopy,* p. 670 (1972).
18. Castles, J. R., Cowley, J. M., and Spargo, A.E.C., *Acta Cryst.*, **A27**, 376 (1971).
19. Watanabe, D., *Intl. Conference on Electron Diffraction and Crystal Defects,* Australian Academy of Science, I-C-3.
20. Watanabe, D., Castles, J. R., Jostons, A., and Malin, U.S., *Acta Cryst.*, **23**, 307 (1967).
21. Watanabe, D., Terasaki, O., Jostons, A., and Castles, J. R., *J. Phys. Soc. Japan,* **25**, 292 (1968).
22. Hilti, E., *Naturwissenschaften,* **55**, 130 (1968).
23. Watanabe, D., Terasaki, O., Jostons, A., and Castles, J. R., *The Chemistry of Extended Defects in Non-Metallic Solids,* L. Eyring and M. O'Keeffe (Eds.), North-Holland Publishing Co., Amsterdam (1970), p. 238.
24. Nakahira, M., and Saekl, M., *7th Intl. Symposium on the Reactivity of Solids,* School of Chemistry, University of Bristol, Bristol, England, Paper 2.1.
25. Morin, F. J., *Phys. Rev. Letters,* **3**, 34 (1959).
26. Austin, I. G., *Phil. Mag.*, **7**, 961 (1962).
27. Kawano, S., Kosuge, K., and Kachi, S., *J. Phys. Soc. Japan,* **21**, 2744 (1966).
28. Steinitz, R., General Telephone and Electronics Laboratory, Inc., Bayside, L.I., New York, unpublished.
29. Ariya, S. M., Bruck, B.J.A., and Vladimirova, V. A., *Vestnik Leningrad University, Ser. Fiz. Khim.*, **22**, 157 (1967).
30. Cheetham, A. K., Fender, B.E.F., and Taylor, R. I., *J. Phys. Chem. Solid State Phys.*, **4**, 2160 (1971).
31. Hayakawa, M., Ph.D. Thesis, *X-Ray Diffraction Studies of Wüstite at High Temperature,* Northwestern University, Evanston, Illinois, June 1973: to be published.
32. Fujiwara, K., *J. Phys. Soc. Japan,* **12**, 7 (1957).
33. Gehlen, P. C., and Cohen, J. B., *Phys. Rev.*, **139**,A844 (1965).
34. Gragg, J. E., Bardhan, P., and Cohen, J. B., *Critical Phenomena in Alloys, Magnets and Superconductors,* R. E. Mills, E. Ascher, and R. I. Jaffee (Eds.) McGraw-Hill Book Company, New York (1971), p. 309.
35. Clapp, P. C., and Moss, S. C., *Phys. Rev.*, **142**, 418 (1966); **171**, 754 (1968); Moss, S. C., and Clapp, P. C., *ibid.*, **171**, 764 (1968).
36. Krivoglaz, M. A., and Hao, T'u, Collection, *Defects and Properties of the Crystal Lattice,* Kive: lzd. "Maukova, Dumka" (1969).
37. Krivoglaz, M. A., *Theory of X-Ray and Thermal Neutron Scattering by Real Crystals,* Plenum Press, Inc., New York (1969).
38. Moss, S. C., *Phys. Rev. Letters,* **22**, 1108 (1969).
39. Tewar, I. S., *Solid State Comm.*, **11**, 1139 (1972).
40. Takeuchi, S., and Suzak, K., *Nippon Kinzoku-Gakkaishi,* **33**, 409 (1969).
41. Norwood, T. E., and Fry, J. L., *Phys. Rev.*, **B2** (2): 472 (1973).
42. Gubernatis, J. E., and Taylor, P. L., *Solid State Comm.*, **12**, 309 (1973).
43. Borie, B., and Sparks, C. J., *Acta Cryst.*, **A27**, 198 (1971).
44. Gragg, J. E., Jr., and Cohen, J. B., *Acta Met.*, **19**, 507 (1971).
45. Gehlen, P. C., *Rev. Sci. Instr.*, **40**, 715 (1969).
46. Schwartz, L. H., Morrison, L. A., and Cohen, J. B., *Applications of X-Ray Analysis,* 7, 281 (1961).
47. Batterman, B. W., Chipman, D. R., and DeMarco, J. J., *Phys. Rev.*, **122**, 68 (1961).
48. Gragg, J. E., Jr., Hayakawa, M., and Cohen, J. B., *J. Appl. Cryst.*, **6**, 59 (1973).
49. *International Tables for X-Ray Crystallography,* The Kynoch Press, Birmingham, England (1965), Vol. III.
50. Tokonami, M., *Acta Cryst.*, **19**, 486 (1965).

51. Carel [*Compt. Rend. Acad. Sci. Parts,* **277**, 69 (1973)] has reported that the data on lattice parameter of Fe_xO in ref. 8 can be interpreted as indicating subphases. He could draw regions of lattice parameter versus x with different parabolic fits through different portions of the data. However he uses only 4 to 5 data points for each region, and it is not clear how the compositions (at which the curvature changes) are chosen. Furthermore with previous data for the lattice parameter of quenched specimens G Carel C., Wengel D., and Vallet, P., *Compt. Rend. Acad. Sci.,* **260,** 4325 (1965), straight lines were employed and the region with the highest slope of a_o vs x was the one with the lowest slope at high temperatures. Changes in thermal expansion of ≈ 20 pct are required as x changes from 0.902 to 0.956. But measurements by the present authors indicate changes of at most 5 percent. There remains no substantive evidence for subphases.

DISCUSSION on Paper by J. B. Cohen

ANDERSON: Concerning the ferrous oxide, is it right to infer that the combined result of your high-temperature X-ray work and Cheetham's high-temperature neutron diffraction agree, and confirm that the rates of concentrations of octahedral vacancies to tetrahedral atoms remain close to the value (3.25) that corresponds to the low-temperature superstructure?

If so, then does the comparison with the computer simulation mean that the clusters are dissociated into smaller units, while maintaining the statistics of site ratios?

COHEN: Our measurements at high temperature indicate tetrahedral occupation because of the differences in intensity of the diffuse supplemental peaks near the 200, 400, and 600 peaks. However, we did not obtain *quantitative* information on the tetrahedral occupation. We must rely on the work by Cheetham and Fender which shows that the ratio is the same at high temperatures as that for quenched specimens (see ref. 30).

ILSCHNER: Is it correct to say that the ordered structures observed in the quenched samples are not "quenched-in" structures but structures formed during the quenching process? If so, should the perfection of the ordered state observed at room temperature not be dependent on cooling rate? (Mind that you have different cooling rates within different zones of any crystal.)

COHEN: The results seem to indicate that the size of the cluster is a bit smaller at temperature than that for a quenched specimen. The regularity of the spacing of these clusters is less. The amount of clustering and, especially, the periodic arrangement or spacing of the clusters is certainly somewhat dependent on cooling rate.

ILSCHNER: Did you obtain information concerning the temperature dependence of the fraction of all cations located on tetrahedral sites?

COHEN: We did not obtain information on tetrahedral occupation at temperature, but this has been determined – see ref. 30.

GOODENOUGH: From a knowledge of site-preference energies, the placement of an Fe^{3+} ion in tetrahedral coordination is quite reasonable. I would be surprised if you find tetrahedral Ni^{2+} or V^{2+} ions. If tetrahedral Ni^{2+} does exist, spin-orbit coupling below the Néel temperature should produce a cooperative Jahn-Teller deformation and large magnetic anisotropy. As for the diffuse scattering, the existence of a Fermi surface is unlikely in wüstite, and in VO_x any band calculation must be consistent with the change from an n-type semimetal for $x < 1$ to a p-type semiconductor for $x > 1$. Are the Fermi-surface calculations to which you refer consistent with this observation?

COHEN: As our work was on NiO-CoO solid solutions, it is possible that all the tetrahedral ions we see are cobalt ions. Further work on pure NiO is needed.

The calculated Fermi surfaces in the literature (see our ref. 41 for example) do not take into account the vacancies in the vanadium and oxygen sublattices. Thus these calculations probably do not provide the kind of information that would explain the p-n transition. This is one of the reasons comparisons of the calculated surface and the actual one will be useful, and it certainly would be useful to have calculations that take into account the vacancies (and their arrangement).

HEUER: How do you know that diffuse intensity is not just due to the geometry of the scattering centers, i.e., how can you be sure the diffuseness does relate to some detail of the Fermi surface?

COHEN: The two are related; if there is an electronic contribution to the ordering or clustering energy, then the arrangements (and hence the intensity) are an expression of this energy.

NOWICK: Could you discuss the structure of the FeO defect and the possibility of dissociation at high temperatures?

COHEN: For quenched specimens with $x = 0.9$ there is, on the average, a cluster consisting of a vacant octahedral cation site and vacancies in its twelve nearest-neighbor octahedral cation sites. This vacancy cluster "occludes" 4 tetrahedral cations. The clusters are arranged periodically on a supercell – 3 x 3 x 3 NaCl cells (approximately). At high temperatures, these clusters are similar in size – a bit smaller, less periodic in arrangement – and fewer in number.

HONIG: Could you supply some further information regarding the construction of the Fermi surface in VO? Whose calculations did you use and how many electrons were used in filling the Brillouin zone?

COHEN: The Fermi surface was taken from the literature. The latest calculations are in ref. 41, and these were made for a variety of valencies for the ions. Our results lie between the calculations for monovalent and divalent ions. But, the calculations do *not* include vacancies.

HAGEMARK: With respect to the ordering that takes place during the cooling of FeO, how far do the atoms have to move?

COHEN: My first guess (from the cluster spacing) would be that this distance is about 12 to 20 Å.

SHEWMON: Would you tell me more about how one infers the caliper of a Fermi surface from diffuse scattering?

COHEN: The diffuse intensity, I_D, is related to the Fourier transform of the ordering energy, V(K), by the equation $I_D \cong 1/1+V(K)$. Electronic energy contributes to this ordering energy, and it has been shown that the electronic portion of V(K) has large singularities for flat or cylindrical portions of a Fermi surface; hence there are regions where I_D is large.

RAPP: To test the relative stabilities (energies) of cobalt or nickel cations in a tetrahedral site of the NaCl lattice compared with octahedral occupancy in the NaCl lattice, could you not gain some insight from the knowledge of the solubilities of NiO and CoO in a zinc blende type of oxide and the values of the partial molar enthalpies of solution of these oxides in the zinc blende oxide? Especially for cobalt, the effect of P_{O_2} on this solubility should demonstrate the effect of changing valence from +2 to +3 with increasing P_{O_2}.

COHEN: This sounds like an interesting experiment, but I do not know of any results.

WELCH: Have you considered the possibility that your observed correlation between calculated Fermi surface geometry and the enhanced diffuse scattering might be due to large static displacements around the defects, of the appropriate wavelength, resulting from "soft" phonons which are a consequence of a strong Kohn anomaly?

COHEN: This could be so, and would be worth further study, but regardless of the cause – soft modes or ordering energy – one is imaging the Fermi surface. That is, the Kohn anomaly is also an image of the surface.

FRANKLIN: Do the clusters give charge balance?

COHEN: As the studies were made with X-rays, there is no way to determine charges. However, the detailed study of the structure of the clusters for quenched Fe_xO indicates that the cations near a cluster are displaced toward the cluster and the nearby anions are displaced away. This *may* be associated with charges on the complex and such charges may be part of the cause of the ordering of the complexes.

O'KEEFFE: With reference to the NiO-CoO data, is it at all possible that the diffraction data could have an alternative explanation in terms of, for example, anharmonic motions of the ions?

COHEN: The details of our work on NiO-CoO can be found in ref. 5; the tetrahedral occupation varied with oxygen pressure and thus an interpretation in terms of anharmonicity is unlikely. Furthermore the highest angle peak we examined was the 420. I don't think this is a high enough $\frac{\sin\theta}{\lambda}$ to see effects beyond our experimental error. Finally the octahedral site is centrosymmetric. It is my recollection that anharmonicity would not lead to any special effects in such a case, but this should be checked.

TANNHAUSER: Your slide showed 3 percent tetrahedral occupancy with 25 percent CoO in NiO and less for 75 percent CoO. Can one extrapolate this to pure NiO? Also, were these measurements done at temperature?

COHEN: The measurements werre made at temperature but the results are not accurate enough to indicate any change in tetrahedral occupation with CoO concentration (see our ref. 5). The differences shown in ref. 5 are within the experimental error. Thus, it is not possible to extrapolate the data to pure NiO. It could well be that all the interstitials at any composition we examined are cobalt ions. The X-ray scattering factors of nickel and cobalt are too similar for us to distinguish this – or the charge on the ion. Thus it might be that there are no interstitials in pure NiO but there certainly should be many in CoO.

TALLAN: If one has several percent of tetrahedral occupancy in the NiO-CoO solid solutions, one would have also several percent vacant sites and these should overshadow the vacancies introduced by departure from stoichiometry. Did you look for an oxygen pressure dependence of the diffusion coefficient for nickel in these solid solutions and did you find one?

COHEN: We did find such a dependence for the occupation. Measurements were made on a given specimen alternately in dried air and in an Ar-O_2 mixture with $P_{O_2} = 5 \times 10^{-3}$ atm. The tetrahedral occupation increased when P_{O_2} increased. This was done with several samples. We did look for a pressure dependence of the diffusivities. As the oxygen pressure decreased, so did the diffusivity [see *J. Am. Ceram. Soc.*, **56**, 119 (1973)].

LIBOWITZ: What is the average cluster size in wüstite at elevated temperature?

COHEN: I think this can be obtained from Fig. 8. The size is of the same magnitude as for a quenched specimen; the clusters are a bit smaller, and much fewer in number.

HIGH-RESOLUTION ELECTRON MICROSCOPY OF DEFECTS AND DISORDER IN CRYSTALS

J. M. Cowley and S. Iijima

Department of Physics
Arizona State University
Tempe, Arizona

ABSTRACT

Recent developments in the techniques of high-resolution electron microscopy have allowed the direct visualization of heavy-atom positions in crystal lattices. The detailed structures of ordered phases and the atom configurations in planar, linear, and point defects have been indicated with a resolution of about 3 Å for a number of oxides of niobium and mixed oxides containing titanium, zirconium, and tungsten. Much the same information is available for a number of minerals. Indications of the mechanism of phase transformation can be derived in some cases from studies on quenched samples or by observation of changes induced by electron irradiation. Some information can be obtained regarding the state of short-range order in relatively simple systems by the use of dark-field imaging techniques in combination with electron-diffraction observations.

1 INTRODUCTION

The recent technical advances in the production of electron microscopes have made available a relatively small improvement in resolution but a large increase in versatility and convenience for high-resolution work. For the study of crystals at the level of atomic resolution (about 3 Å) the requirements include a highly stable goniometer stage and sufficient intensity in images magnified by 500,000X or more on the fluorescent screen to allow the operations of stigmation and focusing to be performed by direct observation.

The necessary level of performance was obtained by Iijima[1] in 1971 using a JEM-100B electron microscope modified for the use of a pointed filament as well as in other ways. Several commercially available microscopes now appear to offer the same capabilities. We may therefore expect that electron micrographs of the type used to illustrate this paper will soon become commonplace, providing a new and valuable means for the study of many aspects of the form and behavior of crystals and their defects.

Until now the application of the electron microscope to the study of crystals has been almost entirely confined to the observation of one-dimensional lattice fringes or to the recording of two-dimensional periodicities with poor resolution[2], but the application of the better resolution under carefully controlled conditions will extend the range of application of the instrument greatly.

An adequate body of theory exists to provide a basis for the interpretation of the contrast observed in these electron micrographs and computer programs have been written for the calculation of both the interaction of electrons with the potential distribution of a crystal and the effects of lens aberrations and other factors on the image intensities.[3-6] The useful approximation that, for small thicknesses, the crystal can be regarded as a planar phase object, has been evaluated and used to define the optimal experimental conditions for obtaining the images.[6-9]

It has been shown, both theoretically and experimentally,[7] that for 100 kev electrons and 3 Å resolution a direct intuitive interpretation of the bright-field images in terms of atom positions is possible if the microscope is underfocused by a distance of about 900 Å (for spherical aberration constant C_s = 1.8 mm), if the crystal thickness is less than about 150 Å, and if the crystal is aligned so that the incident beam is parallel to a principal axis of the crystal with an accuracy of about 2×10^{-3} radian. Under these conditions, the image-intensity variation is proportional to the projected potential distribution of the object only for extremely thin crystals (less than 10 to 20 Å thick). For greater thicknesses than this the relationship is nonlinear but almost monotonic.[6-8] A row of atoms parallel to the incident beam will give a dark spot in the image, but the optical density of the spot will increase very slowly with an increasing number of atoms in the row or with the atomic number of the atoms present. A darker spot will be given if there are two or more rows of atoms closer together than the resolution limit of the microscope.

There are some important limitations to the use of the lattice imaging method. For biological materials, for most organic materials, and for many inorganic crystals, the specimens are either unstable in the vacuum of the microscope or else are destroyed rapidly by the very heavy irradiation dose received ($\sim 10^{12}$ rads) from the incident electron beam under high-resolution imaging conditions. Many minerals and ionic compounds are visibly damaged and tend to sublime away within minutes or seconds but, fortunately, alloys, semiconductors, and many other inorganic materials are unharmed by large amounts of electron irradiation. The requirement that the crystal thickness should be less than about 150 Å imposes experimental difficulties for many specimens. For the oxides and minerals mentioned below, a finely ground specimen was dispersed in solvent and deposited on a holey carbon microgrid. Thin edges of microcrystals protruding over the edges of holes were sought. Perhaps one microcrystal in a hundred proved suitable for imaging. For other materials the fracturing process may well prove much less productive. Techniques for sample thinning by electrochemical or ion bombardment methods have not been adequately tested for such small thicknesses. One encouraging thought is that the useful thickness should increase (approximately with λ^{-1}) as the accelerating voltage is increased. Hence, lattice imaging should provide an important area of application for ultrahigh-voltage electron microscopes (1 MeV or more) when these are developed to the stage of having resolution comparable with that of the current 100-keV instruments.

It is a fundamental limitation of the method that the images obtained are, at best, two-dimensional projections of the crystal structure. In general, the relationship of image intensity to projected potential is nonlinear so that stereoscopic techniques for revealing the third dimension will not apply. In particular cases it may be possible to obtain projections of the structure parallel to more than one principal axis. The oxide systems examined so far have mostly been particularly favorable in that projections can usually be obtained parallel to the short b-axis (3.8 Å) and these projections contain most of the structural features of interest.

At present the observations that can be made are limited by instrumental difficulties to those at room temperature and in the poorly controlled vacuum of the electron microscope. Attempts to control the temperature and the nature and pressure of the gases in the environment of the specimen inevitably lead to a loss of resolution. These difficulties are not fundamental, however, and may be overcome in the near future if the incentive is sufficiently great. For the moment, as we shall see, the normal, or an increased, electron irradiation by the incident beam may serve to provide some evidence regarding possible reaction mechanisms.

2 OBSERVATIONS ON STRUCTURAL DEFECTS

The oxides of niobium and mixed oxides with titanium, zirconium, or tungsten, which we will consider, have structures based on the ReO_3 structure, which may be pictured as being composed of superposed layers of square arrays of corner sharing octahedra of oxygen atoms, each containing a metal atom, with a spacing of about 3.84 Å between centers of octahedra. The electron-microscope image of such a structure would look like the background of Fig. 1, namely a square array of dark spots.

Variations of stoichiometry from the 3:1 ratio are accommodated by the introduction of "crystallographic shear (CS) planes", along which a shearing of the structure translates the octahedra so that they share edges rather than corners.[10] Isolated CS planes, known as Wadsley defects, may occur as illustrated in Fig. 1. A model for the structure is shown in the inset.

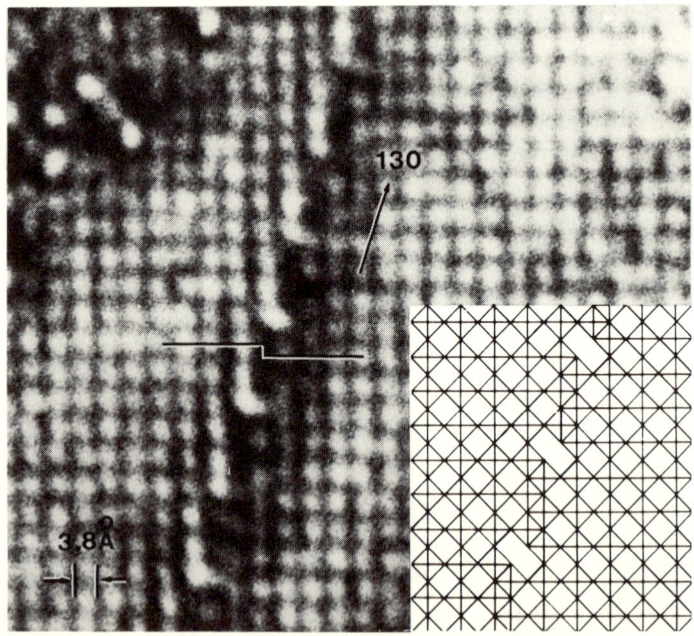

Fig. 1. High-resolution electron-microscope image of a crystal of Nb-W oxide showing a Wadsley-type defect along the (130) direction in a simple ReO_3-type (or WO_3-type) square lattice of corner-sharing octahedra. The inset is a diagram suggesting the nature of the displacements of octahedra across the defect plane.

Periodic arrays of crystallographic shear planes in one or two dimensions frequently occur to produce superlattices having large unit cells in one or two directions and metal-oxygen ratios determined by the relative numbers of corner-sharing and edge-sharing contacts between octahedra. A diagram and an electron-microscope image of the structure $TiNb_{24}O_{62}$ are shown in Fig. 2.[13] The CS planes occur at intervals of 3-1/2 and 2-1/2 octahedra (14.3 and 10.3 Å) in the a- and c-axis directions, giving a structure composed of 3 x 4 blocks of square arrays of octahedra on two levels (drawn light and dark). The characteristic 3 x 2 arrays of white spots in the image show positions having no metal atoms. The high concentrations of projected metal atoms along the shear planes give dark lines in the image.

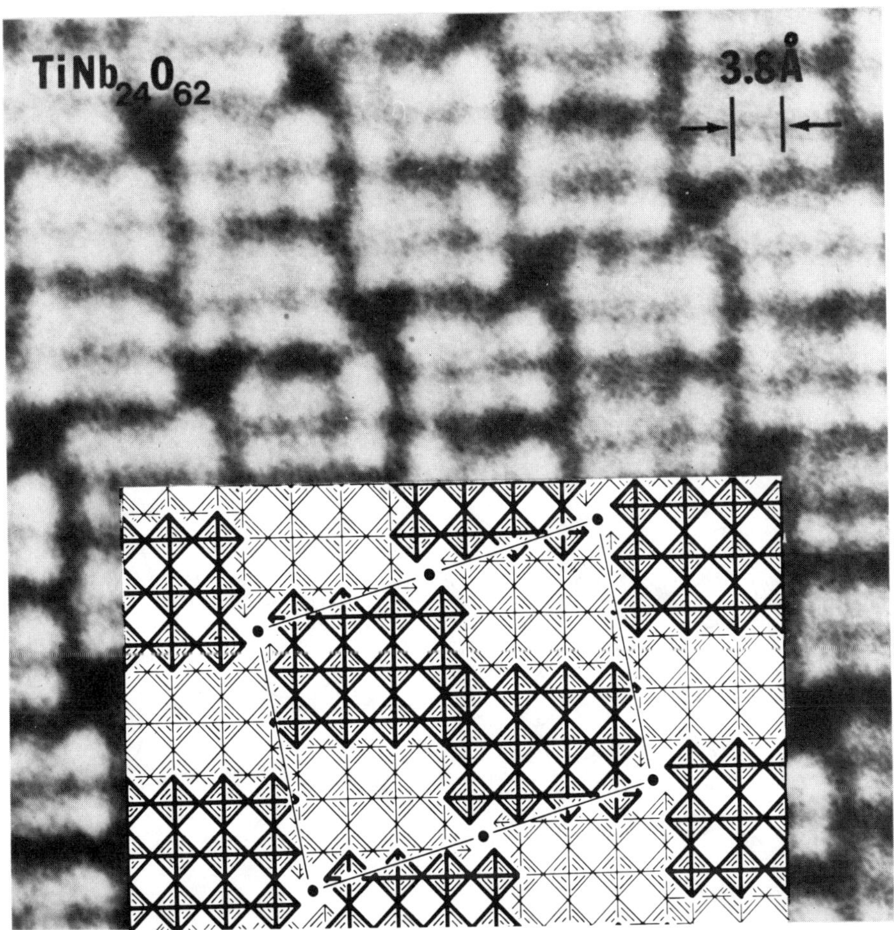

Fig. 2. Electron-microscope image and diagram of the structure of a crystal of $TiNb_{24}O_{62}$.[13] In the diagram, each shaded square represents an octahedron of oxygen atoms containing a metal atom; the black dots represent tetrahedrally coordinated metal atoms.

For many structures, tetrahedral metal-atom sites may occur, particularly at the points of intersection of shear planes. These appear in the images as blacker spots as in Fig. 2 because there is a higher concentration of metal-atom rows about these points.

A very large number of regular superlattice structures, corresponding to well-defined compositions, is now known to be produced by combinations of CS planes and tetrahedral rows. An increasing number have been observed directly by electron-microscope imaging[12-14] including some[14-15] for which the X-ray diffraction results had not been correctly or adequately interpreted. A further class of superlattice structures may be considered to be formed from the basic ReO_3 structure (Fig. 3a) by a different mechanism – a rotation of a column of the structure having four octahedra in its cross section, rather than a shearing along a plane.[16,17] The basic unit of structure, 12 Å in diameter, is illustrated in Fig. 3b. The isolated defects in the square-lattice structure correspond to positions

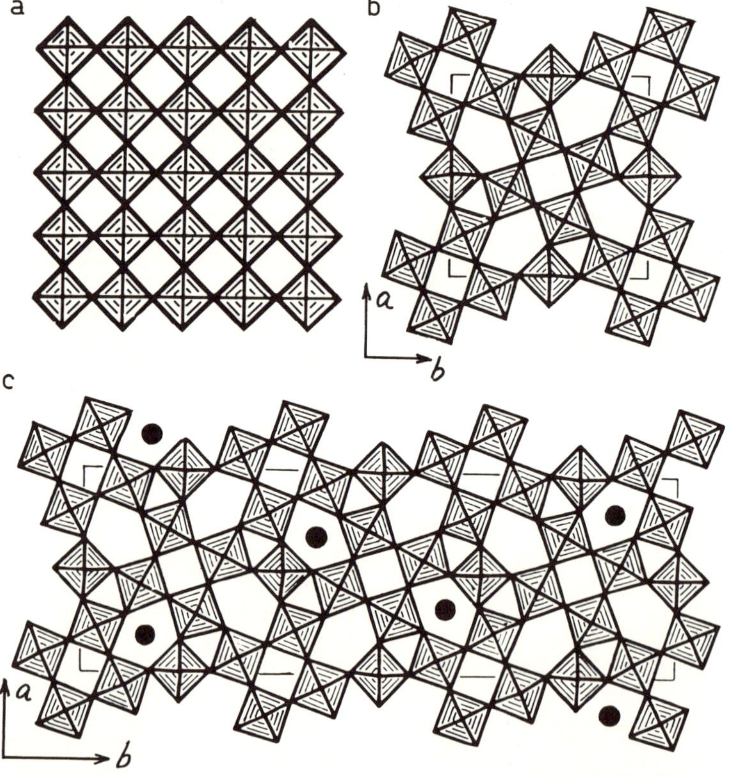

Fig. 3. Diagrams illustrating (a) the ReO_3-type lattice, (b) the transformation of this lattice by rotation of columns of square (2 x 2) octahedra through 45 degrees to form the host tetragonal tungsten-bronze (TTB) structure, and (c) the structure of $4Nb_2O_5 \cdot 9WO_3$ formed by an ordered filling of the pentagonal tunnels of (b) by metal-oxygen atom strings (indicated by black dots).[13]

where four rows of octahedra have been rotated by 45 degrees to give a structure containing a characteristic group of four pentagonal tunnels around one square tunnel through the lattice. These tunnels show up as white dots in the image as at E in Fig. 4.

Periodic arrays of these rotated columns in various modified forms[16] provide superlattices having a variety of compositions. A further feature providing a wide range of variation in composition is the possibility that a varying proportion of the pentagonal tunnels may be filled with metal-oxygen rows of atoms. For a filled tunnel, the image shows a dark spot instead of a light one. A square array of the rotated-block units (12 x 12 Å) with an ordered filling of a fixed proportion of the pentagonal tunnels gives the tetragonal tungsten-bronze (TTB) structure of $4Nb_2O_5 \cdot 9WO_3$ shown in Figs. 3c and 5. Disordered structures, containing a disordered filling of the tunnels in the TTB structure and also the near-random insertion of 4 x 4 square ReO-type blocks of octahedra, have been observed in quenched samples.[15] Also, a new type of ordered phase containing regular arrays of the 4 x 4 square blocks intergrown with a TTB structure has been observed and analyzed.[15]

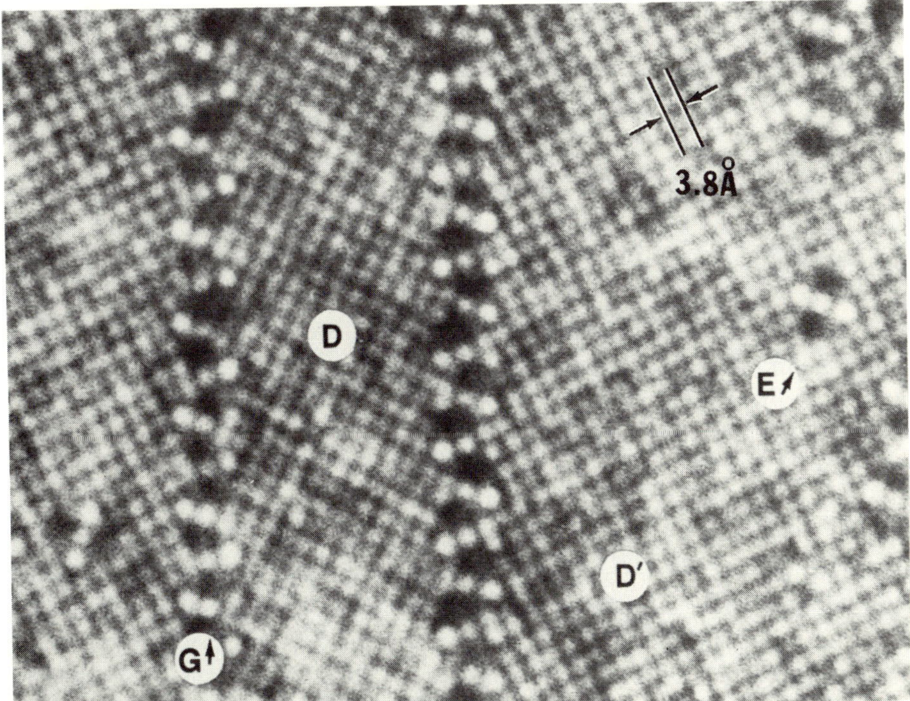

Fig. 4. Electron micrograph showing the growth of the TTB structure from the ReO_3-type structure (D) at isolated points (E) and along twin planes (G) in a crystal of Nb-W oxide of nominal composition $17Nb_2O_5 \cdot 48WO_3$[15]

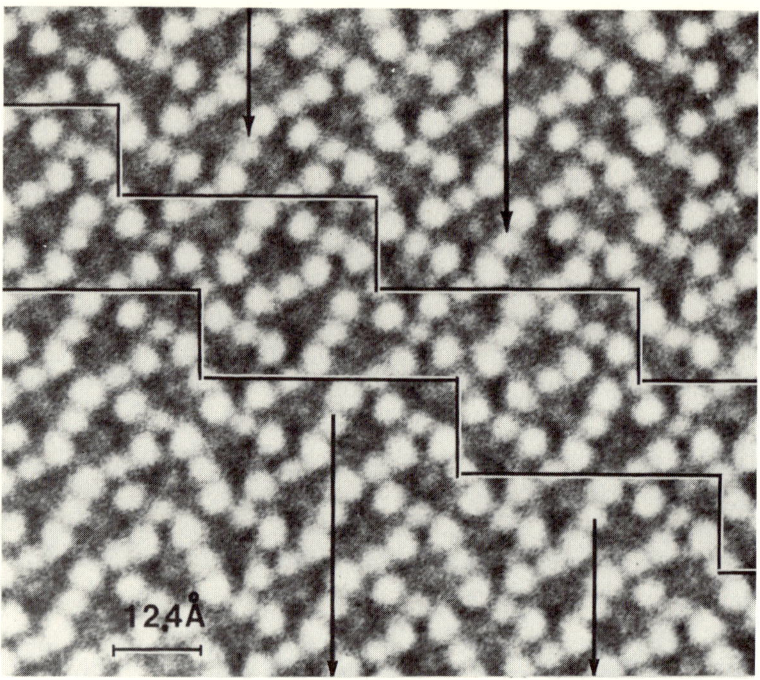

Fig. 5. Electron micrograph of the ordered $4Nb_2O_5 \cdot 9WO_3$ structure (c.f., Fig. 3c) interrupted by an out-of-phase domain boundary across which there is a shift of one-third of the b-axis.

In each of the structures examined, a variety of faults could be seen. These included a high degree of disorder in the superlattice formation for some quenched oxides, isolated planar faults, groups of planar faults occurring at more-or-less regular intervals to give modified superlattice periodicities, linear faults either associated with planar faults or else isolated, and, in a few cases, individual point defects. Since the imaging process is one which gives a point-by-point representation of the projected potential of the object (unlike most lattice fringe patterns which give only an indication of lattice plane spacing with no definable relation to the detailed crystal structure), the resolution of detail within the faults is equal to the resolution of detail in the quasi-periodic, well-ordered regions. Hence it is possible to derive information on the distribution of atoms within the faults, subject only to the restrictions implied by the use of projections of the atomic configuration in the beam direction. For the particular oxide systems under discussion here, the situation is especially favorable since planar faults perpendicular to the short b-axis occur only rarely. The planar and linear faults of interest in the superlattice structure are then parallel to the b-axis and to the incident beam for the optimum imaging conditions.

The form of the isolated CS planes, or Wadsley defects in Fig. 1, may be deduced immediately from the image. When superlattices are formed by a regular spacing of shear planes in one or two dimensions, Wadsley defects may occur in the form of perturbations of the periodicity of repetition of the CS planes and are clearly visible in the images. For example, in Fig. 6b, the principal periodicity seen is that corresponding to the composition $Nb_{22}O_{54}$. The structure contains 4 x 3 blocks of octahedra, 3 x 3 blocks, and tetrahedral metal-atom positions. In the middle of the picture, a planar defect is seen to give a perturbation of the structure. Across the projected line of this defect, the sequence of alternate 4 x 3 and 3 x 3 blocks is interrupted and the tetrahedral sites are eliminated. The region of the fault can be said to form a thin band of a $Nb_{12}O_{29}$ structure which has no 3 x 3 blocks and no tetrahedral positions. The region of such a fault will have a higher metal/oxygen ratio than the rest of the crystal. From observations of the frequency of occurrence of these faults, a deduction can be made regarding the variation of stoichiometry which they induce.

Many types of planar faults have been observed and analyzed by Iijima and co-workers, especially in the H-Nb_2O_5 structure.[12] Some produced an increase and some a decrease in the local metal/oxygen ratio. Others, including some twin planes and some local interpolations of other polymorphs of Nb_2O_5 produced no apparent modification of the stoichiometry.

Similar types of planar defects have been observed to occur in the lattice images of several minerals[18], although the investigation of the structures is as yet in an early stage.

Linear defects of various types have been seen in the images of oxides. Some correspond to kinks in planar defects or to terminations or intersections of planar defects. Very little evidence has been found for the existence of linear faults which could be described as dislocations of the types familiar in metals or simple inorganics.

Point defects may, in general, be expected to give relatively poor contrast and provide only localized intensity fluctuations that might well be confused with the effects of small surface irregularities, contamination of the surface, or statistical fluctuations of electron currents.

However Iijima et al.[19,20] have recently presented strong arguments to suggest that black spots which they observed in images of crystals having a composition near $Nb_{12}O_{29}$ are in fact point defects formed when two interstitial oxygen atoms are introduced to stabilize two metal atoms in tetrahedral instead of octahedral sites. The dark spots appear at particular parts of the unit cell for which such modifications are favorable (Fig. 7).

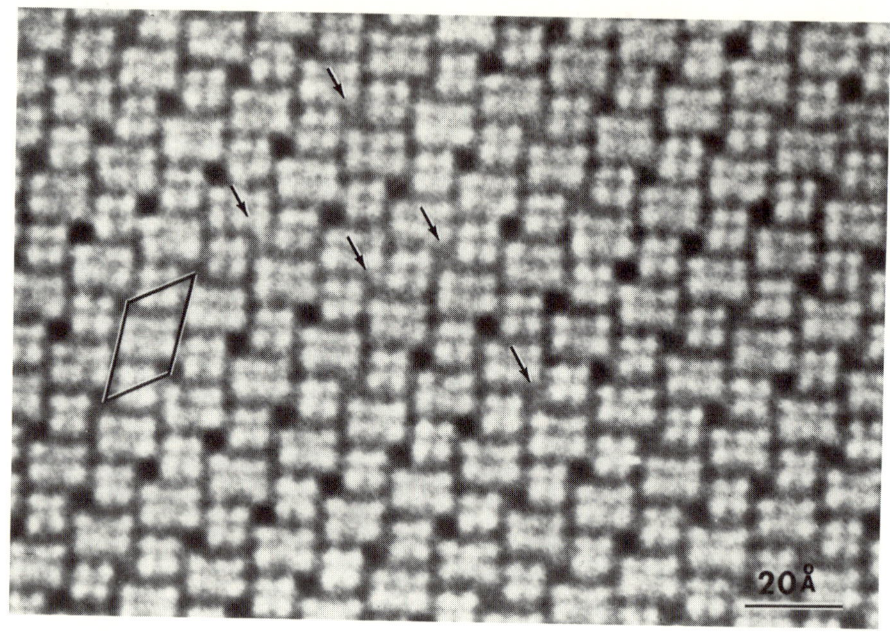

a

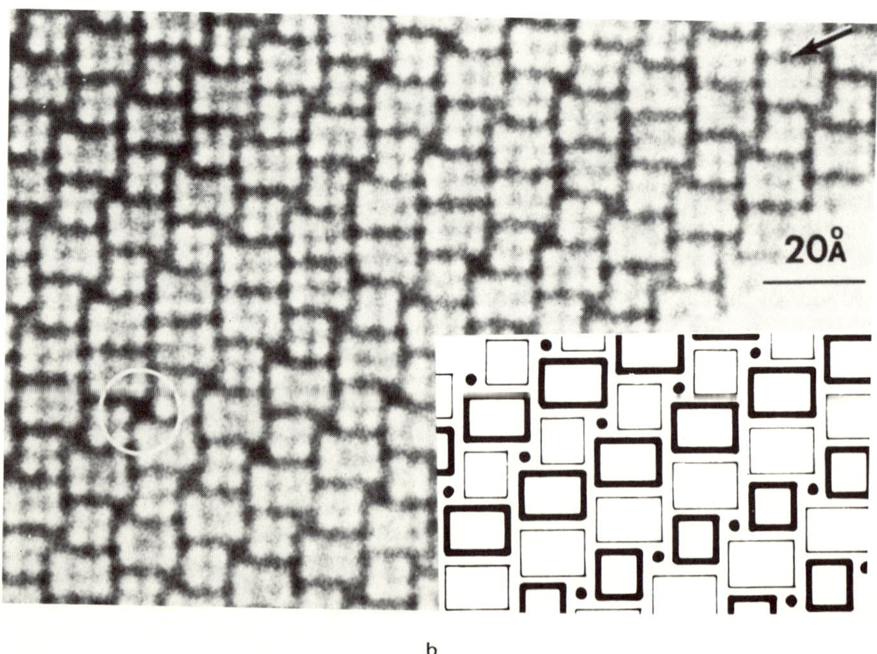

b

Fig. 6. Electron micrographs of a crystal of a niobium oxide close to $Nb_{22}O_{54}$ in composition showing in (a) a variable contrast of the tetrahedral metal-atom rows and in (b) a Wadsley defect of the form suggested in the inset diagram.[20]

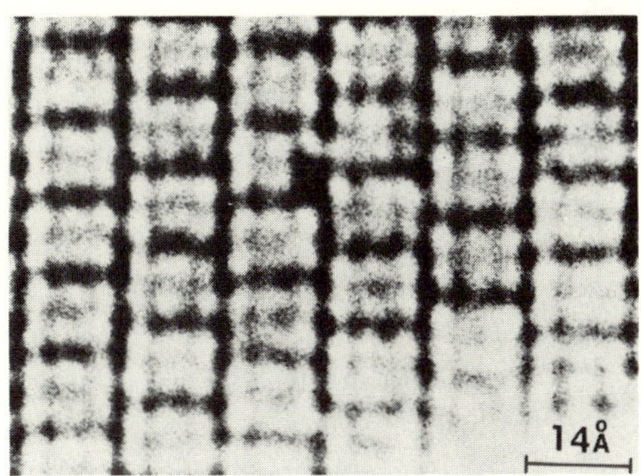

Fig. 7. Image of a thin crystal of composition near $Nb_{12}O_{29}$ showing a black spot considered to be given by a point defect of the type illustrated in the diagram. The open circles in the diagram represent interstitial oxygen atoms. The black circles represent tetrahedral metal atoms which may return to octahedral sites (indicated by arrows) if the excess oxygens disappear.[19]

The arguments favoring the interpretation in terms of point defects rather than lines of atoms parallel to the beam direction are:

(a) On chemical grounds it had been proposed that point defects must exist for these compositions since the number of Wadsley defects is insufficient to account for the deviation in stoichiometry from the ideal structure.

(b) The number of black dots increases linearly with the crystal thickness.

(c) Estimates of the density of the point defects, obtained from the number per unit area in the projection of a known thickness of crystal, give a deviation from stoichiometry close to that deduced from chemical analysis.

(d) The dark spots disappear gradually when irradiated with the electron beam in the microscope. This has not been observed to happen with rows of atoms in tetrahedral sites.

These point defects provide an excess of oxygen in the crystal. Others that provide a defect of oxygen have been postulated to exist in the rows of tetrahedral atoms in the $Nb_{22}O_{54}$ structure giving the variability in the density of the dark spots seen in Fig. 6a.

Detailed calculations of image intensities have yet to be performed to verify that the observed dark spots and contrast variations can be attributed to single point defects. Almost certainly it will be necessary to postulate some localized lattice distortions associated with the excess or defect of atoms.

3 EVIDENCE ON TRANSPORT PROCESSES

It is not possible at present to make high-resolution observations with the electron microscope on oxide systems held at elevated temperatures or in equilibrium with controlled atmospheres of gas. Heating stages are available but usually degrade the resolution to a serious extent. Devices have been made to provide a limited range of gas pressures in the specimen area, but the actual composition and pressure of the gas at the specimen position is not well defined and the additional mechanical complication introduced again limits the resolution. Hence, the studies on oxide systems at an atomic level of resolution have been confined to quenched specimens viewed at room temperature, and the evidence on transport processes have been derived mostly by inference from the configurations of defects observed.

Some modifications of the images have been observed to result from the irradiation by the incident electron beam in the microscope. Part of the effect of the electron beam is to heat the specimen by an amount varying from a few degrees or tens of degrees under normal operating conditions to several hundred degrees when the condenser apertures are removed to produce maximum illumination. The other effect, which is probably the more important in most cases, is the production of local excitations of the crystal electrons, leading to ionization or strong perturbations of the atoms with a consequent enhancement of diffusion processes.

For the oxides of present interest, sublimation under the electron beam is not often observed, but evidence of enhanced diffusion processes has been seen in some cases. The progressive reduction of molybdenum oxides has been studied with lattice fringe resolution by Bursill[21] for example, and Eyring et al.[22] have reported similar changes with irradiation of praseodymium and terbium oxides.

Figure 4 and similar micrographs of Nb_2O_5-WO_3 oxides quenched after incomplete high-temperature annealing provide indications of the processes by which the tetragonal tungsten-bronze structure is formed from the basic WO_3-type structure. Some isolated rotational defects are visible but, for the most part, the propagation of the rotated structure appears to be along planes, seen in projection as lines, and especially along twin planes of the initial structures. Such a growth mechanism leads to a disordered or poorly ordered superlattice structures. No indication of the mechanism by which ordered superlattice structures are generated from these disordered ones has been found.

In a few cases the growth of isolated rotational faults has been induced by electron irradiation. At positions where only a slight perturbation of the primary structure is visible initially, a fully developed square of pentagonal tunnels has appeared after strong electron irradiation. This confirms the inference that can be drawn from the individual micrographs, viz., that propagation of the rotational defects is most rapid along the direction of the tunnels.

The point defects of the type shown in Fig. 7 disappear after prolonged electron irradiation. Individual dark spots become weak and then vanish. If it is assumed that much of the contrast of the defect comes from a distortion of the surrounding lattice, it seems reasonable to suppose that the distortion and the contrast will decrease as the defect approaches the surface of the thin crystal and then vanish as the excess oxygen escapes into the vacuum.

4 OBSERVATIONS ON SHORT-RANGE ORDER

For the many cases where superlattice structures form from a high-temperature disordered state through an ordering reaction, the possibility exists of obtaining considerable information concerning the ordering process, and the energy terms involved, from a study of the short-range ordered state by diffraction methods.[23] It might be supposed that more complete and even more valuable information might be obtained from an actual imaging of the atom positions in short-range ordered materials.

The oxide superstructures we have been considering would appear to provide excellent test cases for this supposition. The ordering is two dimensional and is not complicated by the uncertainties associated with the interpretation of projections of three-dimensional structures. Images showing details of atom distributions in disordered materials have been obtained from quenched samples of various compositions, including Nb_2O_5[12] and the Nb_2O_5-WO_3 compounds[15,24]. From the latter, in particular, electron-diffraction patterns showing a variety of striking diffuse-scattering configurations have been obtained.[15] However, the interpretation of diffuse scattering due to short-range order and the analysis of configurations in terms of ordering energies have not yet been extended to systems having this degree of complexity, and this, together with the limitations on the experimental methods which so far give only rather qualitative results, has prevented any significant analysis of the observations. It has been observed that the short-range ordered state contains microdomains of structures which are recognizable as being those of known ordered phases as well as other regions having no such correlation[12], but the analysis has not proceeded much beyond this point.

High-resolution electron microscopy of some varieties of the mineral tourmaline has revealed ordering of the cations over distances of the order of 100 Å or more[25], even though the X-ray diffraction analyses of these crystal structures suggest complete randomness in the occupancy of the cation sites. It seems likely that this ordering is a result of the enhanced diffusional process induced by electron irradiation.

There are difficulties in the application of the lattice-imaging method to the simpler systems, such as the binary-alloy solid solution for which the theoretical analysis of the short-range ordered state has been the most thoroughly developed and for which the ordering process is best understood, even though the picture is far from being complete. One difficulty is that the relevant interatomic distances are usually about 2 Å in projection and instruments giving point-to-point resolutions of this order for readily interpreted images are not yet available. Another difficulty is that even images of sufficient resolution will give only two-dimensional projections of the three-dimensional atom configurations and the images will therefore have to be interpreted in terms of the statistics of superposition of atoms.

One technique that has been used to obtain information on the nature of the short-range ordered state is to obtain dark-field images using only the diffuse scattering maxima in the electron diffraction pattern. Since the objective aperture of the microscope then excludes all the fundamental Bragg reflections generated by the average lattice, the image contains (as a first approximation) only information concerning the deviations from the average lattice due to the short-range ordering. Such images, obtained from various disordered alloys, have been shown to contain small white dots, 20 to 50 Å in diameter, which are interpreted as representing "microdomains" of the ordered lattice.[26] An analysis of the origins of the contrast produced in relevant idealized cases of dark-field imaging[27] has suggested that such interpretations should not be made too freely since the sizes and shapes of the intensity maxima may be considerably affected by the statistical superposition of microdomain regions and by the square-law dependence of the intensity on the deviation from the average projected potential value.

A case of intermediate complexity, for which it appears possible to apply all techniques and approaches with some degree of satisfaction, is represented by $LiFeO_2$, in which the lithium and iron ions are disordered on one of the face-centered cubic sublattices of the NaCl-type structure. In the disordered state, a striking pattern of diffuse scattering is produced (Fig. 8a). An analysis of the X-ray diffuse-scattering measurements has been made[28] and the short-range order coefficients so found have been used to deduce the nature of the effective interatom potential with limited accuracy[29].

From the configuration of the diffuse scattering it may be concluded that an essential feature of the short-range ordering is the occurrence of linear arrays of alternate iron and lithium concentrations (positive and negative deviations from the average projected potential) with a periodicity

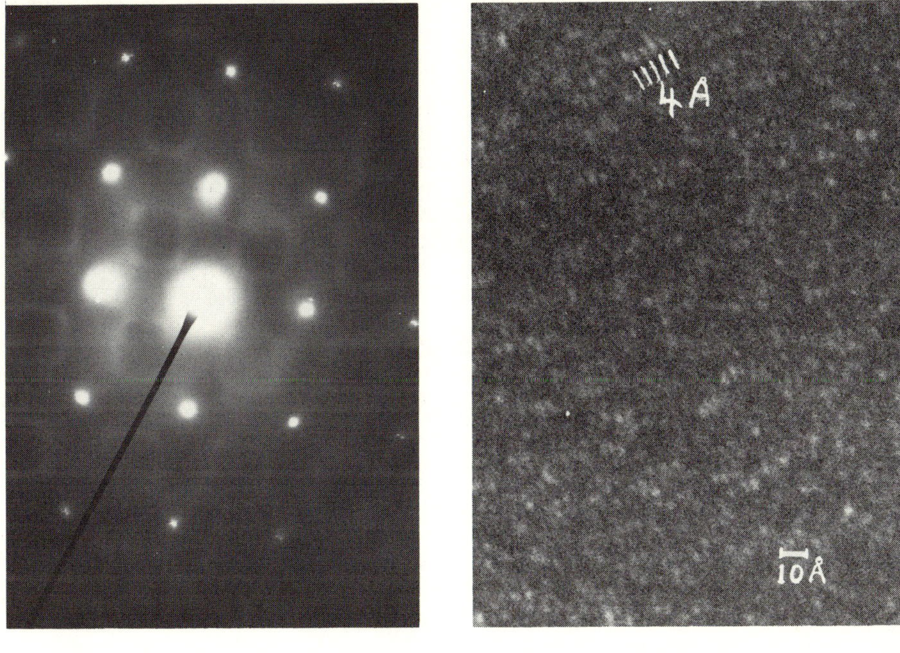

a b

Fig. 8. (a) Electron diffraction pattern of $LiFeO_2$ in (100) direction showing diffuse scattering due to short-range order. (b) Dark field electron micrograph of $LiFeO_2$ crystal obtained from part of the diffuse scattering shown in Fig. 8(a).

of about 8 Å. Dark-field imaging of the materials may then be expected to give lines of spots 4 Å apart, since both positive and negative deviations from the average structure should give intensity maxima.[27] Such lines of spots are seen in the image in Fig. 8b.

In this case, the electron microscopy has given some degree of confirmation of the deductions made from the diffraction pattern. However, the results of all the techniques that have been applied are as yet of limited accuracy and reliability. Each has referred to a different aspect of the problem of the nature of the short-range ordered state. The refinement of the observations and correlation of the results to give a more complete understanding of the ordering process is a matter for future development.

This work has been supported by the National Science Foundation Grants GU3169 and GH36667.

REFERENCES

1. Iijima, S., *J. Appl. Phys.*, **42**, 5891 (1971).
2. Allpress, J. G., and Sanders, J. V., *J. Appl. Cryst.*, **6**, 165 (1973).

3. Goodman, P., and Moodie, A. F., *Acta Cryst.* (In press).
4. O'Keeffe, M. A., *Acta Cryst.,* **A29,** 389 (1973).
5. Antis, G. R., Lynch, D. F., Moodie, A. F., and O'Keeffe, M. A., *Acta Cryst.,* **A29,** 138 (1973).
6. Fejes, P. L., Ph.D. Thesis, Arizona State University, 1973.
7. Cowley, J. M., and Iijima, S., *Z. Naturforsch.,* **27a,** 445 (1972).
8. Cowley, J. M., *Proceedings of 31st Annual EMSA Meeting, 1973,* Claud J. Arcenaux (Ed.), Claitors Publishing Div. (1973), and in press (in extended form).
9. Moodie, A. F., *Z. Naturforsch.,* **27a,** 437 (1972).
10. Wadsley, A. D., and Andersson, S., in *Perspectives in Structural Chemistry,* J. D. Dunitz and J. Albers (Eds.), Wiley, New York (1970), Vol. III, pp. 1-58.
11. Hyde, B. G., and Bursill, L. A., in *The Chemistry of Extended Defects in Non-metallic Solids,* LeRoy Eyring and Michael O'Keeffe (Eds.), North-Holland Publishing Co. (1970), pp. 347-381.
12. Iijima, S., *Acta Cryst.,* **A29,** 18 (1973). See also Anderson, J. S., Browne, J. M., and Hutchison, J. L., *Nature,* **237,** 151 (1972).
13. Iijima, S., and Allpress, J. G., *J. Solid State Chem.,* 7, 94 (1973).
14. Allpress, J. G., Iijima, S., Roth, R. S., and Stephenson, N. C., *J. Solid State Chem.,* **7,** 89 (1973).
15. Iijima, S., and Allpress, J. G., *Acta Cryst.* (in press).
16. Hyde, B. G., and O'Keeffe, M., *Acta Cryst.,* **A29,** 243 (1973).
17. Bursill, L. A., and Hyde, B. G., *Nature Phys. Sci. London,* **240,** 122 (1972).
18. Buseck, P. A., and Iijima, S., *Amer. Mineral.,* in press.
19. Iijima, S., Kimura, S., and Goto, M., *Acta Cryst.,* **A,** in press.
20. Iijima, S., Kimura, S., and Goto, M., in preparation.
21. Bursill, L. A., *Proc. Roy Soc.,* **A311,** 267 (1969); also ref. 11.
22. Skarnulis, A. J., Tai, L. T., Lundberg, M., Kunzman, P., and Eyring, LeRoy, paper presented at Symposium on Solid State Chemistry at 165th National Meeting of the A.C.S., Dallas, Texas, April, 1973.
23. Cowley, J. M., in *Advances in High Temperature Chemistry,* LeRoy Eyring (Ed.), Academic Press, New York (1971), pp. 36-86.
24. Iijima, S., and Allpress, J. G., *Acta Cryst.,* **A,** in press.
25. Iijima, S., Cowley, J. M., and Donnay, G., TMPM, in press.
26. Gaudig, W., and Warlimont, H., *Z. Metallk.,* **60,** 488 (1969); Warlimont, H., and Thomas, G., *Metal Sci. J.,* **4,** 47 (1970).
27. Cowley, J. M., *Acta Cryst.,* **A,** in press.
28. Brunel, M., and Bergevin, F., de, *J. Phys. Chem. Solids,* **30,** 2011 (1969).
29. Cowley, J. M., and Wilkins, S., in *Interatomic Potentials and Simulation of Lattice Defects,* Pierre C. Gehlen, Joe R. Beeler, and Robert I. Jaffee (Eds.), Plenum Publishing Co., New York (1972), pp. 265-280, Wilkins, S., *Phys. Rev.,* **B2,** 3935 (1970).

DISCUSSION on Paper by J. M. Cowley

SELTZER: Do you ever see dislocations in your cold-worked powder specimens? What do they look like?

COWLEY: We have seen very little indication of the existence of anything in the pictures of oxides which can be related to the classical ideas of dislocations. One possible exception is the observation of a relative twist of superimposed crystal portions which suggests a side view of a screw dislocation. Near the thin edge of a crystal of $Ti_2Nb_{10}O_{29}$ structure, the image appears like this (drawing only the dark lines parallel to the c-axis):

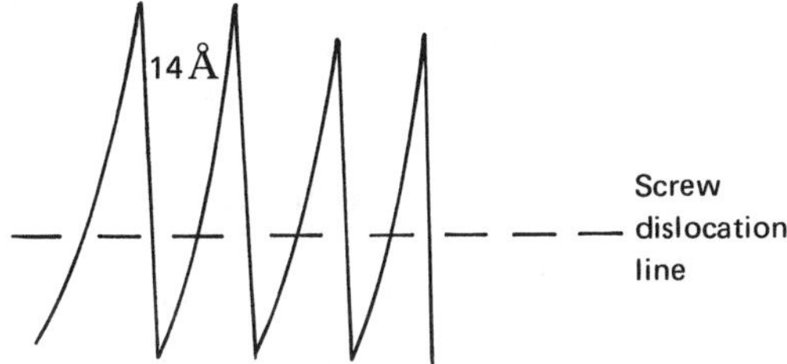

Such a configuration may occur only as a result of severe distortion at a thin crystal edge during fracture of the crystal.

HAGEMARK: Have you looked at zinc oxide?

COWLEY: To my knowledge, no high-resolution observations have been made on zinc oxide. There have been many low-resolution studies on zinc oxide, including observations [Cowley, *Acta Cryst.*, 21, 192 (1966)] on severe modifications of the crystal morphology by selective sublimation or of rapid disintegration of crystals under electron irradiation. Such effects are extremely variable even between neighboring crystals formed together on a specimen grid.

CARO: What is the effect of electron-beam irradiation on ordering, or microdomain formation, or phase segregation? What is the part of temperature effects with respect, for instance, to radiation-induced vacancy mobility?

COWLEY: The contributions of increased temperature and electron-excitation effects are difficult to separate. However, I believe that there is ample evidence for the enhancement of diffusion processes leading to the ordering of amorphous materials or disordered crystals, the sublimation of crystals, and the enhancement of chemical reaction rates under conditions where the increase in temperature cannot be more than a few degrees and so cannot contribute to any important extent. Heating effects will be most important for thick specimens and when the illuminated area of the specimen is large. The processes of electron excitation and the means by which the diffusion processes are assisted do not seem to be well understood.

O'KEEFFE: Further to Caro's comment, it is interesting that apparently stoichiometric $Nb_{12}O_{29}$ can be made in the microscope but not in the laboratory and the evidence is that the stoichiometric compound is unstable. One wonders whether this is due to enhanced diffusion in the electron beam (possibly associated with the large electron current) occurring at a sufficiently low temperature to preclude dissociation to a stable lower oxide.

COWLEY: Indirectly, this may be so. It must be realized that the environment of the specimen in the electron microscope is not well controlled. It usually contains about 10^{-5} Torr of oxygen and nitrogen plus water vapor and carbonaceous materials, with a solid layer (at least a few Angstroms thick) of "contamination" which is supposedly amorphous semigraphitized carbon (with possibly silicon oxides from silicone pump oils). The electron irradiation will tend to produce an equilibrium with this environment. An otherwise unstable structure may be stabilized under these conditions, as in the case of β-BeO [J. M. Cowley, *Acta Cryst.*, 21, 192 (1966)].

VINK: Did you ever see things in your pictures that could be attributed to impurities?

COWLEY: There have been observations of perturbations of the structure of the niobium and mixed oxides which are difficult to explain except in terms of the presence of impurity atoms. Several such cases are discussed in Ref. 12. Some more systematic studies on the localization of impurity atoms have been made by Dr. Anderson and his colleagues [see comment (2) by Anderson].

ANDERSON:

(1) In regard to O'Keeffe's comments – The problem with $Nb_{12}O_{29}$ (ideal composition $NbO_{2.4167}$) is that the deviation from ideal stoichiometry is right on the limit of analytical significance – possibly up to $NbO_{2.418}$.

(2) With regard to Vink's comments – In niobium oxides doped with other cations, it has been seen that crystals have a structure with domains of perfect structure, separated by walls of blocks with different dimensions. From the direct relation between block structure and metal/oxygen ratio, it can be identified that the impurity cations are – to a considerable extent – at least – segregated in these walls.

COWLEY:

(1) The observations of point defects (Fig. 7) were made mostly on specimens of composition $NbO_{2.419}$. From the observed density of spots in the image of a crystal of known thickness it was estimated

that there was one point defect per 30 unit cells, which agrees within the probable limits of accuracy with the figure of 1 defect per 20 unit cells corresponding to the deviation from the ideal stoichiometry.

(2) See comment above in reply to Vink.

SHEWMON: How do you obtain your samples?

COWLEY: See text of paper.

SHEWMON: Could you use this technique to study the defect clusters in FeO or NiO that Prof. Cohen discussed?

COWLEY: It should be possible to obtain useful information concerning the defect clusters in these oxides, although the type of image will not be exactly comparable with the images of the oxides based on the ReO_3 structure. Projections of the basic structure of FeO or NiO will contain metal-atom peaks having a separation of less than 3 Å, whereas for contemporary 100-kev electron microscopes, the least resolvable distances in images which can be directly interpreted (e.g., one black dot equals one row of atoms) are greater than 3 Å. However, the deviations from the average structure could undoubtedly be imaged, especially when the defect clusters are ordered into a regular superlattice. We intend to try this.

NOWICK: Is there any information on reduced rutile which would indicate whether or not there are point defects present, above and beyond the Wadsley-type defects?

COWLEY: There have been extensive electron-microscope observations of reduced rutile, for example by Bursill and Hyde, but none of this work has been done with sufficient resolution or with sufficiently well-controlled orientation or thickness of the crystals to allow point defects to be observed if present.

FRANKLIN: I don't understand the transformation that appeared to be taking place with the tungsten bronzes. Was there a compositional change taking place?

COWLEY: The sample used to obtain Fig. 4 was quenched after an anneal at high temperature, which may have been insufficient to produce uniformity of composition. Then Fig. 4 and similar pictures may be representative of regions having a concentration gradient and indicate the process by which the transformation takes place at high temperature. The alternative explanation, viz., that the transformation was initiated below the annealing temperature during the quench, is not favored (see Ref. 24).

TALLAN: Does the fact that one sees contrast effects corresponding to mismatch of the metal-oxygen polyhedra only in the plane of your figures imply that there is not mismatch or disorder in the thickness direction.

COWLEY: In the examples of electron micrographs used to illustrate this paper, all the superstructure ordering and all the defects have involved mismatches only in the plane of the figures, i.e., along planes parallel to the short b-axis which is the beam direction. This is a characteristic of these types of oxide which makes them particularly favorable for study by high-resolution electron microscopy. Defects involving a disruption of the ordering in the b-axis direction have been observed to occur (see Ref. 12) and give striking contrast effects, which can be interpreted in favorable cases.

VACANCY-RELATED DEFECTS IN THE ALKALINE-EARTH OXIDES

John E. Wertz

Department of Chemistry
University of Minnesota
Minneapolis, Minnesota

ABSTRACT

Cation vacancies profoundly affect the observable electronic properties of alkaline-earth oxides. Their presence serves to compensate the charge of inevitable trivalent ions with which they tend to be associated in a 1:1 or 1:2 ratio. Isolated cation vacancies ordinarily constitute less than 10 percent of the total number. However, neutron irradiation of samples kept at room temperature leads to large numbers of isolated cation vacancies and of anion vacancies. One or two positive holes may be trapped by the former to give the V^- and the V^0 centers, respectively. V^- centers at room temperature show a single isotropic ESR line which is the result of rapid migration of the hole among the six equivalent oxygen ions. Association of a trivalent ion with a cation vacancy prevents such reorientation. Anion vacancies trap one or two electrons to form F^+ or F^0 centers, respectively.

1 INTRODUCTION

Of the many known oxides, those alkaline earths which exhibit the rock-salt structure (MgO, CaO, SrO, and BaO) represent ideally simple systems. The octahedral symmetry of the host lattice may be preserved about an imperfection, or the symmetry may be lowered to tetragonal, or orthorhombic, or to low symmetry. The band gaps decrease progressively from the value of 8.7 ev for MgO; the application of optical, electron-spin resonance, and other techniques permits study of a given defect in hosts of various band gaps. It is primarily the ESR technique applied to these ideally simple host crystals which has permitted the geometry of numerous defects to be unambiguously described.

The nature of the defects in alkaline-earth oxides has been summarized in several reviews.[1-3] A number of significant conclusions have been reached since the most recent of these appeared; these are described here after some of the most significant defects have been introduced for those who have not been concerned with these model materials.

2 POSITIVE ION VACANCIES

2.1. Impurities

If one were asked to specify the one most important defect in all of these oxides, it would unquestionably be the positive-ion vacancy. Even for oxides prepared with great care[4], there are still trivalent iron-group impurities present at the parts-per-million level. The level of such non-paramagnetic impurities as Al^{3+} may be an order of magnitude greater; foreign alkaline-earth ions, e.g., Sr^{2+} in CaO may be another order of magnitude higher in concentration. Ordinarily, the extra charge of the impurity trivalent ions is compensated by half their number of positive-ion vacancies. A variety of experiments indicate that these cation vacancies have a significant mobility at 400°C; one observes a marked tendency of vacancies to associate with other defects to give neutral centers in this temperature range. At room temperature, less than 10 percent of all the cation vacancies in a relatively pure, untreated crystal are isolated. Evidence for this remark will be given later. The majority of these vacancies will be associated with one trivalent ion, leaving a net charge of -1 for the vacancy-impurity pair. A small fraction of vacancies controllable by extended heating of crystals in the vicinity of 400°C, will be associated with two trivalent ions to form a neutral center.[1,2] These vacancies will also serve as traps for monovalent anions, such as OH^- or F^-.[5] As substitutional impurities at the site of an O^{2-} ion, they have an effective charge of +1; they are thus able to trap a wandering cation vacancy, and become associated with it.

We have noted that only a small fraction of all the cation vacancies in an untreated alkaline-earth-oxide crystal are isolated from other defects such as impurities. As an operational concept for "isolation", we require that the vacancy have effectively octahedral symmetry at room temperature. The significance of this definition will shortly become apparent when we consider the isotropic line of the V^- center. One can also study the ESR spectra of numerous substitutional transition-metal ions. Many of the iron-group impurities are able to exist in two or more stable or metastable valence states. For example, iron is readily observed in the +1, +2, and +3 states.[2] One may inquire whether, if most vacancies are found to be associated, the same may be said of the observable transition-metal ions. At least for such ions as Fe^{3+} or Cr^{3+}, one concludes that the majority of these ions are *not* associated with a vacancy.[2,6] The ESR spectra of these ions in octahedral symmetry, i.e., when they are remote from a cation vacancy, are very characteristic and easily distinguishable from the spectra of the same ions associated with a vacancy. The latter show either tetragonal or orthorhombic symmetry.[7] Ti^{3+} may also be observed in tetragonal or orthorhombic symmetry, but it is not observable by ESR techniques in octahedral symmetry.[8] Tetragonal symmetry is displayed by an M^{3+} ion with a cation vacancy along a [100] axis in the next-nearest-neighbor (nnn) position. One may designate the immediate defect as follows:

$$\text{O M}^+ \text{ O [=] O} \qquad [100] \text{ axis} .$$

(Here and subsequently, only *deviations* from normal site charge are shown.) If the vacancy is in the nearest-neighbor (nn) position, the defect shows orthorhombic symmetry, which may be represented as follows:

$$\text{M M}^+ \text{ [=] M} \qquad [110] \text{ axis} .$$

Of course, such diamagnetic ions as Al^{3+} or Ti^{4+} may also participate in association with vacancies. Not only may one detect the association of one trivalent transition-metal ion with a vacancy, but one may also detect the spectra of two trivalent ions-plus-vacancy to give either tetragonal or orthorhombic symmetry:

$$\text{O M}^+ \text{ O [=] O M}^+ \text{ O} \qquad [100] \text{ axis}$$

$$\text{M M}^+ \text{ [=] M}^+ \text{ M} \qquad [110] \text{ axis} .$$

One may refer to these two as doubly associated centers; their spectra indicate a somewhat smaller crystalline electric field than for the singly associated ion. In each of the cases for which one can make ESR comparisons of the ions in associated and in unassociated forms, one observes a marked predominance of the unassociated ions. The difference is usually far more than can be accounted for by the observation that there is only one cation vacancy for every two trivalent ions. One must admit that those ions which find themselves at sites of very low symmetry or in impurity clusters will not be observed in the ESR spectrum. Hence, if all ions were detected, perhaps one-half would be associated with other defects. However, for Cr^{3+}, one may arrive independently at the conclusion that the unassociated ions predominate, for one may observe separately the R-line luminescence – akin to that in ruby – for the unassociated and for the associated ions.[9-12] The optical absorption of associated and unassociated Cr^{3+} has also been detected.[13] Similar luminescence observations may be made for the isoelectronic ion V^{2+}.[14] One should expect somewhat different binding energies to cation vacancies for different trivalent ions. One may infer that the binding energy of Al^{3+} may be somewhat greater than that of such ions as Fe^{3+} or Cr^{3+}, since the great majority of all cation vacancies which trap a positive hole are found to be associated with Al^{3+}.[15,16]

2.2 Trapped Holes

One electron is very readily lost by an O^{2-} ion subjected to any type of ionizing radiation. The resulting positive hole is free to migrate if the oxide ion is at an unperturbed lattice site. This hole may then be trapped at any site that has an effectively negative charge. Two such sites are a cation vacancy or a univalent ion deliberately incorporated into the oxide and substituting for the normal divalent cation.[17] These centers may respectively be designated as follows:

$$\mathrm{O\ Mg\ O^{+}\ [=]\ O} \qquad V^{-}\ \text{center}$$

and

$$\mathrm{Mg\ O^{+}\ M^{-}\ O} \qquad M^{0}\ \text{center}\ .$$

(Such impurity ions as Cr^{2+} or Fe^{1+}, produced by irradiation, may also serve as hole traps, as may the stable Fe^{2+} ion.) Except for the divalent charge of the anion and the cation of the host lattice, the hole trapped adjacent to an isolated cation vacancy is just the model of the V_1 center that Seitz[18] proposed for the alkali halides; the V_1 centers are never observed in these hosts because the hole is immediately trapped at the site of production to form X_2^- molecules. It was anticipated that the hole would be delocalized over the six anions surrounding the cation vacancy. Speaking in the language of electron-spin resonance, one would expect to observe a single isotropic line from such a defect center. This was not initially observed, for reasons which will be detailed. Instead, the observed spectrum of three lines exhibited the tetragonal symmetry of the defect.[19] The components of the spectroscopic splitting were respectively $g_{\parallel} = 2.0032$ and $g_{\perp} = 2.0385$ in MgO. These ESR observations were made at 77°K, at which the lines have a width of about 0.3 G; at room temperature, the line width was a hundredfold greater, and the spectrum was still anisotropic. In CaO, the values are $g_{\parallel} = 2.001$ and $g_{\perp} = 2.0710$. For SrO, they are $g_{\parallel} = 2.0010$ and $g_{\perp} = 2.0703$.[2] To account for the tetragonal symmetry, it is necessary at least that the electron-spin density on the two oxygen ions on the axis be different from those in the equatorial plane about the vacancy. We chose to regard the electron as localized on one oxygen atom. At the time, we considered it likely that there was also a trivalent ion along the defect axis in the next-nearest-neighbor (nnn) position to the vacancy:

$$O^+ \; [=] \; O \; M^+ \qquad V_M \text{ center } (M = Al^{3+} \text{ or other trivalent ion}) \; .$$

Subsequently, we ascribed the tetragonal component of crystalline electric field to the outward relaxation of the O^- ion from the vacancy. The M^{3+} ion was dropped from the model as unnecessary. It was thus presumed that the hole center corresponding to the ESR observations involved an isolated cation vacancy. It was ultimately renamed the V^- center to specify the overall charge state of the defect.[20] However, very recently, electron-nuclear double resonance (ENDOR) experiments have shown that – as ordinarily produced – most of the hole centers give rise to a set of lines from aluminum in the nnn position.[15,16] Such defects have thus been referred to as V_{Al} centers, to indicate both the association with the impurity and the neutral state of the defect. However, it was found that in MgO crystals which had received a low neutron dose while the crystal was kept near room temperatures, no such aluminum ENDOR lines were detected. Remarkably, for both the true V^- center and for the V_{Al} center, the ESR spectrum at 77°K is superposable within the present limits of resolution of the narrow lines. Thus, at least as far as the ESR aspects are concerned, the effect on the tetragonal component of crystalline electric field of the Al^{3+} ion in V_{Al} centers is negligible in comparison with the contribution due to outward relaxation of the O^- ion from the cation vacancy. Thus, ESR spectra at 77°K are unable to distinguish between the V_{Al} and the V^- centers.

However, we have shown that when the spectra are taken at room temperature, such resolution is readily possible.[21] It had been reported that there is no coalescence of the three ESR lines of the center – presumed to be V^- – in this temperature region.[22] This is indeed the case for the V_{Al} centers, and the lack of coalescence is that to be expected. There might indeed be a hopping of the trapped hole from the axial position to one of the equatorial positions, but the hole would then have to return to its original site, since it would be repelled by the positive charge on the Al^{3+} (or any other trivalent) ion lying along the defect axis. The three ESR lines of the V_{Al} center at room temperature are indeed found in the same positions at 300°K as at 77°K, which indicates that any reorientation of the hole from its axial position is insufficient to essentially alter the spectrum. However, the lines are very broad at the higher temperature.

We have shown that for the true V^- center there is a strong isotropic line at room temperature as a result of the rapid reorientation of the hole about the cation vacancy.[21] This line, shown as V_{iso} in Fig. 1, represents the coalescence of the three anisotropic lines, and thus it has considerable intensity, notwithstanding its considerable width (about 9 G in MgO). The position in field of the line is within experimental error precisely at the expected position, i.e., the isotropic g factor g_{iso} is given by:

$$g_{iso} = \frac{1}{3}(2g_{\perp} + g_{\parallel}) = 2.027 \quad .$$

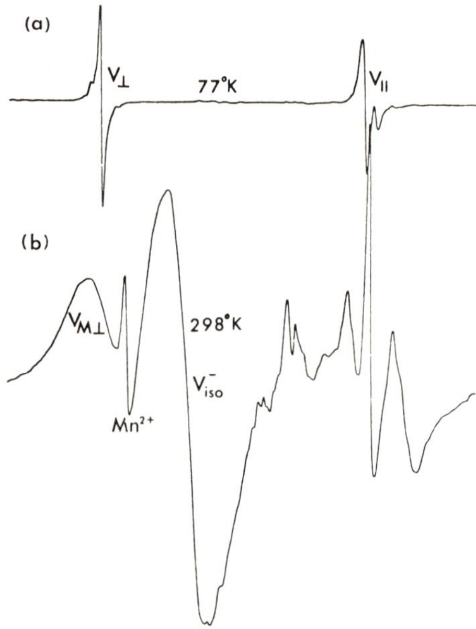

Fig. 1. (a) ESR spectrum at 140 K of an MgO crystal which has been irradiated with fast neutrons and then x-irradiated at 77°K. The lines marked $V_{\parallel}$ and $V_{\perp}$ refer to lines arising from both V^- and V_M centers, with the magnetic field respectively parallel and perpendicular to the tetragonal axes of these defects. (b) Spectrum at 298°K of the sample giving the above spectrum at 140°K. The line marked $V_{M\perp}$ is unchanged in position and it arises only from V_M centers. The line labelled V^-_{iso} is the coincidence of the $V^-_{\perp}$ line shifted to higher field and the $V_{\parallel}$ line shifted to lower field. The region of the $V^-_{\parallel}$ line for the V_M center is complicated by the presence of F^+ center lines.

This isotropic line has been shown to occur both for the V^- centers arising from neutron irradiation and from those centers produced by ultraviolet or X irradiation. In the former case, the cation vacancies are produced randomly, and hence most of these vacancies will be isolated. By keeping the crystals cooled to near room temperature during irradiation, one can be assured that there will be no diffusion to permit association of an impurity with the vacancy. However, even in MgO crystals which have been slowly cooled from the melt, there should be some cation vacancies that are not associated with trivalent ions in crystals of high purity. We did indeed find that there is a weak isotropic line for crystals subjected only to ionizing radiation, thus verifying the existence of the V^- center as a minority defect under ordinary conditions. However, the concentration of the V^- centers after X irradiation can be considerably increased if the crystal has first been heated to about 1000°C to dissociate vacancies from impurities and rapidly quenched so as to minimize the reassociation of vacancies and trivalent ions during the cooling stage.

There is a very striking difference between stability of the ordinarily produced V_M centers and that of the true V^- centers. The former, depending upon the impurity concentration, may show a decay with a halflife of 2 to 15 hours at 20°C. However, the true V^- centers are so stable at room temperature that there is only a small decrease in concentration in 2 years for crystals which have been kept in the dark. Compared with the labile V_M center, the stability of the V^- center is striking. As a matter of fact, in quenched, X-irradiated MgO crystals, one observes a growth in the V^- center isotropic line at the expense of the V_M line when the crystal of MgO is subsequently heated at 50°C. The lack of V_M centers in MgO crystals after neutron irradiation at 32°C and after the subsequent period required for decay of short-lived induced radiation and for transport from a remote reactor facility is ascribed to their more rapid decay. In the course of the irradiation, there will indeed be holes trapped in preexistent vacancies associated with a trivalent ion, or at the smaller number of vacancies associated with impurities which are produced by atomic displacement. These V_M centers merely lose the trapped holes; they may readily be regenerated by subsequent X irradiation. In CaO, the half-life of V^- centers produced by proton irradiation is 2 days, in contrast to 5 minutes for V_M centers.[23] The corresponding half-lives in SrO are 1 minute and a few seconds. The progressively decreasing stability in the series MgO, CaO, SrO is reasonable if there is a relaxation of the O^- ion away from the vacancy by similar distances in the three cases. The marked increase in lattice spacing in this series leads to the lowered stability of the V^- centers in CaO and SrO.

There are various reasons to believe that decay of the V_M center occurs by release of holes.[25] One line of evidence comes from a study of

V^0 centers which have the configuration:

$$\text{Mg O}^+ \ [=] \ \text{O}^+ \text{ Mg} \qquad V^0 \text{ center} \quad .$$

These centers, which have been observed in MgO, CaO, and SrO, can be produced in large numbers ($>10^{17}$ cm^3) by X irradiation at 77°K of crystals which have large numbers of *isolated* cation vacancies.[19,23,24] As noted earlier, these can most readily be produced by neutron or proton irradiation, though some are always present. While the V^- center in MgO is indefinitely stable at room temperature, the V^0 center loses one hole under these conditions. In a series of V-center decay studies at room temperature following X irradiation at 77°K, we observed that after 60 hours, the V_M centers had decayed to a low level, as had the V^0 centers.[24] However, for each V^0 center which decayed, one V^- center was produced. Thus the sum of the V^- and the V^0 center concentrations remained fixed. This gives evidence that the decay at room temperature of both V_M and V^0 centers occurs by loss of a positive hole, which is not retrapped by a V-type center. One had also to consider the possibility that under these conditions, other ions such as Fe^{1+} might lose an electron, which then annihilate some of the positive holes. Such a possible mode of V-center decay has been ruled out by our observations.

In CaO there is some evidence for the breakup of vacancy clusters when the crystal is heated in the vicinity of 600°K.[23] These inferences ensue from ESR observations of a significant increase in the concentration of V^- centers subsequent to X irradiation following a period of heating. While this heating does not alter the number of cation vacancies, it probably aids in the dispersal of some that have been formed close to other defects. It appears clear that there is an increase in the number of isolated cation vacancies after this heating. However, heating in the range of 650 to 700°K serves to reduce drastically the number of isolated cation vacancies available for V^- formation.

In a modified type of V center in which an OH^- ion is associated with a cation vacancy, i.e., the V_{OH} center,[5] the hole is far more readily lost than in the V_M center. In this case, one does observe in MgO the retrapping by V^- (or V_M) centers of holes liberated from V_{OH}.[26]

$$\text{Mg O}^+ \ [=] \ \text{HO}^+ \text{ Mg} \qquad V_{OH} \text{ center} \quad .$$

This same center had been shown earlier to evince a characteristic infrared absorption peak at 3323 cm^{-1} which, upon decay of the V_{OH} center, also decayed, with transfer of its intensity to a peak at 3296 cm^{-1}. The latter peak, which is stable ordinarily, represents the characteristic absorption of an OH^- ion adjacent to a cation vacancy. Experiments with the OD^- ion in MgO show that the absorption of the V^-_{OD} center occurs at 2464.5 cm^{-1}.[3] There is yet another center, presently designated V_{OR}, in which the symmetry of the V^- center is lowered to orthorhombic.[27] It is presumed that a nearby OH^- ion, with the hydrogen in an interstitial position, is responsible for the distortion. Another substituted V center is the V_F center in which an F^- ion substitutes for O^{2-} adjacent to the vacancy:

$$\text{Mg } O^+ \ [=] \ F^+ \quad .$$

This center has the special significance that it was the first in which the substitution of ^{17}O permitted the observation of the hyperfine splitting of the oxygen atom upon which the hole is localized.[28]

In MgO crystals of reasonable purity, ionizing radiation induces a violet color which is associated with a gaussian band centered at 2.33 ev and having a half-width of 1.1 ev at 298°K.[29] The intensity of this band induced by X irradiation of untreated crystals was correlated with that of the ESR spectrum at 77°K of what was believed to be the V^- center. In retrospect, it is apparent that most of the absorption was from V_M centers. However, subsequent studies of the true V^- center, the V_{OH}, and the V^0 centers reveal that all of the trapped-hole centers show nearly identical optical absorption.[24] For example, the peak absorption occurs at 2.21 ev for the V_{OH} center and at 2.37 ev for the double-hole V^0 center. A simpleminded approach presumes that the threefold degeneracy of the P orbitals in octahedral symmetry is partially lifted by the tetragonal crystal field; the expected optical absorption would be from the stabilized P_σ to the P_π orbitals. From a variety of experiments, including studies of the hyperfine interaction of $^{25}Mg^{2+}$ ions adjacent to the trapped hole,[30] it appears that the simple description of the localized hole is appropriate for the ground state. A combined magnetic circular dichroism ESR experiment has unambiguously shown that the optical band at 2.3 ev arises from the V^- centers.[22] However, the shape of the dichroism peaks is not consistent with a P_π description of the excited state. Rather, it appears to demand molecular orbital excited states. This conclusion is reinforced by optical dichroism studies recently completed.[31]

Studies of the self-trapped hole in the alkali halides by means of polarized bleaching and a resulting dichroism have contributed greatly to the understanding of these defects.[32] Earlier attempts to detect dichroism in the alkaline-earth oxides after polarized bleaching were unsuccessful.[2] Recently, dichroism has been detected in MgO after the V^- centers were aligned in an electric field.[31] Increasing dichroism at the largest applied fields is indicative of the existence of large random electric fields due to other defects. Both F^+ centers and linear associated defects presumably are responsible (F^+ centers have one electron trapped at an anion vacancy). To insure a large number of isolated cation vacancies and V^- centers, the MgO was neutron irradiated. This necessarily produces F^+ centers. These stray internal fields help to restrict the hole to a specific site by preventing tunneling at low temperature. The wavelength dependence of the dichroism gives evidence that the 2.3 ev absorption band obscures transitions to widely different excited states, which presumably are molecular orbitals rather than the P_π orbitals expected from simple theory. This conclusion has also been reached in the magnetic circular dichrosim studies which firmly linked the 2.3 ev band to the V^- center.[22]

3 ANION VACANCIES

There have been a large number of studies of negative ion vacancies, singly and in association with other defects in the alkaline earth oxides since first the F^+ center was detected subsequent to neutron irradiation.[33] (Here, one electron trapped at an anion vacancy has an effective charge of +1; the charge state is indicated in the notation for the center.[20]) This same defect, as well as the two-electron F^0 center, may be produced in each of the alkaline-earth oxides by heating in the corresponding metal vapor. The oxide is maintained at about 1800°C, while the metal in a separate furnace is kept at a temperature such that the vapor pressure will correspond to the desired level of doping. Professor Hensley and his co-workers have primarily been responsible for this work.[34]

Little will be said about F^+ or F^0 centers here. However, the center that is formed subsequently upon heating is of interest. The isotropic F^+-center ESR line diminishes in intensity, and in MgO an asymmetric line grows. This has been ascribed to a vacancy pair in which one electron is trapped in the anion vacancy.[35] Such centers have been observed for MgO, CaO, and SrO at 35 GHz, at which frequency the components $g_\parallel$ and $g_\perp$ of these centers can be resolved.[2] The vacancy pairs appear to be stable at 900°C, at which temperature all the isolated anion vacancies have disappeared, for subsequent X irradiation at room temperatures does not give rise to any F^+ centers.

If the oxide samples are not cooled during neutron irradiation in a reactor, there will be numerous defects formed with various numbers of aggregated vacancies. Some of these are akin to the M, N, and R aggregate centers in the alkali halides. A number of the defects show very sharp zero-phonon lines and broader phonon-assisted peaks. The former make it possible to perform stress splitting experiments which allow the symmetry and the orientational degeneracy of the aggregate defects to be determined.[2,3]

4 RÉSUMÉ

This brief review has indicated that whereas one lacks direct means for making observations on vacancies in the alkaline-earth oxides, one may learn much about vacancy behavior by examining the characteristics of impurities, trapped holes, or trapped electrons. In some cases, two of these entities may be found in association with a vacancy. While most of the results have been obtained by using the ESR technique to study systems which are paramagnetic, other techniques are valuable supplements.

REFERENCES

1. Wertz, J. E., *Mass Transfer in Oxides, N.B.S. Spec. Pub.* 296, 11 (1968).
2. Henderson, B., and Wertz, J. E., *Advan. Phys.*, **17**, 749 (1968).
3. Hughes, A. E., and Henderson, B., in *Point Defects in Solids,* J. H. Crawford and L. Slifkin (Eds.), Plenum Press, New York (1972), Chap. 7.
4. Abraham, M. M., Butler, C. T., and Chen, Y., *J. Chem. Phys.*, **55**, 3752 (1971).
5. Kirklin, P. W., Auzins, P., and Wertz, J. E., *J. Phys. Chem. Solids*, **26**, 1067 (1965).
6. Wertz, J. E., and Auzins, P., *Phys. Rev.*, **106**, 484 (1957).
7. Griffiths, J.H.E., and Orton, J. W., *Proc. Phys. Soc. (London)*, **73**, 948 (1959).
8. Davies, J. J., and Wertz, J. E., *J. Mag. Res.*, **1**, 500 (1969); *J. Phys. Chem. Solids*, **31**, 2489 (1970).
9. Schawlow, A. L., Wood, D. L., and Clogston, A. M., *Phys. Rev. Lett.*, **3**, 544 (1959).
10. Schawlow, A. L., Piksis, A., and Sugano, S., *Phys. Rev.*, **122**, 1469 (1961).
11. Schawlow, A. L., *J. Appl. Phys. Suppl.*, **33**, 395 (1962).
12. Imbusch, G. F., Schawlow, A. L., May, A. D., and Sugano, S., *Phys. Rev.*, **140A**, 830 (1965).
13. Larkin, J. P., Imbusch, G. F., and Dravnieks, F., *Phys. Rev.*, **B7**, 495 (1973).
14. Sturge, M. D., *Phys. Rev.*, **130**, 639 (1963).
15. DuVarney, R. C., and Garrison, A. K., *Solid State Comm.*, **12**, 1235 (1973).
16. Unruh, W. P., Chen, Y., and Abraham, M. M., *Phys. Rev. Lett.*, **30**, 446 (1973).
17. Schirmer, O. F., *J. Phys. Chem. Solids*, **29**, 1407 (1968); **32**, 449 (1971); Abraham, M. M., Chen, Y., Kolopus, J. L., and Tohver, H. T., *Phys. Rev.*, **B5**, 4945 (1972).
18. Seitz, F., *Rev. Mod. Phys.*, **26**, 7 (1954).
19. Wertz, J. E., Auzins, P., Griffiths, J.H.E., and Orton, J. W., *Discuss. Faraday Soc.*, **28**, 136 (1959).
20. Sonder, E., and Sibley, W. A., in *Point Defects in Solids,* J. H. Crawford and L. Slifkin (Eds.), Plenum Press, New York (1972), p. 266.
21. Halliburton, L. E., Kappers, L. A., Cowan, D. L., Dravnieks, F., and Wertz, J. E., *Phys. Rev. Lett.*, **30**, 607 (1973).
22. Izén, E. H., Mazo, R. M., and Kemp, J. C., *J. Phys. Chem. Solids,* to be published.
23. Tench, A. J., and Duck, M. J., to be published.
24. Kappers, L. A., Dravnieks, F., and Wertz, J. E., *J. Phys. C.,* to be published in J. Phys. C.
25. Searle, T. M., and Glass, A. M., *J. Phys. Chem. Solids*, **29**, 609 (1968).
26. Kappers, L. A., and Wertz, J. E., *Solid State Commun.*, **9**, 1755 (1971).
27. O'Mara, W. C., and Wertz, J. E., *J. Mag. Resonance*, **8**, 366 (1972).
28. Schoenberg, A., Suss, J. T., Szapiro, S., and Luz, Z., *Phys. Rev. Lett.*, **27**, 1641 (1971).
29. Wertz, J. E., Saville, G., Auzins, P., and Orton, J. W., *J. Phys. Soc. Japan, Suppl.*, **18**, 305 (1963).
30. Halliburton, L. E., Cowan, D. L., Blake, W.B.J., and Wertz, J. E., *Phys. Rev.*, **B8**, 1610 (1973).
31. Rose, B. H., and Cowan, D. L., in press.

32. Kabler, M. N., in *Point Defects in Solids,* J. H. Crawford and L. M. Slifkin (Eds.), Plenum Press, New York (1972), Chap. 6. Many references are given here.
33. Wertz, J. E., Auzins, P., Weeks, R. A., and Silsbee, R. H., *Phys. Rev.,* **107**, 1535 (1957).
34. Hensley, E. B., Ward, W. C., Johnson, B. P., and Kroes, R. L., *Phys. Rev.,* **175**, 1227 (1968); Johnson, B. P., and Hensley, E., *Phys. Rev.,* **180**, 931 (1969).
35. Wertz, J. E., Orton, J. W., and Auzins, P., *Discuss. Faraday Soc.,* **30**, 40 (1961).

DISCUSSION on Paper by J. W. Wertz

KROGER: Have you any idea why ESR is unaffected by a 3+ ion?

WERTZ: Both the difference in stability between

$$O^+ \boxed{=} O\ M^+ \quad \text{and} \quad O^+ \boxed{=}\ O$$

I II

and the fact that the hole may move around the vacancy in II but not in I indicate that the hole experiences the electric field due to M^+. The nonappearance of the isotropic ESR line for the V_{Al} center at 300°K demonstrates the electrostatic repulsion of the hole by the trivalent ion if the hole were to jump to one of the equatorial oxygen positions. The identity of the 77°K spectra demonstrates that the outward relaxation of the O^- ion from the vacancy is primarily responsible for the tetragonal component of the crystalline electric field. In view of the narrowness of the lines, it is surprising that the Al^{3+} ion does not give rise to a detectable contribution. This may be the result of accidental cancellation of an electric-field contribution by Al^{3+} and a smaller one arising from oxygen relaxation in the associated center.

VINK: With respect to the fact that you could not find any difference in the spectrum whether Al^{3+} was present or not, did you try any other trivalent or tetravalent ion, to find out whether in those cases a difference could be found?

WERTZ: The ESR spectra of the V^- and the V_{Al} centers at 77°K are indistinguishable in MgO; at 300°K they are very different. We have not closely examined the 77°K spectra of crystals with a large concentration of trivalent cations other than Al^{3+}. It would be of interest to study crystals with a significant concentration of Ti^{4+}.

GOODENOUGH: What is the experimental evidence that enables you to distinguish, at an isolated V^- center, a hole localized at a single anion from a hole in a three-electron bond across the center?

WERTZ: Recent experiments on the axial ^{25}Mg hyperfine line intensity in the ESR spectrum of the V⁻ center, i.e., for the geometry

$$^{25}\mathrm{Mg}\ \mathrm{O}^{+}\ [=]\ \mathrm{O}$$

show just the expected value for a hole localized on one oxygen. If the hole had been delocalized over the two oxygen atoms on the axis, the intensity would have been a factor of two larger. This hole localization refers to the *ground state* of the defect. In the excited state, the hole is doubtless delocalized and must be described in terms of molecular orbitals.

ILSCHNER: From your discussion, there appears to be a preference for the <100> orientation of defect associations as compared to <110>. Is this backed up by experiment, and, if so, could you comment on the possible reasons? <110> seems to be favored by the shorter distance, increasing electrostatic attraction.

WERTZ: For both Cr^{3+} and Fe^{3+}, the total intensity of the ESR lines from tetragonal vacancy-impurity defects is clearly greater than for those with orthorhombic symmetry. Calculations on the alkali halide defects made by Dr. Lidiard's group show that small changes in local conditions can make either the nnn or the nn vacancy-impurity pair the more stable. In the alkaline-earth oxides, the former is the one which predominates and which therefore is presumably the more stable.

FRANKLIN: Have there been any independent experiments, e.g., dielectric relaxation, to observe the reorientation of the hole around the cation vacancy in the simple V^- center?

WERTZ: Some years ago a group at the Illinois Institute of Technology sought unsuccessfully to observe the relaxation of the Cr^{3+}-nnn vacancy defect. As far as I am aware, there has been no published work on dielectric relaxation of the V^- center. However, it has been only recently (early 1973) that we showed the reorientation in the V^- center by observation of the isotropic ESR line. Dielectric relaxation in both these types of defects ought to be observable.

LIDIARD: Can you, from your experiments involving impurity – cation vacancy pairs, give some idea of the activation energy for cation-vacancy motion? The reason I ask the question is the implication of our interpretation of the magnesium self-diffusion in MgO that the cation-vacancy activation energy is as high as 2.5 ev.

WERTZ: Experiments on both impurities and trapped hole centers indicate a significant cation-vacancy motion at 350 to 400°C. This mobility is consistent with a cation-vacancy activation energy of 2 to 3 ev.

DEFECT CHARACTERIZATION BY INDIRECT EXPERIMENTAL TECHNIQUES

N. M. Tallan

Aerospace Research Laboratories
Wright-Patterson Air Force Base, Ohio

ABSTRACT

The use of measurements of the oxygen pressure dependence of the defect concentrations in metal oxides to determine the nature of the predominant point defects present is reviewed. Difficulties arising from defect-structure transitions are discussed and an analysis of such data is given for the case of NiO.

1 THE OXYGEN PRESSURE DEPENDENCE OF DEFECT CONCENTRATIONS

To determine whether the predominant point defects in a metal oxide are vacancies or interstitials, whether they are on the cation or anion sublattice, and whether they are fully or partially ionized, one may, in principle, measure as a function of oxygen partial pressure and temperature some property of the material that is directly proportional to one or more separable defect concentrations. This approach is based upon the fact that if the defect-formation process is treated as a quasi-chemical reaction, then under certain limiting assumptions, the defect concentrations can be related directly to temperature and oxygen pressure through mass-action equations.[1] This leads to expressions for the defect concentrations of the form $[D] = KP(O_2)^n$, where K is exponentially temperature dependent and n will have a constant, rational fraction value over any range of temperature and pressure in which one pair of charge-balancing defects clearly dominates over all others.

Unfortunately, the number of cases where this simple, rational fraction value of n is found is rather small. Frequently, one finds instead that n varies continuously as a function of temperature and oxygen partial pressure because the defect structure under investigation is in a transition state between two extreme simple cases. In fact, it is often found that the temperature and oxygen partial pressure ranges available experimentally are too narrow to permit clear observation of the individual processes contributing to the transition. In these cases the transition can be, and often has been, mistaken for an entirely different and incorrect simple defect structure. For example, one might in this way misinterpret a transition between singly and doubly ionized vacancies as evidence for the predominance of interstitials in one particular charge state, and undoubtedly many related examples have been reported in the literature.

A number of these transitions, arising from the competition between various defect species and effects, have been treated analytically. Kevane[2], for example, treated the case of electronic conduction arising from the presence of both singly and doubly ionized oxygen vacancies. Blumenthal and co-workers[3] reported extensive studies of the conductivity of ceria, and analyzed their results using a computer and the assumption of a transition between mixed-valence defects. Transitions can also arise between defect concentrations fixed by an allovalent impurity content or an intrinsic defect-formation process (Schottky or Frenkel disorder) and pressure-dependent defect concentrations due either to these same processes or to departures from stoichiometry. For example, Lasker and Rapp[4] and Bransky and Tallan[5] have treated the case of the transition from pressure-independent ionic conduction to pressure-dependent electronic conduction in ThO_2, and Kofstad[6] has treated a number of transitions between intrinsic and impurity-dominated behavior. One may also clearly have transitions between isolated point defects and associated defects at higher concentrations. The pressure dependence of the defect concentrations will also vary continuously at large departures from stoichiometry because of the failure of the assumptions involved in the use of simplified mass-action laws. In these cases, one must consider the activities of the defects and the available lattice sites, and also the possibility of "site blocking", as shown by Libowitz and Lightstone[7], by

Kofstad and Hed[8], and others.

Abrupt changes in the pressure dependence of the conductivity or some similar property have often been attributed to a transition in defect structure.[9-11] In fact, their abruptness would seem to preclude this explanation, since a common feature of all of the transitions described above is the gradual nature of the change in pressure dependence. Such transitions should be observed over many orders of magnitude of oxygen partial pressure. For example, in a binary metal oxide, MO, containing singly and doubly ionized metal vacancies, V_M' and V_M'', the electroneutrality condition describing the transition region is given by

$$p = [V_M'] + 2[V_M''] \tag{1a}$$

$$p = \frac{K' P(O_2)^{1/2}}{p} + \frac{2K'' P(O_2)^{1/2}}{p^2} \quad . \tag{1b}$$

The pressure dependence of the electron-hole concentration varies from 1/6 at low pressures, where doubly ionized vacancies predominate, to 1/4 at high oxygen pressures, where singly ionized vacancies predominate. The pressure dependence in the transition region is given by

$$\frac{d \log p}{d \log P(O_2)} = \frac{1}{4}\left[\frac{[V_M'] + 2[V_M'']}{[V_M'] + 3[V_M'']}\right] \quad . \tag{2}$$

If the transition region is considered to be bounded by the slopes 1/4.2 and 1/5.8 in a log p vs log $P(O_2)$ plot (the departure of these slopes from the simple 1/4 and 1/6 values should be experimentally observable), then the transition region would extend over 9-1/2 orders of magnitude in oxygen pressure. Over much of this range, the experimental results could easily be misinterpreted as a simple 1/5 pressure dependence. These difficulties in the interpretation of data in transition regions severely limit the usefulness of indirect techniques for defect-structure determination.

2 EXPERIMENTAL TECHNIQUES

Over the past several years, there have been both refinements of older experimental techniques for measuring properties proportional to the defect concentrations present in an oxide specimen and the appearance of several new techniques. The measurement of electrical conductivity, even in cases where the total conductivity must be carefully separated into electronic and ionic components by galvanic-cell or polarization experiments, is still the single most sensitive and reliable indirect technique available. Its use has been described at length in several recent reviews.[6,12]

Thermogravimetric measurements are extremely valuable because, in addition to being proportional to the defect concentration, they can be used to obtain the actual number of defects present if the correct defect structure model is known. Unfortunately, the need to correct or compensate for thermomolecular flow, gas flow, and buoyancy effects and currently available balance sensitivities limit their usefulness to departures from stoichiometry of at least 100 ppma. An analysis technique alleviating some of these difficulties by the use of only isobaric weight-change measurements has been reported by Osburn and Vest.[13]

Seebeck coefficient and diffusion measurements are rarely precise enough to permit accurate determinations of the oxygen-pressure dependence, but have on occasion proven useful for this purpose. For example, Le Brusq et al.[14] showed $\pm 1/6$ dependences for the electron and hole concentrations in MnO by thermal emf measurements, and Volpe and Reddy[15] utilized cation self-diffusion measurements to test defect structure models in NiO.

Several less frequently used techniques have also been reported. Tretyakov and Rapp[16], for example, used coulometric titration to determine the pressure dependence and absolute magnitude of the defect concentrations in NiO. Greskovich and Schmalzried[17] have reported a gas volumetric analysis that appears to be particularly promising. The technique is based on the measurement of pressure changes in the sample chamber accompanying the incorporation or liberation of oxygen by the sample as the oxygen pressure is varied. Using a differential micromembrane manometer[18,19] that transforms pressure differences into an electrical output, the membrane generally serving as one plate of a capacitor, defect-concentration changes at least one to two orders of magnitude smaller than those observable in thermogravimetric measurements can be followed. Another interesting technique that has been used effectively is thermoelectronic emission, which can be used to follow the electron concentration even at oxygen pressures where ionic defects or positively charged electron holes would dominate the conductivity. Loup and Anthony[11] applied this technique to ZrO_2.

3 APPLICATIONS

The transition-metal monoxides, NiO, MnO, CoO, and FeO are good examples of materials in which many of the transitions described earlier occur and are easily observed experimentally. Libowitz[20] has shown that conductivity data and oxygen activities calculated from galvanic-cell data for FeO cannot be treated by simple mass-action equations written in terms of defect concentrations; at the high defect concentrations involved, the pressure dependences of both isolated defects and defect complexes are continuously varying functions of the departure from stoichiometry. The existence of defect complexes at high defect concentrations in MnO and CoO have also been proposed, although in MnO there appears to be a sudden transformation between the isolated and associated defect regions rather than a gradual transition[21] and in CoO there is not enough of the associated defect region observable experimentally to follow the transition[22]. Although large concentrations of Frenkel defects in NiO, CoO, and their solid solutions have recently been proposed on the basis of X-ray measurements[23], these large concentrations have never been observed by the kind of indirect measurements discussed here. One finds instead, for example, strongly pressure-dependent diffusion measurements, more characteristic of nonstoichiometric disorder than of intrinsic thermal disorder (unless an argument involving immobile complexes is invoked). NiO, in fact, has been one of the most thoroughly studied transition-metal monoxides, and recent results provide a good example of the analysis of indirect measurements in terms of detailed defect models.

The presence of singly and doubly ionized cation vacancies in NiO at temperatures above 1000°C and oxygen partial pressures above about 10^{-5} atm has been suggested for many years in conflicting reports. Using conductivity, weight change, coulometric titration, and diffusion measurements, some investigators found a nearly pure 1/6 oxygen pressure dependence characteristic of doubly ionized nickel vacancies while others found a 1/4 dependence characteristic of singly ionized nickel vacancies. In an early attempt to reconcile the discrepancy, Meier and Rapp[24] proposed that doubly ionized vacancies predominate in NiO, that the 1/6 dependence is characteristic of sufficiently pure material, and that the 1/4 dependence frequently observed is due to the doubly ionized vacancy concentration being fixed by an allovalent impurity contaminant. Volpe and Reddy[15], on the other hand, interpreted their nickel self-diffusion and conductivity results on the basis of a pressure dependence intermediate between 1/6 and 1/4 and a defect model involving the presence of both singly and doubly ionized nickel vacancies in comparable concentrations. Recent results support this latter view.

Figure 1 shows the results of very careful and very detailed conductivity measurements on high-purity (99.999 percent pure starting powder) polycrystalline specimens.[25] The apparently linear plot, at least on the scale used, would at first suggest a simple defect model involving one positive and one negatively charged predominant defect. Since the observed pressure dependence is 1/5.33, a more complicated defect model must be examined.

If we assume the presence of singly and doubly ionized nickel

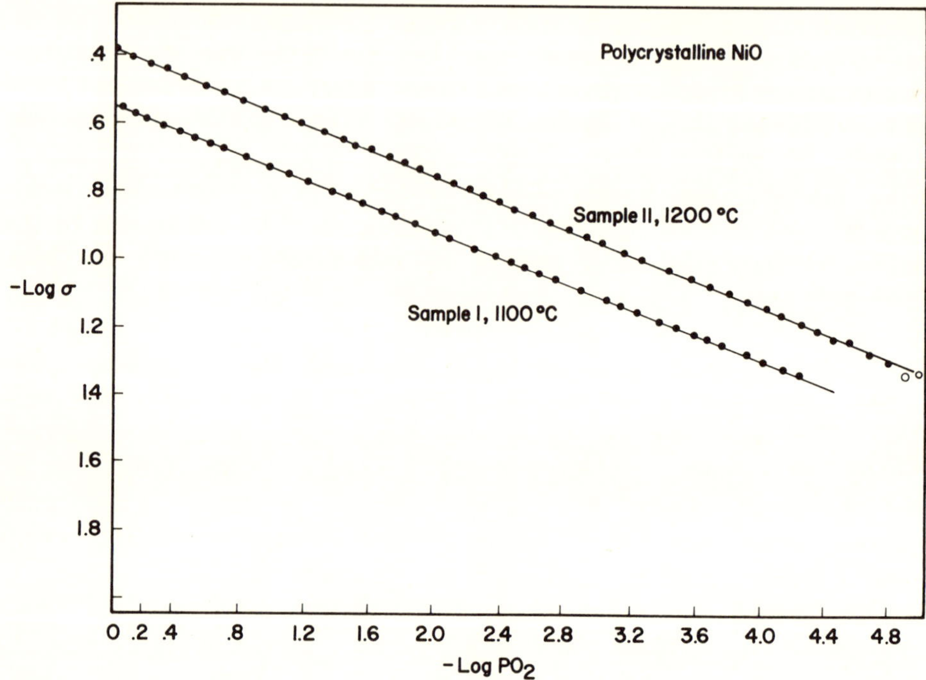

Fig. 1. Electrical conductivity of NiO as a function of oxygen partial pressure. Solid lines represent the fit of Eq. 6, using the constants shown in Table I, to the data.

vacancies, a pressure-independent concentration of singly ionized impurities, and electron holes, then the electroneutrality condition is

$$[V_M'] + 2\,[V_M''] = p + [F_M] \quad . \tag{3}$$

If the vacancy formation reactions are written

$$\tfrac{1}{2}O_2 \rightleftharpoons O_o^x + V_M' + p \tag{4a}$$

$$\tfrac{1}{2}O_2 \rightleftharpoons O_o^x + V_M'' + 2p \quad , \tag{4b}$$

then at low defect concentrations, the mass-action equations are

$$[V_M'] = \frac{K_{V'}\,P(O_2)^{1/2}}{p} \tag{5a}$$

$$[V_M''] = \frac{K_{V''}\,P(O_2)^{1/2}}{p^2} \quad . \tag{5b}$$

If the hole mobility is essentially independent of the oxygen pressure, then the conductance G is related to the hole concentration by a proportionality constant, $p = \alpha G$, and Eq. 3 can be written as

$$\left(\frac{\alpha K_V{'}}{2K_V{''}}\right) G - \left(\frac{\alpha^2 [F_M]}{2K_V{''}}\right) \frac{G^2}{P(O_2)^{1/2}} - \left(\frac{\alpha^3}{2K_V{''}}\right) \frac{G^3}{P(O_2)^{1/2}} + 1 = 0 \ . \qquad (6)$$

A computer-determined least-squares fit of the data of Fig. 1 to Eq. 6 was used to obtain the values of the constants shown in Table I. Separate specimens were used by Stroud et al.[25] to obtain the 1100°C and 1200°C data, so little can be inferred about the temperature dependence of the constants. It is interesting to note that the effective impurity content, calculated on the basis of this defect model, is only on the order of 4 to 14 ppm. This is certainly very much less than the total impurity content of the specimens, as determined by mass spectrographic analysis. This tendency for the effective impurity level to be much less than the total apparent excess allovalent impurity content was also noted by Tripp and Tallan[27] in thermogravimetric measurements.

Table I. Equilibrium Constants for NiO

T, °C	[F] (a), ppma	$K_V{'}$(a), atm$^{-1/2}$	$K_V{''}$(a), atm$^{-1/2}$	$K_V{'}$(b), atm$^{-1/2}$	$K_V{''}$(b), atm$^{-1/2}$
1100	3.71	2.38×10^{-8}	2.81×10^{-12}	7.03×10^{-8}	6.16×10^{-12}
1200	14.1	5.04×10^{-8}	1.37×10^{-11}	2.04×10^{-7}	2.48×10^{-11}

(a) Values reported by Stroud, Bransky, and Tallan[25].
(b) Equilibrium constants calculated from the values of K_{ox}, K_1, and K_2 given by Osburn and Vest[26].

The equilibrium constants obtained by fitting the data of Fig. 1 to Eq. 6 are also compared (in Table I) to values based on the constants calculated by Osburn and Vest.[26] Osburn and Vest used their calculated constants and a five-term form of the electroneutrality condition to generate theoretical conductivity curves, particularly useful at temperatures below the point at which the NiO sample ceases to equilibrate with the atmosphere. The agreement they obtained with their experimental results was striking. The agreement between the results of Stroud et al.[25] and Osburn and Vest[26] shown in Table I is rather good, considering the differences in specimens and approaches to the determination of the constants.

It is interesting to note that, with either set of constants, the concentrations of singly and doubly ionized nickel vacancies are quite comparable over most of the experimentally observed ranges of temperatures and oxygen partial pressures. For example, using the equilibrium constants in Table I obtained by Stroud et al.[25], the singly to doubly ionized metal vacancy-concentration ratio in pure NiO at 1200°C would be 0.25 at 10^{-4} atm oxygen and 1.3 at 1 atm oxygen pressure. Inserting these concentration ratios in Eq. 2, it is clear that the pressure dependence of the hole concentration in pure NiO ought to vary from 1/5.8 to 1/5.2 over the experimental pressure range 10^{-4} to 1 atm oxygen. Clearly it should not be possible to see *either* a 1/4 or a 1/6 dependence at 1 atmosphere oxygen in a pure sample. The effects of impurities, however, should not be overlooked. With an effective trivalent impurity content of 14 ppm, the pressure dependence at 1200°C would vary from 1/5.4 to 1/5.1 over this pressure range, as observed by Stroud et al.[25] With an effective monovalent impurity content of 100 ppm, the pressure dependence at 1 atm oxygen would be 1/5.8, but the impurities present in this case would nearly fix the hole concentration at lower oxygen pressures, and a dependence of 1/14.6 would be observed at 10^{-4} atm oxygen.

The intentional addition of allovalent impurities has frequently been used to at least qualitatively distinguish between alternative defect models. O'Keeffe and Valigi[28] have recently applied a detailed analysis of intentional-doping experiments to determine the equilibrium concentrations of the defects in MnO. They showed that in cases where an electronic conductivity minimum is observed, determination of the oxygen activity at the minimum as a function of dopant concentration permits the calculation of the defect concentrations and departure from stoichiometry of the pure compound. A particularly complete analysis is possible if the ratio of electron to hole mobility is also known. Where this technique can be applied, it should be a particularly valuable adjunct to thermogravimetric or coulometric-titration measurements.

Indirect defect-structure studies have made many contributions to our understanding of some binary oxides. There are many others, however, in which competing defect concentrations, alternative defect models, and limited ranges of the experimental variables have combined to prevent any conclusive determination of the defects present. In the case of ZrO_2, HfO_2, and ThO_2, for example, it is still not really clear whether the pure materials exhibit any degree of oxygen excess below 1 atm of oxygen pressure and whether intrinsic thermal disorder in these materials is Schottky or Frenkel in nature. Clarification of these cases may well require more direct studies, perhaps by precise X-ray diffraction analysis of the occupancy of interstitial sites[23] or by electron-spin-resonance techniques. Certainly, further studies of these materials based on the interpretation of the oxygen-pressure dependence of the defect concentrations will have to apply more sophisticated analyses of the results to separate the defect contributions and to test alternative models than those generally used in the past.

REFERENCES

1. Kroger, F. A., *Chemistry of Imperfect Crystals,* John Wiley & Sons, Inc., New York (1964), pp. 245-57.
2. Kevane, C. J., *Phys. Rev.,* **133**, 1431 (1964).
3. Blumenthal, R. N., Lee, P. W., and Panlener, R. J., *J. Electrochem. Soc.,* **118**, 123 (1971).
4. Lasker, M. K., and Rapp, R. A., *Z. Physik. Chem. (Frankfurt),* **49**, 198 (1966).
5. Bransky, I., and Tallan, N. M., *J. Amer. Ceram. Soc.,* **53**, 625 (1970).
6. Kofstad, P., *Nonstoichiometry, Diffusion, and Electrical Conductivity in Binary Metal Oxides,* Wiley-Interscience, New York (1972), pp. 38-45.
7. Libowitz, G. G., and Lightstone, J. B., *J. Phys. Chem. Solids,* **28**, 1145 (1967).
8. Kofstad, P. K., and Hed, A. Z., *J. Electrochem. Soc.,* **115**, 102 (1968).
9. Fisher, B., and Tannhauser, D. S., *J. Chem. Phys.,* **44**, 1663 (1966).
10. Hed, A. Z., and Tannhauser, D. S., *J. Chem. Phys.,* **47**, 2090 (1967).
11. Loup, J. P., and Anthony, A. M., *Compt. Rend. Acad. Sci. Paris,* **268**, 772 (1969).
12. Tallan, N. M. (Ed.), *Electrical Conductivity in Ceramics and Glass,* Marcel Dekker, Inc., New York, in press.
13. Osburn, C. M., and Vest, R. W., *J. Phys. Chem. Solids,* **32**, 1331 (1971).
14. Le Brusq, H., Oehlig, J. J., and Marion, F., *Compt. Rend. Acad. Sci. Paris,* **266**, 965 (1968).
15. Volpe, M. L., and Reddy, J., *J. Chem. Phys.,* **53**, 1117 (1970).
16. Tretyakov, Y. D., and Rapp, R. A., *Trans. AIME,* **245**, 1235 (1969).
17. Greskovich, C., and Schmalzried, H., *J. Phys. Chem. Solids,* **31**, 639 (1970).
18. Baumbach, H. H., and Wagner, C., *Z. Phys. Chem.,* **B24**, 59 (1934).
19. Meurer, H., Thesis, Clausthal University, 1969.
20. Libowitz, G. G., "Mass Transport in Oxides", *N.B.S. Special Publication* 296, pp. 109-18 (1968).
21. Bransky, I., and Tallan, N. M., *J. Electrochem. Soc.,* **118**, 788 (1971).
22. Bransky, I., and Wimmer, J. M., *J. Phys. Chem. Solids,* **33**, 801 (1972).
23. Stiglich, J. J., Whitmore, D. H., and Cohen, J. B., *J. Am. Ceram. Soc.,* **56**, 211 (1973).
24. Meier, G. H., and Rapp. R. A., *Z. Phys. Chem.,* **B74**, 168 (1971).
25. Stroud, J. E., Bransky, I., and Tallan, N. M., *J. Chem. Phys.,* **58**, 1263 (1973).
26. Osburn, C. M., and Vest, R. W., *J. Phys. Chem. Solids,* **32**, 1343 (1971).
27. Tripp, W. C., and Tallan, N. M., *J. Am. Ceram. Soc.,* **53**, 531 (1970).
28. O'Keeffe, M., and Valigi, M., *J. Phys. Chem. Solids,* **31**, 947 (1970).

DISCUSSION on Paper by N. M. Tallan

TANNHAUSER: In her thesis, B. Fisher fitted her elementary results for CoO to a curve calculated for the transition region. The fit was good, indicating that her data actually pertained to a transition from singly to doubly ionized vacancies.

WAGNER: What concentration of interstitial nickel ions would be necessary to alter appreciably your very good fit of the experimental data to the model involving singly and doubly ionized cation vacancies in nickel oxide?

TALLAN: At the temperature of our measurement, the total vacancy concentration varies from about 10 ppm at low oxygen pressures to about 100 ppm at high oxygen pressures. The free-nickel interstitial content would have to be within an order of magnitude of these concentrations to alter the analysis I have given here.

RAPP: A potentially very important transition in the defect state upon changes in pO_2, especially when one is concerned about transference numbers in solid electrolytes, is the change in valence of one or more impurities which enter the electrical-neutrality condition. Have you been able to make a quantitative evaluation for the change in valence (or the change in defect type) with change in pO_2 for an impurity in any given compound?

TALLAN: In the case I have described here for NiO, the concentration of effective impurities is considerably smaller than the hole concentration and any change in valence of the impurity would be a second-order effect. We do get slightly different values for the equilibrium constants if we leave out the impurity concentration entirely, but it is not a very large effect.

I have not specifically analyzed a transition involving the valence change of an impurity, but any such transition dominating the neutrality condition should have a reasonable width and should be easily observable experimentally.

LIBOWITZ: If the uncertainty in the NiO data is large enough to initially cause one to draw a straight line with 1/5.3 slope, is it valid to do a three-parameter analysis of the data and attach any significance to the values of the constants ($K_{V_M}{}'$ $K_{V_M}{}''$ K_{F_M}) so obtained?

TALLAN: There is not much uncertainty in the NiO conductivity data. The data are really very good. The curvature, on the scale plotted, is very small and one can plot a straight line through the points which is, visually, a good fit to the data. The least-squares fit to the equation given shows that in fact the data fit the theoretical curvature very well.

KROGER: It should be remembered that the Brouwer approximation was proposed to help us see how to relate effects to be expected under various conditions. It was never intended to be used for quantitative analysis of data.

Our own experience is that an analysis based on the results from a variety of different experiments is much better than analysis from one type of experiment. In Dr. Tallan's example, for instance, the values of the constants K' and K'' found could be compared to what is to be expected on the basis of low-temperature values of ionization constants. In work on CdS and CdTe, we have been successful in combining

Hall-effect studies with tracer self-diffusion studies – all as function of partial pressure, temperature, and doping. In work on Al_2O_3-Co, we have used conductivity combined with quantitative spin-resonance, optical absorption, and electromotive-force measurements.

TALLAN: I did not mean to imply that the Brouwer approximation is anything less than an extremely valuable means of simplifying the mapping of the defect concentrations and, in fact, telling us where the transitions I have described are centered. Too often, however, the more complete neutrality condition describing the transition region itself has not been used to assess the width of the transition in oxygen partial pressure.

In addition to the conductivity measurements described here, we have done thermogravimetric and Seebeck-coefficient measurements on NiO. I agree that the more correlated measurements we can make, the better our analysis of the data is likely to be.

HAGEMARK: It seems to be a common phenomenon that the electrically active impurity concentration is considerably less than the total impurity concentration. Have you speculated on how the impurities are tied up in NiO so they become electrically inactive?

TALLAN: Even in the relatively pure NiO we have used, chemical analysis indicates that a variety of both lower and higher valent impurities are present. We don't know where these levels really lie and what the effective net impurity concentration ought to be. We are also concerned by the fact that these analyses are reported as a volume average, although the impurities may be distributed very nonhomogeneously.

FRANKLIN: The situation described for hole conduction in NiO as a function of pO_2 is analogous to that for ionic conduction in the alkali halides. In both cases the phenomena observed are the result of the sum of a number of contributions. In the alkali halides, early attempts to fit the log σ versus T^{-1} curve by computer to evaluate the different contributions proved to be unreliable. It was found instead that the different contributions had to be isolated by doping in such a way that a given contribution would dominate the behavior, and a measurement of σ or of a diffusion coefficient would give a reliable estimate of this contribution. These then could be reassembled to give the overall expression for the conductivity in the more general, pure case and the whole checked against the measured data. Perhaps an equivalent technique would be of value for, e.g., NiO.

TALLAN: The doping experiments you describe are indeed a very valuable way of separating contributions to a transport process, either conductivity or diffusion. For example, lithium-doping in NiO has been very valuable for assessments of the concentration and mobility contributions to the activation energy for conduction in NiO. In the work I have described here, however, we are looking at the isothermal, oxygen-pressure dependence of the conductivity to evaluate the equilibrium constants for defect formation. I do not think, in this case, we could get the defect concentrations for pure NiO from doping experiments.

AGENDA DISCUSSION: THEORY AND CHARACTERIZATION OF DEFECTS AND TRANSPORT IN OXIDES

*LeRoy Eyring**

Department of Chemistry
Arizona State University
Tempe, Arizona

*Alan L. Dragoo***

U.S. Bureau of Standards
Washington, D. C.

Three short presentations began the agenda session. Each was followed by a discussion expected to range over related topics covered in the introductory lectures and those of Session II. A more general discussion then ensued.

The discussion as it appears later is, in all cases, an edited paraphrase of the recorded proceedings. The chairman assumed responsibility for editing the transcription won from the tape with great labor and skill by the secretary. The prepared contributions appear as submitted by the respective contributors.

*Chairman

**Secretary

Rhéal Turcotte had agreed to describe his adventures in calculating lattice energy changes for various models of nonstoichiometric UO_{2+x}.

1 LATTICE ENERGY AND DEFECT STRUCTURES IN NONSTOICHIOMETRIC UO_{2+x}*

1.1 Introduction

Since its introduction[1] in 1919, the Born model of ionic solids has been utilized to correlate chemical and physical properties of a wide variety of substances. The early contributions (in the period 1920-1940) by Born and Mayer[2], Huggins[3-5], Pauling[6], and others were concerned primarily with the simplest and most nearly ionic crystals (alkali halides) and established a basis from which applications to more difficult problems could be examined. That extension has been relatively slow because of difficulties relating to the necessity of incorporating the interaction of many ions in order to approach the correct limiting electrostatic sum. This is no longer true because of the availability of high-speed computers, and the original concepts are now undergoing a renewed application and further extension.

Considering the many successful applications of the simple ionic model and the need for even a qualitative correlation between defect structures and energy in nonstoichiometric compounds, an investigation of the UO_{2+x} system has been undertaken. The primary objective was to investigate possible correlation of the local defect structure to a minimum lattice energy.

1.2 Method of Calculation

The crystal lattice energy (U_L) is defined as the sum of electrostatic interactions (Madelung energy, U_m), Born repulsive energy (E_r) and lesser polarization, van der Waals, etc., corrections. The complex nature of the compounds of interest and large uncertainty in virtually all corrective terms have necessitated consideration of only the Madelung and Born repulsion terms in the present work.

Two computer codes capable of the required summations have been obtained and modified appropriately. One, based on Evjens method[7], was supplied by Boeyens and Gaffner[8]. The second, which uses a counting scheme related to Evjen's but is based on the Ewald method[9], was developed by Van Gool and Piken[10,11]. Both of these methods have been well covered by Tosi[12] and will not be discussed in depth here. Although the Evjen code is not capable of computing noncentrosymmetric struc-

*This paper is based on work supported by Battelle Memorial Institute internal funding.

tures, it is accurate to any desired level, depending on the number of shells included, and is a useful check on the Ewald-based program which is used most extensively. The two methods agree with literature values to within ±0.3 kcal/mole for the test cases considered, which include CaF_2, CaCl, $AlCl_3$, UO_2, and U_4O_9.

The repulsive contribution is calculated in the present work as by Huggins and Mayer[3]. The evaluation is made directly as the electrostatic sum is computed with a provision to limit the inclusion of long-range repulsions since repulsions due to ions beyond r > 7 Å make insignificant contributions. The lattice energies were therefore obtained by the equation

$$U_L = U_m + \sum_{i,j} c_{ij}\, b \exp\left[(r_i + r_j - r_{ij})/\rho_{ij}\right] \quad ,$$

where c_{ij} are the Pauling coefficients[5], b is the energy constant, r_i are the so-called basic radii, r_{ij} is the ionic separation, and ρ_{ij} is the hardness parameter for the ion pair considered.

Several problems arise in applying the repulsive corrections to any materials not part of a thoroughly studied family. For the actinide elements, for example, there is an unresolved question as to how to treat the inner-shell f electrons. The true electronic configurations are not known in the solid, and several possibilities exist. For UO_2, the 7s shell may or may not contain electrons. Possible sets of Pauling coefficients for $C_{\pm}$, $C_{\ddagger}$, $C_{=}$ are 2.75, 5, 0.5 and 1.25, 2.0, 0.5. These are based on the two configurations suggested by Benson et al.[13] for U^{+4} and others are possible. There are obviously different possibilities for U^{+5}. For the fluorite structure where $r_{o^{-2}}$ = 1.36 Å and $r_{U^{+4}}$ = 1.00 Å, the metal-metal repulsions are very small and the net effect of the Pauling coefficients is to fix the relative importance of M-O and O-O repulsions.

In the usual treatments, having fixed the Pauling coefficients and the energy parameber b, a knowledge of unit cell dimensions for a family of salts (alkali halides for example) and the compressibility allows determination of a self-consistent set of radii and ρ through the equation

$$r(dU_L/dr) = (3T/N\beta)\,(\partial V/\partial T)_P \quad ,$$

where β is the compressibility. Unfortunately, the necessary experimental data are not available for the actinide dioxides (which could be a suitable family), and considerable uncertainty exists even for UO_2 which is perhaps the most studied of all solid oxides. For this reason, the calculation of cohesive and surface energies of UO_2 and ThO_2 by Benson et al.[13] was carried out over a wide range of conditions related to uncertainty in the compressibility factor. The energy parameter in this work was given a value of 2×10^{-12} ergs which falls within the possible range suggested by Benson as between 0.6×10^{-12} and 2.7×10^{-12} ergs.

As the interest in the present study was comparative rather than absolute, the more serious of the above-mentioned problems lies in setting relative importance of the repulsive terms (the Pauling coefficients) rather than their absolute magnitude. We have therefore varied these constants appropriately, as pointed out later. The hardness parameter ρ was fixed at ρ = 0.30. Geometric radii were used assuming O-O contact in UO_2 and U_4O_9, with a_o = 5.4700 for UO_2 and 5.4410 for U_4O_9. These give radii (Å) - U^{+4} = 1.001, U^{+5} = 0.001, and O^{-2} = 1.368 (UO_2) or O^{-2} = 1.360 (U_4O_9).

1.3 Results and Discussion

The fluorite lattice and defect models proposed by Willis[14] from neutron diffraction experiments are shown in Fig. 1.1. It is well known that uranium dioxide can take up additional oxygen to a ratio O/U = 2.25 with retention of the basic cubic symmetry. Only very weak superstructure lines appear in the X-ray patterns, a finding long taken as evidence that the metal positions are unchanged. This relative immobility of the metal-ion lattice (also found by Willis) is observed in virtually all of the fluorite-related systems. Cation relaxations out of cubic sites therefore have not been considered in the present work.

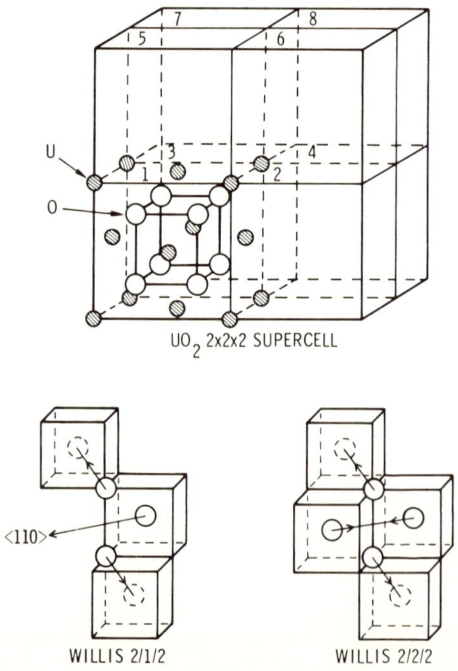

Fig. 1.1 UO_2 supercell and Willis defect models.

Considering the oxidation of $UO_{2.0}$, a simple model would be for additional oxygen to occupy the body-center position in the fluorite cell with relaxation of neighboring oxygen ions in the <111> directions. The results of calculations for this model, with U^{+5} placed at the (0.5, 0, 0.5) and (0, 0.5, 0.5) positions, are compared with relaxation in the parent UO_2 structure (see Fig. 1.2) The minimum in lattice energy at u = 0.25 for the dioxide, shifts to u = 0.232 for U_4O_9, an entirely reasonable result from bond-length considerations.

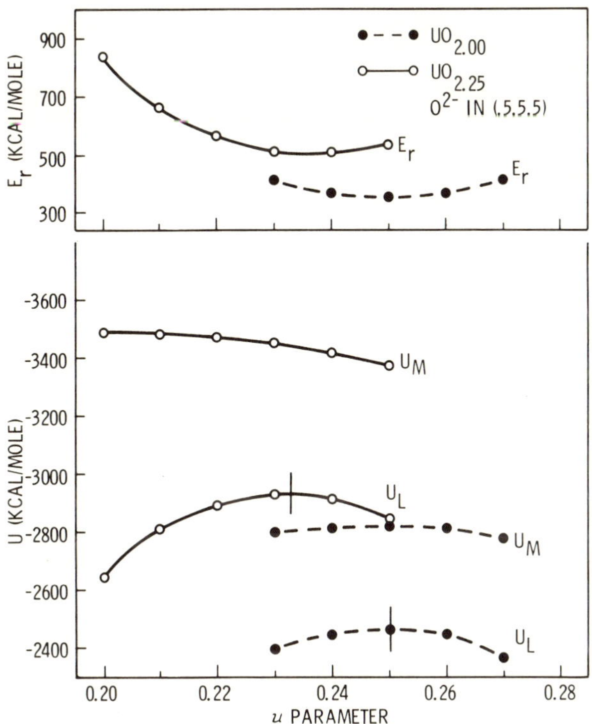

Fig. 1.2. Energy changes in UO_2 and body-center U_4O_9 as a function of u in (u, u, u).

Turning now to comparison of this simple defect model with defect clusters of lower symmetry, we show (in Fig. 1.3) the energy changes associated with a regular structural transformation to the Willis 2/1/2 model. The displacement involves motion of the body-center interstitial in a <110> direction and a coordinated shift of two of the fluorite oxygens in <111> directions. The starting structure at u = 0.5 is the relaxed interstitial model, and for this set of calculations with the Pauling coefficients in the ratio $C_\pm/C_= = 1.3/1$, the simple model is favored by 30 kcal/mole over the 2/1/2 model. The same calculation but with the Pauling coefficients in the ratio $C_\pm/C_= = 5/1$ gives even lower relative stability for the Willis model, as shown in Fig. 1.4. As would be expected, coefficients in the ratio $C_\pm/C_= = 1/5$ impart substantial apparent stability

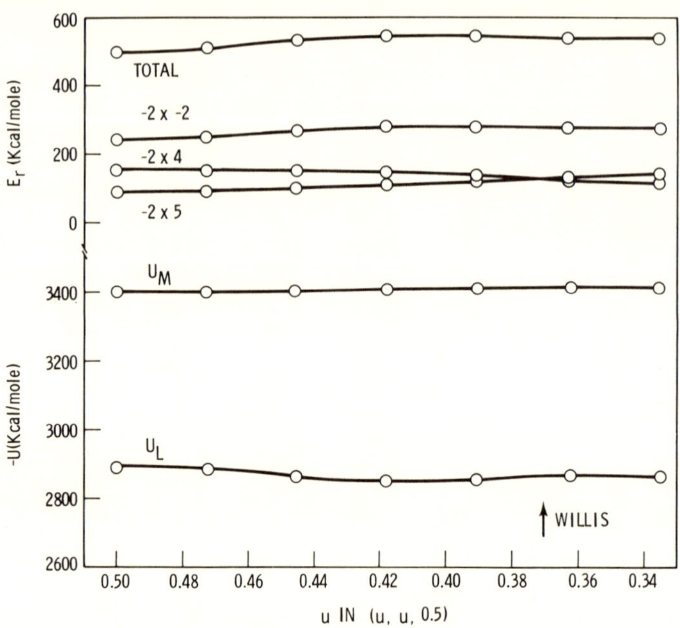

Fig. 1.3. Energy changes with formation of Willis 2/1/2 model and $C_{\pm}/C_{=} = 1.3$.

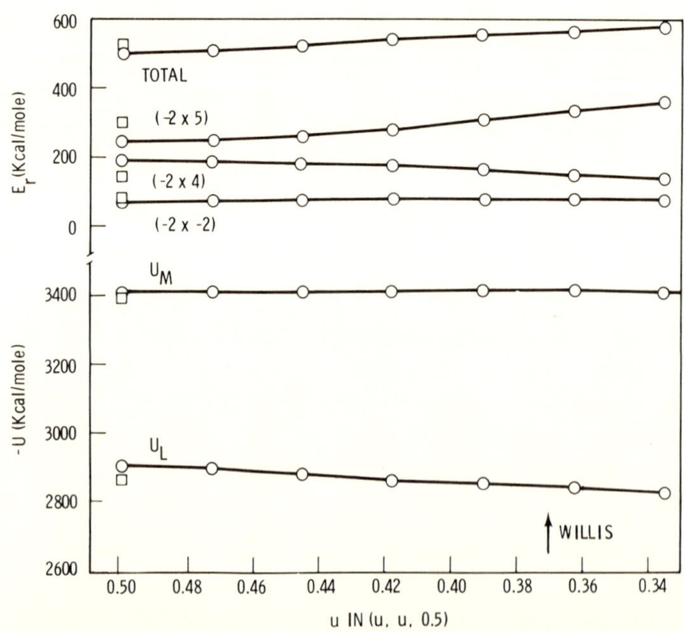

Fig. 1.4. Energy changes with formation of Willis 2/1/2 model and $C_{\pm}/C_{=} = 5$.

to the 2/1/2 defect. The results given in Fig. 1.3, however, do represent midrange values, and show only a metastable stability for the complex defect, since intermediate u values (0.42) give higher lattice energies than either the interstitial or 2/1/2 models.

These results suggest clearly that energy differences between related defect structures are very small, with a subsequent high demand for precise computational modeling. Although reasonable lattice relaxation appears to coincide with lattice-energy minima for simple defects, any quantitative structural understanding of complex clusters does not appear likely, within the framework of the Born ionic model.

ACKNOWLEDGMENT

The author is grateful to Drs. Boeyens and Gaffner and Drs. Van Gool and Piken for providing computer programs for calculating Madelung energies, and to Dr. T. D. Chikalla for helpful discussion.

REFERENCES

1. Born, M., *Verh. d. D. Phys. Ges.*, **21,** 13 (1919).
2. Born, M., and Mayer, J. E., *Z. Physik,* **75,** 1 (1932).
3. Huggins, M. L., and Mayer, J. E., *J. Chem. Phys.*, **1,** 643 (1933).
4. Huggins, M. L., *J. Chem. Phys.*, **5,** 143 (1937).
5. Huggins, M. L., *J. Chem. Phys.*, **15,** 212 (1947).
6. Pauling, L., *Z. Krist.*, **67,** 377 (1938).
7. Evjen, H. M., *Phys. Rev.*, **39,** 675 (1931).
8. Boeyens, J.C.A., and Gafner, G., *Acta Cryst.*, **A25,** 411 (1969).
9. Ewald, P. P., *Ann. Physik.*, **64,** 253 (1921).
10. Van Gool, W., and Piken, A. G., *J. Mater. Sci.*, **4,** 95 (1969).
11. *Ibid.*, p. 105.
12. Tosi, M. P., *Solid State Phys.*, **16,** 1 (1964).
13. Benson, G. C., Freeman, P. I., and Dempsey, E., *J. Am. Ceram. Soc.*, **46,** 43 (1963).
14. Willis, B.T.M., *IAEA Symposium,* Vienna (1964).

GROUP DISCUSSION

Lidiard pointed out that the absence of polarizability in the model would be a serious defect in looking at any unsymmetrical structures. Turcotte replied that if you could not handle the simple repulsive term then there would be no point in including the polarizability. He justified this response by pointing out the variability in the compressibility data which, even for UO_2, shows a very large spread. Lidiard pointed out that the elastic constants for UO_2 were now quite firm. Turcotte noted, however, that this is not true for U_4O_9 and continued by saying that distortions occurring at high concentrations are near the composition of U_4O_9 and not near UO_2 and that elastic constants were needed for these highly defective systems. Lidiard agreed that the Willis observations were for UO_{2+x} where the smallest x was 0.13 but that the calculations which Catlow carried out on the Catlow-Norgett model showed the unsymmetrical structure for the isolated complex of only one interstitial and two electron holes. Lidiard said further that although someone should do those experiments at lower degrees of nonstoichiometry he thought the Willis experiments and the calculations were now in complete agreement with respect to the 2/1/2 Willis structure. Turcotte asked Lidiard if he had attempted the calculations for the 2/2/2 structure. Lidiard replied that the 2/2/2 defect was not neutral and added that the calculations were attempted with one hole, a U^{+5} ion, and one interstitial and this gave surprisingly little binding. However, two holes will produce just the Willis structure.

Goodenough raised a question about the stability of the outer electron on an interstitial O^{-2} ion where the Madelung potential could no longer stabilize it, and how the charge distribution is handled when the electron is distributed over so wide an area. Lidiard replied that the Madelung potential at the center of a cube is repulsive to a negative charge and that Catlow's calculations considered carefully the relative stabilities of O^{-2} and the O^-, and found that O^{-2} was stable against the possible dissociation, namely, taking an electron off and putting it somewhere else. Goodenough then asked if there was a sizable radial distribution for the outer wave function. Lidiard emphasized that the calculation was within the framework of a classical model so that there is no redistribution except that described by polarizability terms, such as in the shell model and so on.

Franklin recalled that experiments had been done for the alkaline-earth fluorides at less than 1 mole percent of the added rare earth using electron spin resonance, and that the experiments had not shown the kinds of symmetries that you would expect based on the Willis model, although something seems to appear in the data at the higher concentration ranges. Kröger asked how one can get information about UO_2 by looking at calcium fluoride. Franklin replied that Norgett and Catlow had done the calculations for calcium fluoride and found structures that occur both in calcium fluoride and UO_2. On this point Lidiard remarked that Cheetham, Fender, and Cooper, carried out neutron scattering experiments down to 6 percent excess anion in calcium fluoride and that it may be possible to do UO_{2+x} at lower concentrations. Furthermore, W. Hayes at the Clarendon Laboratory, Oxford, looked with ESR at some of the specimens provided by Fender containing ~ 5 percent trivalent rare-earth ions but had been unable to observe any meaningful ESR spectrum. Thus there is still a gap between what has been seen with neutron diffraction and what is measurable by ESR, but it may nevertheless be possible to bridge this gap and observe the low symmetry of these clusters directly by ESR.

O'Keeffe critized the defect calculations in oxides. He noted that in perfect oxides, the oxygen ion always sees about the same potential and behaves in about the same way, but in the presence of a defect, the oxygen sees large differences in potential, large fields and large field gradients which are not fully taken into account in the calculations. There are theoretical calculations of how the polarizability of an oxygen ion would change, as the potential it sees changes, and this is a dramatic effect. In the limits where it sees no potential at all, this polarizability obviously is infinite as it ionizes. Also, if there are large fields and large field gradients, then other parameters such as the hyperpolarizability of the ion and the quadrupole polarizability should be included in the model. This will in effect introduce three-body forces. Goodenough offered the example of a color center. Lidiard replied that there are indeed these effects, but the important theoretical question is how big they are and secondly, the more important practical question is how far can you get with these relatively simple empirical models. He said that when you are faced with all these theoretical complexities it is necessary to adopt the strategy of seeing how far you can get with a simple model, and finally challenged O'Keeffe to provide him with a better potential function. O'Keeffe replied that one can feed in more information – there are theoretical calculations [e.g., E. Paschalis and A. Weiss, *Theor. Chim. Acta,* **13**, 381 (1969)] and experimental data which show, for example, how the polarizability might depend on the potential felt by the ion. Franklin responded to O'Keeffe by pointing out the classical idea that the polarizability of the oxygen changes from oxide to oxide comes from Tessman, Kahn and Shockley (TKS), who calculated the electronic polarizabilities first for the ions in alkali halides and then used these to obtain polarizabilities for ions in other compounds, eventually extracting the oxygen polarizabilities which they found to vary from oxide to oxide. However, several years ago, Boswarva showed that if you start with the alkaline earth chalcogenides rather than with the alkali halides and do exactly the same thing, you get just as nice a set of single ion polarizabilities for the alkaline earth chalcogenides as TKS got for the alkali halides. O'Keeffe replied that this comes about because you are always looking at oxygen in about the same potential in normal oxides. Lidiard stated that the shell model puts in the effects of other forces. O'Keeffe responded that the shell model posed another problem in that all its parameters are fitted to data for a perfect crystal and one wonders how relevant they are for oxygen in a very unusual environment, such as in the vicinity of a defect.

Lidiard said that you can wonder but he was simply trying to show that you can use a relatively simple model, that this is the first time that any such calculations had got anywhere. He continued by saying that such criticisms had been made for the last 20 years and that calculations based on polarizable point ion models (TKS or otherwise) had divergences and other gross failures, but this was the first time that successful calculations in UO_2 had been made. Lidiard concluded by saying that such criticism must ultimately be considered, but you don't take everything into account in the first step.

Professor Anderson agreed to present data he had which amplified some comments made earlier by John Cowley.

2 ELECTRON MICROSCOPY OF THE HEXAGONAL LAYER LATTICE BARIUM FERRITES

2.1 Some Experimental Observations

Cowley outlined some directions in which high-resolution lattice-imaging microscopy is being extended to structures less tractable and more densely packed than the niobium oxides. In this context, work done on the barium ferrites during the last few months in Oxford, by Dr. J.D.M. McConnell, is of some interest.

A very large number of closely related materials can be obtained when barium ferrite is made by flux growth in the presence of divalent transition-metal ions. All are based on the close packing of oxygen ions, in mixed cubic and hexagonal stacking, and the structural principles were worked out by X-ray diffraction and some ingenious electron microscopy of etch-pit structure by Kohn and Eckart.[1,2] In brief, all the structures can be resolved into subunits: a spinel block (S) of cubic stacking with the composition Fe_6O_8 or $Me_2Fe_4O_8$ (Me = divalent transiton metal), a three-layer hexagonal block (R), $BaFe_6O_{11}$, and a four-layer hexagonal block (T), $Ba_2Fe_8O_{14}$. These fit in perfect coherence between the (0001) plane of the hexagonal subunits and the (111) plane of the spinel subunit and combine so as to maintain electroneutrality. The simplest structures are those of the combination RS, the magnetoplumbite structure $BaFe_{12}O_{19}$ (Braun's phase M)[3], and TS, $Ba_2Me_2Fe_{12}O_{22}$ (Braun's phase Y).

These are the end members of a numerous series of intergrowth compounds M_nY_m, of which many representatives may be formed simultaneously in crystal-growth experiments.[4] The members of this series identified by diffraction methods can be formulated as

M_2Y_m with m = 1, 2, 3 . . . . 13, 15, 18, 21

M_4Y_m m = 5, 8, 9, 10, 13, 15, 33.

M_6Y_{14}

M_8Y_{27}.

Since the dimensions of one subunit along the c-axis direction are 11.6 Å for M and 14.5 Å for Y, it is evident that the c-axes of the hexagonal unit cells – and even more so for the rhombohedral polymorphs – are extremely long. M_4Y_{33} and M_8Y_{27} have c-axes in excess of 1500 Å.

Several questions suitable for attack by electron microscopy are apparent: (1) the preferred stacking sequences, out of the many possible sequences provided by the more complex formulas; (2) the regularity of structures with long repeat distances – stacking errors with or without changes in composition, the size of well-ordered domains, changes in stacking sequence within one macroscopic crystal; and (3) the existence of other intergrowth compounds not recognized by X-ray diffraction.

These questions can be tackled by one-dimensional lattice fringe imaging, but many-beam images from thin crystals in the right orientation showed two-dimensional resolution (Fig. 2.1). A model of the structures (Fig. 2.2) shows that there are tunnels, approximately 2.5 x 4.5 Å in size and lying obliquely to the c-axis, running through the structures parallel to $[1\bar{1}00]$. Across the barium plane of the M phase, they are related by a mirror plane; in Y they slope in the same direction. No image calculations have yet been made, but by using the crudest approximation to the projected charge density – blocking in the coordination polyhedra round the heavy cations – we obtain a highly schematic representation that can be superimposed with some success on the highly magnified lattice image (Fig. 2.3). There are some electron-optic features of the lattice images that are not yet understood, and that can hinder simple fringe-to-fringe measurements of the 11.6 and 14.5 Å subunits; resolution of the lattice has helped in identifying stacking sequences. Thus, the structure in Fig. 2.4 can be identified as the stacking structure . . .$MYMY_3MYMY_3$. . . .; in Fig. 2.1 (from which Fig. 2.3 is taken), the sequence is . . . .MY_5MY_5. . . . .

Simpler stacking sequences may repeat over long distances, as far as can be judged from the small crystal flakes examined in the microscope. Abrupt changes from one well-ordered structure to another – e.g., from . . . .MY_4MY_4. . . . to . . . . MY_2MY_4 . . . . – have been observed and, not surprisingly, faults in the form of individual lamellae of "wrong" structure are found. Sometimes these are self-correcting – the crystal keeps its stoichiometry and overall periodicity right – as, for example, where two

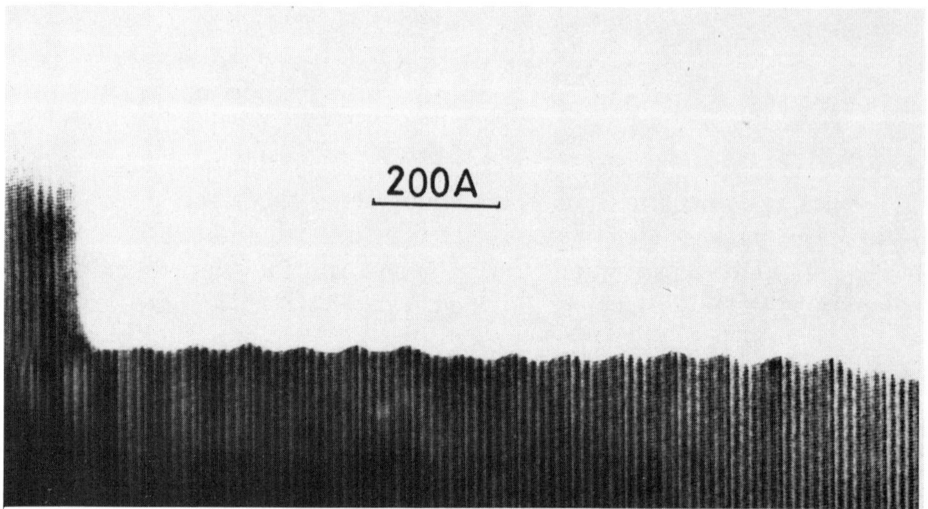

Fig. 2.1 Lattice image of the ferrite M_2Y_{10}. Vertical fringes demarcate the (0001) layers. Note the two-dimensional fine structure of the image.

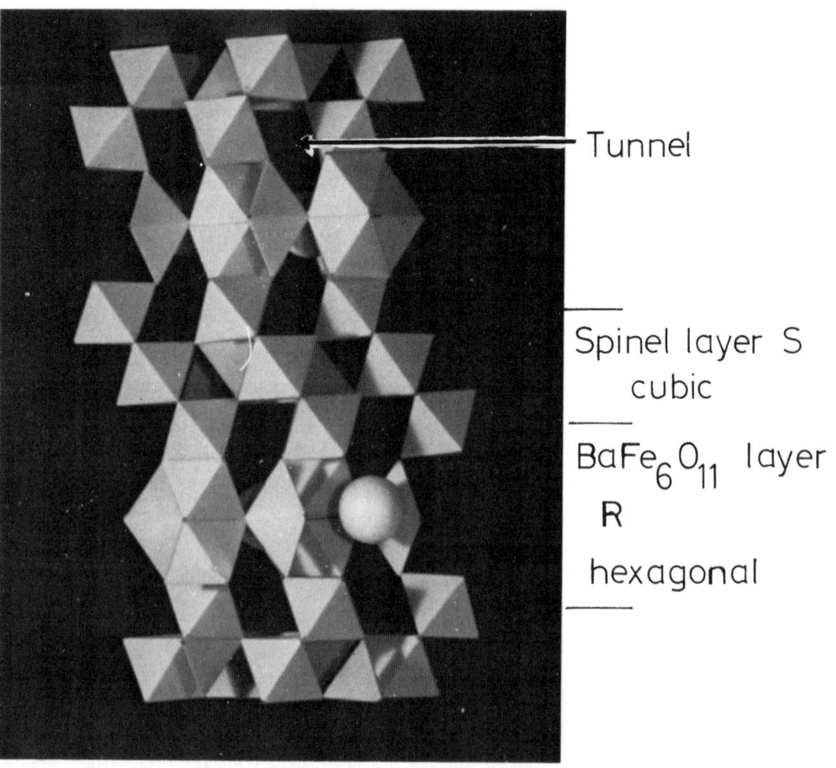

Fig. 2.2. The M-phase structure: $BeFe_{12}O_{19}$. Viewed along $[1\bar{1}00]$, showing pentagonal tunnels through the structure.

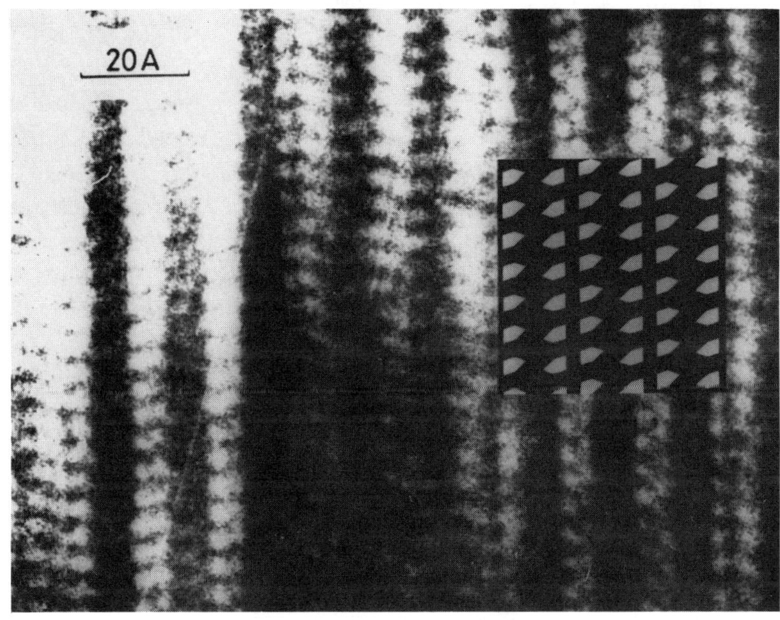

Fig. 2.3. A section from Fig. 1, greatly magnified, with the crude projection of scattering-center density superimposed.

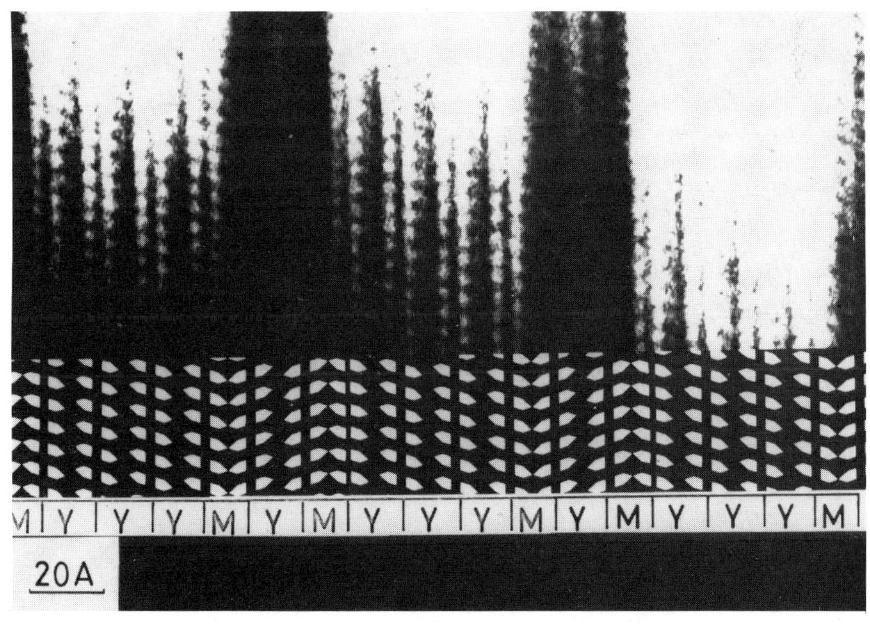

Fig. 2.4. Identification of the layer sequence in a crystal of M_2Y_4 in terms of the projected scattering-center density.

lamellae, $MYMY_6 \cdot MYMY_4$, have been found interpolated in the $MYMY_5 \cdot MYMY_5 \cdot MYMY_5 \ldots\ldots$ structure. Some of these faults could be formally described as the deposition of M and Y layers concurrently, but occasionally the wrong way up. The whole is, indeed, a kind of one-dimensional site-preference problem, i.e., of the preferred environment for depositing and covering an M layer.

Results to date only scratch the surface of a large subject, but draw attention to many problems of crystal chemistry. The morphology of the crystals leaves little doubt that the long c-axis is the slow growth direction. One must enquire how each successive sheet of structure is nucleated and what interactions and information steer the sequence in which the structurally distinct layers are deposited in a regular fashion.

Figure 2.1 reveals two distinct kinds of crystal fracture. At the left is a natural cleavage. In a crystal that has cracked along the cleavage, we have observed that the fracture appears to be atomically smooth, and the two bits of the broken stacking sequence, on either side, exactly make up the normal lattice period. At the right of Fig. 2.1 is a fracture across the natural cleavage, which shows a feature that we have repeatedly observed. The conchoidal fracture edge is rippled, and the ripple has exactly the periodicity of the crystal structure. In the micrograph shown, each peak coincides with an M layer in the crystal.

This work is continuing.

REFERENCES

1. Kohn, J. A., and Eckart, D. W., *Z. Krist.*, **119**, 454 (1964); **124**, 69 (1967).
2. Kohn, J. A., and Eckart, D. W., *Mater. Research Bull.*, **2**, 55 (1967).
3. Braun, P. B., *Phillips' Res. Reports*, **12**, 491 (1957).
4. Savage, R. O., and Tauber, A., *Mater. Research Bull.*, **2**, 469 (1967).

DISCUSSION OF PRESENTATION
By J. S. ANDERSON

To Lidiard's question about the nature of the dislocations observed, Anderson replied that the dislocation is a rearrangement of blocks producing a Burger's vector either one or some multiple of a half-octahedron diameter, and it may involve displacements in both the x direction and y direction vertically.

Honig asked about the numerical significance of the sequence of crystal layers in which mistakes had occurred. Anderson replied that sometimes where the crystal makes mistakes, it puts down two layers at a time, but the wrong way around.

Vink asked whether the repeat units of extended defects should be viewed as phases in equilibrium or as composites, adding that composites would not require equilibrium and would only have a surface energy, which might explain the numeralogy commented on by Honig. Anderson pointed out that one still comes back to the fact that certain ordered sequences of these phases or composite units seem to have a thermodynamic stability, and then we are back to the old question of how to define a phase. In many of these relatively simple sequences, very long repeat units are seen which are perfectly regular. That is, although the sequences can be regarded as a buildup of composites of two pseudo-components, their ordering is something which is thermodynamically required in terms of the chemical potentials of the component atoms present in the flux melt from which the crystals are grown. He added that his crystals were flux grown in the prescribed way and were single phases, as identified by their diffraction patterns. In conclusion, Anderson said that ultimately we must come back to a closer definition of what is meant by a phase, what is meant by a compound, and how we define our fundamental structural units.

Franklin asked about the correspondence between the rippled pattern of the conchoidal fracture and the repetition of holes in the electron micrographs. Anderson replied that the peak of every hump coincides with an M unit and that in some cases the fracture runs obliquely to the plane. In every case, they are sinusoidal and the periodicity is practically that of the lattice. He suggested that this might have something to do with the periodicity of a strong phonon which runs as a shock wave through the structure in the act of cracking.

Lindenmeyer asked if the type of dislocation observed was the only kind in these oxides. He noted that in metals, for example, and in polymers even, images of both screw and edge dislocations are observed. Anderson recalled that Cowley and Iijima have identified dislocations, and that their identifications come out much the same way; that is, when one moves around the crystal one finds a gap or an overlap in the Burger's vector, or in terms of a two-color map, you've gone wrong. There may be a variety of such things but they cannot be staged and must be taken as they come. Lindenmeyer then asked if they saw the image of a dislocation line. Anderson replied that because they are working at such high magnification they do not see these.

Ilschner asked to return to the question of what a "phase" is in these systems and referred to the theory of the buildup of concentration fluctuations in unstable solid solutions as described by Hillert and by Cahn and Hilliard. In this case also, one has regions of the crystal having slightly different compositions. The energy of the single crystal is the sum of energies of the regions plus the energy (an interfacial energy) of transition between them. If the different regions are not considered as phases, then this is not really an interfacial energy, but is the energy of transition of composition or the energy of transition of structural patterns. He concluded that he thought it was really a matter of philosophy whether one talks about phases or about regions of different structures. Anderson agreed that this would be a fruitful line of theoretical reasoning and that Dr. J.D.C. McConnell, a mineralogist at Cambridge, has been working in this direction. In conclusion, Anderson said that one needs to bring into our ideas of complex-structure stability the notions of periodic fluctuations in composition and in structure.

Next, Welch discussed shell-model calculations he and Dienes were doing at Brookhaven.

3 THE SIMPLE SHELL MODEL AND INTERATOMIC FORCES IN IONIC CRYSTALS AND MOLECULES*

3.1 The Calculations

In calculating the energy of formation and migration of point defects in ionic crystals, it is essential to include contributions due to ionic polarizability, since neglect of this effect results in a large overestimate of the energy.[1] Similar problems are found in calculating the phonon spectrum of nonpolarizable ionic crystals[2] and the properties of ionic molecules[3]. However, the inclusion of polarization interactions can result in various theoretical "catastrophes". This is most easily seen by considering the energy of interaction at distance r of a pair of ions with charges ± Ze and polarizabilities $\alpha_{\pm}$:[4]

$$E(r) = E_r(r) - \frac{Z^2e^2}{r} - \frac{Z^2e^2}{2r}\left[\frac{(\alpha_+ + \alpha_-)}{r^3} + \frac{4\alpha_+\alpha_-}{r^6}\right]\left[1 - \frac{4\alpha_+\alpha_-}{r^6}\right]^{-1}, \quad (1)$$

where $E_r(r)$ is the repulsive interaction. It is easily seen that as r approaches $(4\alpha_+\alpha_-)^{1/6}$, the energy diverges.

When the polarizability of one of the ions (usually the cation) is negligibly small, a catastrophe of this type cannot occur. A second type of catastrophe can result, however, if the repulsive energy is of the commonly used exponential form, $E_r(r) = A \exp(-br)$. If the remaining polarizability, say α_-, is sufficiently large, there will be no value of r for which Eq. (1) has a minimum. This catastrophe occurs when:

$$\alpha_- > \frac{4}{b^3}\left\{\frac{4\exp(-2)\,e^2}{Z^2\,Ab} - 1\right\} \equiv \alpha_c, \quad . \quad (2)$$

*Work performed under the auspices of the U. S. Atomic Energy Commission.

Values of α_-[5] and α_c for several salts in which $\alpha_+ \cong 0$ are listed in Table I. It may be seen that the stability criterion is violated in each of the cases listed. (If an inverse-power-law description of the repulsive energy is used, there is always a minimum, but it occurs at a very small value of r and yields an enormous polarization energy.) Polarization catastrophes of this nature do not generally occur in defect calculations for alkali halides[6] and alkaline-earth oxides[7] of the rock-salt structure, primarily because of the symmetry of the crystal structure, but they have been encountered in calculations of the surface energy[8], where the ionic environment is less symmetric, and in calculations of the properties of diatomic molecules[3,11]. Furthermore, such a catastrophe results for Al_2O_3 even in the perfect crystal.[9]

Table I. Values of Crystal Anion Polarizabilities and Critical Polarizabilities for Several Salts

Salt	α_-, Å^3 [5]	α_c, Å^3
LiF	0.64	0.09
NaCl	2.96	1.78
Al_2O_3	1.34	0.11
MgO	1.65	0.33

Polarization catastrophes of the type described above can be avoided by the use of the simple shell model, originated by Dick and Overhauser.[10] The essence of this model is that the repulsive interaction between ions is determined primarily by the same part of the electron distribution which displaces during polarization. This suggests that the repulsive interaction is a function of both the internuclear separation and the dipole moments on the ions. The simplest representation of this effect is to use an effective interionic separation in the expression for the repulsive energy, and to make this effective separation proportional to the dipole moments $\vec{\mu}$:

$$\vec{r}^{\,eff} = \vec{r}_{ij} + \frac{\vec{\mu}_i}{Q_i} - \frac{\vec{\mu}_j}{Q_j}, \tag{3}$$

where $\vec{r}_{ij}$ is the vector from nucleus i to nucleus j. Applying this approximation to the energy of interaction of the pair of ions described above yields:

$$E(r) = -\frac{Z^2e^2}{r} - \frac{Z(\mu_+ + \mu_-)}{r^2} - \frac{2\mu_+\mu_-}{r^3} + \frac{\mu_+{}^2}{2\alpha_+{}^\circ} + \frac{\mu_-{}^2}{2\alpha_-{}^\circ}$$

$$+ A \exp\left[-b\left(R + \frac{\mu_+}{Q_+} - \frac{\mu_-}{Q_-}\right)\right] \quad , \tag{4}$$

where the $\alpha_\pm{}^\circ$ are the polarizabilities of the free ions. The values of $\mu_\pm(r)$ are obtained from the simultaneous solution of the coupled (generally nonlinear) equations obtained from the condition that $\partial E/\partial\mu_+ = \partial E/\partial\mu_- = 0$. Similar equations can be obtained for the energy and equilibrium dipole moments of a crystal, either perfect or containing defects. We have recently[11] applied this approach to obtain values of $Q_\pm$ for the alkali halides and alkaline-earth oxides by fitting experimental data for crystals (static dielectric constant, refractive index, transverse optic mode frequency) and molecules (binding energy, equilibrium separation, vibration frequency, and dipole moment.) The values obtained are listed in Tables II and III. For the alkali halides, a single value of Q could be found for each anion and cation, independent of the salt, such that the properties mentioned above can be described with reasonable accuracy, as indicated in Table II. (This is important in the calculation of impurity-ion properties.) However, it was found that the value of Q for oxygen varies substantially, depending upon the salt, as indicated in Table III.

Table II. Shell-Model Parameters ($Q_\pm$) and Free-Ion Polarizabilities for Alkali Halides, Obtained from Crystal and Molecular Properties

(Values are independent of the salt.)

Cations	$\alpha_+{}^\circ$, Å^3	Q_+	Anion	$\alpha_-{}^\circ$, Å^3	Q_-
Li	0.00	∞	F	1.10	1.35
Na	0.18	5	Cl	3.42	2.65
K	0.83	10	Br	4.67	2.55
Rb	1.40	5	I	7.15	2.50
Cs	2.44	5			

Note: The standard deviations of the fit are: for molecular properties, 9 percent; for crystal properties, 16 percent; for combined properties, 14 percent.

The usefulness of the shell-model approach for defect calculations in oxides is illustrated by the successful calculation of the energies of Schottky and Frenkel defects in Al_2O_3, a material which has a polarization catastrophe even in the perfect crystal. (It was found in this calculation that aluminum Frenkel defects are energetically favored.)

Table III. Shell-Model Parameters and Free-Ion Polarizabilities for Some Oxides, Obtained from Crystal and Molecular Properties

Salt	α_+°, Å^3	α_-°, Å^3	Q_+	Q_-
MgO	0.094	3.88	∞	2.00
CaO	0.47	3.88	20	2.75
SrO	0.86	3.88	20	3.50
BaO	1.55	3.88	28	4.00
Al_2O_3[(a)]	0.00	3.88	∞	2.24

Note: The standard deviations of the fit are: for molecular properties, 13 percent; for crystal properties, 7 percent; for combined properties, 11 percent.

(a) Q_- deduced from refractive index only.

REFERENCES

1. Mott, N. F., and Gurney, R. W., *Electronic Processes in Ionic Crystals,* 2nd Ed., Chapter II, Oxford University Press (1948).
2. Bilz, H., "Lattice Vibrations", in *Optical Properties of Solids,* S. Nudelman and S. S. Mitra (Eds.), Plenum Press, New York (1969).
3. Tosi, M., and Doyama, M., *Phys. Rev.,* **160,** 716-18 (1967).
4. Moelwyn-Hughes, E. A., *Physical Chemistry,* 2nd Revised Edition, Pergamon Press, New York (1961), p. 448.
5. Tessman, J. R., Kahn, A. M., and Shockley, W., *Phys. Rev.,* **92,** 890 (1953).
6. Barr, L. W., and Lidiard, A. B., in *Physical Chemistry,* W. Jost (Ed.), Academic Press, New York (1970), Vol. X, p. 151.
7. Boswarva, I. M., *J. Phys., C5,* **15** (1972).
8. Anderson, P. J., and Scholz, A., *Trans. Faraday Soc.,* **64,** 2973 (1968).
9. Dienes, G. J., Fischer, C. R., Hatcher, R. D., Lazareth, O., Samberg, M., and Welch, D. O., to be published.
10. Dick, B. G., Jr., and Overhauser, A. W., *Phys. Rev.,* **112,** 90 (1958).
11. Welch, D. O., and Dienes, G. J., to be published.

DISCUSSION OF D. O. WELCH'S PRESENTATION

Franklin asked whether Welch had taken the parameters for the oxygen interaction and the oxygen-aluminum interaction, that is, the Born-Mayer parameters, out of the Wedepohl calculations rather than developing them by fitting to empirical data. Welch replied that this was true, and noted that the parameters of the unit cells which are a little puckered, come out nicely. The contents of the unit cell are in about the right position. He said that they intend to do things such as calculate the elastic constants next. Franklin noted that this is the kind of thing that

C.R.A. Catlow had been doing in calculating the oxygen-oxygen interaction (as reported by Lidiard). Welch said that his calculation, however, does not have the attractive component that Lidiard talked about, that it is purely a Wedepohl calculation. In response to Franklin's observation that this might be because in his problem the oxygens are well separated, he replied that the oxygens were relatively far apart in the alumina structure, but they were doing an oxygen interstitial for which results had not been obtained.

Anderson questioned whether, in view of the variability of the theoretical calculations, there was any real distinction between the Schottky and the Frenkel solutions. Welch replied that so far the distinction between these defect types is small and one ought not to take these numbers and calculate defect concentrations. In any case, the high energy for formation of Schottky defects means that the intrinsic defect concentration is very low; there is evidence, however, that the aluminum Frenkel pairs are favored. The experimental observations are probably extrinsic.

Huggins inquired whether Welch had found a systematic fit only so long as he did not include lithium salts and Welch replied affirmatively. Huggins then commented that lithium has been shown by a number of experiments not to sit in the center of its hole in materials such as the chloride and hence he was not surprised that it didn't go along with Welch's calculations and fit in with the results for the several oxides. Welch replied that Tosi and Fumi had calculated the ionic radii and the stiffness parameters of the alkali halides and they also found fairly poor agreement for lithium salts. He said that he hoped that the shell model would work and that it would produce a consistent set of parameters so that one could make calculations on impurities and on off-center ions without knowing everything about a particular salt. If that doesn't work, then the theoretician is faced with a far more difficult task of calculating these things using some kind of quantum chemical techniques. Welch observed that this question will probably have to be faced as people get further into the periodic table.

Kröger said that in Al_2O_3 the conduction is often determined by foreign atoms. In addition, these foreign atoms often turn out to be acceptors such as Mg^{+2} and Fe^{+2}. He also said that they had shown that the counter ion which is responsible for conduction is an interstitial aluminum, a result that would fit in with the Frenkel idea. Welch remarked that they were trying, so far unsuccessfully, to answer the question of which is the most mobile ion. Kröger was puzzled by the fact that the Frenkel interstitials in calcium fluoride and UO_2 are the ions of low charge and from this it might be expected that oxygen would be the interstitial in Al_2O_3. Welch remembered that the structure of Al_2O_3 is different, but beyond that he couldn't comment since they had not finished their calculations for the oxygen interstitial.

O'Keeffe asked whether the interstitial aluminum is in octahedral or tetrahedral surroundings. Welch replied that the interstitial was the hole between the pair of aluminum ions going up the c-axis, which is the octahedrally coordinated site, and that they had not done the calculation for the tetrahedrally coordinated site yet.

Anderson commented that although Welch and co-workers had found justification for the alkali halide model by looking at the alkali halide molecules whose vibrational spectrum has been analyzed, they were in trouble when they came to aluminum oxide because Al_2O_3 molecules are a minor vapor species and their spectra have not been measured. Excluding the alkaline-earth oxides, he continued, the less ionic oxides generally have molecular species that do not correspond to the solid molecular unit either in atomic ratio or in ionic charge; for example, the vapor specie over Al_2O_3 is Al_2O or Al_4O_2 and that over SnO_2 is polymers of SnO. Welch replied that they are interested in molecular species having the same charge state as in the solid.

4 GENERAL DISCUSSION

The preceding sessions had illustrated the elaborate models and techniques that have been evolved for characterizing or calculating defect energies in relatively simple ionic oxides, and these were juxtaposed against powerful experimental methods that reveal, sometimes in a most graphic way, the details of extended defects and the texture of solids in the range of 5 to 100 Å in oxide systems not notably ionic in character. No comprehensive theory for these latter cases has been put forward.

The following general discussion afforded an encounter between persons fresh from the battle of observing and explaining the expanding list of point defects and those developing precise descriptions of materials having extended defects, but where the experimental characterization has far outstripped the development of theory. Obviously the discussion had taken into account the chemical and physical differences between oxides of the alkali and alkaline-earth elements of fixed oxidation state and low electronegativity on the one hand and of the transition and inner transition elements of possible varied oxidation state on the other.

The basic questions remaining to be discussed were: (1) To what extent would the successful experience with point defects carry over to the exploration and description of extended defects? (2) What types of models utilizing the new information on real solids can be put forward as a basis for understanding the practical problems of solid-state reactions?

Stringer commented on the application of the point-defect model in the understanding of the oxidation of metals. He then raised the question of how the extended defect model might be applied to the interpretation of the kinetics of the oxidation of metals. He said that the growth of an oxide layer takes place by the migration of defects through the layer and that an understanding of this process will allow its control. Over the years, the "point-defect" model of defect oxides, following the early work of Wagner, has been very successful in the interpretation of oxide layer growth. For example, measurements of the oxygen-pressure dependence of the oxidation rate can be correlated with those of the oxygen-pressure dependence of conductivity in oxides. The point-defect model is used to predict the effect of alloy elements on the oxidation rate by regarding them as doping the oxide, following the work of Wagner and Hauffe. Now, Stringer said, many find the extended defect model very difficult to reconcile with the earlier models and would welcome the advice of those at this meeting on how the details of the oxidation process should be interpreted and, in particular, the effect of oxygen pressure and of alloying elements on the oxidation rate. Anderson replied (1) that a general extended defect model does not exist, (2) that for many oxide systems the defect concentration is probably low enough that any type of analysis which Kröger developed is applicable, and (3) that for more exotic oxides, such as those of tungsten, tantalum and niobium, problems do exist if extended defects are neglected. Stringer replied that in that case his work and that of Kofstad on tantalum and niobium might be in deep trouble, and that a problem also exists in the interpretation of the oxidation of iron to wüstite and in the oxidation of cobalt and nickel because of the high defect concentrations in the oxides. Eyring raised the question of whether the properties of wüstite and other oxides showing extended defect structures have really been adequately treated using a point defect model. Cohen replied that the fundamental property measurement of diffusion in wüstite is all fouled up. Shewmon asked Cohen whether he thought that people didn't know how to measure a diffusion coefficient in that system or that they didn't know how to understand it. Cohen replied that he thought they didn't know how to measure it properly.

Stringer, noting that electron micrographs shown by Anderson and Cowley were of oxides at nearly room temperature, asked whether the extended defect structures observed are the same as those that exist at high temperatures. He further commented that in the case of "Ta_2O_5" and "Nb_2O_5", measurements of mechanical properties show an abrupt softening at about 600°C, e.g., the hot hardness drops from around 1000 Vickers to around 200 in a very short temperature range.* He asked

*I. I. Korobkov et al.; *Symposium on Metallurgical and Physical Properties of Pure Metals,* **4** 93 (1965); **7**, 158 (1968); *Fiz. Metal i Metalloved.,* **25**, 85 (1968).

whether this phenomenon might be related to the disordering of the low-temperature extended defect structure which has been observed. In reply to a question by Anderson, Stringer said that the measurements he was quoting were for the bulk oxides and not for oxide films on the metals. Anderson then noted that there had been about 18 polymorphs of Nb_2O_5 reported. Stringer stated that although Nb_2O_5 was not very well characterized, both Nb_2O_5 and Ta_2O_5 had been examined and both showed the same effect. Anderson suggested that the Nb_2O_5 probably had the low temperature Ta_2O_5 structure which is very complicated and in which the structure multiplicity is a function of temperature.

Returning to earlier comments of Cohen, Anderson said that the effect of the extended defect model is that point defects responsible for mass transport in nonstoichiometric oxides cannot be calculated from a simple arithmetic relation, and that a relation does exist but that it involves, in some complicated way, the internal dissociation mechanisms and these have yet to be worked out. Summarizing, Anderson said that in the observations of high-resolution electron microscopy there do not appear to be any measurements to answer questions on the role of extended defects in mass transport.

Wagner then raised the question of what measurements of properties other than very elegant measurements of lattice parameters or of conductivity, diffusion, hardness, etc., need be made to sort out the mechanisms of oxidation in these materials. Lidiard suggested that from Stringer's point of view of oxidation, the influence of extended defects on electronic transport needs to be measured. Goodenough said that an important question is how these defect structures are transformed into one another and suggested that measurements of the dynamics of domain-boundary motion would be useful.

Huggins raised the question of the integrity and compactness of the oxide film of transition metals and cited photographs taken by Gulbransen of Westinghouse a number of years ago which showed very complicated oxide film patterns – "whiskers" and "butterfly wings". He questioned whether oxide-film growth takes place by transport through an oxide layer as assumed or by growth of whiskers from below or some other mechanism, and whether the observed phenomena had much to do with transport through the oxide film. Stringer stated that he would comment further on this in his talk on Friday.

Kofstad, commenting on the oxidation of titanium, recalled Anderson's comments from the previous day that the nonstoichiometry of TiO_x had to be moved back from $x = 1.96$ to $1.99 \pm .004$; recalling also another paper by Anderson, he noted that at 1200°C, 24 hours were required to develop these structures. On the other hand, an oxidation process such as that Stringer might observe would be finished 23 hours earlier. He then asked how these structures are related to the oxidation phenomena, i.e., whether oxidation is associated with quasi-equilibrium and not with extended defect structures. Eyring stated that in substances which have extended defects, although the number of mobile defects is much lower than one would calculate in a simple way from the departure from stoichiometry, it is possible to account for oxygen diffusion in a rare-earth oxide in terms of a simple point-defect model, even though the calculated defect concentration is far from the actual defect concentration and even though it is far off from equilibrium. He suggested that many times this growth of defective oxides can be treated by more classical models which do not really apply.

Libowitz contrasted the observations of oxidation and diffusion at high temperatures with the observations of extended defects at low temperatures. He then noted a remark of Cohen's made at the afternoon session that the defect clusters in FeO seem to be smaller as the temperature rises. On the basis of these two things, he asked whether defects at high temperatures were the same as those observed at low temperatures and, further, whether at sufficiently high temperatures a point-defect regime might be realized. Cohen characterized this as a kind of wishful thinking that we all would like to engage in. Continuing, he said that although they had made quantitative measurements of the FeO system up to 840°C, they had made qualitative observations up to 1150°C and had found indications of extended defects even at those high temperatures. Libowitz said however that the defects were changing and that more studies such as those Cohen had done were needed. Cohen disagreed, arguing that the measurements of diffusion and electrical conductivity suggested by the theoreticians had already been made by many investigators and the problem now was that of relating existing measurements to known properties such as the p-n transition in wüstite which occurs at high temperatures with the strange behavior of the diffusivity in the system. Furthermore, no careful tie-in with defect structure has been made other than a very qualitative "hand-waving" one.

Eyring then asked Cowley about the prospects for the incorporation of a hot stage into the high-resolution electron microscope. Cowley said that the resolution which can be achieved with a hot stage is usually only about 20 Å, which is far short of what is required for good imaging of crystal structure. Anderson remarked that they had gotten 17 Å at 800°C. Cowley added that he thought better resolution could be achieved but that it might take a couple of years to do it.

Lindenmeyer spoke of the analogy between the extended defect question in oxides and what has happened theoretically in polymer research. In regard to defects in oxides, he said that people were happy with point defects because they could be treated with conventional statistical mechanics using a random distribution but they were very discouraged when it came to extended defects because they are much more complicated. Similarly, in the polymer field, people were happy with rubber elasticity because they could apply Gaussian statistics and the theory worked very nicely for many properties. However, they began looking closely at the experimental evidence and found that systems are much more ordered than they had been led to believe by the success of past theories. That a similar experience is being encountered in oxides he considered to be an encouraging sign rather than a discouraging one because if one can understand how these materials are disordered and hence explain them in depth using something more definite than random statistics, one would have a much better chance of relating them to properties than would otherwise be possible. Concluding, he said that both fields are beginning to mature now because we are beginning to see how things are arranged rather than how we would like to have them arranged for mathematical convenience.

Hagemark, commenting on the periodicity of the extended defect structures observed by Anderson and Cowley, suggested that their influence upon the lattice phonon spectrum and upon optical properties sensitive to the lattice spacing might be measured. He also suggested that materials could be doped with paramagnetic defects and the influence of the different extended-defect environments upon the EPR spectra observed. He asked whether any such measurements had been made. Anderson mentioned experimental measurements under way at the University of Oxford, in particular, infrared studies of the niobium oxides. In regard to doping, he said that one must be very careful because one has no idea where the doped-in atoms will go. However, from neutron crystallography and from calculations of site potentials, it is known that there are very strong differences between various cation sites. Anderson emphasized the desirability of measuring the phonon spectrum.

Tannhauser returned again to the question of phases in crystals with several defect structures. He asked Anderson to comment on the point that a phase can be uniquely defined thermodynamically according to the phase rule if equilibrium exists, whereas Anderson has said that the definition of a phase is not clear anymore. Further, he asked what experiments would be needed to figure out whether one has phases or one continuous phase at high temperatures. Replying to the second question first, Anderson said that the complex defect structures which had been observed are the high-temperature-equilibrium configurations and are not the low-temperature product of ordering processes. Turning to Tannhauser's first question, he said that in the usual discussion of the phase rule, one ordinarily thinks in terms of the following intensive quantities: chemical potential, temperature, and molar volume. He went on to say that this leaves out the question of coherent intergrowth between structures. Regardless of the coherence of intergrowth, there is a change in lattice potential in going from one structure to another. Consequently, there is an energy term in passing from one structure to another even though there is a perfect geometric coherence. This introduces an extra degree of freedom to our thermodynamics and we have thereby lost our ordinary definition of a phase.

Franklin commented on the observed independence from oxygen partial pressure of the oxygen diffusion coefficient for rutile. He suggested that the defects in the bulk might react with shear structures in the oxygen sublattice of the rutile as if it were a two-phase system, i.e., at a special plane the missing oxide ions act as sources and sinks for point defects so that the bulk oxygen point-defect concentration does not change as the oxygen partial pressure changes, and, accordingly, the transport of oxygen does not change with oxygen partial pressure.

In regard to the two defect domains observed in hexagonal ferrites, Kröger raised the question of the feasibility of growing the pure macroscopic structures and then studying the various processes when the two structures were juxtaposed as in experiments on bicrystals, which have been used for careful studies of grain boundaries. Anderson suggested that although this would be an interesting experiment it is unlikely to prove practicable. The flux growth methods used to grow these crystals almost invariably produce a whole range of the compounds concurrently. To grow a pure macroscopic crystal of one end structure, and then to change conditions so as to continue growth of the other end member, would be exceedingly difficult.

Caro commented on the uncertainties of doping with paramagnetic ions and the problems associated with the interpretation of the influence of structure on the optical spectra. The rare-earth ion exhibits a series of discrete crystal field energy levels, which by itself is a message giving precise information about the ligand distance and the site symmetry of the position where the ion is located. The site may not be exactly the one derived from crystallographic considerations. For instance the rare earth ion seldom prefers a site having a center of inversion. Take the example of the pyrochlore structure $Ln_2M_2O_7$. If europium is used as an optical structural probe, it can be determined whether the rare earth is at the D_{3d} site from the absence of the $\Delta J = 2$ lines (absorption $^7F_0 \rightarrow {}^5D_2$, emission $^5D_0 \rightarrow {}^7F_2$) which being dipolar electric in nature should be forbidden if the ion is at a center of inversion. The spectrum then shows only the dipolar magnetic lines ($\Delta J = 1$). Examples are $Eu_2Sn_2O_7$ and $Eu_2Ti_2O_7$. But, if the $\Delta J = 2$ lines are present, the site is distorted with respect to the one in true pyrochlore, for example in $Eu_2Zr_2O_7$.

In addition, the broadness of a line is a sure indication of the degree of disorder. If one has a very complicated structure, but with numerous, well defined sites, the complex spectrum will still exhibit narrow lines from which the number of different sites can be determined. However, lines indicate disorder, characterized by various distances and angles to the ligands: the crystal field parameters, which depend on the distance, being modulated from one ion to another. For example, a disordered microphase can be identified in samples of $Eu_2Zr_2O_7$.

There are several examples of segregation of rare earth ions introduced as dopants in one phase, this occurs through syntactic intergrowth, or epitaxial microdomains. Such can be clearly detected from the optical signal, but not always from X-ray crystallography.

Lidiard, commenting on a "phase", said that from a statistical mechanical point of view one has to ask what kind of phase space the system is averaged over in thermodynamic equilibrium. These structures, such as the shear structures, are essentially fixed and are not thermodynamically averaged; that is, although there is a sizable energy of ordering [A. M. Stoneham and P. J. Durham, *J. Phys. Chem. Solids,* (December, 1973)], there is no entropy, i.e., the system is not capable of taking itself over an average of those structures. Similarly, in a metal with dislocations one does not average over dislocation distributions when one performs statistical thermodynamics. In this sense, you have another variable which is outside the thermodynamic averaging. One, therefore, has a distinct phase for every given structure, but insofar as the existence of that structure does not affect the rest of the crystal very much, it does not affect the average thermodynamic properties, just as extensive defects (dislocations or voids) do not affect the average thermodynamic properties. Nevertheless, in kinetic processes, the interplay between point defects and these extended defects (dislocations, voids, shear-plane structures) can be overwhelmingly important. Furthermore if one frames the questions about extended defects and point defects in these oxides in terms of equilibrium distributions and the mechanisms of approach to equilibrium, then some of these paradoxes about phases will vanish. It is, however, obviously important to know what happens to these extended structures at high temperatures.

Smeltzer raised the question of whether extended defects, in particular the shear structures, could be utilized in the interpretation of the parabolic to linear transition in the kinetic oxidation curve of tungsten, citing an observation dating from the 1950's which noted an inner layer, identified as W_4O_{11}, called the "diffusion membrane" and an outer layer of WO_3 which is completely porous to oxygen. He suggested that the observed transition in mechanical properties of the oxides might be related to the shear structures and that a question could be asked about the role of the shear structures in the mechanical breakdown of the oxide when the chemical stoichiometry is altered. Anderson expressed uncertainty stating that the layered tungsten oxides have a very complicated chemistry and their structures often show rotary shear, and some of these oxides exhibit tunnel structures such as those which Cowley talked about during the afternoon session. There is, then, a series of shear structures leading up to WO_3. Although the reduction of WO_3 has been studied, work has not progressed in the other direction. Smeltzer then asked if these structures would show similar mechanical properties. Anderson said that they are all related to the WO_3 structure and that there should be a continuous transformation between them.

Smeltzer next brought up the problem of the interpretation of oxygen self-diffusion in niobium pentoxide. He pointed out that, apart from the difficulty of getting suitable materials for study, diffusion measurements have shown that D does not depend upon the degree of nonstoichiometry, but diffusion is highly anisotropic, preferring a direction corresponding to the open structures in the crystal. A point-defect argument would suggest a strong dependence upon stoichiometry. Investigators who have interpreted their results in terms of point defects have suggested either a vacancy mechanism or atomic oxygen going down the channels. Perhaps only the residual vacancies between the shear planes are important. Smeltzer asked Anderson for his views on the possible mechanisms. Anderson replied that they had studied the oxidation of $Nb_{12}O_{29}$ to Nb_2O_5 and that they had concluded that the oxidation may involve the diffusion of oxygen down channels so that the anisotropy of diffusion is very high and the rates are very fast. $Nb_{12}O_{29}$ crystals one to several tenths of microns in size can be oxidized readily at 400°C to Nb_2O_5, which implies a very free diffusion. Thus, the mechanism is very facile and probably does not depend upon the concentration of point defects.

Smeltzer inquired about the relationship between the defect clusters and shear structures observed in the electron microscope and the intergrowth of two phases observed when the solution is cooled below the eutectic point. Anderson replied that the electron-microscope work was not done on quenched specimens (except that on metals), but on crystals grown at low temperatures by chemical-vapor deposition or by solid-state reactions. Smeltzer asked whether the electron microscope would distinguish between defect clusters and the clusters due to the presence of a second solid phase. He cited the work on wüstite by Roth [W. L. Roth, *Acta Cryst.,* 13, 140 (1960)], who thought he had observed clusters of magnetite in the wüstite structure since the spinel structure of magnetite differs from that of wüstite in that the iron occupies octahedral rather than tetrahedral positions. The question is then whether the observed quenched structures of single crystals are due to eutectic decomposition which does not propagate owing to strain energy or some other reason.

Cohen replied that, first, measurements have all been made above the eutectic and, second, that Roth's interpretation was wrong; what he thought was magnetite was cold-work broadening of the supplementary peaks.

Smeltzer then raised the question of the structure of the oxide phase that forms on iron during the early stages of oxidation. From the LEED technique, investigators have argued that the structure is wüstite; however, the phase diagram would lead you to expect magnetite. By taking the spinel structure and altering the occupancy of the tetrahedral sites, thereby altering the stoichiometry, you can get reflections for a spinel phase which correspond very closely with what you would get for wüstite when some of the ions go into tetrahedral positions. Cohen responsed that the structure of spinels and the structure of wüstite are entirely different. Putting an ion into the tetrahedral hole of wüstite would not give you a spinel.

Cohen then commented on the definition of a phase by saying that a copper crystal having a grain or coherent twin boundary would be considered a single-phase system. He asked, however, that we consider a metallic alloy such as copper-gold or copper-molybdenum which has a long-period superlattice and a periodic antiphase boundary much like the ones seen in oxides. This would be considered a one-phase system with an occasional antiphase boundary. He then suggested that the question of a phase description might revolve around the question of a change in the symmetry of the structure rather than around the occurrence of regions, such as a stacking fault in a simple hexagonal metal that is cubic. Anderson replied that there are several things that must be borne in mind. The main feature of a crystal is that the structure repeats itself periodically in a definite number of translational steps. We can therefore define a compound crystallographically because a solid has a crystallographically repeating unit containing a certain number of atoms. When we come to a definition of a phase, we mean something different from a structure and we must speak thermodynamically. Where an argument can arise is where you have a crystal with one structure, periodic in itself, passing over coherently into another structure. If you reduce Nb_2O_5 at 1400°C, as shown by Cowley, you have thermodynamic equilibrium conditions maintained, but you can have thermodynamic pseudoequilibrium conditions if you reduce Nb_2O_5 at 1200°C. Partially reduced oxides will show domains of perfect Nb_2O_5 and of the next phase, the $Nb_{25}O_{62}$ structure, which is the first structural product of reduction. There is no step in the thermodynamic-activity composition curve. You go through what looks like a true nonstoichiometric curve for small displacements from equilibrium and it is reversible. According to X-ray examinations, you would say that you are dealing with one phase of variable composition. According to the structure, you are dealing with a crystal that has the domain texture with two different structures within it. Anderson queried whether this is a compound with a one-phase structure or a two-phase structure? Cohen said that he would call that a two-phase structure. Anderson said that he agreed.

Lindenmeyer called attention to the thermodynamics of small systems, pointing out that in such systems you cannot disregard surfaces. When you include surface effects, the problem of whether you have one phase or two phases completely disappears. Since you are dealing with a small system, you must put in the correction for the surface involved and everything else corrects itself very nicely.

Kröger asked for a clarification from Anderson on an apparent contradiction. Anderson had said, he recalled, that diffusion in Nb_2O_5 occurs by migration down channels and later that point defects are not involved. Anderson said that by point defects he meant vacancies, and that diffusing interstitials would, of course, be a point defect while in the channel.

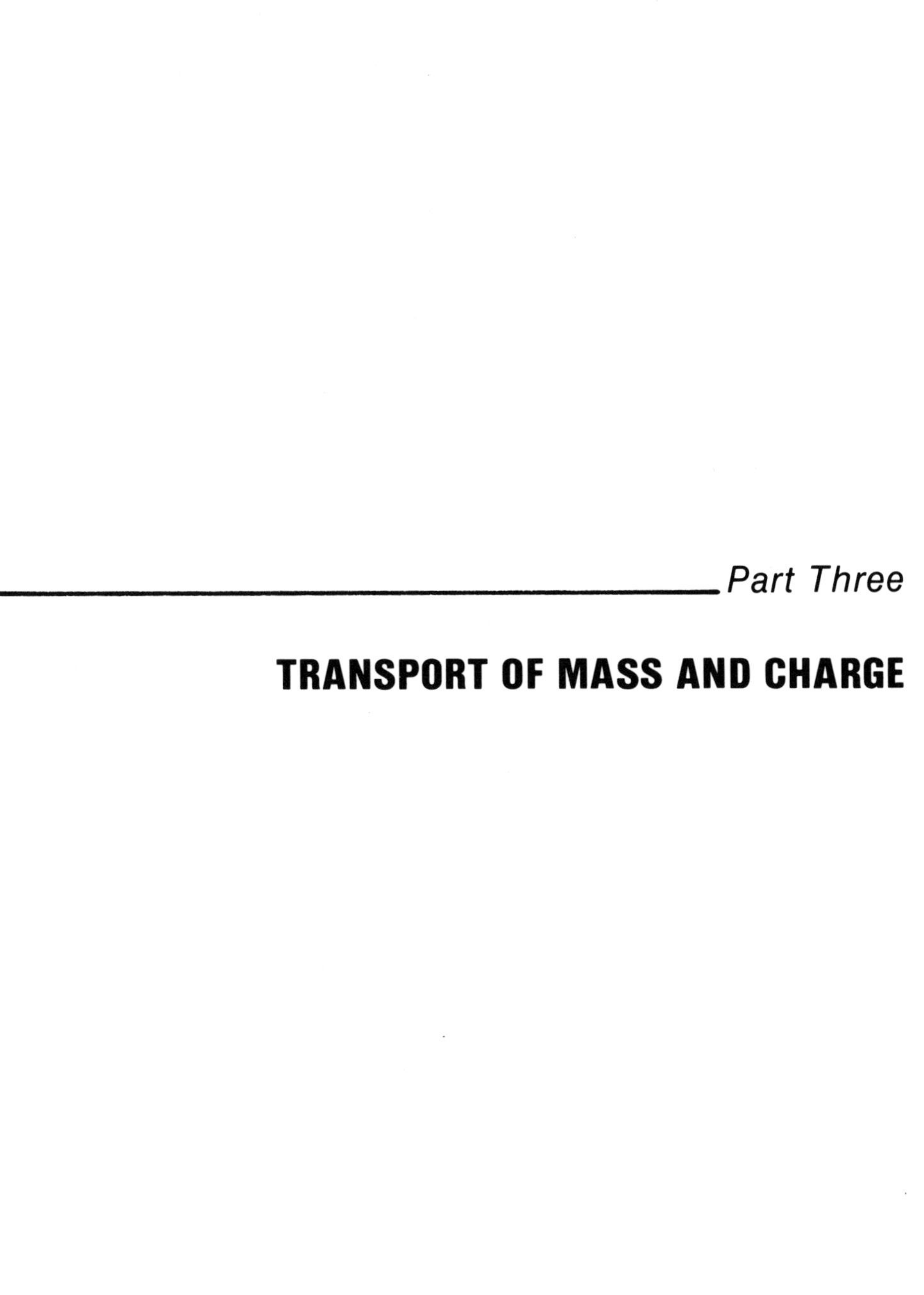

Part Three

TRANSPORT OF MASS AND CHARGE

ELECTRICAL CONDUCTIVITY, DIFFUSION, AND MINORITY DEFECTS IN SOME TRANSITION-METAL OXIDES

J. Bruce Wagner, Jr.

Materials Research Center
and
Department of Materials Science
Northwestern University
Evanston, Illinois

ABSTRACT

Electrical conductivity and diffusion data for some transition-metal oxides are discussed in terms of the majority and minority defects existing in the oxides. The rate of transport and the transport mechanism for sulfur and chlorine are discussed with particular reference to NiO.

1 INTRODUCTION

The purpose of this paper is to show that the type of minority defect may be inferred from electrical conductivity and diffusion data and also to present data for the diffusion of anion impurities, sulfur and chlorine, in some transition-metal oxides.

Relatively many data have been accumulated on the transport of cations in oxides but few on the transport of anions, especially anion impurities. In the transition-metal oxides, NiO, CoO, Al_2O_3, and Cr_2O_3, and in some temperature and oxygen pressure ranges, also Fe_2O_3, it has generally been assumed that the oxide ion is relatively large and immobile compared with cations. By inference, it has been suggested that anion impurities migrate slowly in these oxides, and therefore these oxides might offer resistance as barrier layers in carious applications. Within the last half dozen or so years, evidence has been accumulating that anion impurities somehow migrate rapidly through these oxides. A noteworthy example has been the corrosion of alloys used in turbines. When the engines were exposed to atmosphere containing salt (over seawater), the combination of the sulfur in the jet fuel plus oxygen resulted in a condition known as "hot corrosion". In some cases, a layer of sulfide was found underneath an oxide layer at the metal-oxide interface, and a question arose as to the rate and mechanism of transport of sulfur in oxides.[1] Another example is provided by the corrosion of nickel-base alloys by chlorine-containing atmospheres.[2]

Neither the rate of transport, the transport mechanism, nor the solubility of anion impurities in oxides is well understood. Presented herein are the results obtained for the transport of radiotracers sulfur-35 and chlorine-36 in some transition-metal oxides, along with a comparison of known data on the transport of the cations and oxygen ions.

2 EXPERIMENTAL

The general procedure has been to prepare a single crystal or a polycrystalline compact and to give the sample a prediffusion anneal under a fixed oxygen pressure to fix the composition and hence the type and number of defects in the materials. Most samples were annealed in closed quartz tubes containing air, and the oxygen pressure was calculated using the ideal gas law. For the dependence of diffusivity on oxygen pressure, two-phase mixtures of oxides (Cu/Cu_2O or Fe_3O_4/Fe_2O_3) were sealed in the quartz tubes in order to provide a known oxygen pressure (the isopiestic method).[3] Details can be found elsewhere.[4-6]

The sample is then quenched and the radiotracer is applied to one surface. The sample is then given a diffusion anneal at the same temperature and oxygen pressure as the prediffusion anneal. At the end of the diffusion anneal, the sample is quenched and then mounted in plastic. The radioactivity is determined by serially sectioning the sample and counting the freshly exposed surface (residual activity method).

Because of the high vapor pressure of the sulfur tracer (applied to the sample as elemental sulfur in benzene which was allowed to dry) and of the chlorine tracer (applied to the sample as a solution of $NiCl_2$ in ethanol which was also allowed to dry), the tracers vaporized and the solution to Fick's second law for a constant source into a semi-infinite crystal was used. It is:

$$I = I_o \, [1-\mathrm{erf}(x/2\sqrt{Dt})] \quad , \tag{1}$$

where I_o is the activity of the diffusant at the original surface (x=0), I is the activity at any penetration depth, x, D is the appropriate diffusion coefficient in cm^2/sec and t is the time. Absorption corrections[4] were made for both the sulfur-35 (0.167 Mev beta) and chlorine-36 (0.714 Mev beta).

3 RESULTS AND DISCUSSION

The diffusion of nickel[7,19,20] and of oxygen[8] has been studied in NiO. The reported data show the diffusivity of oxygen to be much less than that for nickel. Likewise, the diffusivities of cobalt[9] and oxygen[10] have been studied in cobaltous oxide with a similar large difference in diffusivities of the cation and oxygen ion. These large differences might be expected on the basis of the radii of the respective ions (see Table I).

For NiO single crystals annealed under virtually constant (0.77 to 1.0 atm) oxygen pressure, the diffusivity of sulfur was found to be[5,11]

$$D(S \rightarrow NiO) = 1.38 \times 10^{-4} \exp(-48{,}600/RT) \quad . \tag{2}$$

Preliminary data on the diffusivity of chlorine in single crystalline NiO

Table I. Comparison of Some Ionic Radii (Å)

	Goldsmidt[27]	Pauling[28]	Nickless[29]
Ni^{2+}	0.78	0.69	
Co^{2+}	0.82	Co^{+2}	
Fe^{3+}	0.67	–	
O^{2-}	–	1.40	
S^{2-}	1.74	1.84	
S^{o}	–	1.04	
S^{-}	–	–	1.22
Cl^{-}	1.81	1.81	

were found to be[12]

$$D(Cl \rightarrow NiO) \approx 1.8 \times 10^{-7} \exp(-38{,}230/RT) \ . \qquad (3)$$

These data are shown in Fig. 1 along with the diffusion data for nickel and for oxygen. The surprising result is that the diffusivity of sulfur lies between those of nickel and oxygen. Because both $S^{=}$ and Cl^{-} are larger than $O^{=}$, it was initially expected that oxygen would migrate more rapidly than these impurities. A detailed study was carried out on nickel oxide to determine the most probable mechanism and state of charge on sulfur. For the mechanism, the oxygen pressure dependence of the diffusion of sulfur was studied.[4,5] Results are presented for sulfur in Fig. 2. Increasing the oxygen pressure decreased the diffusivity of sulfur. These results can be explained as follows. The majority of lattice defects in NiO under the conditions of temperature and oxygen pressure employed here are doubly ionized cation vacancies which are compensated by electron holes. The equation may be written

$$\tfrac{1}{2} O_2 = V_{Ni}'' + 2h^{\bullet} + O_O \qquad (4)$$

whence

$$K_3 = [V_{Ni}''] \, [h^{\bullet}]^2 / p_{O_2}^{1/2} \ . \qquad (5)$$

The electroneutrality condition requires that

$$2[V_{Ni}''] = [h^{\bullet}] \ . \qquad (6)$$

Hence,

$$[V_{Ni}''] = \text{constant } p_{O_2}^{+1/6} \ . \qquad (7)$$

Because the sulfur diffusivity decreased with increasing oxygen pressure, we rule out the possibility of interstitial diffusion; moreover, the oxygen-pressure dependence was -1/6.9 (see Fig. 2). If the diffusion of sulfur occurred via jumps into anion (oxygen) vacancies,

$$D(S \rightarrow NiO) \propto [V_O^{\bullet\bullet}] \ , \qquad (8)$$

the diffusivity would be proportional to the -1/6th root of the oxygen pressure because there is a Schottky constant,

$$K_S = [V_{Ni}''] \, [V_O^{\bullet\bullet}] \ ; \qquad (9)$$

and using Eqs. (7), (8), and (9) yields

$$D(S \rightarrow NiO) \propto [V_O^{\bullet\bullet}] \propto p_{O_2}^{-1/6} \ . \qquad (10)$$

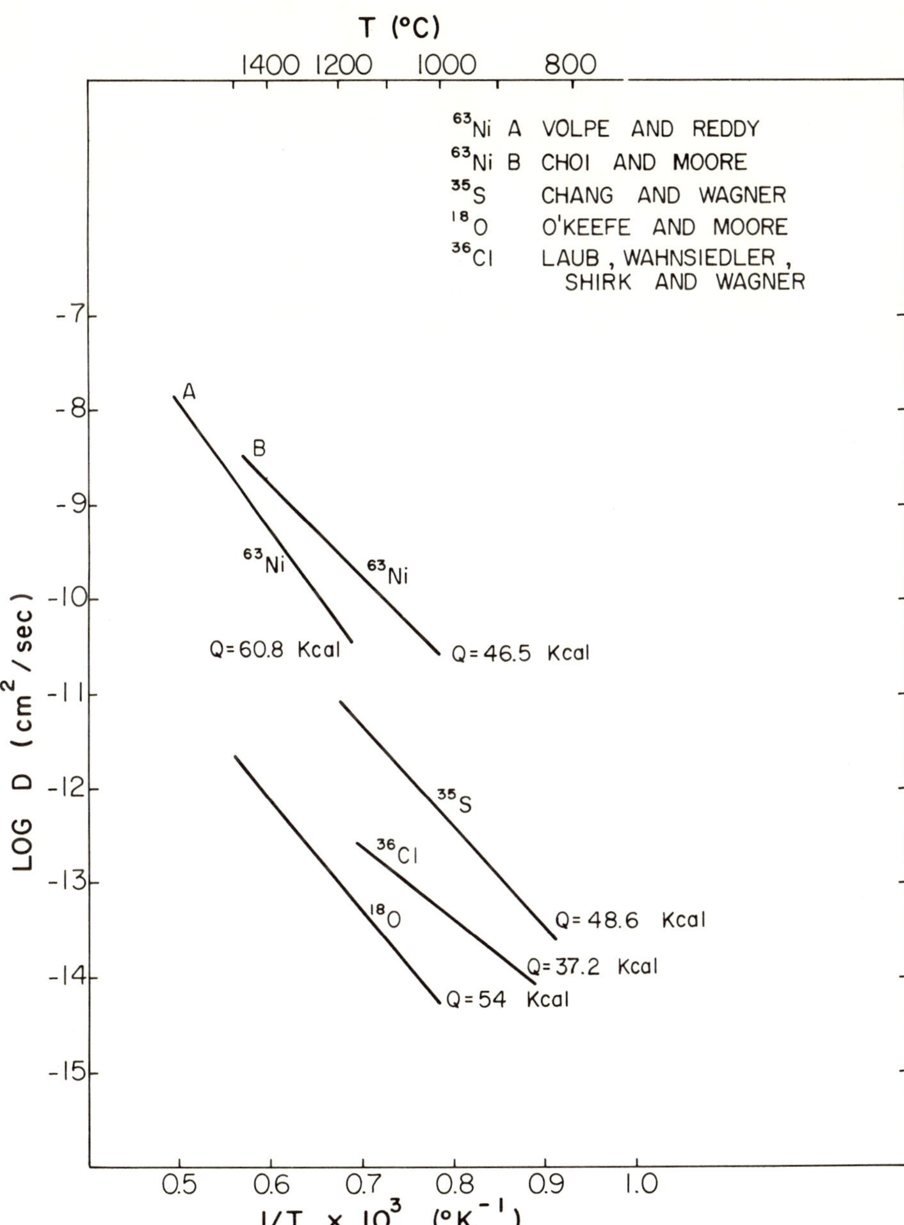

Fig. 1. Arrhenius plot of the diffusivities of nickel, sulfur, chlorine, and oxygen in NiO.

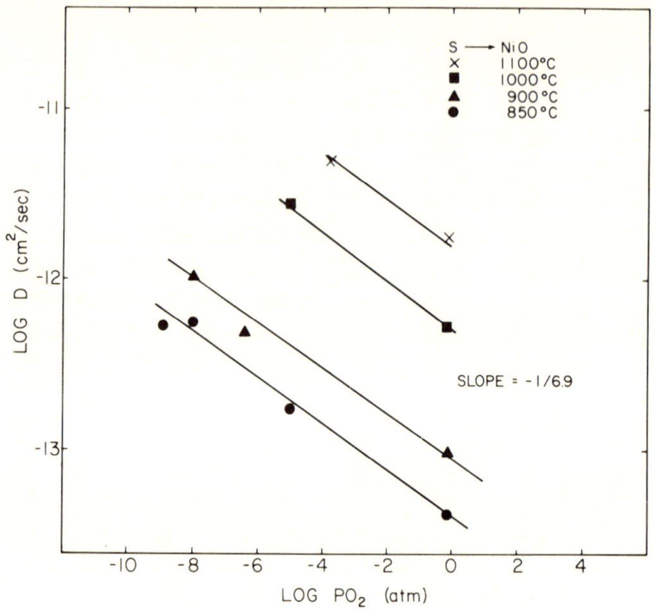

Fig. 2. Plot of the diffusivity of sulfur-35 in single-crystalline NiO as a function of oxygen pressure.[5]

The experimental value was -1/6.9, which suggests that sulfur migrates via anion vacancies. The possibility of a vacancy-impurity complex is discussed later.

The analogous experiments for chlorine-36 have not yet been carried out. The oxygen pressure dependence will be difficult to establish because the Gibbs free energy for the displacement reaction,

$$NiO\ (s) + Cl_2\,(g) = NiCl_2\ (s) + \tfrac{1}{2}O_2\,(g)\quad , \tag{11}$$

amounts to only a few kilocalories in the temperature range of interest. Thus the pressures of oxygen and of chlorine are of the same order of magnitude (under equilibrium conditions). Although this reaction will be limited by the small amount of $NiCl_2$ used, the oxygen pressure will be difficult to control. An alternative method is to use doped crystals. For example, chromium is a donor and doping additions should increase the cation-vacancy concentration according to

$$Cr_2O_3 = 2Cr^{\bullet}_{Ni} + V''_{Ni} + 3O_O\quad . \tag{12}$$

Because the concentration of cation vacancies is increased, the concentration of anion (oxygen) vacancies will be decreased according to Eq. (9). Consequently, if the chlorine migrates via anion vacancies, the diffusivity in the chromium-doped crystals will be smaller than that in an undoped crystal. Such studies are under way at this time.[12] A comparable study has been carried out by Chang[4] for sulfur diffusing into undoped and chromium-doped NiO. Results are presented in Fig. 3. Migration of sulfur via anion vacancies is inferred because of the above argument.

Consider next the solubility limit and state of ionization of sulfur dissolved in NiO. Because divalent sulfur is about the same size as divalent oxygen, as a first approximation one would expect the diffusivities to be about the same. Experimentally, sulfur migrates one and a half orders of magnitude more rapidly than oxygen in NiO. Further, Konev et al.[13], on the basis of electrochemical cell measurements, suggested earlier that sulfur dissolved in NiO as a singly charged species. If sulfur dissolves in NiO as a singly ionized ion, it will act as a donor and thus reduce the electronic conductivity which is due to electron holes [see Eq. (4)]. The appropriate equation can be written as

$$SO_2(g) = S_O^{\bullet} + O_2(g) + e' \quad . \qquad (13)$$

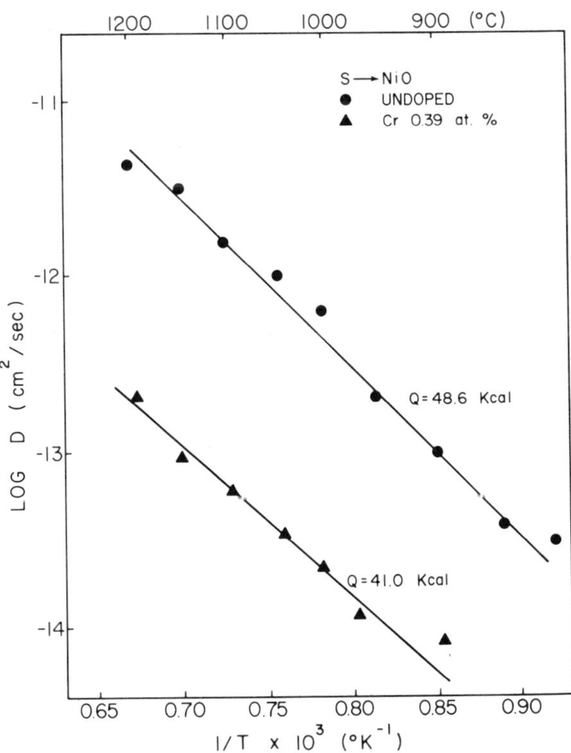

Fig. 3. Arrhenius plot of the diffusivity of sulfur in single-crystalline undoped and chromium-doped nickel oxide under virtually a constant oxygen pressure (0.77 to 1.0 atm). Because the diffusivity is smaller in the chromium-doped crystals, migration via anion vacancies is inferred by Chang.[4] See text.

Accordingly, Chang measured the electronic conductivity of NiO crystals doped with sulfur. His procedure was to first measure the electronic conductivity of the undoped crystal under a 20 percent oxygen-80 percent argon mixture. Thereafter, a mixture of oxygen, sulfur dioxide, and argon, was admitted to dope the crystal. Again the electronic conductivity was measured, and the doped crystal was then submitted for mass spectrometric analysis. Results are shown in Fig. 4. Because sulfur decreases the electronic conductivity, it is inferred that sulfur enters the NiO lattice as a singly ionized species according to the model presented in Eq. (13).

The solubility limit of sulfur in NiO was estimated using two methods. In both cases, a single crystal was heated in a closed quartz tube containing excess sulfur. A thin layer of a golden-colored nickel sulfide formed on the surface. In one set of experiments, the X-ray microprobe of a JEOL JSM-U3 scanning electron microscope was used to measure the concentration of sulfur across a cleaved crystal. For a crystal annealed at 1000 to 1100°C, the concentration just under the NiS_y layer was less than or equal to 10^{18} atoms/cc or 0.01 at.% sulfur.[11] In the second set of experiments, a radiotracer was added to the excess sulfur. After nickel sulfide formed on the NiO surface, the crystal was quenched and the sulfide layer was ground off using metallographic paper. The concentration

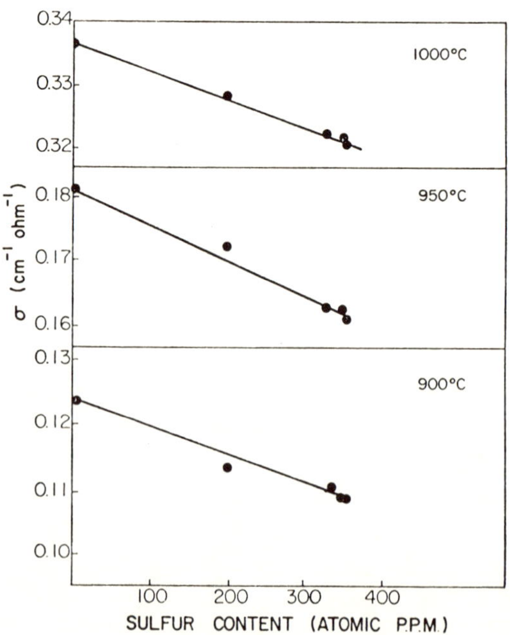

Fig. 4. D-c conductivity of undoped and sulfur-doped single crystals of NiO as a function of sulfur doping.[4] Because sulfur decreases the electrical conductivity, it is inferred that sulfur acts as a donor and probably enters the NiO lattice as a singly ionized species.

of sulfur was estimated to be between 0.05 to 0.1 at.% at 900°C using the radioactivity emanating from the resulting NiO interface.[4] These data are substantially lower than others reported in the literature.[14]

The majority defects in NiO have been inferred to be cation vacancies compensated by electron holes. This model has been confirmed by Seebeck coefficient data[15], electrical-conductivity data[16-18], and radiotracer diffusion studies[7,19,20], as well as by kinetic measurements of the rate of oxidation of nickel metal to nickel oxide[21]. The diffusivity of oxygen in NiO has been suggested to occur via interstitial jumps because O'Keeffe and Moore[8] found the diffusivity of oxygen increased with increasing oxygen pressure. On the other hand, Chen and Jackson, using doped and undoped CoO crystals, report that oxygen migrates in CoO via an anion vacancy.[10] This latter interpretation is consistent with Chang's results for sulfur.[4] A detailed study is needed for NiO and another oxide in this series, perhaps MnO, to test these apparently conflicting modes of migration for oxygen.

If one considers a comparison of only sulfur and oxygen migrating in NiO, the results are consistent for sulfur migrating as an S^- ion based on a size argument. Although the singly charged chloride ion diffuses less rapidly than sulfur, as would be predicted by their relative sizes, the oxygen ion is much smaller than the chloride ion and the steric argument fails. Because, presumably, both chlorine and sulfur diffuse as singly charged ions, the defect equilibria in NiO may be changed as these species enter the lattice. Both ions should act as donors, i.e., both singly ionized chlorine and singly ionized sulfur when substituted for oxygen on a normal site should behave electrically as chromium substituted for a nickel ion on a normal nickel-ion site. Thus both the in-diffusing chlorine and sulfur should result in an increasing concentration of cation vacancies and a *decreasing* concentration of oxygen ion vacancies. If the concentration of diffusant were large enough* to decrease the concentration of oxygen ion vacancies as suggested above, the diffusivities of chlorine and of sulfur should be correspondingly decreased over that for oxygen, especially in view of the size of the chloride ion. This is not observed. The penetration profiles of both the chlorine and sulfur appear to follow a normal complementary error-function form as given in Eq. (1). Doping a nickel oxide crystal with 0.39 at.% chromium decreases the diffusivity of sulfur as expected. But in this case and in the case of initially undoped crystals, there was no evidence of the type of penetration profiles observed by

*Approximately 10^{-4} mole of radiotracer ion are enclosed in the tube. This corresponds to about 10^{19} atoms. If all these ions were incorporated in the surface layer, a change in stoichiometry could result. Chang[4] has calculated the sulfur concentration in the surface layer to be 10^{16} atoms/cc and Shirk et al. estimate the concentration of chlorine to be less than this. This concentration is too low to affect the native defects in NiO and CoO.

Seltzer[22] and by Perkins and Rapp[23] for the diffusion of chromium into initially undoped NiO crystals. Chen, Peterson, and Robinson[24] have recently reported that small concentrations of chromium diffusing into very pure NiO crystals did not exhibit the self-doping effect reported by Seltzer[22], and furthermore, the activation energy was *larger* (67.4 kcal) than the highest value yet reported for nickel in NiO. Further experiments are needed to test the concentration dependence by diffusing sulfur into crystals previously doped with known sulfur concentrations.

The activation energy for the diffusion of nickel into NiO has been reported to range from about 44 to 61 kcal/mole. It is probable that the larger values represent diffusion in the purer crystals. The activation energies for the diffusion of both iron[19] and chromium[22] into NiO have been reported to be lower than that for nickel in pure NiO, and explanations have been advanced that the diffusion may have been due to extrinsic impurities in the crystals. The analyses for the NiO and the CoO crystals used in the present studies are presented in Tables II and III.

Table II. Impurity Content of NiO

(Parts per Million Atomic)

Element[(a)]	Detection Limit[(b)]	NiO
Li	0.1	N.D.[(c)]
F	0.1	3.4
Na	0.1	14
Al	0.7	N.D.
Si	0.3	2.5
S	7	38
Cl	0.7	0.92
K	0.1	0.21
Ca	0.7	10
Mn	1	N.D.
Cr	0.3	N.D.
Fe	1	12
Co	3	46
Cu	0.3	N.D.
Zn	3	7.6

(a) No analysis was made for hydrogen. Analyses for tantalum and gold are not given since tantalum slits are used in the mass spectrometer and the samples were sparked against high-purity gold counterelectrodes. Other elements not listed were not detected and have concentrations less than 10 ppm atomic.

(b) Determined for 3×10^{-8} coulomb exposure.

(c) N.D. – not detected.

Table III. Impurity Content of CoO
(Parts per Millino Atomic)

Element[a]	Detection Limit[b]	CoO Single Crystal
C	0.1	20
N	0.1	N.D.[c]
F	0.2	N.D.
Na	0.1	N.D.
Mg	0.5	17
Si	2	15
P	0.7	1
S	7	12
Cl	2	5
K	0.3	79
Ca	1	120
Cr	2	450
Fe	3	150
As	5	21

(a) No analysis was made for hydrogen. Analyses for tantalum and gold are not given since tantalum slits are used in the mass spectrometer and the samples were sparked against high-purity gold counterelectrodes. Other elements no listed were not detected and have concentrations less than 10 ppm atomic.
(b) Determined for 3 x 10^{-8} coulomb exposure.
(c) N.D. – not detected.

An alternative explanation is that the diffusion occurs via an anion vacancy – impurity ion complex such as has been advanced by Lidiard[25] and invoked for cadmium into silver bromide. At this time, there are insufficient data to make a test of this model.

Sulfur diffuses more rapidly than oxygen in the analogous crystal, CoO. Results for single crystalline CoO heated under virtually constant oxygen pressure (0.77 to 1 atm) are shown in Fig. 5. Presumably, sulfur migrates via the same mechanism as in NiO. Studies are under way in this laboratory to test the oxygen-pressure dependence of the diffusivity in single crystalline CoO.

A few data have been obtained[4] for sulfur diffusing into single-crystalline Al_2O_3 and Fe_2O_3, and these are given below:

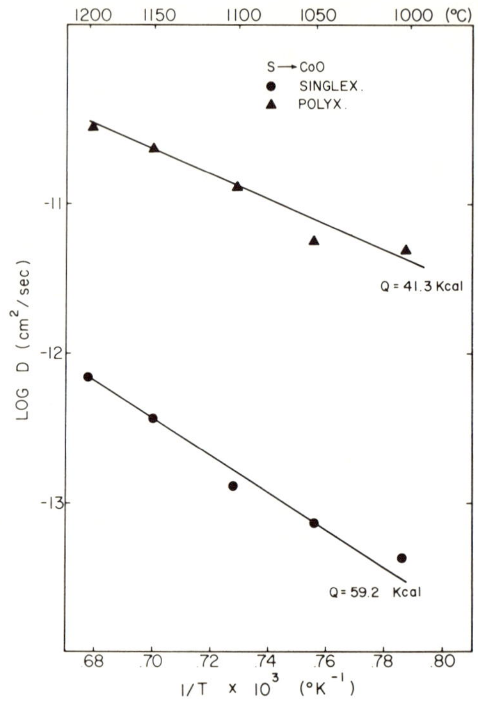

Fig. 5. Arrhenius plot for the diffusivity[5] of sulfur-35 in single-crystalline and in polycrystalline CoO heated under virtually constant oxygen pressure (0.89 to 1 atm). The polycrystalline sample had a density of 95.6 percent of theoretical and an average grain size of 10 to 12 microns.

Crystal	T, °C	p_{O_2}, atm	D, cm^2/sec
Al_2O_3	1300	1.1	$\sim 10^{-15}$
Fe_2O_3	1200	~1	$\sim 2.1 \times 10^{-13}$

Again the sulfur appears to migrate more rapidly than oxygen. However, these are preliminary results and must be treated as such.

Some studies were also carried out to determine the kinetics of transport of sulfur on polycrystalline compacts. The samples were pressed and sintered. They were then given a prediffusion anneal, followed by a diffusion anneal as described above for the single crystalline samples. The data were analyzed in the same way. For comparative purposes, the data obtained from the penetration profiles were treated as being due to bulk diffusion [Eq. (1)]. Data for polycrystalline NiO, CoO, Al_2O_3, and Cr_2O_3 are shown in Figs. 5 through 8. It is obvious that diffusion down surfaces is extremely rapid compared with bulk diffusion. The diffusion of sulfur in single-crystalline Al_2O_3 at 1300°C is three to five orders of magnitude less than that in polycrystalline material. Most surprising were the data on chromia. Chang's data[5] are in good agreement with Seybolt's[26]. However the activation energy is very low (17.8 kcal) – lower than polycrystalline NiO of the same density and grain size. Diffusion studies of sulfur in single-crystalline Cr_2O_3 are under way to provide a comparison.

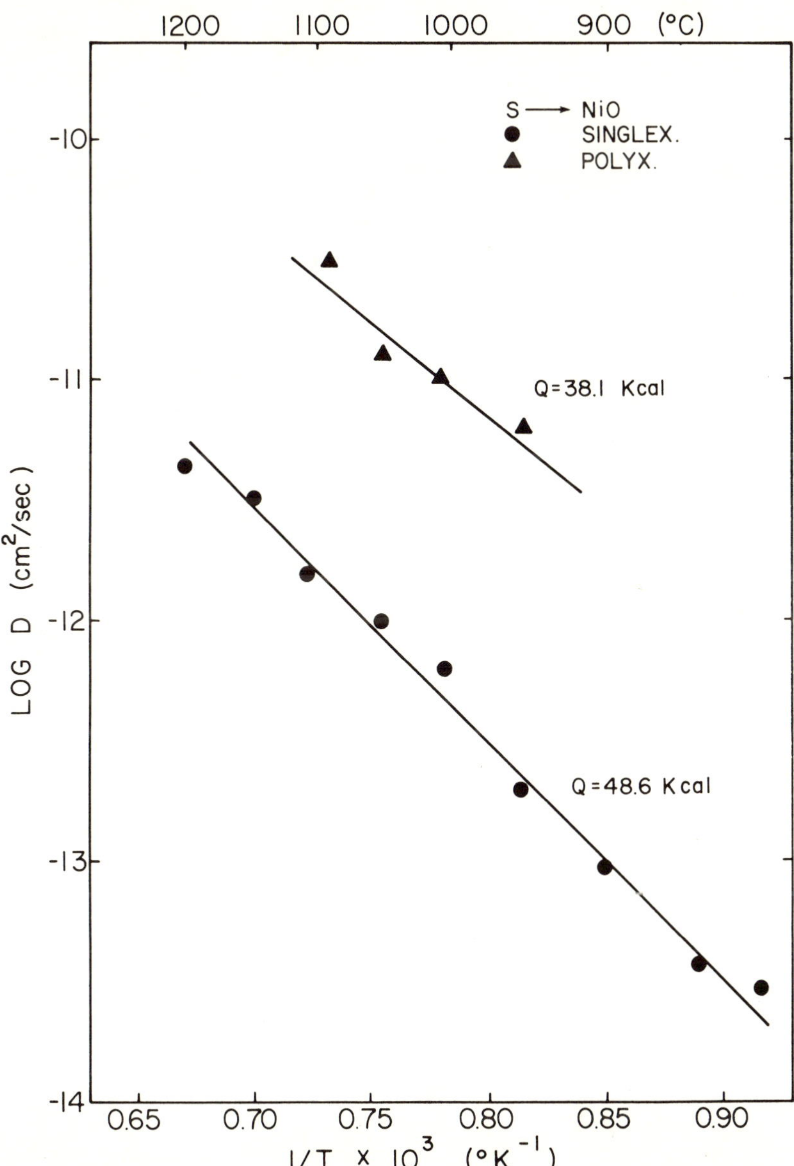

Fig. 6. Arrhenius plot for diffusion[6] of sulfur in single and polycrystalline NiO heated under virtually constant oxygen pressure (0.77 to 1 atm). The polycrystalline sample had a density of 93.1 percent of theoretical and a grain size of 10 to 12 microns.

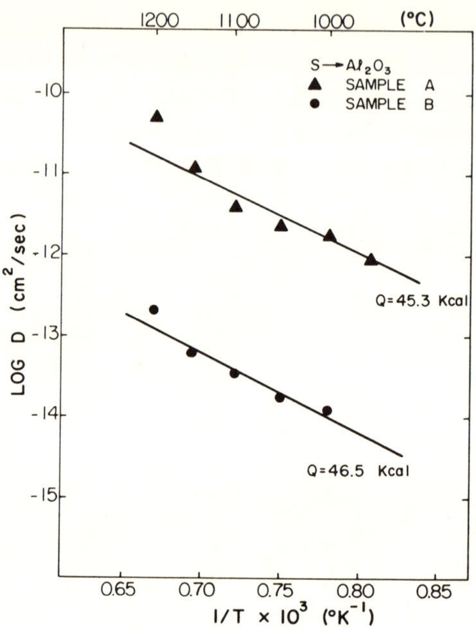

Fig. 7. Arrhenius plots of the diffusion of sulfur in polycrystalline Al_2O_3 samples heated under virtually constant oxygen pressure (0.86 to 1.0 atm). Sample A had a density of 94 percent of theoretical, while the density of Sample B amounted to 97.2 percent of theoretical. Sample A had an average grain size of 3 microns, while Sample B had an average grain size of 10 to 12 microns.

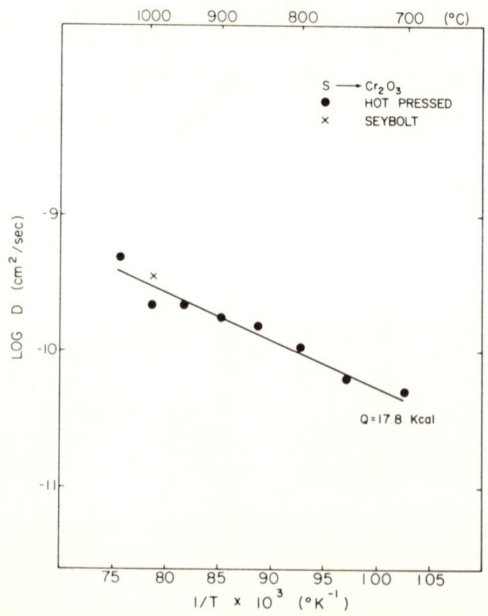

Fig. 8. Arrhenius plot of the diffusion of sulfur in polycrystalline Cr_2O_3. The density was 92.7 percent of theoretical and the average grain size was 3 microns.[5] The datum of Seybolt[26] is also plotted.

4 SUMMARY AND CONCLUDING REMARKS

Measurements of radiotracer diffusion coupled with electrical conductivity have been shown to be useful in determining the role of minority defects in oxides. Sulfur has been shown to migrate via atom jumps into anion vacancies in NiO. Further studies are needed to determine whether the impurity-anion vacancy complexes play a significant role in the transport process. Sulfur was also shown to behave as a donor, i.e., sulfur enters the NiO lattice as a singly charged ion. On the basis of preliminary data, it is speculated that chlorine may behave similarly to sulfur.

In spite of the relatively large sizes of chlorine and sulfur ions, they diffuse more rapidly than oxygen in NiO. Sulfur also diffuses more rapidly than oxygen in CoO and Al_2O_3.

ACKNOWLEDGMENT

These studies were supported in part by grants from the National Science Foundation and the Army Research Office – Durham.

REFERENCES

1. *High Temperature Metallic Corrosion of Sulfur and Its Compounds,* Z. A. Foroulis (Ed.), The Electrochemical Society, Princeton, N. J. (1970).
2. Hauffe, K., and Hinrichs, J., "Simultaneous Oxidation and Chlorination of Nickel", in *Oxidation of Metals and Alloys,* American Society for Metals, Metals Park, Ohio (1971).
3. Jannick, R. F., and Whitmore, D. H., *J. Phys. Chem. Solids,* **27**, 1183 (1966).
4. Chang, R. H., "Sulfur Diffusion in NiO, CoO, α-Al_2O_3 and Cr_2O_3", Ph.D. Thesis, Northwestern University (1973).
5. Chang, R. H., Nemoto, R., and Wagner, J. B., Jr., submitted for publication to *Trans. AIME,* June 21, 1973.
6. Chang, R. H., Nemoto, R., and Wagner, J. B., Jr., submitted for publication to *Trans. AIME,* July 2, 1973.
7. Choi, J. S., and Moore, W. J., *J. Phys. Chem.,* **66**, 1308 (1962).
8. O'Keeffe, M., and Moore, W. J., *J. Phys. Chem.,* **65**, 1438 (1961); Notes and Corrections, *ibid.,* **65**, 2277 (1962).
9. Carter, R. E., and Richardson, R. D., *J. Metals,* **6**, 1244 (1954).
10. Chen, W. K., and Jackson, R. A., *J. Phys. Chem. Solids,* **30**, 1309 (1969).
11. Chang, R. H., Stewart, W., and Wagner, J. B., Jr., *Proc. 7th International Symposium on the Reactivity of Solids,* Chapman and Hall, London (1972), p. 231.
12. Laub, L., Shirk, A., and Wahnsiedler, W., research in progress, Northwestern University.
13. Konev, V. N., Chebotin, V. N., Suntsov, N. V., and Stavtseva, L. I., *Zaschita Metallov.,* **6**, 448 (1970).
14. Alcock, C. B., Hocking, M. G., and Zador, S., *Corrosion Sci.,* **9**, 111 (1969).
15. Bosman, A. J., Daal, H. J., van, and Knuvers, G. F., Phys. Letters, **19**, 372 (1965); Bosman, A. J., and Crevecoer, C., *Phys. Rev.,* **144**, 763 (1966); Daal, H. J., van, and Bosman, A. J., *Phys. Rev.,* **158**, 736 (1967).

16. Baumbach, H. H., von, and Wagner, C., *Z. Physik. Chem.*, **B24**, 59 (1934).
17. Eror, N. G., and Wagner, J. B., Jr., *Phys. Stat. Solidi,* **35**, 641 (1969).
18. Osburn, C. M., and Vest, R. W., *J. Phys. Chem. Solids,* **32**, 1331 (1971).
19. Crow, W. B., "Diffusion of Cobalt, Nickel, and Iron in Cobalt Oxide and Nickel Oxide", Ph.D. Thesis, The Ohio State University, 1969.
20. Volpe, M. L., and Reddy, J., *J. Chem. Phys.*, **53**, 1117 (1970).
21. Fueki, K., and Wagner, J. B., Jr., *J. Electrochem. Soc.*, **112**, 384 (1965).
22. Seltzer, M. S., *J. Electrochem. Soc.*, **118**, 802 (1971).
23. Perkins, R. A., and Rapp, R. A., *Trans. AIME*, **4**, 193 (1973).
24. Chen, W. K., Peterson, N. L., and Robinson, L. C., *J. Phys. Chem. Solids,* **34**, 705 (1973).
25. Lidiard, A. B., *Phil. Mag.*, **46**, 815 (1955); *ibid.*, **46**, 1218 (1955).
26. Seybolt, A. U., *Trans. AIME,* **242**, 752 (1968).
27. Goldsmidt, V. M., *Chem. Ber.*, **60**, 1263 (1927).
28. Pauling, L., *J. Am. Chem. Soc.*, **49**, 765 (1927); *The Nature of the Chemical Bond,* 3rd Ed., Cornell University Press, Ithaca, N.Y. (1969).
29. Nickless, G. (Ed.), *Inorganic Sulphur Chemistry,* Elsevier Publishing Co., Amsterdam (1968).

DISCUSSION on Paper by J. B. Wagner, Jr.

JOHNSON: Did you examine the Cr_2O_3 specimens for open porosity? At 93 percent of theoretical density, there may be pores open to the atmosphere.

WAGNER: No, we examined the polished surfaces for average grain size but we did not examine the samples for interconnecting pores.

GOODENOUGH: Your observation of a decrease in the number of mobile holes with sulfur concentration does not necessarily imply the existence of an S^- ion. Strong covalent mixing occurs between the nickel 3d orbitals and the sulfur 3p orbitals. The stronger covalent mixing has two effects: (1) it lifts the energy of the acceptor level at a sulfur site, thereby trapping the holes and removing them from the pool of mobile holes, and (2) it spreads the hole out over the sulfur and its nearest-neighbor cations. If this latter effect enhances the sulfur diffusibility, then there should be interesting differences for sulfur diffusion in other oxides where there is no equivalent covalent mixing between cation d and sulfur p orbitals.

WAGNER: Thank you for this explanation. One additional type of information supports the model of a singly charged sulfur ion substituting for a doubly charged oxygen ion in NiO. Workers in England have reported that the parabolic kinetics of a layer of NiO growing on a pure-nickel sheet are increased when sulfur is incorporated in the growing NiO layer. The inference is that sulfur enters the NiO lattice under the formation of additional cation vacancies. Because NiO grows on nickel by migration of nickel ions via cation vacancies, the parabolic rate of oxidation is increased by the presence of sulfur. Thus, to a first approximation, sulfur acts as a donor in NiO, similar to chromium which also creates cation vacancies.

VINK: In view of the small solubility of sulfur in NiO that you did deduce from your experiments, one could doubt that a possible donor activity of sulfur could explain the decrease in the p-type conductivity of your NiO samples.

Did you consider the possibility that sulfur actually reduces the NiO by the formation of SO_2, and in that way does lead to a decrease in the p-type conductivity of NiO?

WAGNER: Thank you for this suggestion. I wish to make distinctions between the three experimental procedures. For the radiotracer-diffusion studies, the samples were first preannealed in a closed quartz tube containing oxygen but no sulfur. Then a thin layer of radiotracer sulfur was applied to one surface of the crystal and it was again sealed in a quartz tube containing initially the same oxygen pressure as for the prediffusion anneal. At the elevated temperature, the sulfur presumably evaporated to form SO_2. Calculations showed this amount of sulfur to be too small to affect the defect equilibria in the crystal, i.e., to cause reduction.

For the solubility studies, crystals were heated in a tube containing oxygen *and an excess of sulfur.* In this case, the crystal was reduced and a layer of nickel sulfide formed on the surface of the nickel oxide. The solubility limit was estimated from the concentration of sulfur in the NiO just under the nickel sulfide layer.

In the case of the electrical-conductivity measurements, the crystals were first preannealed in a gas mixture of 20 percent oxygen-80 percent nitrogen. This provided a reference point for the undoped crystals. A four-probe d-c method was used. In order to dope the samples, the crystals were heated for a period of 1 to 2 months in the following gas atmospheres:

20% O_2, 2% SO_2, 78% Ar
20% O_2, 5% SO_2, 75% Ar
20% O_2, 7% SO_2, 73% Ar
20% O_2, 10% SO_2, 70% Ar.

These gas compositions maintain nickel oxide in its homogeneity range. The crystals then were heated again in the 20 percent oxygen-80 percent nitrogen mixture for a period of about 1 week and the electrical conductivity was measured. Then the crystals were quenched and submitted for mass spectrometric analysis. This procedure is analogous to measurements on chromium-doped or lithium-doped nickel oxide in which the presence of the small amount of ternary constituent is neglected in fixing the chemical potentials coexisting with the specimen.

KROGER: Chromium in NiO increases the V_{Ni}'' concentration to a fixed level, with $[V_{Ni}''] \approx 1/2\ [Cr_{Ni}^{\bullet}]$ = constant. From this the V_O'' concentration is found through $[V_O''] = K_S/[V_{Ni}''] = K_S$/constant. Therefore $[V_O'']$ varies with temperature just as K_S does. The slope of sulfur diffusion should therefore be the sum of that of mobility + H_S.

WAGNER: You are correct. As far as I am aware there is no reported value for the enthalpy of formation of the Schottky pairs in NiO. Therefore the migration enthalpy for sulfur diffusing in NiO cannot be calculated at this time.

We plan to study crystals doped with an acceptor, e.g., lithium, in order to increase the concentration of anion (oxygen) vacancies. At this time, however, we can state only that increasing the concentration of cation vacancies in NiO results in a smaller diffusivity of sulfur in NiO than in undoped NiO, which is consistent with the results of studies on the oxygen-pressure dependence of the diffusivity.

KOFSTAD: The source of radioactive sulfur is a mixture of O_2 + SO_2 (as the deposited sulfur film is oxidized during the diffusion anneal). When the partial pressure of oxygen is reduced, the sulfur activity is increased. Has this aspect been taken into account in the interpretation of the results (showing increased sulfur diffusion with decreasing partial pressure of oxygen)? Correspondingly, the sulfur activity changes with temperature. Has this been taken into account in the interpretation of the activation energy for sulfur diffusion?

WAGNER: For the diffusion studies, the crystals or polycrystalline compacts were sealed in a closed quartz tube containing oxygen. The only source of sulfur was from the thin layer of radiotracer on the surface. It is most probably true that the sulfur vaporized and formed SO_2. However, the amount of SO_2 was too small to affect appreciably the partial pressure of oxygen in the tube. The presence of this small amount of SO_2 was not taken into account in my discussion.

SOCKEL: Is it possible to give an estimate of the enthalpy of formation of Schottky defects in NiO with the experimental and other available data?

WAGNER: In my judgment, there are insufficient data for one to calculate a value of the enthalpy of formation of Schottky defects in NiO. Many workers have studied properties of NiO treated under constant oxygen pressure. Such data, when plotted on an Arrhenius plot, yield slopes which involve the partial molar enthalpy of dissolution of oxygen in nonstoichiometric NiO plus a term for the migration of the species under study. As an example, the data for electrical conductivity of NiO treated under constant oxygen pressure yield an activation energy

$$Q = \left[\frac{\Delta \bar{H}_O{}^M}{(m+1)} + \Delta H^{\ddagger} \right] ,$$

where $\Delta \bar{H}_O{}^M$ denotes the partial molar enthalpy of solution, $\Delta H^{\ddagger}$ the migration enthalpy of electron holes and m the degree of ionization of the vacancies. From thermodynamic data, $\Delta \bar{H}_O{}^M$ is known and the migration enthalpy of electron holes (or of cations if nickel self-diffusion is studied) may be calculated. As far as I am aware, the enthalpy for formation of Schottky pairs has not been estimated from other data and it cannot be inferred from our data.

NOWICK: I wish to question your statement that a large ion, like $S^=$ in NiO, would not diffuse easily enough. I see no reason why the activation energy for migration of a large ion should not be reasonably low. Rather, if the ion is too large, its *solubility* will be very limited, and this may be the reason that sulfur prefers to enter the lattice as S^- plus an electron.

WAGNER: Thank you for your comment. I would have predicted a much lower diffusivity for sulfur in oxides than has been found. If sulfur enters NiO as a divalent ion substituting for an oxygen ion, there should be no change in the electrical conductivity to a first approximation. Because sulfur decreased the electrical conductivity, we inferred that it entered the lattice as S^- plus an electron.

DIFFUSION-PATH NETWORKS IN THE WURTZITE LATTICE

Ralph H. Condit

University of California
Lawrence Livermore Laboratory
Livermore, California

ABSTRACT

A description is given of the network of cation jump possibilities between interstices in structures where the anions are in hexagonal close packing. The discussion focuses primarily on the wurtzite lattice (BeO and ZnO), where the activation energies for diffusion (D) in the a- and c-directions have been experimentally found to be the same. Two types of vacancy mechanisms give D_c/D_a = 1.3 and 1.5, and an interstitialcy mechanism gives 1.0; interstitial diffusion can be ruled out since it would be strongly anisotropic.

1 INTRODUCTION

Atomic diffusion takes place as a result of a random-walk sequence of jumps between normally occupied sites. In the case of self-diffusion by a vacancy mechanism in a close-packed cubic structure, the jump is probably simple, consisting of a direct passage through a single energy barrier. In more complicated structures, the jump path may include several metastable stopping points along the way and branch points from which the atom may take one of several alternative routes. This is certainly the case in oxides. Since one purpose of diffusion research is to understand the mechanisms of the basic jump process, we must identify the component jump parts of the paths in the hope that we can find ways of differentiating between them and assigning each a value based on experiment. Thus, determining the network of diffusion paths should provide a bridge between experiment and theory.

Structures where the anions are in hexagonal close packing with the cations nestled between them seem to be well suited to studies in which theory and experiment might be compared, because one has the handles of anisotropy of diffusion, possibilities for measuring the correlation coefficient, and comparisons between analogous structures. In many oxides the cations are much smaller than the anions, and can fit between them without much distortion of their close packing. The interstices in an anionic close-packed lattice can be identified as one octahedral interstice (an o-site) and two tetrahedral interstices (t-sites) per anion. The radii of cations that can fit into these interstices without distortion are 0.414 and 0.22 of the anionic radii for the o- and t-sites, respectively, using a hard-sphere model. In the compounds to be discussed in this paper, we have the radius ratios 0.42 in PbI_2 (about 0.9/2.16), 0.41 in FeS (0.76/1.84), and 0.38 in NiS (0.70/1.84), which are compounds where the cation is in the o-sites, and 0.22 in BeO (0.31/1.40) and 0.53 in ZnO (0.74/1.40), which are compounds where the cation is in t-sites. Zinc oxide is the one deviant from the picture of easy fit of the cation into the anionic interstices.

The arrangement of the interstices within an hcp unit cell is shown in Fig. 1. Anion positions are not shown, but it may be noted that each anion is surrounded by six o-sites and six t-sites. The shapes of the o- and t-sites are roughly cubic and tetrahedral, respectively, as shown in Fig. 1. The interstices are connected by lines which we regard as plausible jump paths for the cations; they pass through an orifice formed by the three anions that touch on each of the two connected interstices. While there will obviously be local distortions of a lattice as an ion passes through an orifice, we will assume that no other jump paths than these are opened up. We will apply this model to ZnO along with the other compounds, but with somewhat less assurance.

In the hcp matrix, the o-sites extend in chains along the c-direction, while the t-sites are coupled in pairs. In the basal plane, the sites are arranged in a hexagonal net of o-sites with a t-site in the center of each triangle of o-sites, alternate t-sites being a little above and below the plane of o-sites (see Fig. 2). It can be seen that diffusion in the c-direction is possible by a sequence of direct-o-o- jumps or -o-t-t-o- jumps, while diffusion in the a-direction can be by -o-t-o- or -o-t-t-o- jumps. The diffusion of an interstitial atom in an hcp matrix has been discussed previously[1], and that discussion can serve as an introduction to some aspects of the present approach.

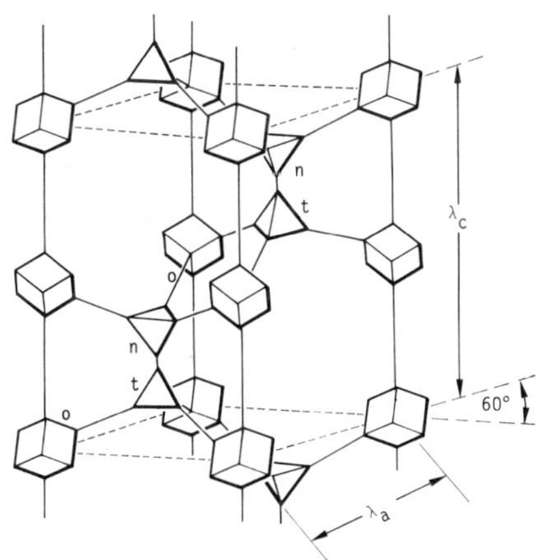

Fig. 1. The interstice network within the hcp structure. A unit cell containing two o-sites (octahedral coordination) and four t-sites (tetrahedral coordination) is shown. In the wurtzite structure the upper one of each pair of t-sites is normally filled; this is called an n-site.

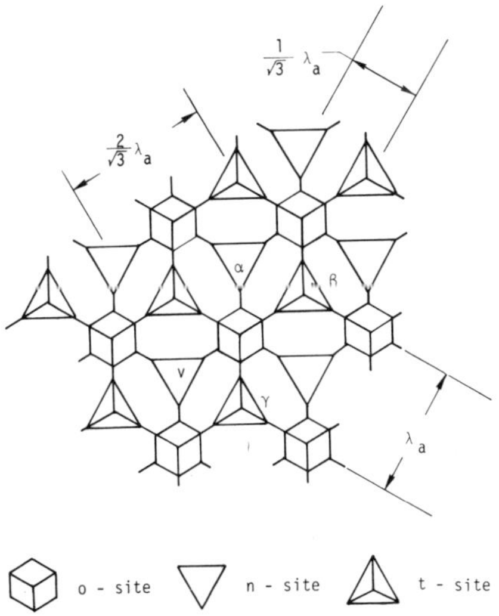

Fig. 2. Top view of the interstice network in the hcp structure. If one n-site (normally occupied) is vacant (v), there will be three types of site from which it can be filled, an α-, a β-, or a γ-position (see Table I).

2 DISCUSSION

We will now remark on the structures of several compounds in which the anions are in hexagonal close packing. It is interesting to compare PbI_2 (CdI_2 structure), FeS and NiS (NiAs structure), and BeO and ZnO (wurtzite structure). In the first two cases, the cations occupy the o-sites. In CdI_2, only every second plane of o-sites (counting in the c-direction) is occupied, while in NiAs, all of the o-sites are occupied. In the wurtzite structure, none of the o-sites and only half of the t-sites are filled. As noted above, the t-sites are paired; the lower one is normally empty. This crystal asymmetry gives rise to piezoelectric and related properties. There seem to be no structures in which both of the t-sites are filled in the hcp matrix. In Ni_2In, the In atoms provide the hcp matrix, and the Ni atoms are distributed evenly among the o-sites and the t-site pairs; in a t-site pair, the Ni atom resides at the midpoint between the two sites.[2] This suggests that in the hcp matrix the t-t midpoint energy barrier is low and, in at least this one case, the t-t midpoint is even the stable position. In BeO, there is a form stable above 2350 K, β-BeO, in which the anions are retained in an hcp structure (although with some distortion) but the cations on alternate pairs of t-sites move down to the lower site.[3] The heat of this transformation is about 8 kJ/mole, a small amount compared with most diffusion energies. These two observations suggest that the pair of t-sites, the normally occupied and the normally unoccupied, not only have a low barrier between them but also have similar energy levels.

In turning to the discussion of diffusion mechanisms, it is appropriate to clarify terms. For example, while interstitial diffusion has a well-defined meaning for carbon migration in iron, it may not be clear when we are talking about cation diffusion in an anionic matrix. An actual diffusion jump might be contingent on there being a vacancy on a normal site to receive the cation, even though the cation may have passed through several interstices within the anion lattice. Azaroff[4] has suggested the term *voidal diffusion* for this process; the mechanism as generally understood might be a vacancy or interstitial mechanism, but the path will take the cation through what he calls voids, our o- and t-sites.

For PbI_2, only ionic conductivity data are available[5], and these are open to several interpretations. However, it is known that the activation energy for migration in the c-direction is about three times that for migration in the a-direction. If transport is by lead ions, this suggests that the -o-o- jumps are not the easy ones for the cations. However, since diffusion in the c-direction could possibly involve an interstitialcy mechanism with a high activation energy, the differentiation between energies of individual jump components is not straightforward.

In both FeS[6] and NiS[7], it has been established that (a) cation diffusion is by a vacancy mechanism, (b) the activation energy for diffusion is the same in the a- and c-directions, (c) the effective frequency of jumps in the c-direction is about twice that in the a-direction, and (d) the correlation coefficient is not greatly different from 1.0, i.e., the diffusion coefficient one derives from the rate of the metal-sulfur reaction is close to the value obtained in tracer studies. Knowledge about the correlation coefficient gives important information about the rate-limiting step. Inspection of the structure reveals that if the -o-o jumps were relatively easy, a vacancy might run up and down the chain of o-sites without contributing to a net tracer-atom displacement; the correlation factor would be zero. In this case, an interchange of cation positions would require one to jump out from the chain, an o-t jump, to permit interchange of positions within the chain. In this way, it would work out that the rate-limiting step for tracer diffusion would be the same for the a- and c-directions, namely an o-t jump, even though the o-o jumps were more frequent. The fact that cations do not stream down this -o-o channel during the metal-sulfur reaction eliminates this possibility from consideration. This leaves us with the mechanism being one where the vacancy is filled from a neighboring site by an o-t-o or an o-t-t-o jump, with the rate-limiting jump component being the o-t barrier. That the D_c/D_a ratio is near 2 rather than the expected 1 can be explained only if jumps into a vacancy cannot occur from sites which are coplanar with it. The relaxation of anions about vacancies on normal cation sites seems to affect the strength of binding between coplanar neighboring cations and their sites, and these cations are less likely than other neighboring cations to jump out from their sites. If this binding is about 6 kJ/mole, the energy of vacancy ordering in Fe_7S_8 at lower temperatures, the theory and experiment would agree. It is interesting to note that below the ordering temperature in FeS, the D_c/D_a ratio becomes nearly 1.0.[6]

For the wurtzite structures, we have the following data: (a) diffusion in BeO is by a vacancy mechanism[8,9], (b) the correlation coefficient is close to 0.78[10], (c) the activation energy for diffusion is the same in the two directions, and (d) the D_c/D_a ratio is 1.3 ± 0.1. In ZnO, the mechanism is still under discussion, but the diffusion activation energy is the same for the two directions and D_c/D_a is nearly 1.0.[11]

The network of sites and paths in the wurtzite lattice includes o-sites (normally unoccupied), t-sites that are normally unoccupied, and t-sites that are normally occupied, which we will call n-sites. Let us first consider a vacant n-site and ask how it might be filled from any one of the neighboring filled n-sites. If we assume that all jumps of the same type have the same energy barrier regardless of where they lie relative to a vacancy, it turns out that it is impossible to derive a jump sequence having a rate-limiting step common to both a- and c-directions. This argument runs through the successive elimination of possibilities: (a) If the o-o barrier were small, c-diffusion would be rapid. In fact, this voidal diffusion route would be available for both vacancy and interstitial diffusion. Since this is not the experimental observation, the o-o barrier must be large. (b) Since the o-o barrier is large, the n-t barrier must be smaller than any of the remaining barriers. If it were not, diffusion in the a-direction rather than in the c-direction would be preferred. (c) This leaves us with a choice between the n-o and the t-o barriers as being rate limiting. Inspection of the structure in Fig. 1 shows that diffusion in the c-direction requires passage through these two barriers in succession (n-t-o-n or n-o-t-n jumps), while diffusion in the a-direction requires passage through either one or the other (n-o-n or n-t-o-t-n jumps). Thus, since the highest barrier limits diffusion in one direction while the lowest limits diffusion in the other, the activation energies for diffusion in the a- and c-directions would differ. It would be an unlikely coincidence for the two barriers to be the same. Moreover, if they were, it could be shown that there should be no anisotropy of diffusion by using the same arithmetic as in Eqs. (10), (11), and (12) below.

Let us consider the possibility that the barriers have different energies depending on their positions relative to the vacancy. The t-o barrier might be smaller than the n-o, but the path n-t-o-t-n might be forbidden as a way a vacancy could be filled from a neighboring coplanar site. Actually, inspection of the structure reveals that a cation taking this latter path will find itself, when in the o-site, very close to three adjoining filled n-sites, and the electrostatic repulsion energy would be high. The remaining path for a-direction diffusion would be the n-o-n sequence, which includes the same barrier as for the c-direction. In this case, the moving cation would still pass through an o-site, but only two of the three neighboring n-sites would be filled.

The logic of arguing that cation-cation repulsion is important leads us to consideration of what we will call a key-lock mechanism. We know that a cation on an n-site can move easily to the adjoining t-site. This stepping back of a cation from the immediate neighborhood of an o-site should make that site more open to the passage of another cation. A similar proposal has been offered by Slifkin[12] to explain the fact that in AgCl the activation energies for Cl^-, Br^-, and I^- ion diffusion are the same; an Ag^+ ion shifting back from its normal o-site into a t-site could open up an orifice large enough to permit passage of any one of the three anions with equal ease.

The interstitial mechanism cannot explain diffusion data in either BeO or ZnO because no path for voidal diffusion around normally filled sites can be found having a common rate-limiting step for the a- and c-directions.

The interstitialcy mechanism does appear plausible. An extra cation in the wurtzite lattice must reside in an o-site (somewhat as in Ni_2In). For the purposes of this discussion, the n-site and t-site of a pair may be treated as a single unit, an nt-site. Each nt-site is surrounded by six o-sites and each o-site by six nt-sites. After a cation has been ejected into an o-site by an interstitial next to it, it can jump into any of the six nt-sites around it, one of which would be its original position. Since there is only one jump mechanism that can lead to a- or c-diffusion, the activation energies for migration in these two directions would be the same.

Other diffusion mechanisms in the wurtzite structure, such as ring or other multiatom processes, might be considered, but the entropy factors would not seem to favor them enough to justify discussion at this time.

Let us now count up the contributions of various jumps to the net displacement and incorporate a weighting factor for the number of ways a given displacement may occur. It can be seen from Figs. 1 and 2 that a vacancy can be filled from sites having three types of position relative to it: (1) an n-site coplanar with the vacancy (an α-path); (2) an n-site one plane above or below the vacancy and more or less colinear with the vacancy and the o-site between them, really a next nearest neighbor as measured by hcp indexing (a β-path); and (3) an n-site that is one plane above or below the vacancy and is a nearest neighbor as measured by hcp indexing (a γ-path). Some of the properties of these paths are given in Table I.

Table I. Properties of the α, β, and γ Displacement Paths

Site Type	Number of Such Sites	Jump Frequency	Path Type	Displacement, X_i a-Direction	Displacement, X_i c-Direction
α	6	Γ_α	n–o–n	a	0
β	6	Γ_β	n–t–o–n n–o–t–n	$(2/\sqrt{3})$ a	c
γ	6	Γ_γ	n–t–(o, o)–n n–(o, o)–t–n	$(1/\sqrt{3})$ a	c

Noting that the diffusion coefficient is

$$D = \frac{1}{2} \Sigma g_i \Gamma_i X_i^2 , \qquad (1)$$

where g_i is the weighting factor, Γ_i is the jump frequency, and X_i is the displacement vector, we can show that

$$D_a = a^2 \left(\frac{3}{2} g_\alpha \Gamma_\alpha + 2 g_\beta \Gamma_\beta + \frac{1}{2} g_\gamma \Gamma_\gamma\right) \quad (2)$$

and

$$D_c = \frac{c}{4}^2 \left(3 g_\beta \Gamma_\beta + 3 g_\gamma \Gamma_\gamma\right) \quad . \quad (3)$$

With this framework, we now turn to the first case proposed above, vacancy diffusion with the n-t-o-t-n path being forbidden. The rate-limiting step will be the n-o jump, and we can designate its frequency as ϕ_{no}. The weighting factors will be $g_\alpha = 1$, $g_\beta = 1$, and, since there are two paths leading from a γ-type site to the vacancy, $g_\gamma = 2$. It follows that

$$D_a = (9/2)\,\phi_{no}\,a^2 \quad , \quad (4)$$

$$D_c = 9\phi_{no}\,(c^2/4) \quad , \quad (5)$$

and, since c/a = 1.63 in BeO,

$$D_c/D_a = 1.33. \quad (6)$$

This is in agreement with the experimental value.[8,9]

Considering the key-lock mechanism, we propose that the necessary condition for passage of one cation through an o-site is that at least one adjoining cation on an n-site step back to its t-site. Given this condition, the rate-limiting step would be the jump through the n-o barrier. The weighting factors will be $g_\alpha = 1$, $g_\beta = 2$, and $g_\gamma = 4$. In the γ-path case, there are two branches which the cation may take and each has a double chance of being opened, since there would be two filled n-sites next to each. We find that

$$D_a = 7.5\phi_{no}\,a^2 \quad , \quad (7)$$

$$D_c = 18\phi_{no}\,(c^2/4) \quad , \quad (8)$$

and

$$D_c/D_a = 1.6 \quad . \quad (9)$$

Correction for correlation effects reduces the values given by Eqs. (6) and (9) to 1.28 and 1.5, respectively.

This key-lock mechanism has been justified on the assumption that all cations about a vacancy have an equal probability of stepping into their t-sites. This may not be true, since the same ordering forces which, in β-BeO, select those cations that shift to t-sites may also tend to eliminate some cations from their operation as keys. This effect of this might be judged by selecting as an example the effect of a change in g_γ from 4 to 2. In this case, $D_c/D_a = 1.22$ after correction for correlation effects. It would appear that the details of a possible key-lock mechanism should be studied further.

The interstitialcy mechanism can be handled by Eqs. (2) and (3) also. The weighting factors are $g_\alpha = 2$, $g_\beta = 1$, and $g_\gamma = 2$. The α-value is twice the value for vacancy diffusion because the n-t pair is treated as one site and a neighboring coplanar site can be reached via either the upper or the lower o-site between them. Taking the rate-limiting jump frequency to be ϕ without specifying it in more detail gives

$$D_a = 6\phi a^2 \quad , \tag{10}$$

$$D_c = 9\phi(c^2/4) \quad , \tag{11}$$

and, taking $c/a = 1.602$ as in ZnO,

$$D_c/D_a = 0.98 \quad . \tag{12}$$

Calculation of the correlation coefficient does not change this isotropy. This value seems to agree with the measured data on ZnO.[11]

Any conclusions to be drawn from agreements among calculations [Eqs. (6), (9), and (12)] and experimental data can be only as valid as the model. The model should be especially good for BeO, but less certain for ZnO.

3 CONCLUSIONS

From the data on the wurtzite structures and comparisons with others in which the anions are also in hexagonal close packing, it appears to be generally the case that the energy barriers for migration are, in order of decreasing height: o-o, o-t (or for wurtzite either o-t or o-n), and t-t (or n-t). Taking the activation energies for diffusion to be the same in the a- and c-directions, calculations of diffusion anisotropy for a vacancy, a vacancy-key-lock, and an interstitialcy mechanism lead to the suggestion that one of the first two is able to describe diffusion in BeO, while the third is appropriate for ZnO.

This discussion has been based on a very simple hard-sphere interpretation of the packing of the anions and the possible jump paths between them. While it may not be applicable in all the examples selected – ZnO in particular deviates from the close anion packing which has been assumed – it should be a useful first analysis.

ACKNOWLEDGMENTS

This work was performed under the auspices of the U.S. Atomic Energy Commission. Many of the ideas were developed while the author was a guest of Prof. H. Schmalzried at the Institut für Theoretische Hüttenkunde, T.U., Clausthal-Zellerfeld, W. Germany, and his advice was much appreciated. I owe special thanks to Dr. M. W. Guinan of the Lawrence Livermore Laboratory for valuable discussions and for presenting this discussion in my absence at the Colloquium.

REFERENCES

1. Condit, R. H., *Materials Science Research,* Vol. 4, T. Gray and V. D. Frechette (Eds.), Plenum Press, New York (1969), p .284.
2. Kjekshus, A., and Pearson, W. B., *Progress in Solid State Chemistry,* Vol. 1, H. Reiss (Ed.), The Macmillan Co., New York (1964), p. 83.
3. Smith, D. K., Cline, C. F., and Austerman, S. B., *Acta Cryst.,* **18,** 393 (1965).
4. Azaroff, L. V., *J. Appl. Phys.,* **32,** 1658, 1663 (1961).
5. Seith, W., *Z. Elektrochem.,* **39,** 538 (1933).
6. Condit, R. H., Hobbins, R. R., and Birchenall, C. E., "The Self-Diffusion of Iron and Sulfur in Ferrous Sulfide", UCRL-75215, University of California, December 18, 1973.
7. Klotsman, S. M., Timoveyev, A. N., and Trachtenberg, I. Sh., *Phys. Metals Metallog.,* **16,** 92 (1963).
8. Condit, R. H., and Hashimoto, Y., *J. Am. Ceram. Soc.,* **50,** 425 (1967).
9. Cline, C. F., and Newkirk, H. W., *J. Chem. Phys.,* **49,** 3496 (1968).
10. Cline, C. F., Newkirk, H. W., Condit, R. H., and Hashimoto, Y., "Mass Transport in Oxides", J. B. Wachtman and A. D. Franklin (Eds.), *U.S. Nat. Bur. Standards Special Publ. 296,* p. 177 (1968).
11. Kim, K. S., and Wuensch, B. J., *Bull. Am. Ceram. Soc.,* **50,** 373 (1971); Ph.D. Thesis of Kee Soon Kim, Massachusetts Institute of Technology, 1971.
12. Batra, A. P., and Slifkin, L., *J. Phys. Chem. Solids,* **30,** 1315 (1969).

DISCUSSION on Paper by Ralph H. Condit

FRANKLIN: The suggestion was made in the talk that in the NiAs structure, the correlation coefficient can give information as to the path taken during the vacancy-tracer exchange. However, it is worth noting that the correlation coefficient is determined in vacancy-controlled self-diffusion purely by the geometry of the lattice upon which the vacancy and tracer move. Thus, it can give no information about the path by which the exchange takes place.

CONDIT: This is an important clarification. One might say that the value of the correlation coefficient is to suggest the location of a second normal site occupied by the mobile atom relative to its first. Given this relationship, one may then try to deduce the route taken between the normal sites through the possible metastable sites. The correlation coefficient does not give direct information about which of the possible routes was taken.

ANDERSON: In the diffusion network postulated for BeO, one step is the jump of a cation from an occupied tetrahedral site to an unoccupied one. Is there any evidence from crystal-structure determinations that the Debye-Waller factor for the cations is anisotropic, and large in the c-axis direction? This might be expected if the postulated step takes place easily.

CONDIT: I have not been able to find any direct experimental data.

E ASPECTS OF ELECTRICAL
PERTIES OF METAL OXIDES

Department of Chemistry
Purdue University
West Lafayette, Indiana

ABSTRACT

Earlier reviews concerning electrical properties of highly conducting metal oxides (RuO_2, ReO_3, and NbO) and of oxides that undergo a metal-insulator transition (Ti_2O_3 and V_2O_3) have been brought up to date. An attempt has been made to correlate the known physical properties of these materials with their respective band structures.

1 INTRODUCTION

While it has been known for a long time that metal oxides display an astonishing variety of electrical characteristics, ranging from highly insulating to highly metallic properties, systematic investigations of the metallic metal oxides have only recently been pursued. Among the most interesting materials are oxides that undergo a semiconductor-metal transition; because of their possible applications, this class has been studied extensively over the last decade.

Review articles on the subject of conducting oxides and their transitions are fairly numerous[1-12]; for this reason, the present discussion will be limited exclusively either to aspects not covered earlier or to items that should be brought up to date.

2 HIGHLY METALLIC OXIDES: RuO_2, ReO_3 NbO

One of the best oxide conductors is RuO_2 whose resistivity is approximately 2 x 10^{-9} ohm-cm at 4.2 K[13-16]; however, at room temperature this value is 4 x 10^{-5} ohm-cm. These rather large positive temperature coefficients of resistivity seem to be a general characteristic of highly conducting oxides.

The standard prototype material which is highly conducting and whose metallic characteristics are readily explained is ReO_3. Several recent resistivity measurements on this material are displayed in Fig. 1[17-19] and are related to the electrical characteristics of copper. The resistivity of ReO_3 depends to a large extent on sample stoichiometry and purity, but is not too far removed from that of low-purity copper.

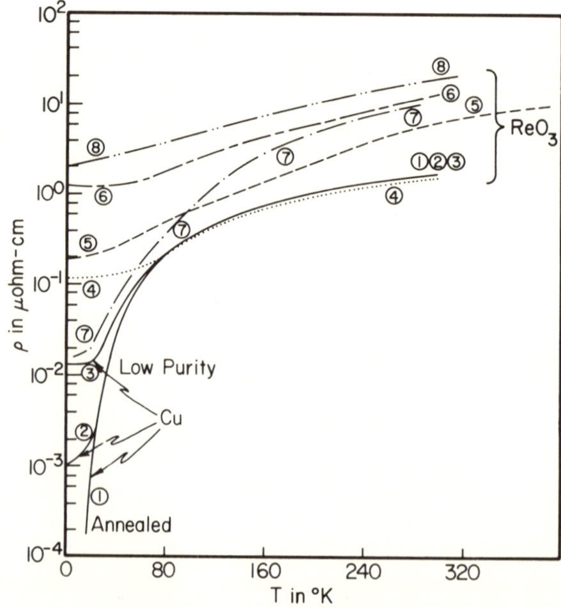

Fig. 1. Resistivity of ReO_3 compared to that of copper. Curves 1, 2, 3: copper metal annealed, drawn, and low purity, respectively. Curves 4-8: measurements on ReO_3 according to ref. 17 (Curves 4, 5, and 6 for different specimens), ref. 18 (Curve 7), and ref. 19 (Curve 8).

The metallic nature of ReO_3 is also manifested by other properties. Thus, the material is Pauli paramagnetic[20] and has a small Seebeck coefficient[21]. Plots of C/T versus T^2, where C is the heat capacity and T the temperature, are linear up to 6°K[19], and display a nonzero intercept. The reflectivity spectrum has been carefully studied[22], and the plasma resonance frequency, characteristic of metals, identified. Studies of the Fermi surface have been carried out; these involve measurements of the de Haas – van Alphen effect[13,18] and its dependence on pressure[23], as well as magnetothermal oscillations[24]. Experiments on positron lifetimes have been reported.[25] Finally, a direct examination of occupied portions of the band structure by use of X-ray photoelectron spectroscopy has been undertaken.[26,27] This type of investigation reveals bands, scvcral cv in width, intersected by the Fermi level.

The above findings may be explained at a qualitative level using a procedure pioneered by Goodenough.[1-3] The explanation offered below represents a slight revision and simplification of a scheme advanced and reviewed earlier.[1-3,7,8] One begins with the atomic structure of the perovskite unit cell [shown in Fig. 2(a)] and with the listing of atomic levels corresponding to valence shells for rhenium and oxygen, as shown on the left and right in Fig. 2(c). The splitting of rhenium 5d and of oxygen 2p states is occasioned, respectively, by the octahedral disposition of the nearest-neighbor anions surrounding each cation and by the presence of two nearest-neighbor cations surrounding each anion. As shown in Fig. 2(b), the e_g, s, and p orbitals of rhenium point in the direction of the

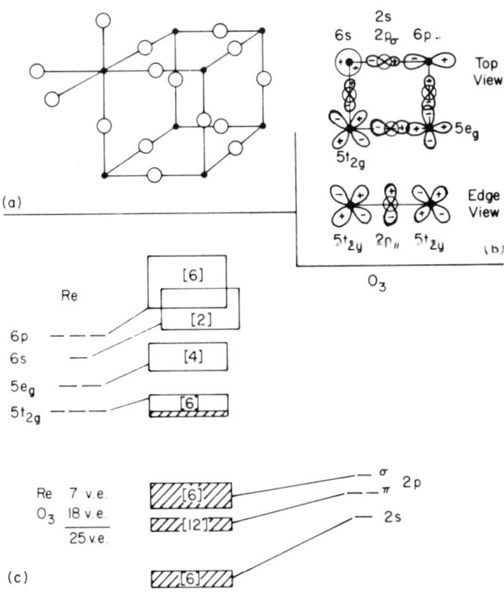

Fig. 2. Crystal structure (a), band structure (b), and types of bonding (c) in ReO_3[1-3].

anions, and the s and p_σ orbitals of the anions point in the direction of the cations. Upon turning on the interactions between these, the respective atomic levels are shifted away from each other; also, because bonding is continuous through the entire lattice, the resulting molecular energy levels broaden into bands. These are shown in the center of the diagrams.

The t_{2g} and p_π lobes are in relative orientations that permit π bonding to occur over the entire lattice network [see Fig. 2(b)]; the corresponding molecular levels also broaden into bands, as shown in the center of Fig. 2. The total electron capacity of each band per ReO_3 formula unit is indicated in brackets. Of 25 valence electrons for ReO_3, 24 can be accommodated in the lower lying bands derived from primarily anionic energy states, but the 25th must be placed in the next higher lying band of primarily t_{2g} character. This immediately rationalizes the metallic properties of ReO_3.

While the Goodenough scheme is delightfully attractive because of its simplicity, it obviously cannot be expected to be correct in detail. Indeed, this is verified by comparing Fig. 2 with the sophisticated calculations based on the APW procedure, as carried out by Mattheiss[28,29] and commented upon elsewhere[7,8]. One manifestation of this difference is the shape of the Fermi surface which must be accurately known in order to predict fine details in transport properties. According to the Goodenough scheme, the Fermi surface should essentially be representable as a simple sphere in reciprocal space, since the e_π* band is filled to only 1/6 of its capacity. The Fermi surfaces predicted by Mattheiss are shown in Fig. 3; the differences arise from two factors: the spin-orbit coupling effects in rhenium, neglected by Goodenough, split the e_π* band into three units (except along lines of high symmetry in reciprocal space), so that one encounters three distinct bands of e_π* character, rather than one threefold degenerate set. Each of these close-lying bands is intersected by the Fermi energy; correspondingly, one obtains three Fermi surfaces, of which two

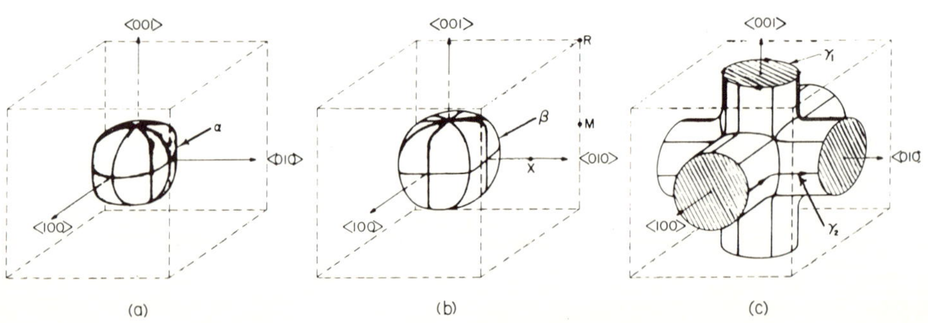

Fig. 3. Three Fermi surfaces of ReO_3[28,29].

are very nearly spherical. The severe departure from sphericity of the third surface is due to the fact that along the <100> directions in reciprocal space the energy of the third band changes very little with wave vector, **k**, while the ϵ-**k** variation along other directions in reciprocal space is still nearly parabolic at the Fermi energy.

The experimental facts cited earlier are very well matched by the above-mentioned theoretical calculations of Mattheiss; however, it must be pointed out that some semiempirical adjustments of parameters were found to be necessary to bring theory and experiment into good agreement.

A third example in the category of highly conducting oxides is furnished by NbO, which has the metallic luster of noble metals whose properties it matches. Although the compound was synthesized over 100 years ago, its electrical properties have been only sporadically investigated. A typical resistivity curve[30] is entered in Fig. 4, along with comparable data on platinum. The residual resistivity below 50°K of NbO is quite

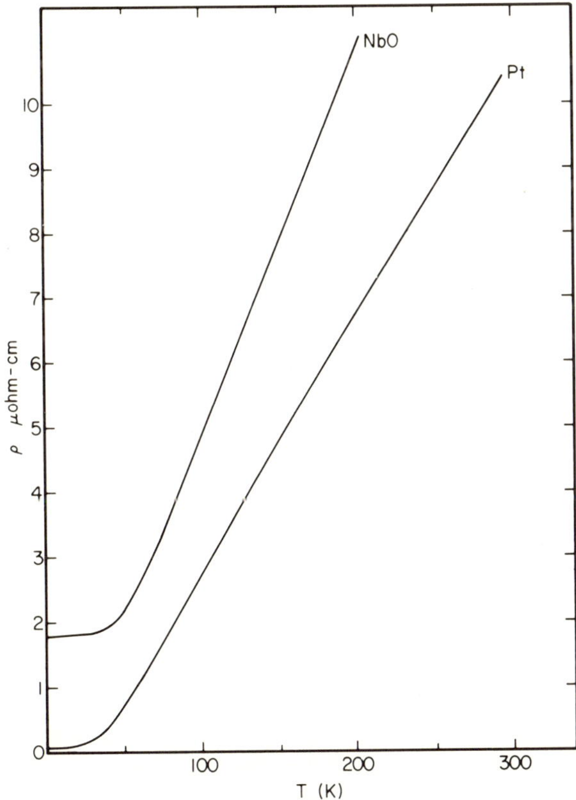

Fig. 4. The resistivity of NbO compared with that of platinum. NbO data by G. V. Chandrashekhar, unpublished; see also ref. 31.

high, presumably because of the presence of impurities, defects, and strains that inevitably develop in the rapid growth of single crystals by arc-melting techniques.[30,31] It should be noted that the straight-line portion beyond 75°K in NbO is maintained, in conformity with the Grüneisen law, up to 1000°K. Again, the slope is somewhat larger than that encountered in conventional metals. Optical reflectivity data[32] indicate the existence of a plasma edge at about 4.3 ev. Magnetoresistance effects reported in the literature[30] point to magnetic breakdown effects; the Hall coefficient in NbO is of variable sign.

The above data are rationalized in terms of Fermi surfaces constructed by the Harrison technique[33] according to the prescription that the radius of the Fermi sphere k_f is subject to the condition $k_f = (3\pi^2 n)^{1/3}$, where n is the number of electrons in unfilled bands. With n = 3, these spheres are considerably larger than the primitive unit cells of reciprocal space they enclose; for this reason one obtains the rather complex patterns, illustrated in Fig. 5, when the various appropriate portions of the Fermi sphere outside the central unit cell of reciprocal space

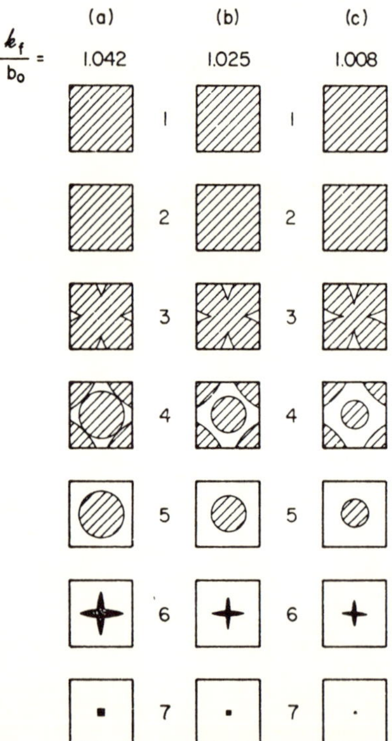

Fig. 5. Fermi surfaces of NbO_x projected onto the 100 plane of the reciprocal lattice. Parts (a), (b), (c) refer to x = 0.98, 1.00, 1.02. k_f/b_o represents the ratio of the Fermi sphere radius to the reciprocal lattice vector for the indicated composition; seven Brillouin zones are shown.

are reassembled within this unit cell. Outside two completely filled zones, hole pockets appear in a third and fourth zone, and electrons sparsely occupy three additional zones. The alteration of sample stoichiometry (in NbO_x, $0.98 \leqslant x \leqslant 1.02$) significantly affects the electrical properties since x and n are proportional, thereby altering k_f. While these diagrams are undoubtedly an oversimplified version of what would be obtained with a more sophisticated calculation, they do provide a zero-order explanation for the experimental observations. In particular, one can readily rationalize the sensitivity of Seebeck and Hall coefficients to oxygen stoichiometry, and the peculiarities in magnetoresistance phenomena which are thought to be manifestations of magnetic breakdown effects, when the electrons switch orbits or tunnel through narrow energy gaps at the Brillouin zone boundaries.

3 OXIDES UNDERGOING SEMICONDUCTOR-METAL TRANSITIONS: Ti_2O_3 AND V_2O_3

Because of obvious commercial applications and because of their intrinsic scientific interest, oxides undergoing a semiconductor-metal transition have been extensively investigated. The scientific literature in this area is still growing rapidly and early reviews of this area[1,2,5-7] are by now seriously out of date.

A case in point is Ti_2O_3. Originally, there was considerable confusion as to the cause of the transition which is represented[34] by the top curve of Fig. 6. It now seems generally accepted that the transition is caused by

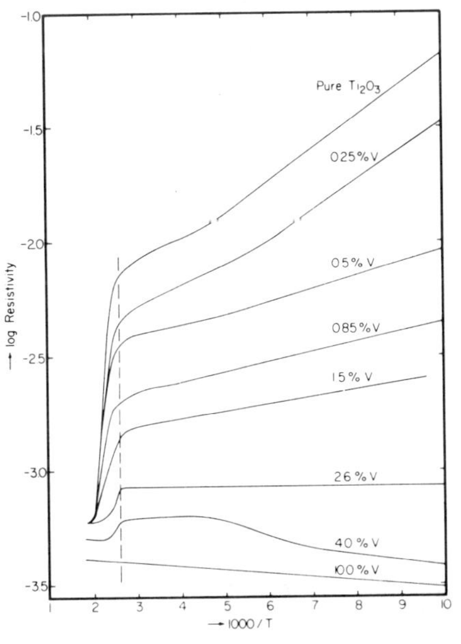

Fig. 6 Resistivity of pure and of vanadium-doped Ti_2O_3 as a function of reciprocal temperature.[34]

the gradual disappearance with rising temperature of an initially small band gap, separating two bands of primarily d-type character.[35] This is ascribed to a distension, over the same temperature interval, of the unit cell, involving the lengthening of the *c* axis and contraction of the *a* axis without change in lattice symmetry as the temperature is raised.[36,37] Very recent measurements[38,39] have shown the atomic positions to scale reasonably accurately with lattice parameters. The band structure of Ti_2O_3 is shown in Fig. 7. The above-mentioned changes in band structure are an immediate consequence of the indicated atomic motions: the lower lying band of a_1 character in Fig. 7 is very sensitive to the Ti-Ti spacing along the *c* axis of the rhombohedral unit cell and moves upward in energy as the spacing is increased, while the higher lying band of e_π character is sensitive to Ti-O spacings along the basal plane and broadens as the *a* axis contracts.

More information concerning the motion of these two bands has recently become available through measurements of elastic constants[40], specific heat[41], and electrical properties[42] which correct or supplement earlier thermodynamic[43-45] and electrical measurements[46-48].

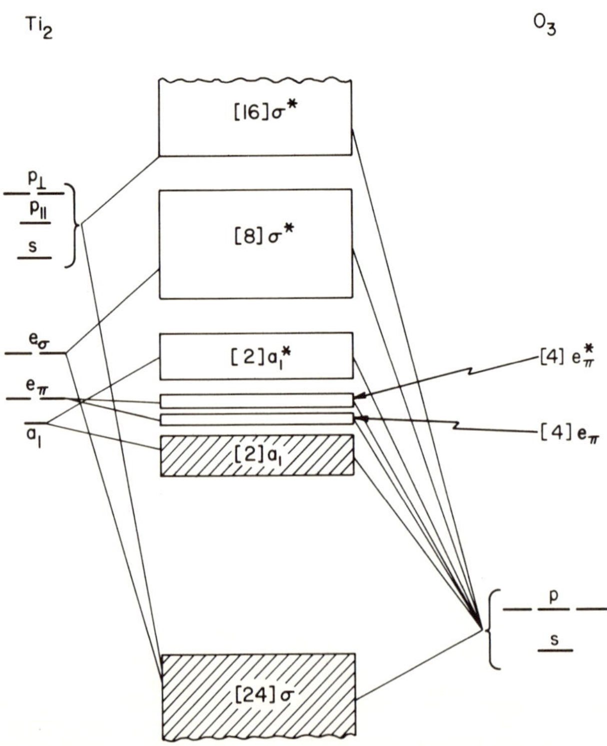

Fig. 7. Schematic band structure diagram for Ti_2O_3.[1-3,35] Numbers in brackets represent electron capacity per Ti_2O_3 formula unit.

Examination of Seebeck and Hall coefficient data has led[40,41] to a tentative assignment of 1.0 and 0.3 ev as bandwidths to the lower a_1 and upper e_π bands, respectively. The band gap was then adjusted to $\epsilon_g = 0.13$ ev at 300°K and allowed to vary with temperature in such a manner as to optimize agreement between computed and observed values of the specific heats[41], as described below. The details of the band motion with temperature were then worked out so as to match the specific heat measurements reproduced in Fig. 8, which were ascribed to electronic effects. It is not immediately obvious that this should be correct because the various lattice-vibration frequencies detected by Raman scattering experiments[49-50] soften (with the exception of one mode that undergoes a partial recovery) over the temperature range of the electrical transition.

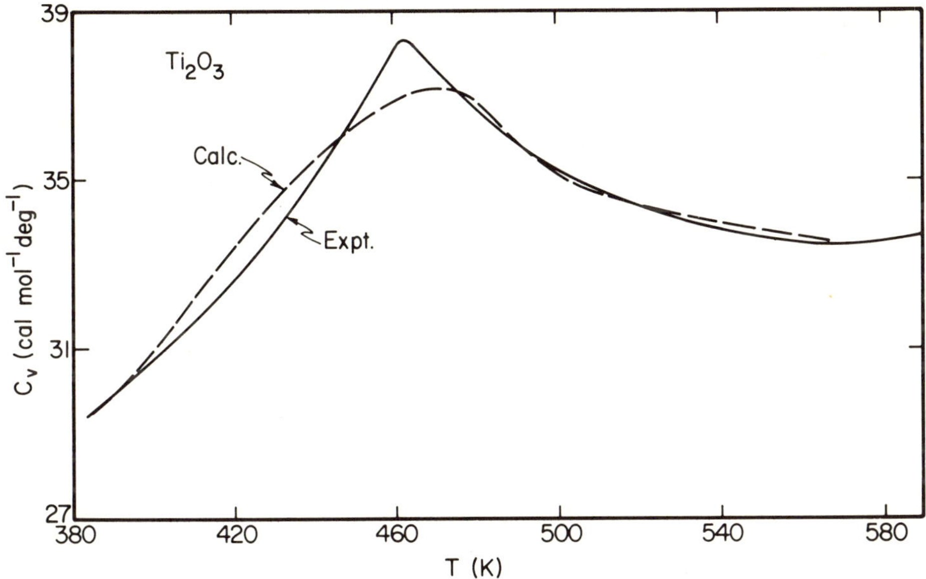

Fig. 8. Calculated and observed specific heat anomaly in the electrical transition range of Ti_2O_3.[41]

However, it was ascertained that the decrease in heat capacity from the peak of the anomaly to the merging with the baseline could not be reproduced with any reasonable variation of lattice vibration frequencies, On the other hand, the anomaly may readily be explained as being due to electronic excitations: as the band gap decreases with rising temperature, more electrons are needed to excite the phonon spectrum and becomes manifest as an excess in heat capacity. However, the band gap ultimately becomes so small that the excitation energy approaches zero; furthermore, after the onset of band overlap, degeneracy sets in. Both factors contribute to a rapid decline of the electronic heat capacity, C_e, toward zero, at which point the anomaly disappears.

Quantitative determinations of C_e may be carried out by standard calculations of the electronic energy U_e of the two-band system and differentiating the result numerically with respect to temperature. A good fit to C_e is obtained with the shift of band position shown in Fig. 9. Double differentiation of $U_e(T)$ with respect to strain components yields elastic constants in good agreement with the various measured values over the 300-degree temperature interval of the experiments. The model employed in these computations is thus self-consistent. However, the results displayed in Fig. 9 are undoubtedly only rough approximations to reality. In the first place, the band shapes in Fig. 9 were chosen to be symmetric about the midpoint of the bandwidth; such a density-of-states

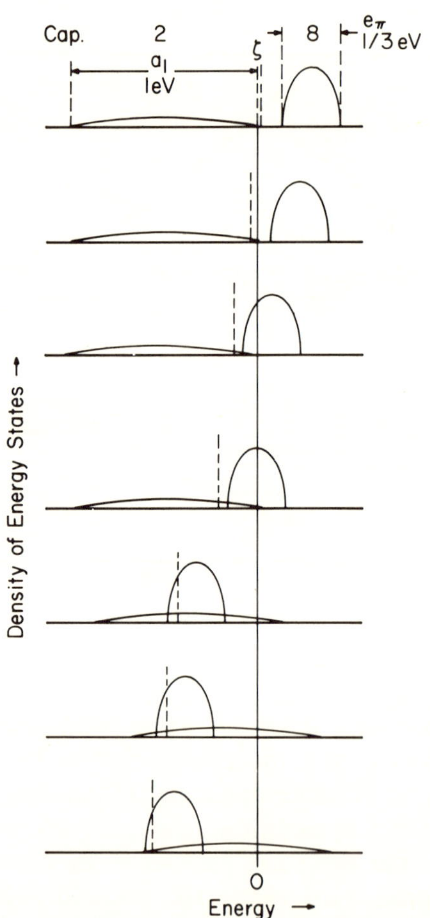

Fig. 9. Shift of a_1 and e_π bands of Ti_2O_3 with temperature. Capacities for electrons indicated by numerals 2 and 8. Dashed line: position of Fermi level. Zero of energy originally coincides with top of valence band at low temperatures. Successive diagrams show motion of bands on energy scale with rising temperature through the transition.

curve is highly unrealistic. On the other hand, since this quantity appears only as an integrand, the final results are unlikely to be strongly dependent on the details of the density-of-states function. Second, the bandwidths undoubtedly are also functions of temperature since they are very sensitive to variations in interatomic spacings in the lattice. Formally, this feature may be taken into account by allowing the integration limits in the standard expression for U_e to vary with temperature. However, this refinement merely introduces further parameters into the problem (thereby presumably further increasing the agreement of theory with experiment), and does not really pin down the actual variation of bandwidth with temperature; hence this particular refinement was not considered further. In conclusion, then, the combination of elastic-constant and heat-capacity measurements has made it possible to obtain a reasonably accurate picture of the band motion in Ti_2O_3 as a function of temperature.

To understand the measurements referred to below it is necessary to backtrack to Fig. 6, and to take note of the effect of adding V_2O_3 to Ti_2O_3. It is seen that this process lowers the size of the transition without altering the temperature range where it occurs, and that for alloys containing more than 4 at.% vanadium, the transition disappears altogether, rendering the material metallic over the entire temperature range. X-ray studies[51] show that the *c* axis expands and the *a* axis contracts on addition of V_2O_3 to the host lattice. Incremental doping of Ti_2O_3 with V_2O_3 thus has the same effect as taking the host material through the temperature range of the transition.

Low-temperature measurements of the heat capacity of vanadium-doped Ti_2O_3 have been carried out by Sjöstrand and Keesom[52-54]; the electronic contribution has been interpreted by them on the basis of a one-dimensional electron band model [recall that $(V_xTi_{1-x})_2O_3$ remains metallic for $x \gtrsim 0.04$]. It has been pointed out[55-57], however, that this interpretation is at variance with the known electrical properties of Ti_2O_3; calculations have just been completed[57] which reinterpret these results in terms of an overlap of a narrow vanadium-type impurity band with the a_1 band.

A very large body of literature exists (reviewed elsewhere[1-12,58-61]) concerning the electrical properties of V_2O_3; most of the interest in this material arises from the sharp, low-temperature electrical transition and the less marked, high-temperature transition which are encountered (see Fig. 10). While there is general agreement on the experimental results, considerable controversy surrounds the mechanism responsible for the transition. The situation is complicated by experimental results reported during the last few years on $(Cr_xV_{1-x})_2O_3$ with $0 < x < 0.05$. No less than 12 papers on this subject have been published, principally by workers

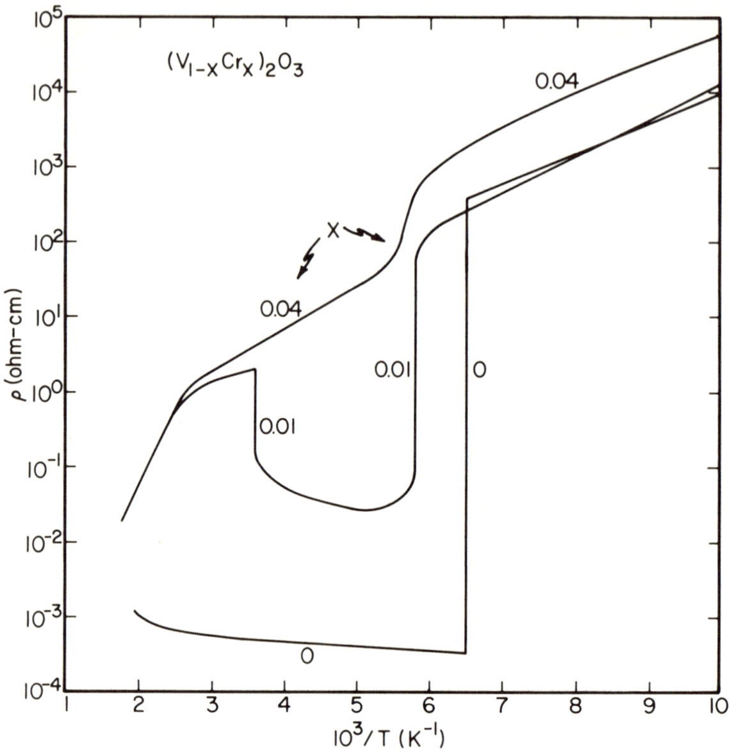

Fig. 10. Electrical properties of pure and chromium-doped V_2O_3. Earlier measurements by McWhan et al. (See refs. 62-73).

at Bell Telephone Laboratories.[62-73] Their findings on chromium-doped V_2O_3 with $x = 0.01$ are readily summarized by reference to Fig. 10, showing that there now exist two first-order transitions: one close to the temperature where the first-order transition occurs in pure V_2O_3, and a second near 350°K. This fact, and related work on Hall coefficient, thermoelectric, thermal conductivity, NMR, magnetic susceptibility, thermodynamic, and structural studies have been repeatedly linked to the Mott transition as the mechanism responsible for the dramatic alterations of physical properties in doped V_2O_3 and, by implication, in pure V_2O_3.

That this idea has merit is indicated by the following qualitative argument: a material which is to undergo a transition from a band metal to a hopping-type insulator must be delicately poised near a critical interatomic separation, so that relatively small changes in temperature might suffice to push the material past the critical separation distance which permits the Mott transition to occur. The probability of encountering this situation in any material undergoing a transition is slight. Nevertheless, V_2O_3 is expected to be a reasonable candidate for this category; what is more, the addition of Cr_2O_3 expands the lattice, so that even if pure V_2O_3 were not in the edge of the critical atomic separation, chromium-doped V_2O_3 might be. The emergence of a new first-order phase transition in the doped sesquioxide appears to vindicate this position.

Sporadically, doubts have been raised concerning the adequacy of this viewpoint and other mechanisms of the transition have been proposed.[5,7,8,11,59,61,69,74] These doubts appear to be substantiated by recent measurements in the author's laboratory.[75] Using single-crystal specimens of $(Cr_{0.01}V_{0.99})_2O_3$ with a total impurity content less than 150 ppmw, the results displayed in Fig. 11 were obtained. These indicate that while the earlier work of Fig. 10 could be reproduced in essentials upon rapidly cycling new samples through the temperature range 300 to 600°K, markedly different phenomena were encountered on slowly passing these specimens through this range. In particular, the anomalous rise in resistivity could be greatly reduced and ultimately nearly suppressed (Curves 2, 3, and 4 in Fig. 11). A similar phenomenon was encountered in studies of the Seebeck coefficient as a function of temperature (Curves 5, 6, and 7 of Fig. 11). The reduction in the anomaly was easily accom-

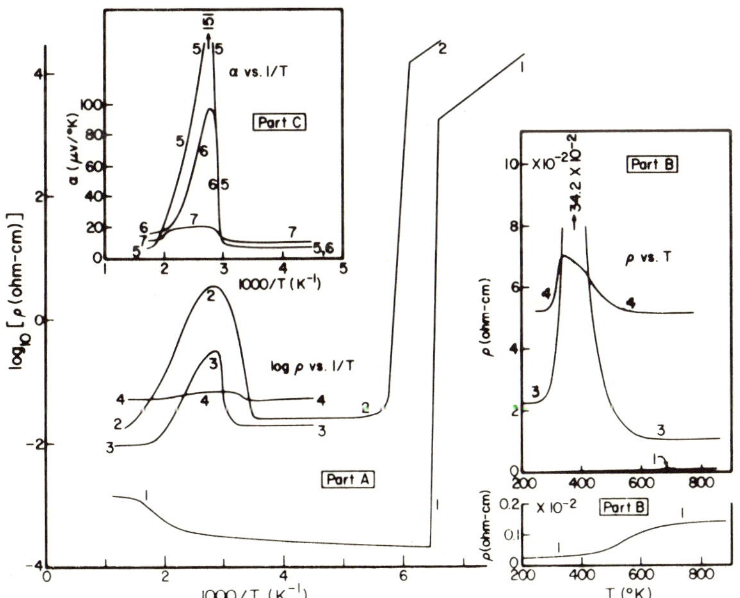

Fig. 11. Recent measurement of electrical properties of pure and chromium-doped V_2O_3.[75] Part A: Plot of $\log_{10} \rho$ vs $10^3/T$. Part B: Plot of ρ vs T. Part C: Plot of α vs $10^3/T$, where α is the Seebeck coefficient. Curve 1, pure V_2O_3; Curves 2 and 5, virgin samples subjected to rapid cycling; Curves 3 and 6, same specimens taken through transition slowly; Curves 4 and 7, specimens which were repeatedly cycled through the transition region.

plished in samples that were repeatedly cycled. Furthermore, two rhombohedral phases were identified by X-ray crystallography in the region of the anomaly, namely an α phase for which $a_H = 4.957$ Å, $c_H = 13.99$ Å, $\alpha_H = 53° 53'$, and a β phase for which $a_H = 4.995$ Å, $c_H = 13.93$ Å, and $\alpha_H = 54° 22'$. On raising the temperature, the latter grows at the expense of the former; the resistivity peak is encountered when both phases are present in approximately equal amounts.

The anomaly was therefore attributed[75] to an extrinsic effect, such as scattering by or trapping of charge carriers in boundaries between the two coexisting phases. There are four experimental findings which buttress this claim. First is the observation that even samples that undergo as sharp a high temperature change in ρ as shown for $x = 0.01$ in Fig. 10 fail to exhibit any anomaly in heat capacity, so that the enthalpy change accompanying this change in electrical properties is zero within experimental error.[75] Obviously, it is very difficult to reconcile a bona fide first-order transition with this observation. Second is the finding, as shown in Curves 3 and 4 of the insert (Part B of Fig. 11), that for $T > 600°$K the high temperature phase is a very good conductor, for which ρ is somewhat smaller than the resistivity of the low-temperature metallic phase encountered at $170 < T < 270°$K. If one were to postulate that $(Cr_{0.01}\text{-}V_{0.99})_2O_3$ is a semiconductor between 350 and 600°K, yet another semiconductor-metal transition would have to be invoked to explain the properties above 600°K. Third is the observation that the Seebeck coefficients shown in Part C of Fig. 11 run parallel to the electrical measurements. This observation and the fact that the α values are very small in the ranges 170 to 300°K and above 600°K cannot be readily reconciled with a transition of the type invoked by earlier workers. Finally, there is the fact that the electrical anomaly is never exactly reproducible, that it is affected by thermal cycling, annealing, and the rate of heating or cooling of specimens, all of which affect the degree of interdispersion of the two phases and, hence, produce effects attributable to the presence of phase boundaries.

Obviously, the above conclusions pertain only to $(Cr_xV_{1-x})_2O_3$ with $x \approx 0.01$; since Cr_2O_3 is a hopping-type semiconductor, there may well exist a composition x at which the material does undergo a Mott transition. However, if the above revision in viewpoint is sustained in essentials, then the electrical anomaly for $x \approx 0.01$ cannot be interpreted in terms of a Mott-type transition.

REFERENCES

1. Goodenough, J. B., *Bull. Soc. Chim. France,* p. 1200 (1965).
2. Goodenough, J. B., *Czech. J. Phys.,* **B17**, 304 (1967).

3. Goodenough, J. B., *Progress in Solid State Chemistry,* H. Reiss (Ed.), Pergamon, Oxford (1971), Vol. 5, pp. 145 ff.
4. Goodenough, J. B., *Mat. Res. Bull.,* **2,** 37, 49 (1967).
5. Adler, D., *Solid State Physics,* **21,** 1 (1968).
6. Adler, D., *Rev. Mod. Phys.,* **40,** 714 (1968).
7. Honig, J. M., *IBM J. Res. Dev.,* **14,** 233 (1970).
8. Vest, R. W., and Honig, J. M., *Electrical Conductivity of Ceramics,* N. M. Tallan (Ed.), Dekker, New York, in press.
9. Honig, J. M., *Proceedings of the Conference on Fast Ion Transport,* held at Belgirate, Italy, September 5-15, 1972, W. van Gool (Ed.), North-Holland, Amsterdam (1973), p. 311.
10. Frederikse, H.P.R., *IBM J. Res. Dev.,* **14,** 295 (1970).
11. Mott, N. F., *Phil. Mag.,* **20,** 1 (1969).
12. Austin, I. G., and Mott, N. F., *Science,* **168,** No. 3927, 71 (1970).
13. Marcus, S. M., *Phys. Letters,* **26A,** 191 (1968).
14. Marcus, S. M., and Butler, S. R., *Phys. Letters,* **26A,** 518 (1968).
15. Ryden, W. D., Lawson, A. W., and Sartain, C. C., *Phys. Letters,* **26A,** 209 (1968); *Phys. Rev.,* **B1,** 1494 (1970).
16. Butler, S. R., and Gillson, J. L., *Mat. Res. Bull.,* **6,** 81 (1971).
17. Honig, J. M., and Strauss, J. A., Unpublished research.
18. Phillips, R. A., and Shanks, H. R., *Phys. Rev.,* **B4,** 4601 (1971).
19. King, C. N., Kirsch, H. C., and Geballe, T. H., *Solid State Comm.,* **9,** 907 (1971).
20. Quinn, R. K., and Neiswander, P. G., *Mat. Res. Bull.,* **5,** 329 (1970).
21. Ferretti, A., Rogers, D. B., and Goodenough, J. B., *J. Phys. Chem. Solids,* **26,** 2007 (1965).
22. Feinleib, J., Scouler, W. J., and Ferretti, A., *Phys. Rev.,* **165,** 765 (1968).
23. Schirber, J. E., Morosin, B., Anderson, J. R., and Stone, D. R., *Phys. Rev.,* **B5,** 752 (1972).
24. Graebner, J. E., and Greiner, E. S., *Phys. Rev.,* **185,** 992 (1969).
25. Noguchi, M., Chiba, T., Mitsuhashi, T., and Tsuda, N., *J. Phys. Soc. Japan,* **31,** 306 (1971).
26. Wertheim, G. K., and Hüfner, S., *Phys. Rev. Letters,* **28,** 1028 (1972).
27. Kasowski, R. V., and Riggs, W. M., *Phys. Letters,* **39A,** 387 (1972).
28. Mattheiss, L. F., *Phys. Rev.,* **181,** 987 (1969).
29. Mattheiss, L. F., *Phys. Rev.,* **B2,** 3918 (1970).
30. Honig, J. M., Wahnsiedler, W. E., and Eklund, P. C., *J. Solid State Chem.,* **6,** 203 (1973).
31. Chandrashekhar, G. V., Moyo, J., and Honig, J. M., *J. Solid State Chem.,* **2,** 528 (1970).
32. Rao, C.N.R., Wahnsiedler, W. E., and Honig, J. M., *J. Solid State Chem.,* **2,** 315 (1970).
33. Harrison, W. A., *Pseudopotentials in the Theory of Metals,* Benjamin, New York (1966), p. 79.
34. Chandrashekhar, G. V., Choi, Q. W., Moyo, J., and Honig, J. M., *Mat. Res. Bull.,* **5,** 999 (1970).
35. Van Zandt, L. L., Honig, J. M., and Goodenough, J. B., *J. Appl. Phys.,* **39,** 594 (1968).
36. Pearson, D., *J. Phys. Chem. Solids,* **5,** 316 (1958).
37. Rao, C.N.R., Loehman, R. E., and Honig, J. M., *Phys. Letters,* **27A,** 271 (1968).
38. Simonyi, F., and Raccah, P. M., *Bull. Am. Phys. Soc.,* **18,** 339 (1973).
39. Robinson, W. R., *J. Solid State Chem.,* in press.
40. Chi, T. C., and Sladek, R. J., *Phys. Rev.,* **B7,** 5080 (1973).
41. Barros, H. L., Chandrashekhar, G. V., Chi, T. C., Honig, J. M., and Sladek, R. J., *Phys. Rev.,* **B7,** 5147 (1973).
42. Shin, S. H., Chandrashekhar, G. V., Loehman, R. E., and Honig, J. M., *Phys. Rev.,* **B8,** 1364 (1973).
43. Naylor, B. F., *J. Am. Chem. Soc.,* **68,** 1077 (1946).
44. Shomate, C. H., *J. Am. Chem. Soc.,* **68,** 310 (1946).
45. Nomura, S., Kawakubo, T., and Yanagi, T., *J. Phys. Soc. Japan,* **16,** 706 (1961).
46. Morin, F. J., *Phys. Rev. Letters,* **3,** 34 (1959).
47. Yahia, J., and Frederikse, H.P.R., *Phys. Rev.,* **123,** 1257 (1961).
48. Honig, J. M., and Reed, T. B., *Phys. Rev.,* **179,** 1020 (1969).
49. Mooradian, A., and Raccah, P. M., *Phys. Rev.,* **B3,** 4253 (1971).
50. Aggarwal, R. L., Shin, S. H., Lax, B., and Honig, J. M., *Bull. Am. Phys. Soc.,* **18,** 370 (1973).

51. Loehman, R. E., Rao, C.N.R., and Honig, J. M., *J. Phys. Chem.*, **73,** 1781 (1969).
52. Sjöstrand, M. E., and Keesom, P. H., *Phys. Rev. Letters,* **27,** 1434 (1971).
53. Sjöstrand, M. E., and Keesom, P. H., *Phys. Letters,* **39A,** 147 (1972).
54. Sjöstrand, M. E., and Keesom, P. H., *Phys. Rev.,* **B7,** 3558 (1973).
55. Eklund, P. C., and Van Zandt, L. L., *Phys. Rev.,* **B7,** 1454 (1973).
56. Eklund, P. C., and Van Zandt, L. L., *Phys. Letters,* **42A,** 237 (1972).
57. Van Zandt, L. L., To be published.
58. Rice, T. M., and McWhan, D. B., *IBM J. Res. Dev.,* **14,** 251 (1970).
59. Zinamon, Z., and Mott, N. F., *Phil. Mag.,* **21,** 881 (1970).
60. Mott, N. F., and Zinamon, Z., *Rep. Progr. Phys.,* **33,** 881 (1970).
61. Doniach, S., *Adv. Phys.,* **18,** 819 (1969).
62. McWhan, D. B., Rice, T. M., and Remeika, J. P., *Phys. Rev. Letters,* **23,** 1384 (1969).
63. McWhan, D. B., and Remeika, J. P., *Phys. Rev.,* **B2,** 3734 (1970).
64. Jayaraman, A., McWhan, D. B., Remeika, J. P., and Dernier, D. P., *Phys. Rev.,* **B2,** 3751 (1970).
65. Menth, A., and Remeika, J. P., *Phys. Rev.,* **B2,** 3756 (1970).
66. Barker, A. S., Jr., and Remeika, J. P., *Solid State Comm.,* 8, 1521 (1970).
67. Dernier, P. D., *J. Phys. Chem. Solids,* **31,** 2569 (1970).
68. Wertheim, G. K., Remeika, J. P., Guggenheim, H. J., and Buchanan, D.N.E., *Phys. Rev. Letters,* **25,** 94 (1970).
69. Rubinstein, M., *Solid State Comm.,* **8,** 1469 (1970).
70. Rubinstein, M., *Phys. Rev.,* **B2,** 4731 (1970).
71. Gossard, A. C., Menth, A., Warren, W. W., Jr., and Remeika, J. P., *Phys. Rev.,* **B3,** 3993 (1971).
72. McWhan, D. B., Menth, A., Remeika, J. P., Brinkman, W. F., and Rice, T. M., *Phys. Rev.,* **B7,** 1920 (1973).
73. Pettifer, R. F., Pitt, G. D., Thomas, D., and Vyas, M.K.R., *Mat. Res. Bull.,* **8,** 87 (1973).
74. McWhan, D. B., Rice, T. M., and Remeika, J. P., *Phys. Rev. Letters,* **23,** 1384 (1969).
75. Honig, J. M., Chandrashekhar, G. V., and Sinha, A.P.B., *Phys. Rev. Letters,* **32,** 13 (1974).

DISCUSSION on Paper by J. M. Honig

WELCH: How was the density of states obtained for the calculation of the specific heat anomaly?

HONIG: Quite arbitrarily we used an analytic expression which made each band parabolic, with an extremum about its midpoint. This is not so bad as might appear at first sight, because the final result for the electronic energy, involving as it does an integral over the energy, is quite insensitive to the choice of the density–of–states. Our main objective was to ascertain whether the anomaly can indeed be correlated with electronic excitation across a shrinking gap.

COHEN: You mentioned that VO underwent a resistivity transition until 1967 – but with careful preparation it now no longer does so. I'm aware that Drs. Reed and Banus have suggested that the reason for the occasional appearance of this transition is grain-boundary precipitation of V_2O_3, which does have such a transition. But is there really any experimental evidence for this? Wouldn't the film have to be continuous and quite thick? Perhaps the transition is sensitive to heat treatment, but for other reasons – such as ordering or clustering of vacancies. For $VO_{1.17}$ an ordering reaction is accompanied by large changes in electrical properties (see the conference last year in Bristol on "Reactivity in Solids").

HONIG: One of the problems is that in the early measurements, the investigators did not adequately characterize their crystals. As far as I am aware, no one has demonstrated that grain-boundary precipitation of V_2O_3 has taken place. For VO_x prepared with reasonable care, the material has metallic properties for $x < 1$, and semiconducting properties for $x > 1$, but with x fixed, no change from one regime to the other takes place.

HEUER: If the transition in $(V_{1-x}Cr_x)_2O_3$ is of the nucleation and growth type, then it should be quenchable. Has such an experiment been conducted?

HONIG: No. In fact, at present very little is known about the nature of the two phases.

KOFSTAD: Dr. Honig has shown that there is a marked effect on conductivity of chromium-doped V_2O_3 by annealing the specimens in air at 1600°C for several days. Cr_2O_3 evaporates rapidly as CrO_3 at high temperatures and high partial pressures of oxygen. Does the heat-treatment of chromium-doped V_2O_3 at 1600°C lead to loss of chromium and a corresponding chromium-gradient in the specimens?

HONIG: We have not checked this point. However, we did ascertain the chromium content for each specimen by means of the electron microprobe, so that the quoted chromium content represents what was measured for each sample after the anneal.

GROWTH OF PURE CRYSTALS OF NiO AND MEASUREMENT OF THEIR DRIFT MOBILITY*

D. S. Tannhauser

Department of Physics
Technion – Israel Institute of Technology
Haifa, Israel

ABSTRACT

Very pure crystals of NiO were grown by an epitaxial process. The drift mobility of holes was measured in these crystals by the time-of-flight method. The drift mobility fits a large polaron model. A new analysis of the results gives fairly good agreement with polaron parameters previously calculated by Bosman and van Daal.

1 INTRODUCTION

This paper summarizes work done recently on NiO which has contributed materially to the understanding of the conduction process in this substance.

*Partially supported by the Volkswagen Foundation.

In the course of this work we have grown very pure single crystals of NiO[1], and in these the drift mobility of holes was determined[2].

NiO is a p-type semiconductor. Its properties have recently been reviewed by Bosman and van Daal.[3,4] Knowledge of the drift mobility and its temperature dependence is needed to settle the controversy regarding small polaron conduction versus band conduction for holes in NiO and is of help in elucidating the band structure.

Previous to this work, the value of the drift mobility in NiO has been determined only indirectly through the conductivity in lithium-doped samples which were supposed to be in the exhaustion range at 1200°K.[3,4] Room-temperature values were deduced from conductivity and thermoelectric-power measurements and were between 0.5 and 5 cm^2/v-sec. The range reflects different assumptions that were made regarding the density of states in the valence band. Bosman and van Daal deduced from these measurements that the free charge carriers in NiO are large polarons.

2 EXPERIMENTAL PROCEDURES

The most direct method of measuring the drift mobility is the time-of-flight measurement of carriers which are generated at one surface of a thin sample by a pulse method.[5] This technique requires samples which are pure enough not to trap too many carriers before they reach the opposite surface. Crystals available previous to this work were probably not pure enough for meaningful measurements.

Pure crystals were grown in the apparatus shown in Fig. 1 by epitaxial deposition of NiO on MgO, a process developed in 1959 by Cech

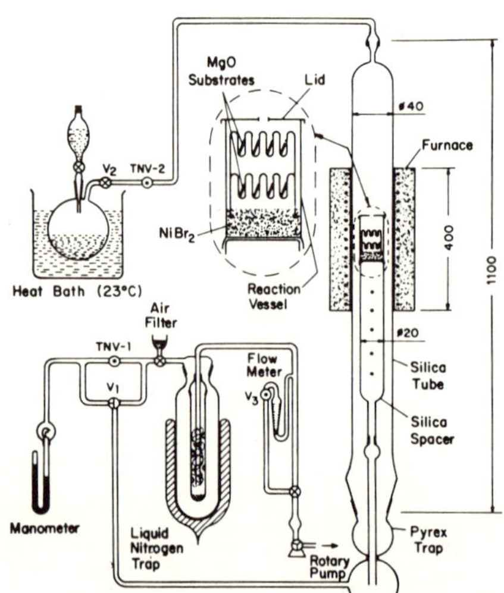

Fig. 1. Scheme of apparatus for expitaxial growth of pure NiO crystals. TNV are Teflon Needle Valves which regulate the flow of water vapor.

and Alessandrini.[6] The main refinements over previous applications of this process were as follows: $NiBr_2$, the starting material for the growth process, was prepared from the purest nickel sponge available (less than 10 ppm metal impurities) and very pure HBr (less than 1 ppm metal impurities). The heart of the growing apparatus, which is maintained at about 700°C during the crystal growth, was constructed completely of pure quartz. The handling procedures for substrates and starting material were refined as much as possible to eliminate sources of impurities during the growth process.

3 RESULTS

Crystals were grown in batches of five to ten, with a thickness of about 100 microns and an area of about 1 cm^2. One crystal from each of the more successful batches was analyzed. Results are shown in Table I. The consistency of the results from batch to batch as well as the improvement in purity with time, as our procedure was refined, can be seen clearly. The analysis, which was done by emission spectroscopy by a commercial laboratory, raises of course the questions familiar to every crystal grower:

1. Are these the impurities?
2. Are these all the impurities?
3. Are you sure?

Table I. Impurities Detected in Epitaxially Grown NiO, PPM by Weight

	Batch									
Element	3-71	5-71	6-71	7-71	12-71	2-72	3-72	9-72	22-72	25-72
Aluminum	10	10	3	3	1	1	3	3	15	5
Calcium	–	<1	<1	<1	<1	<1	<1	<1	<1	7
Iron	20	30	30	30	5	10	10	10	7	10
Magnesium	3	7	3	<1	<1	<1	2	1	<1	1
Silicon	20	30	20	10	5	3	5	5	7	5
Sodium	1	3	<1	1	–	–	–	–	–	–
Boron	–	10	–	3	–	–	–	–	–	–
Copper	–	<1	–	–	<1	<1	<1	<1	<1	<1
Silver	–	–	–	–	–	–	<1	<1	–	–
Manganese	–	–	–	–	–	–	–	–	<1	–
Total cations	54	90	56	47	11	14	20	19	39	28
Bromine	27		114	<1	24					

A partial, if somewhat confusing, check was obtained through mass spectroscopic analysis done at two different laboratories. The results of one laboratory confirmed the emission spectroscopy results for metals within a factor of about two, but another laboratory reported very high impurity concentrations, including hundreds of ppm of both chlorine and sodium.

Since the starting material for the crystal growth is $NiBr_2$, the bromine content of some batches was determined by neutron activation. It varied between 1 and 100 ppm by weight.

Hole-drift-mobility measurements proved to be possible on some of these crystals, but not on all of them. The best samples came from batch 7-71 which contained less than 1 ppm bromine. NiO is normally p-type because of excess oxygen, and bromine probably substitutes for oxygen and acts as a donor in this material. The above result, then, suggests that crystals containing bromine are partially compensated, and, therefore, some of the acceptor levels are empty and can act as traps or recombination centers for holes. The dependence on bromine concentration is not conclusive; some samples with a fairly large concentration gave a measurable signal. It is possible that some of the bromine which shows up in the analysis is included as $NiBr_2$ in the samples and has no influence on the electrical properties.

The drift mobility (μ_h) was found on five samples to be 20 to 50 cm^2/v-sec at room temperature; temperature dependence of drift mobility was determined on the samples from batch 7-71 (see Fig. 2). The drift mobility rises, with a negative activation energy of 0.055 ev, as the temperature is lowered, and reaches a value of 600 cm^2/v-sec at 125°K.*

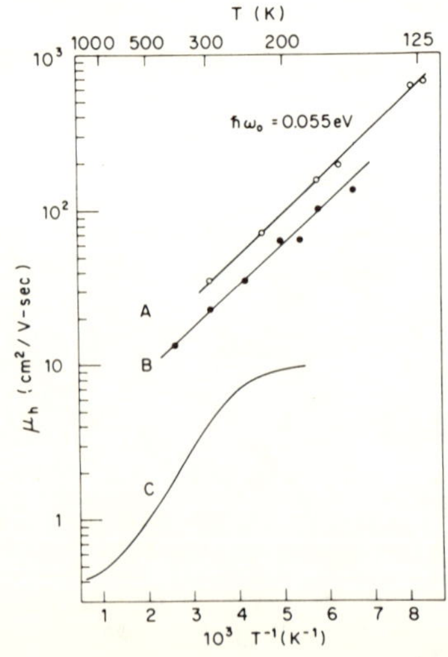

Fig. 2. Temperature dependence of the hole drift mobility for two specimens from batch 7-71 (lines A and B). Temperature dependence of the hole drift mobility calculated in ref. 3 from direct-current measurements (line C).

*The measurements were done in Prof. Spears Laboratory at the University of Dundee.

Figure 2 also shows the mobility values of Bosman and van Daal, calculated with the assumption that the density of states varies as $T^{3/2}$. It should be noted that the temperature dependence of the mobility reflects, in their case, the very small difference between the activation energy of the conductivity (0.30 ev) and that of the thermoelectric power (0.31 ev) and can therefore not be considered as exact.

4 DISCUSSION

The large mobility values and the temperature dependence observed support the model that the charge carriers in NiO are large polarons, with longitudinal optical phonon scattering limiting the mobility. The observed activation energy of $\hbar\omega_0 = 0.055$ ev is somewhat smaller than the zone-center ($k = 0$) longitudinal optical phonon energy of 0.071 ev in NiO. The difference is in the right direction, since phonons with $k \neq 0$ presumably participate in the scattering process and will have lower energies than those at $k = 0$.

In ref. 2, the value and the temperature dependence found for the mobility were analyzed according to the equation of Low and Pines[7]:

$$\mu = \frac{1}{2\alpha\omega_0}\frac{e}{m^*}\left(\frac{m^*}{m_p^*}\right)^3 f(\alpha) \exp\frac{\hbar\omega_0}{kT} , \tag{1}$$

where α is the coupling constant, which can be calculated from the dielectric constants of NiO according to

$$\alpha = \frac{0.33}{(\hbar\omega_0)^{1/2}}\left(\frac{m^*}{m_0}\right)^{1/2} ,$$

$m_p^* = (1 + \frac{1}{6}\alpha)$, m^* is the polaron effective mass, and $f(\alpha) \simeq 1$. The following values were obtained: $m_p^* = 1.5\ m_0$, $m^* = 1.2\ m_0$, and $\alpha = 1.6$. It was pointed out that our results for the coupling constant and the polaron effective mass differ considerably from those of Bosman and van Daal.[3] The starting point of these authors was to obtain $m_p^* = 6$ from an estimate of the density of states in the valence band. By using $\hbar\omega_0 = 0.071$ ev, they then obtained $\alpha = 2.5$. Their analysis was independent of the value of the mobility; if they had used $\hbar\omega_0 = 0.055$ ev it would have given $\alpha = 2.8$.

Langreth[8] has criticized Eq. (1) and shown that the correct equation for the mobility is

$$\mu = \frac{1}{2\alpha\omega_0}\frac{e}{m^*}\frac{m^*}{m_p^*} f(\alpha)\, e^{\hbar\omega_0/kT} . \tag{2}$$

An analysis similar to the one used in ref. 2 but using Eq. (2) leads to the following values

$$m_p^* = 3.4\ m_0,\ m^* = 2.5\ m_0,\ \alpha = 2.2\quad ;$$

i.e., the agreement with Bosman and van Daal is now much better.

One feature of the drift-mobility measurements requiring an explanation is the very low efficiency of generation: about 200 kev is needed for the generation of one mobile hole. It seems likely that two kinds of holes are generated in two overlapping bands: a low-mobility d band and a high-mobility p band. Adler's[9] model of conduction in a wide p band would thereby be supported. This model of two overlapping bands would also explain qualitatively the difference between the drift mobility of 0.4 cm^2/v-sec at 1200°K determined by Bosman and van Daal from direct-current measurements and the value of about 5 cm^2/v-sec extrapolated from our measurements to the same temperature. [The latter value must be regarded as a lower limit since the exponential temperature dependence in Eqs. (1) and (2) is only valid for $\hbar\omega_0/kT >> 1$. At $\hbar\omega_0/kT << 1$, the relation $\mu \propto T^{-1/2}$ is valid].[10,11]

The model implies that in thermal equilibrium the charge carriers excited from the lithium ions are distributed between the two bands proportionally to the effective densities of states. From the above mobility values it follows, then, that the ratio between the densities of states at 1200°K is about 12 in favor of the d band, which is a reasonable figure.

The very low generation efficiency in the drift-mobility experiments cannot of course be explained through the factor 12 in the density of states alone. We have to assume that excitation to the d band is a much more efficient process than excitation to the p band, and that very little thermalization occurs between the two bands during the transit time (about 10^{-7} sec) of the charge carriers.

REFERENCES

1. Lubezky, I., and Tannhauser, D. S., *J. Crystal Growth,* **17,** 162 (1972).
2. Spear, W. E., and Tannhauser, D. S., *Phys. Rev.,* **B7,** 831 (1973).
3. Bosman, A. J., and van Daal, H. J., *Adv. Phys.,* **19,** 1 (1970).
4. van Daal, H. J., *Proceedings of the Second International Conference on Low Mobility Materials,* Taylor and Francis, Eilat (1971), p. 17.
5. Spear, W. E., *J. Non-Cryst. Solids,* **1,** 197 (1969).
6. Cech, R. E., and Alessandrini, E. I., *Trans ASM,* **51,** 56 (1959).
7. Low, F. E., and Pines, D., *Phys. Rev.,* **98,** 414 (1955).
8. Langreth, D. C., *Phys. Rev.,* **137A,** 760 (1965).
9. Adler, D., and Feinleib, J., *Phys. Rev.,* **B2,** 3112 (1970).
10. Baltz, R. V., and Birkholtz, U., *Festkoerperprobleme,* **XII,** 233 (1972).
11. Appel, J., *Solid State Physics,* Academic Press, New York (1968), Vol. 21.

DISCUSSION on Paper by D. S. Tannhauser

HONIG: Recent measurements of Seebeck coefficients (α) on NiO samples with total impurity contents of less than 100 ppmw are essentially independent of temperature between 250 and 600°K. This can be interpreted as indicating a hopping regime for electrons in the d band.

Is there any really compelling evidence that the oxygen 2p band overlaps the nickel 3d band? Contrary band models have been proposed in the literature.

TANNHAUSER: The following argument is based on the data of Bosman and Van Daal. The effective density of states in the 2p band is much lower than that in the 3d band, since the 2p band is much wider. If now the 3d band lies more than kT ($\cong$ 0.1 ev at 1000°C) above the 2p band, we would expect it to act as a source of holes for the 2p band when all the lithium is ionized. The number of holes in the 2p band would then be temperature dependent. In the present paper we assume that $\mu\,(2p) >> \mu\,(3d)$, and in the above situation, we would therefore never get saturation of the conductivity as observed at high temperature by Bosman and van Daal.

DEFECT STRUCTURE AND ELECTRICAL CONDUCTIVITY OF CRYSTALLINE FERROUS SILICATE

H. G. Sockel

Erlangen University
Erlangen, West Germany

ABSTRACT

The possible disorder types in iron orthosilicate, $2Fe_{1-y}O{\cdot}SiO_2$ are discussed on the basis of knowledge of its crystal structure and the defect structure of transition-metal oxides. The defect structure has been investigated by electrical-conductivity and gravimetric measurements in CO_2-CO atmospheres in the temperature range 1000 to 1150°C. The results are consistent with a disorder type of doubly ionized iron vacancies and an equivalent number of electron holes. In comparison with $Fe_{1-y}O$, the point-defect concentrations in the iron orthosilicate are found to be lower by a factor of 10^2. For the interpretation of conductivity, the same models as for transition-metal oxides have been used. The p_{O_2} dependence and Seebeck coefficient indicate a p conduction. The calculated mobility of electron holes of 6×10^{-3} cm^2/v-sec as well as the activation enthalpy for motion of 5.8 kcal/mole suggest a small polaron mechanism.

1 INTRODUCTION

Point defects in binary and ternary oxides influence or control a series of physical properties, for instance the electrical, magnetic, optical, and mechanical behavior. In particular, the influence on the rate of diffusion is pertinent to a variety of technically important processes such as high-temperature deformation, formation of compounds in the solid state, sintering, precipitation, and crystallization.

Despite the technical importance of the silicates, the investigations of ternary compounds have been restricted mainly to ferrites, titanates, chromites, and aluminates.[1-6] Except for some papers which have been concerned with the disorder of $2MgO{\cdot}SiO_2$[7] and other orthosilicates[8-10], there has been relatively small progress in this field during the last years. The reason for this is the lack of radioactive tracers for diffusion experiments with oxygen and silicon, and, further, the difficulties in observing the small deviations from stoichiometry assumed to occur in many silicates.

This investigation was conducted to study the defect structure and electrical conductivity in iron orthosilicate, $2FeO{\cdot}SiO_2$. This silicate, with a relatively low melting point of 1205°C[17], has been chosen for a detailed investigation for two reasons. First, point defects in wüstite, "FeO", have been studied in a series of works, and though there are still problems left in understanding details, some facts are well established by experimental investigations.[11] From results on nonstoichiometry of other ternary compounds[1,3,9-11] and from the large range of homogeneity of "FeO", i.e., $Fe_{1-y}O$,[11-13] one could expect for the iron orthosilicate, in comparison with other orthosilicates, a relatively large deviation from stoichiometry dependent on oxygen partial pressure. Second, a question of general interest is whether there is a strong decrease in conductivity and perhaps a change in conduction mechanism when the transition-metal oxide "FeO" is "diluted" in the compound $2FeO{\cdot}SiO_2$ by "SiO_4" complexes.

For the presentation of the results of electrical-conductivity measurements and thermogravimetric investigations, and further for the following discussions, it seems necessary to review briefly the possible disorder types and the probable mechanism for electrical conduction based on the knowledge of the crystal structure of $2FeO{\cdot}SiO_2$ and of defect structure and electrical conductivity in transition-metal oxides.

2 DISORDER TYPES AND ELECTRICAL CONDUCTIVITY IN $2FeO{\cdot}SiO_2$

2.1 Crystal Structure of $2FeO{\cdot}SiO_2$

All silicate compounds contain SiO_4^{4-} tetrahedra in which the distance between silicon and oxygen is about 1.63 ± 0.03 Å.[14] The SiO_4 group has a negative charge of four and the bonding in the complex has been found to be about 50 percent ionic and the other 50 percent homopolar.[15]

The numerous different silicate structures are characterized by the arranging and linking of the SiO_4 tetrahedra in the lattice.[14] In the case in which the tetrahedra are separated and are connected only by cations, the structure is called an isle silicate. Iron orthosilicate, $2FeO{\cdot}SiO_2$, crystallizes in this type of structure.[16]

A projection of the structure in the (100) plane is shown in Fig. 1. Disregarding a slight deformation of the lattice, the structure can be described by a nearly close-packed oxygen lattice in which the cations are distributed regularly on the tetrahedrally and octahedrally coordinated sites. The close-packed layers of anions with pseudohexagonal symmetry are oriented parallel to the (100) plane. In this figure, the SiO_4 tetrahedra, indicated by the triangles, point alternately in different directions perpendicular to the (100) plane. It is evident from the projection that the iron ions are surrounded by six oxygen ions, the arrangement of which is nearly octahedral. By a slight deformation, the lattice loses its hexagonal symmetry and becomes orthorhombic. The nearest-neighbor distance of the iron ions depends on the lattice site of the ion and on the position of the surrounding nearest neighbor. The two occurring distances are 3.10 and 4.05 Å, and the number of nearest neighbors varies from 8 to 12. In comparison with these data, 12 nearest neighbors with a distance of 3.06 Å are found in the cationic lattice in "FeO".

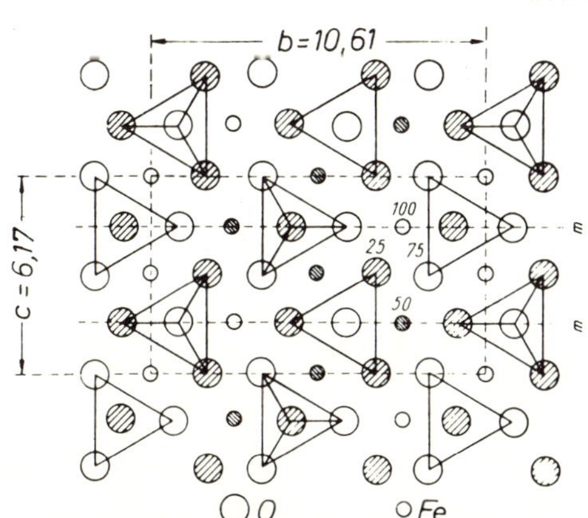

Fig. 1. Projection of the crystal structure of $2FeO{\cdot}SiO_2$ in the (100) plane. SiO_4 tetrahedra, indicated by triangles, point alternately in perpendicular to the (100) plane.

2.2 Disorder Types in $2FeO \cdot SiO_2$

The crystal structure of $2FeO \cdot SiO_2$ and those of the transition-metal oxides "FeO", CoO, and NiO are similar with respect to the nearly close packing of the oxygen sublattice and to the occupation of the octahedrally coordinated sites in the anionic structure by the divalent cations.[18] Taking into account this similarity as well as the knowledge of the disorder types in the above-mentioned transition-metal oxides and of the crystal structure of the iron silicate, one is led to the following restricting assumptions for possible point defects in the orthosilicate:

1. Point defects arising from anions in the cationic sublattice, and vice versa, can be excluded on account of the strong Coulomb interaction. Owing to the relative size of the ions and the high packing density in the anionic sublattice, it is unlikely that interstitial anions will occur.

2. Because of energetic reasons, it can be concluded that Si^{4+} ions prefer the tetrahedrally and the Fe^{2+} ions the octahedrally coordinated interstices of the anionic sublattice.

3. In all silicates the SiO_4 tetrahedra appear to be quite stable structural units; furthermore, SiO_2 as a compound shows an extremely small deviation from stoichiometry. Therefore, it is assumed that the vacancy concentration in the silicon sublattice remains so low that it can be neglected.

With these assumptions, the following point defects may be proposed for the iron orthosilicate lattice structure:

$Si_{i,T}^{\cdot\cdot\cdot\cdot}$	silicon ion on an interstitial site with tetrahedral coordination
$Fe_{i,o}^{\cdot\cdot}$	divalent iron ion on an interstitial site with octahedral coordination
$V_A^{\cdot\cdot}$	anion vacancy
$Fe_{Fe,o}^{\cdot}$	trivalent iron ion on a regular site with octahedral coordination in the iron sublattice
$V_{Fe,o}^{''}$	vacancy in the iron sublattice.

Apart from the four positively charged defects, $V_{Fe,o}^{''}$ has been taken into consideration as the only negatively charged lattice defect.

If a certain sort of positively charged defect dominates, then, owing to the condition of electroneutrality, the following pairs of positively and negatively charged defects may be formulated as possible disorder types:

$$(1)\quad Fe_{i,o}^{\cdot\cdot},\ V_{Fe,o}^{''} \qquad (2)\quad Si_{i,T}^{\cdot\cdot\cdot\cdot},\ V_{Fe,o}^{''}$$

$$(3)\quad V_A^{\cdot\cdot},\ V_{Fe,o}^{''} \qquad (4)\quad Fe_{Fe,o}^{\cdot},\ V_{Fe,o}^{''}$$

Type (1) represents the Frenkel disorder in the iron sublattice and can occur in the stoichiometric as well as in the nonstoichiometric compound with excess of SiO_2 or FeO. At constant composition, the defect concentrations depend only on temperature. In contrast to this, for types (2) and (3), an excess of SiO_2 with respect to the stoichiometric composition is necessary. Because of energetic reasons it is assumed that $Si_{i,T}^{\cdot\cdot\cdot\cdot}$ is more easily formed than $V_A^{\cdot\cdot}$; therefore, type (2) seems more probable for a composition with an excess of SiO_2. The occurrence of an SiO_2 deficiency with $V_{Si,T}^{''''}$ and $Fe_{i,o}^{\cdot\cdot}$ can be excluded from assumption 3. This is consistent with analogous experimental results for $2MgO{\cdot}SiO_2$.[7]

Disorder type (4) includes electronic defects $Fe_{Fe,o}^{\cdot}$, and an electron hole, $\dot{h}$, besides the formerly mentioned ionic point defects. For charge compensation, $Fe_{Si,T}^{''}$ and $V_{Si,T}^{''''}$ in addition to $V_{Fe,o}^{''}$ have not been taken into consideration in view of assumptions 2 and 3. While the defect concentrations of types (2) and (3) are controlled only by composition and temperature, in type (4) the oxygen partial pressure enters as an additional variable of state. In the case of completely ionized point defects, their concentrations should exhibit a characteristic p_{O_2} dependence. The interaction of gaseous oxygen with $2Fe_{1-y}O{\cdot}SiO_2$ can be expressed by the equation

$$\frac{1}{2}\,O_2(g) + 3Fe_{Fe,o} = 2Fe_{Fe,o}^{\cdot} + V_{Fe,o}^{''} + FeO \quad . \tag{1}$$

The corresponding equilibrium condition under the assumption of negligible defect interaction reads

$$(V_{Fe,o}^{''})\,(Fe_{Fe,o}^{\cdot})^2\,a_{FeO}/p_{O_2} = \exp(-\Delta G_1^{\circ}/RT) \quad , \tag{2}$$

where ΔG_1° is the standard free-energy change for reaction (1). The activity of FeO dissolved in $2FeO{\cdot}SiO_2$, a_{FeO}, depends on the activities of SiO_2, $2FeO{\cdot}SiO_2$, and the free energy of formation of $2FeO{\cdot}SiO_2$ from the oxides. Therefore, at constant temperature, a_{FeO} remains constant as long as $a_{SiO_2} = 1$ can be provided experimentally. Since a ternary compound has 4 degrees of freedom, the thermodynamic state is determined by the variables temperature, pressure, oxygen partial pressure, and, in this case, by the activity of SiO_2. Under this condition, the enthalpy of formation of about -8 kcal/mole[19,20] is responsible for a slight change in a_{FeO} with temperature.

Taking this into account and introducing the relationship for electrical neutrality, $(Fe_{Fe,o}^{\cdot}) = 2(V_{Fe,o}^{''})$, the following p_{O_2} dependence of the defect concentrations is obtained:

$$(Fe_{Fe,o}^{\cdot}) = 2(V_{Fe,o}^{''}) = K(T)\,p_{O_2}^{1/6} \quad . \tag{3}$$

If the disorder type (4) comes in competition with types (2) and (3), $V''_{Fe,o}$ changes only slightly with p_{O_2} or remains constant. In the latter case, the defect electron concentration will obey a $p_{O_2}^{1/4}$ dependence. The same variation in the defect concentrations would be expected if $[Fe^{\cdot}_{Fe,o}, V''_{Fe,o}]'$ complexes are the predominant defects.

2.3 Electrical Conductivity in $2FeO{\cdot}SiO_2$

Crystal- and defect-structure similarities between $2FeO{\cdot}SiO_2$ and the transition-metal oxides FeO, NiO, and CoO encourage interpreting the electrical conductivity of the silicate in the same terms used for understanding the electrical behavior of the transition-metal oxides.

Electronic-transport phenomena of transition-metal oxides have been investigated in numerous works and have been reviewed in several papers.[21-24] For the interpretation of carrier mobility, Seebeck- and Hall-coefficient data, localized electron transfer, and narrow band models were used. The hopping model considers the transfer of a polaron – a localized electron with a surrounding lattice deformation – between ambivalent cations, and predicts a thermally activated diffusion process with a corresponding exponentially temperature-dependent mobility. According to the narrow band model, the mobility is limited by scattering in the band; with increasing temperature, a decrease in band width and in carrier mobility should be expected. Following the theories of small polarons[25-27], the mechanisms should be responsible, in different temperature regions, for the mobility of the carrier. At low temperatures the carrier mobility decreases with temperature owing to tunneling in the crystal, as described by band theory. The band model breaks down at mobilities lower than about 0.1 cm^2/v-sec; and at temperatures higher than $\sim 1/2\ \theta_D$, where θ_D stands for the optical Debye temperature, a description of the carrier motion by a hopping model is more adequate.

It has been shown in the preceding chapters that the iron orthosilicate can be treated roughly as "FeO" diluted by SiO_4 complexes. It is thus of interest for the consideration of the conduction mechanism to compare the carrier mobility data for $2FeO{\cdot}SiO_2$ with data for transition-metal oxides in general and for pure FeO and solid solutions of FeO with MgO in particular.

3 EXPERIMENTAL DETAILS

The compound $2FeO{\cdot}SiO_2$ had to be prepared by solid-state reaction. Appropriate amounts of high-purity silica and ferric oxide were weighed after annealing at 800°C for 1 day in air, well mixed, and pulverized. Pellets were made from the mixture by pressing in steel dies at a pressure of about 3 to 4 kp/cm^2. The pellets were heat treated between 1150 and 1170°C in a CO_2/CO atmosphere for several days for reaction and sintering of the oxides. Grinding, pressing, and annealing of the same mixture was repeated several times with additions of SiO_2 or Fe_2O_3 until a stoichiometric compound with no unreacted FeO was obtained. Material produced by this method showed a density of greater than 93 percent of theoretical and a grain size varying between 30 and 100 μm. Material prepared by melting and solidifying or by hot pressing the molten material had the same density but a grain size of about 100 to 300 μm. Before final sintering, about 1 percent SiO_2 in excess was added to fix a_{SiO_2} in the compound. For conductivity measurements samples with final dimensions 10 x 2 x 2 mm^3 were cut from the sintered pellets.

For the conductivity measurements, a two-terminal arrangement with platinum contacts was used. The a-c measurements were carried out with a bridge (Wayne Kerr) operating at 1592 Hz.

The temperature could be kept constant within ±1°C. A CO_2/CO mixture of purified gases was used to establish the oxygen partial pressure. The composition of the gaseous mixture was varied in the range 20 to 95 percent CO_2 and controlled with an accuracy better than 1 percent by infrared absorption measurements.

4 RESULTS

The electrical conductivity of $2Fe_{1-y}O{\cdot}SiO_2$ is plotted in Fig. 2 as a function of 2 log $p_{CO_2}/p_{CO} \propto \log p_{O_2}$ at several temperatures in the range from 1000 to 1150°C. A dependence of conductivity on grain size has not been observed in the samples prepared by different methods. The times necessary to establish equilibrium after a change of p_{O_2} ranged from 3 to 5 hours. These facts indicate that the measured data can be attributed to a volume conduction. The data from different samples showed agreement within ±10 percent. The reproducibility of data from the same sample after temperature and p_{O_2} variations was better than ±5 percent. Investigation of the Seebeck effect at 1100°C and $p_{CO_2}/p_{CO} = 2$ showed that the Seebeck coefficient is positive and that the thermoelectric power equals about 300 μv/°K.

Results from the thermogravimetric investigation are shown in Fig. 3. The weight of the sample at low p_{O_2} near the boundary Fe-$2FeO{\cdot}SiO_2$ has been chosen as reference point for the weight change as a function of

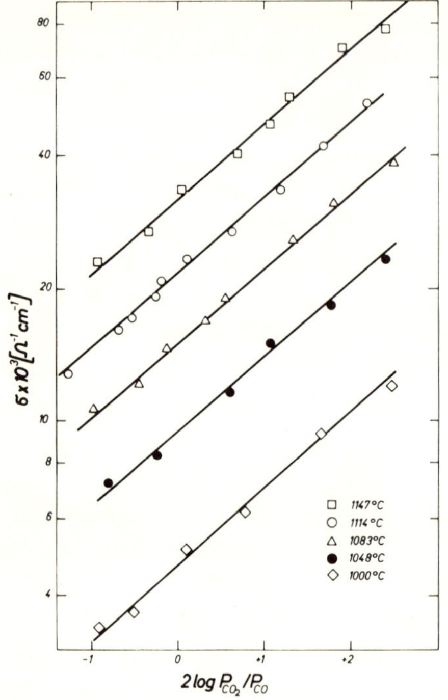

Fig. 2. Electrical conductivity σ of $2Fe_{1-y}O \cdot SiO_2$ vs 2 log p_{CO_2}/p_{CO} as a function of temperature.

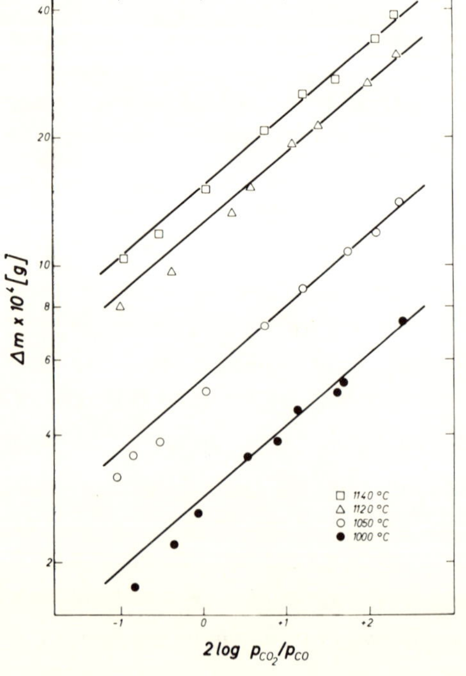

Fig. 3. Weight change Δm vs 2 log p_{CO_2}/p_{CO} as a function of temperature. The weight of the sample at low p_{O_2} near the boundary $Fe\text{-}2FeO \cdot SiO_2$ is chosen as reference point.

p_{O_2}. The weight change Δm for a sample of 10 g $2Fe_{1-y}O \cdot SiO_2$ is plotted as a function of 2 log p_{CO_2}/p_{CO} at several temperatures. At constant temperature and low p_{O_2}, deviations to steeper slopes, d log Δm/d log p_{O_2}, are observed.

The weight change Δm is related to the oxygen excess, z, in $2FeO_{1+z} \cdot SiO_2$ by

$$z = \Delta m \; M_s / 2 \; M_o \; m_s \quad , \tag{4}$$

where m_s, M_o, and M_s equal the actual weight of the sample, the atomic weight of oxygen, and the molecular weight of $2FeO \cdot SiO_2$, respectively. The cation deficiency in $2Fe_{1-y}O \cdot SiO_2$ can be calculated by

$$y = z \; / \; 1 + z \quad . \tag{5}$$

For $z \ll 1$, the oxygen excess, z, is approximately equal to y. A plot of y as a function of log p_{CO_2}/p_{CO} for various temperatures is shown in Fig. 4. Because there is a deviation at lower p_{O_2} from the dependence at high p_{O_2} for the weight changes but not for the electrical conductivity, this effect must arise from certain amounts of iron vacancies and electron holes present in the compound at low p_{O_2} near the phase boundary Fe-$2FeO \cdot SiO_2$. This has not been taken into account in Eqs. (4) and (5). The p_{O_2} dependence of the electrical conductivity σ shows no deviation because σ is proportional to $(Fe_{Fe,o}^{\cdot}) \propto p_{O_2}^{1/6}$. However, the weight change $\Delta m(p_{O_2})$ with a reference weight of the sample at p_{O_2} near the phase boundary Fe-$2FeO \cdot SiO_2$ is proportional to $\left\{ (V_{Fe,o}'')_{p_{O_2}} - (V_{Fe,o}'')_{O} \right\}$, the difference in the vacancy concentrations at p_{O_2} and at p_{O_2} near the phase boundary Fe-$2FeO \cdot SiO_2$. Because only $(V_{Fe,o}'')$ is proportional to $p_{O_2}^{1/6}$, the weight change Δm will show deviations from this p_{O_2} dependence at low defect concentrations on account of $(V_{Fe,o}'')_O$. From the observed deviations, values of $y_o \times 10^5$ at the phase boundary Fe-$2FeO \cdot SiO_2$ of 5,7,8,8 at 1000, 1050, 1120, and 1140°C, respectively, are calculated and the y data have been corrected with these values.

The gravimetric measurements could be carried out within an accuracy and reproducibility of ±5 percent. Within the same error, an agreement among data from different samples was obtained.

The conductivity and y data have been plotted as a function of 1/T for various p_{O_2} in Fig. 5. The difference in the slopes of the curvatures are due to the activation enthalpy of motion of the electron holes.

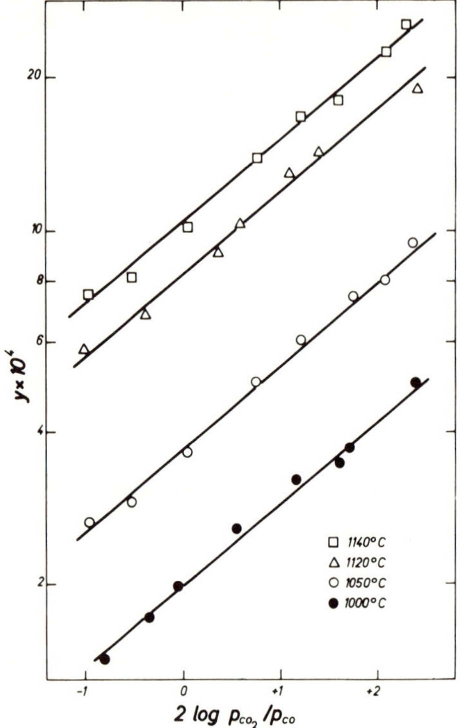

Fig. 4. $y = (V''_{Fe,o})/2$ vs 2 log p_{CO_2}/p_{CO} as a function of temperature.

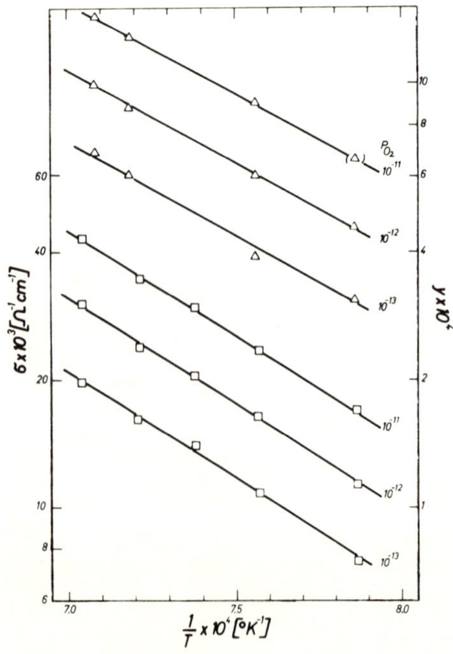

Fig. 5. y (Δ) and σ (□) vs $1/T$ as a function of oxygen partial pressure.

5 DISCUSSION

The specific conductivity σ is given by

$$\sigma = n_{h^{\cdot}} \cdot u_D \; e \quad , \tag{6}$$

where $n_{h^{\cdot}}$ is the concentration of electron holes, u_D is their mobility, and e is the electronic charge. The linearity of the plots in Fig. 2 indicates that at constant temperature the mobility of electron holes and the degree of ionization remains constant with changing defect concentration. This justifies the assumption of ideal behavior of the defects. The slope of d log σ/d log p_{O_2} leads to a $p_{O_2}^{1/6}$ dependence for the concentrations of electron holes and cation vacancies. This fact, together with the results from gravimetric investigation mentioned below, indicates a disorder type with completely ionized iron vacancies and electron holes as majority defects.

For this disorder type, the defect concentration

$$y = \left(V''_{Fe,o}\right)/2 = \left(Fe^{\cdot}_{Fe,o}\right)/4 \tag{7}$$

can be calculated from the gravimetric data. The observed $p_{O_2}^{1/6}$ dependence for higher p_{O_2} at constant temperature emphasizes again the result from conductivity measurements concerning the defect structure. In comparison with $Fe_{1-y}O$[11-13, 28-30], the cation vacancy concentration in the silicate is lower by a factor of 10^2. The same defect structure and nearly the same factor for the corresponding defect concentrations have been found for cobalt orthosilicate.[9-11, 31, 32]

When investigating such low concentrations of defects in the silicates, one has to ensure that the samples are free of FeO as a separate phase. Even small amounts, of the order of 1 percent, would produce an effect which would outweigh that of the silicate. Therefore, the substance has been analyzed by both chemical and optical investigations, and, in addition, small amounts of SiO_2 were added to the stoichiometric substance. Further, in the gravimetric investigation, the presence of FeO would cause a steplike increase in the weight change with increasing p_{O_2} at the FeO-Fe and FeO-Fe_3O_4 phase boundaries. Since such steplike changes are not observed, it can be concluded that the contribution to the weight change and correspondingly to the defect concentrations can be neglected.

According to Eq. (2), the dependency of the vacancy concentration on temperature and p_{O_2} can be described by

$$(V''_{Fe,o}) = a_{FeO}^{-1/3} \, p_{O_2}^{1/6} \exp(-\Delta G^{\circ}_1/3RT) \quad . \tag{8}$$

It has been mentioned that a_{FeO} with a_{SiO_2} = 1 shows a slight temperature dependence. Taking this into account, the standard change of enthalpy for reaction (1), ΔH_1°, is calculated to be 66.5 ± 4 kcal/g-mole oxygen. For the binary cation-deficient oxides "FeO"[13] and NiO[33] ΔH_1° has been found to be nearly equal to the absolute value of the enthalpy of formation from the elements. As the enthalpy of formation of 2 $FeO{\cdot}SiO_2$ from the oxides is relatively small, about -8 kcal,[19,20] it seems consistent with the results for the mentioned binary oxides that for the occurring defect structure, the enthalpy ΔH_1° for $2FeO{\cdot}SiO_2$ lies near the absolute value of the enthalpy of formation for "FeO", 63.7 kcal.[34] A similar result has been obtained for solid solutions (Mg,Fe)O. For comparison, ΔH_1° for "FeO"[13] and the solid solutions (Mg,Fe)O[62] has been found to be 66 kcal/mole and 62 to 70 kcal/mole, respectively.

The specific electrical conductivity in the range of 10^{-2} ohm^{-1} cm^{-1} indicates a semiconducting behavior. Both the p_{O_2} dependence of conductivity and the positive thermoelectric power are consistent with a p conduction. The defect concentration together with the conductivity data enable one to calculate the mobility of electron holes according to Eq. (6). The obtained mobilities of about 6 x 10^{-3} cm^2/v-sec and the activation enthalpy for motion lead to the assumption that a small polaron mechanism is responsible for electrical conduction.

The mobilities of electron holes for the transition-metal oxides FeO[39], CoO[40], NiO[40,41], MnO[40], a solid solution of "FeO" with MgO[42], and the iron silicate are plotted in Fig. 6 as a function of

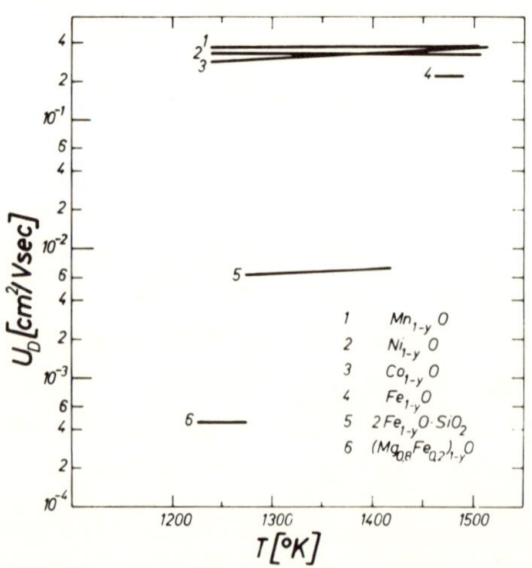

Fig. 6. Mobility of electron holes vs temperature for "FeO"[39], CoO[40], NiO[40,41], MnO[40], $(Mg_{0,8}Fe_{0,2})O$[42], and $2FeO{\cdot}SiO_2$ (this work).

temperature. Mobilities for the transition-metal oxides have been found to be in the range from 0.1 to 0.4 $cm^2/v\text{-sec}$ between 900 and 1200°C. In the solid solution (Fe, Mg)O, a substitution of FeO by MgO leads to a decrease in the electron-hole mobility.[39,42] For 31.7 percent MgO and 1200°C, a value of 8.6×10^{-2} $cm^2/v\text{-sec}$[39] has been obtained, and for 80 and 95 percent MgO and temperatures between 900 and 1000°C, values of 4.5×10^{-4} and 1.5×10^{-6} $cm^2/v\text{-sec}$[42], respectively, have been measured. The data for iron orthosilicate lie between the values for highly diluted "FeO" (80 to 95 percent MgO) and those of the solid solution (Fe, Mg)O containing 31.7 percent MgO. In comparison with pure "FeO", there is a greater decrease in the mobilities in the silicate than in the solid solution (Fe, Mg)O with 31.7 percent MgO by a factor of about 10. This may be due to the fact that more than 50 percent of the nearest-neighbor distances have increased in the silicate from 3.1 Å – nearly the same as in "FeO" – to 4.05 Å.

Activation enthalpies of motion for electron holes for "FeO" and solid solutions (Fe, Mg)O have been reported in the literature in the range from 1.5 to 16 kcal/mole.[39,42,43] For "FeO"-rich solid solutions, an increase of enthalpy with decreasing O/Fe ratio (0 to 31.7 percent MgO) has been observed. Data cover the range from 1.5 to 9 kcal/mole.[39] For solid solutions containing 10 and 20 percent FeO activation enthalpies of about 9 kcal/mole have been obtained[42], while other experimenters obtained enthalpies of 4 to 10 kcal/mole[43]. For concentrations of FeO < 10 percent the observed values vary between 9 and 16 kcal/mole.[42,43]

The activation enthalpy evaluated for $2FeO{\cdot}SiO_2$ equals 5.8 ± 1.8 kcal/mole. In comparison with the data for (Fe, Mg)O solid solutions, the measured enthalpy seems to be a reasonable value.

6 CONCLUSION

On the basis of knowledge of defect structures in transition-metal oxides, the possible point defects in the stoichiometric compound $2FeO{\cdot}SiO_2$ with orthosilicate structure are considered. Measurements of the electrical conductivity as a function of temperature and p_{O_2} and thermogravimetric analysis indicate a disorder type with completely ionized cation vacancies in the iron sublattice and electron holes. The point-defect concentrations in the silicate lie lower than in FeO by a factor of 10^2. An attempt was made to apply the same models for electrical conduction as in transition-metal oxides. A mobility of electron holes of 6×10^{-3} $cm^2/v\text{-sec}$ and an activation enthalpy of 5.8 kcal/mole have been found. The corresponding mobility in wüstite is about 0.2

cm^2/v-sec. Both the low carrier mobility and the increase with temperature clearly suggest a hopping mechanism for the electrical conductivity.

ACKNOWLEDGMENT

I wish to thank Prof. Dr. B. Ilschner for helpful criticism and suggestions, Dr. J. Varner for assistance in language difficulties, and the Deutsche Forschungsgemeinschaft for financial support of this work.

REFERENCES

1. Schmalzried, H., *Prog. in Solid State Chem.*, **2**, 265 (1965).
2. Tretjakov, J. D., and Schmalzried, H., *Ber. Bunsenges. Phys. Chem.*, **69**, 396 (1965).
3. Schmalzried, H., and Tretjakov, J. D., *Ber. Bunsenges. Phys. Chem.*, **70**, 180 (1966).
4. Jagodzinski, H., *Problems of Nonstoichiometry*, A. Rabenau (Ed.), North Holland Publishing Company, Amsterdam (1970), p. 131.
5. Joubert, J. C., Berthet, G., Bertaut, E. F., *ibid.*, p. 179.
6. Reijnen, P.J.L., *ibid.*, p. 219.
7. Pluschkell, W., Dissertation, Clausthal, 1968.
8. Borchardt, G., and Schmalzried, H., *Z. Phys. Chem. N.F.*, **74**, 265 (1971).
9. Schwier, G., Dieckmann, R., Schmalzried, H., *Ber. Bunsenges. Phys. Chem.*, **77**, 402 (1973).
10. Greskovich, C., and Schmalzried, H., *J. Phys. Chem. Solids*, **31**, 639 (1970).
11. Sockel, H. G., and Schmalzried, H., *Ber. Bunsenges. Phys. Chem.*, **72**, 745 (1968).
12. Lehmann, G., *Ber. Bunsenges. Phys. Chem.*, **73**, 349 (1969).
13. Engell, H.-J., *Arch. Eisenhüttenwes.*, **28**, 109 (1957).
14. Liebau, F., *Naturwiss.*, **49**, 481 (1962).
15. Pauling, L., *The Nature of Chemical Bonding*, Ithaca, New York (1960).
16. Kleber, W., *Einführung in die Kristallographie*, 7th edition, VEB Verlag Technik, Berlin (1963), p. 223.
17. Bowen, N. L., and Schairer, J. F., *Am. J. Sci.*, 5th Ser. **24**, 200 (1932).
18. Flörke, O. W., *Ber. Deut. Keram. Ges.*, **40**, 451 (1963).
19. Darken, L. S., and Schwerdtfeger, K., *Trans. Met. Soc. AIME*, **236**, 201 (1966).
20. King, E. G., *J. Am. Chem. Soc.*, **74**, 4446 (1952).
21. Gerthsen, P., Kauer, E., and Reik, H. G., *Festkörperprobleme*, F. Sauter (Ed.), Friedr. Vieweg & Sohn, Braunschweig (1966), Vol. 5, p. 1.
22. Adler, D., *Solid State Physics, Advances in Research and Applications*, F. Seitz, D. Turnbull, and H. Ehrenreich (Eds.) Academic Press, New York (1968), Vol. 21, p. 1.
23. Appel, J., *ibid.*, p. 193.
24. Austin, I. G., and Mott, N. F., *Adv. Phys.*, **18**, 41 (1969).
25. Holstein, T., *Ann. Physique*, **8**, 343 (1959).
26. Sewell, G. L., *Phys. Rev.*, **129**, 597 (1963).
27. Friedmann, L., *Phys. Rev.*, **135A**, 233 (1964).
28. Swaroop, B., and Wagner, J. B., *Trans. Met. Soc. AIME*, **239**, 1215 (1967).
29. Darken, L. S., and Gurry, R. W., *J. Am. Chem. Soc.*, **68**, 798 (1946).
30. Bransky, I., and Tannhauser, D. S., *Trans. Met. Soc. AIME*, **239**, 75 (1967).
31. Fischer, B., and Tannhauser, D. S., *J. Electrochem. Soc.*, **111**, 1194 (1964).
32. Gleitzer, C., *Bull. Soc. Chim.*, p. 75 (1962).
33. Tretjakov, J. D., and Rapp, R. A., *Trans. Met. Soc. AIME*, **245**, 1235 (1969).
34. Rossini, F. D., et al., *Selected Values of Chemical Thermodynamic Properties*, U.S. Dept. of Commerce, National Bureau of Standards (February 1, 1952).

35. Berthet, A., Ph.D. Thesis, University of Nancy, France, 1963.
36. Schmahl, L. G., Frisch, B., and Stock, G., *Arch. Eisenhüttenw.*, **32,** 297 (1961).
37. Brynestad, J., and Flood, H., *Z. Elektrochem.*, **62,** 953 (1958).
38. Alcock, C. B., and Iyengar, G.N.K., *Proc. Brit. Ceram. Soc.*, **8,** 219 (1967).
39. Levin, H. A., Kassner, T. F., and Wagner, J. B., *Z. Phys. Chem. N.F.*, **74,** 331 (1971).
40. Bransky, I., and Tallan, N. M., *Modern Oxide Materials,* B. Cockayne and D. W. Jones (Eds.), Academic Press, London (1972), p. 67.
41. Sockel, H. G., and Ilschner, B., *Z. Phys. Chem. N.F.*, **74,** 284 (1971).
42. Iyengar, G.N.K., and Alcock, C. B., *Phil. Mag.*, **21,** 293 (1970).
43. Hansen, K. W., and Cutler, I. B., *J. Am. Ceram. Soc.*, **49,** 100 (1966).

DISCUSSION on Paper by H. G. Sockel

ANDERSON: I may have misunderstood some part of the defect description. If it is right to take it that you have only one Fe_2SiO_4 phase present, the ratio of iron atoms to silicon atoms in the condensed phase is fixed and immutable. All that can be done is to push oxygen atoms in or out of the solid; the formulation must be $Fe_2O_{2-y}\cdot SiO_2$. If oxygen atoms are removed, this would leave silicon only 3-coordinated unless silicon shifts to the alternative tetrahedral site to form a $Si_2O_7{}^{6-}$ anionic unit. The defect description would then be much more complex.

SOCKEL: During the discussion of the reaction for the incorporation of oxygen from the gaseous phase into $2FeO\cdot SiO_2$ it was mentioned that a_{SiO_2} had to be fixed experimentally in order to keep a_{FeO} constant apart from the influence of nonstoichiometry at constant temperature. This is possible by adding small amounts of SiO_2 as a separable phase. In this case, the defect concentrations depend only on p_{O_2} at constant temperature, and the state of the compound is thermodynamically fixed by the last variable.

SCHMALZRIED: Let me try to answer Dr. Anderson's question about the metal to oxygen ration in Dr. Sockel's experiment. For coexistence with SiO_2 ($a_{SiO_2} \simeq 1$), the fayalite changes composition with regard to the ratio Fe/Si if one changes the oxygen potential. This is so because SiO_2 coexisting with Fe_2SiO_4 is either a source or a sink for silicon ions. One could do the change in oxygen potential also, with no coexisting phases such as SiO_2 or "FeO". But in this kind of experiment, the defect equilibrium of which can also be worked out, one never is really sure as to how much the surroundings will react with the components and thus disturb the defect equilibria. Personally, I feel that a predetermination of all independent thermodynamic variables is the safest way to operate in using defect chemistry.

ANDERSON: As I now understand, uncombined SiO_2 is present in the condensed phase as an additional 2nd-phase constituent, to react with the "FeO" formed as postulated in the basic defect-forming equation. This deals with the point.

HEUER: Isn't it true that if one had a stoichiometric single crystal of Fe_2SiO_4 and one heated this crystal in a high p_{O_2}, any oxygen takeup by the crystal would cause the formation of defects in both the iron and silicon sublattices with the charge compensation being oxidation of Fe^{2+} to Fe^{3+}.

SOCKEL: This may be possible. The difficulty is that without SiO_2 as a separate second phase, the state of the compound is not fixed thermodynamically. Furthermore, it can be shown that the p_{O_2} dependence in this case will be different from the $p_{O_2}^{1/6}$ relationship for the defect concentrations.

WAGNER: Would you explain why the plot of the logarithm of weight change versus logarithm of the oxygen pressure (from CO_2-CO mixtures) was not linear at low oxygen pressures?

SOCKEL: There is a deviation at lower p_{O_2} from the dependence at high p_{O_2} for the weight changes, but not for the electrical conductivity in the log Δm vs log p_{O_2} plots. This effect must arise from certain amounts of iron vacancies and electron holes present in the compound at low p_{O_2} near the phase boundary Fe-2FeO·SiO_2. Because $\sigma \propto (Fe^{\bullet}_{Fe,o}) \propto p_{O_2}^{1/6}$, there is no deviation in the p_{O_2} dependence of the electrical conductivity. However, the weight change Δm with respect to a reference weight at the phase boundary Fe-2FeO·SiO_2 is proportional to $\left\{(V''_{Fe,o})p_{O_2} - (V''_{Fe,o})_O\right\}$, the difference in the vacancy concentrations at p_{O_2} and at p_{O_2} near the phase boundary Fe-2FeO·SiO_2. For the log Δm vs log p_{O_2} plot, it has been assumed that $(V''_{Fe,o})_O = 0$. On account of the deviation, $y \times 10^5$ must lie between 5 and 8 for different temperatures.

AGENDA DISCUSSION: TRANSPORT OF MASS AND CHARGE

*P. Kofstad**

Central Institute for Industrial Research
Blindern, Oslo, Norway

*J. M. Wimmer***

Aerospace Research Laboratories
Wright-Patterson Air Force Base, Ohio

Kofstad opened the session with several remarks on lattice self-diffusion in binary oxides. He first posed the question of how the presence of shear structures or defect clusters affect lattice diffusion and also brought up the question of the stability of these structures at high temperature. He then stated that if one looks at actual data rather than comparing separately D_0's or activation energies as is often done, the state of diffusion measurements was not in such bad shape as was sometimes inferred. There are systematic trends and variations for certain classes of oxides. For example, for some of the NaCl structure monoxides, there is

*Chairman
**Secretary

fairly good agreement among various investigators that the cation-tracer diffusion coefficient decreases as the extent of departure from stoichiometry decreases i.e., in the order FeO, CoO, NiO, MgO/CaO. In addition, if one subtracts out the compositional dependence from the tracer diffusion data, then the activation energy associated with the mobility of the cation defect is 30 to 35 kcal/mole in the cases where data are available.

Cohen questioned the statement about diffusion data being in good shape, especially with respect to FeO where the compositional dependence of the cation-tracer and chemical diffusion coefficients are opposite.

In response to this question, Wagner presented a summary of our understanding of diffusion in FeO. The properties that appear reasonable and well behaved are the tracer diffusion of iron and the variation of lattice parameter with deviation from stoichiometry. If we take the data of various investigators and plot the tracer diffusion coefficient as a function of the O/Fe ratio, the curves are linear and all extrapolate back to approximately zero (on the scale of $D_{Fe} \rightarrow FeO \cong 10^{-7}$ to 10^{-8} cm^2/sec) at the stoichiometric composition. This behavior does not rule out clusters, but suggests that the same general diffusion mechanism is operative across the entire wustite phase field. The lattice parameter decreases linearly with O/Fe ratio both at high temperature and on quenched samples.

One of the properties that does not exhibit this linear behavior is electronic conductivity at elevated temperatures. The conductivity varies with $P_{O_2}{}^{1/6}$ at low iron vacancy concentrations, but at higher defect concentrations (approximately in the middle of the phase field) the dependence starts to decrease, and this results in a very small oxygen pressure dependence. Something else peculiar takes place in the middle of the phase field, there is a p to n transition, as determined by Seebeck coefficient measurements. On the low-oxygen-excess side of the phase field, one can make a reasonable argument that there the classical case of the dissolution of oxygen to yield cation vacancies and holes obtains. The change in sign of the Seebeck coefficient indicates that this is far too simple a model, and in fact none of the defect models proposed for FeO take into account this change. The final puzzle concerns the chemical diffusion coefficient obtained from the kinetics of isothermal equilibration after a step change in oxygen pressure. This chemical diffusion coefficient, being a defect diffusion coefficient, should be easily related to the tracer diffusion coefficient, but they have the opposite dependence, the chemical diffusion coefficient decreasing with increasing O/Fe ratio.

O'Keeffe pointed out that the more concentrated the defects become, the more difficult it is for them to move, and since the chemical diffusion coefficient is intuitively understandable as a defect diffusion coefficient, the observed dependence is reasonable. The real point at issue is the failure of the Darken equation to correctly relate the chemical and tracer diffusion even though good thermodynamic data on FeO are available. Kofstad asked if the use of the Wagner equation would not be more appropriate. Wagner responded that both equations had been used and neither gave good correlation between the tracer diffusivities and the chemical diffusivities.

Seltzer stated that our understanding of anion diffusion is in fairly good shape for some groups of compounds. For example, sulfur diffusion data for PbS, PbTe, PbSe, Cd salts, and Hg salts are all consistent, and show the same 1/2 partial pressure dependence; also, a number of laboratories agree that the anions are diffusing by an interstitial mechanism, probably as a neutral species. There are a number of oxide systems for which similar data are being obtained.

Stringer agreed with Kofstad's remark that the actual diffusion data should be compared rather then the D_0's or activation energies used to describe the data. Earlier in the day, several people had used the word reasonable in reference to certain diffusion data and Stringer asked what were reasonable values for the various parameters used to describe diffusion in oxides.

When one is trying to use diffusion data to predict, for example, oxidation rates, one has difficulty in deciding which data in the literature are reasonable and which are not. For substitutional self-diffusion in metals, there are reasonable values for the pre-exponential factor and activation energies, and if these are not observed, then alternative transport mechanisms are invoked. The consensus was that metals, being monoatomic, are much simpler systems than compounds. There are different classes of compounds and each has its own reasonable parameters.

Shewmon voiced the opinion that accurate diffusion data are just beginning to be reported now. There are older data in the literature which are often unreliable, but which were publishable at the time because they were the first obtained. He recommended looking at the newer data being published by the better laboratories.

Huggins stated that there are really three different types of materials and what is reasonable varies for all three. In the first type, we have isolated defects either clustered or not, such as are found in the monoxides. In the second type, we have liquid sublattice materials and all the parameters including the language used to describe defects are different. The third type are the extended-defect-structure materials.

Kröger suggested that isotherms were more useful in studying diffusion than were temperature dependencies. Often one finds that over a range of activities, the isotherms clearly indicate the operation of certain mechanisms much more so than the temperature dependence. Transitions between mechanisms are also easier to analyze from the isotherms.

Nowick suggested that rather than deciding in advance what is reasonable, the user of the data should examine them closely, for example, by looking at penetration plots to substantiate a volume diffusion mechanism and by noting the impurity content of the materials for which measurements were made.

Dragoo stated that he was suspicious of data obtained by O^{16}/O^{18} exchange using ground-up crystals since the crystals are undoubtedly damaged to a significant depth. Heuer agreed, pointing out that for low diffusivity materials, the penetration can easily be of the order of only 10 μ and unless special precautions are taken, high dislocation densities in the near surface region might be expected.

O'Keeffe presented some recent data on CaF_2 and PbF_2 where the electrical conductivity in the solid is of the order of the molten-salt conductivity and there is no discontinuity in conductivity at the melting point. In this case, essentially every anion is in motion and one speaks of a molten anion sublattice. In such a case one needs to examine carefully the idea of extended defects or clusters. For both of these compounds the entropy of fusion is much lower than for normal salts because the anion sublattice is already disordered. From the magnitude of the conductivity one can calculate a diffusion coefficient and, assuming a jump distance, the time between jumps. The result is 2.3 picoseconds, which is of the order of a lattice vibration period. On the other hand, a simple gaseous calculation gives numbers of the right order.

Lidiard asked about the experimental conditions, since hydrolysis of the surface leads to highly conducting surface layers.

O'Keeffe replied that purified argon was used for CaF_2 but in the case of PbF_2, both purified argon and air gave the same results and at high temperature the results are very reproducible.

Eyring questioned the doubts that still exist on the stability of shear structures at high temperature and made the following points. First, the most stable of the intermediate phases in the lanthanide and actinide oxides is M_7O_{12}, and in each case this phase is stable to about half the melting point. Second, Dr. Rustum Roy's phase diagram of the TiO_x-O_2 system shows that the homologous Ti_nO_{2n-1} oxide series members melt congruently or are at least stable to very near the melting point. Third, the elegant high-resolution micrographs shown by Dr. Cowley were taken under an intense electron beam from a pointed filament, a condition which causes many substances to demonstrate behavior typical of high temperature. From these qualitative observations, it is clear that the ordered phases observed in quenched systems are indeed present also at high temperature in many materials.

O'Keeffe agreed entirely. He wanted only to emphasize that in some materials at high temperature, there is very expensive disorder, so not in every instance does one need to worry about complicated defects. He then asked Anderson if it were not true that sometimes shear planes vanish in the microscope when the material is heated, and then reappear when it is cooled.

Anderson commented that he thought that this was an oxidation-reduction reaction involving backstreaming pump oil and not a dissolution of the shear planes.

Huggins pointed out that many materials show the very strange, unusually high ionic transport at surprisingly low temperatures and that there are three striking features of these materials. One, the magnitude of the conductivity; two, the extremely low activation energies; and three, the very low D_0 values, typically on the order of 10^{-4} cm^2/sec. One of the interpretations of a low D_0 is that a small number of species are contributing to the transport, but for many reasons it appears that nearly every ion in the sublattice is contributing and this seems to be a contradiction. The only reasonable explanation seems to be that the jump attempt frequency for these very mobile ions is extremely low.

The discussion then turned to electronic transport and Nowick presented recent results on reduced ceria (CeO_{2-x}). Electronic conduction in ceria seems to provide one of the clearest cases for the small polaron model. The strongest evidence for this model is that at fixed composition the Seebeck coefficient is found to be independent of temperature, a result which is consistent with a hopping model and not a band model. Other evidence is the activation energy for conduction of 0.3 eV (which represents the hopping activation energy) and the magnitude of the mobility, which is about 10^{-2} $cm^2/v\text{-}sec$ at 1000°C. The data indicate that the doubly ionized oxygen vacancy is the major defect close to stoichiometry, and that a transition to singly ionized vacancies occurs at larger departures from stoichiometry. Goodenough remarked that one would expect a small polaron in CeO_2 because the electrons are going into a $4f^1$ level in the gap.

Kofstad asked if one should not expect clusters, pairs, triplets, etc., at these high defect concentrations. Assuming the correct charge on the clusters, the observed pressure dependencies could be explained. There was discussion of this point, but Nowick did not believe that pairs, etc., existed in the range of defect concentration in which he was working.

Rapp concluded the presentations by suggesting an alternative method to the one presented earlier by Sockel for examining the defect structure of ternary oxides. This method consists of equilibrating the ternary compound with different oxygen activities in a sealed chamber so that the ratio of the two cations remains fixed. This fixes the system thermodynamically without the addition of a second phase, although there is no real advantage to either method.

Part Four

DEFECT DEPENDENT PROCESSES I
(Creep and Sintering)

NONSTOICHIOMETRY AND DOPING IN THE SINTERING OF OXIDES*

D. Lynn Johnson

Department of Materials Science
Northwestern University
Evanston, Illinois

ABSTRACT

Doping and nonstoichiometry influence sintering through their effects on volume, grain-boundary, and surface diffusion, as well as through their effects on grain growth. Carefully executed sintering experiments can determine the mechanisms of sintering and the diffusion coefficients of the species that has the lowest net mobility, taking into account both volume and grain-boundary diffusion. However, the volume diffusion coefficient appropriate for sintering may be different from the bulk value since the transport during sintering takes place in a narrow regional parallel to the grain boundaries. A space charge is likely to influence transport in this region. Results for initial- and intermediate-stage sintering of CoO may be explainable in terms of near-surface diffusion.

*Supported by the National Science Foundation.

1 INTRODUCTION

The fundamental mass-transport phenomenon during the sintering of a few metals is fairly well understood[1-5], although details of the microstructural evolution are not adequately documented. The situation is much less satisfactory in the case of the sintering of oxides. Far fewer definitive experiments have been performed and there are fewer data on diffusion coefficients and other parameters significant to sintering. Moreover, impurities and deviations from stoichiometry alter the point-defect concentrations and, thus, the diffusion coefficients. Nevertheless, there is a continuing need to understand the fundamentals of sintering in order to help optimize the process.

Besides being useful in characterizing sintering itself, fundamental sintering studies can provide very useful insights into fundamental mass-transport phenomena in materials. This is particularly so in the case of oxides, where so few diffusion data are available. The cation and anion volume self-diffusion coefficients have been measured on only a few oxides. There are even fewer data on surface diffusion, and grain-boundary self-diffusion data are almost entirely lacking. Carefully performed sintering experiments can be used to measure volume, grain-boundary, and surface diffusion coefficients if these are operative during sintering.[1,6]

It must be borne in mind at all times in interpreting sintering results that the transport takes place along or within a short distance of the surface and grain boundaries. Whereas in pure metals the diffusivities near the surfaces and grain boundaries are the bulk values, this may not be true in the case of compounds where near-surface space-charge effects can alter the defect concentration, and thus the diffusion coefficients, for rather extended regions near the interfaces.[7-9] Sintering may become a very useful tool in studying these near-surface transport phenomena.

Studies of the initial sintering of compacts of uniformly sized spherical particles have been successful in delineating the contributions of volume, grain-boundary, and surface diffusion to sintering.[1-3,10] The diffusion coefficients calculated from these data are in good agreement with the values determined directly, even though more than a single mechanism has been operating. Alpha iron, for instance, sinters by a combination of volume, grain-boundary, and surface diffusion.[1] Silver particles in the range of 40 to 150 μm in diameter also sinter by all three mechanisms, but surface diffusion is significant only at lower temperatures.[2-5] The ionic compound LiF sinters primarily by volume diffusion, with an initial enhancement due to plastic deformation by dislocation creep.[10] In this case the sintering rate was controlled by the diffusion of the lithium ion. The success of the sintering models in delineating sintering mechanisms in metals and LiF indicates that meaningful results should be obtainable in the case of oxides.

2 SINTERING MODELS

Curvature variations from point to point within a powder compact result in chemical potential gradients for the atoms or ions that make up the solid. The atoms or ions respond to these chemical potential gradients by moving via volume, grain-boundary, and surface diffusion on or within the solid phase, and by vapor transport in the gaseous phase. (Similar gradients also exist in the liquid phase in liquid-phase sintering, which is beyond the scope of this paper.)

The shrinkage rate of a compact of uniform-size spherical particles in which the average number of contacts per sphere is unchanging with sintering is related to the instantaneous geometry and the volume and grain-boundary diffusion coefficients, provided dislocation motion and particle rearrangement are negligible. The following equation was derived under these assumptions and under the additional assumption that the chemical potential gradients for volume diffusion and grain-boundary diffusion are identical:

$$-\frac{X^3 R}{X+R}\,\frac{dL}{L_o dt} = \frac{2\gamma\Omega D_v}{\pi kT\, a^3}\cdot\frac{A_v}{X} + \frac{4\gamma\Omega b D_b}{kT\, a^4}\,, \qquad (1)$$

where

X = fractional neck radius

R = minimum radius of curvature of the neck surface divided by the sphere radius

$-dL/L_o dt$ = fractional linear shrinkage rate

L_o = initial length of compact

γ = surface tension

Ω = atomic volume

D_v = volume self-diffusion coefficient

b = grain-boundary-diffusion width

D_b = grain-boundary self-diffusion coefficient

a = sphere radius

A_v = neck surface area divided by a^2

and

kT has its usual meaning.

The shrinkage, shrinkage rate, and neck radius are measured experimentally on a series of compacts sintered for various times at the same temperature, and R and A_v are calculated from the measured values of X and the shrinkage.. Then the left-hand side of the equation is plotted versus A_v/X, with the slope containing the volume diffusion coefficient and the intercept containing the product of the grain-boundary width and the grain-boundary self-diffusion coefficient. Surface diffusion, volume diffusion from the sphere surfaces to the neck surface, and vapor transport will influence the instantaneous geometry and, therefore, the sintering rate, but do not invalidate the use of Eq. (1), since the geometry is measured explicitly. The surface-diffusion and vapor-transport contributions can be obtained by successive approximations using a computer simulation of the sintering curves.[1,11] Although no systems in which vapor transport is significant have yet been studied, the surface-diffusion contributions have been determined quantitatively for the case of iron, as mentioned above[1], and for α-Fe_2O_3.[12]

Equation (1) is valid for the initial stage of sintering, i.e., when the necks on a given sphere are all independent of each other. During the intermediate stage of sintering, which follows as the necks grow beyond this, the following equation has been proposed[6]

$$\frac{\overline{G}}{\overline{H}\,L_V}\,\frac{dL}{Ldt} = \frac{5\gamma\Omega D_v}{kT}\,\frac{S_V}{L_V} + \frac{5\gamma\Omega b D_b}{kT}\,, \tag{2}$$

where

$\overline{G}$ = mean grain size

$\overline{H}$ = mean surface curvature of the gas-solid interface

L_V = grain-boundary perimeter per unit volume at the gas-solid interface

S_V = pore surface area per unit volume

L = instantaneous length of the compact.

Note that the densification rate is normalized to the instantaneous length, rather than to the initial length as was done in the initial sintering model. This eliminates one of the problems inherent in initial-stage studies: the need to quantify nondiffusional shrinkage produced by plastic flow and particle rearrangement. The geometric parameters of Eq. (2) can be obtained by simple counting techniques on random polished sections of the compacts as follows[13-17]:

$$\overline{G} = 1.56\,\overline{\ell} \tag{3}$$

$$S_V = 2N_L \tag{4}$$

$$L_V = 2N_A \tag{5}$$

$$\bar{H} = \frac{\pi T_{A\ net}}{N_L}, \tag{6}$$

where

$\bar{\ell}$ = mean grain intercept of a random line on a plane section

N_L = number of intersections a random line makes with solid-vapor interfaces, per unit length of the line

N_A = the number of grain boundary – pore intersections per unit area on a plane section

$T_{A\ net}$ = number of times a test line is tangent to convex solid-pore interfaces, minus the number of times it is tangent to concave interfaces, per unit area, as the test line is moved normal to itself over the plane of polish.

While the preceding equations are appropriate for metals, they must be altered for the sintering of compounds since one must account for the transport of all species in the compounds. Under the requirement that the material that is transported during sintering has the same composition as the starting material, equations similar to Eqs. (1) and (2) can be derived with the following results:

$$-\frac{X^3 R}{X+R}\frac{dL}{L_o dt} = \frac{2\gamma\Omega}{\pi kT\, a^3 X} M \tag{7}$$

$$\frac{\bar{G}}{\bar{H}\, L_V}\frac{dL}{L dt} = \frac{5\gamma\Omega}{kTL_V} M' \tag{8}$$

where M and M′ are mobility terms defined as follows for the compound $A_\alpha B_\beta$:

$$M \equiv \frac{\alpha\left(aA_v D_{vA} + 2\pi X b D_{bA}\right)\left(aA_v D_{vB} + 2\pi X b D_{bB}\right)}{a\left[\alpha\left(aA_v D_{vA} + 2\pi X b D_{bA}\right) + \beta\left(aA_v D_{vB} + 2\pi X b D_{bB}\right)\right]} \tag{9}$$

$$M' \equiv \frac{\alpha\left(S_V D_{vA} + L_V b D_{bA}\right)\left(D_V D_{vB} + L_V b D_{bB}\right)}{\alpha\left(S_V D_{vA} + L_V b D_{bA}\right) + \beta\left(S_V D_{vB} + L_V b D_{bB}\right)}, \tag{10}$$

where the self-diffusion coefficients are subscripted to indicate the species involved. Gordon[18] has derived similar relationships for the case of diffusional creep.

It can be seen from these equations that the species with the lowest mobility, taking into account both volume and grain-boundary diffusion, will be rate controlling. Equations similar to Eqs. (1) and (2) are obtained from Eqs. (7) and (8) if one of the terms in parentheses in Eqs. (9) and (10) is much larger than the other. This could result, for instance, in the sintering being controlled by volume diffusion plus grain-boundary diffusion of the cation, even though the volume diffusivity of the anion might be less than the volume diffusivity of the cation, provided the grain-boundary diffusivity of the anion is sufficiently large. If the volume-diffusion term for the cation and the grain-boundary diffusion term for the anion predominate their respective terms in Eqs. (9) and (10) for all values of the shrinkage, Eqs. (7) and (8) can be rearranged into the following:

$$-\frac{X^3 R}{X+R}\,\frac{dL}{L_o dt} = \frac{2\pi\beta b D_{bB}}{\alpha a\, D_{vA}} \cdot \frac{X^4 R}{A_v(X+R)}\,\frac{dL}{L_o dt} + \frac{4\gamma\Omega b D_{bB}}{kT\, a^4} \tag{11}$$

$$\frac{\bar{G}}{\bar{H} L_V}\,\frac{dL}{Ldt} = -\frac{\beta b D_{bB}}{\alpha D_{vA}} \cdot \frac{\bar{G}}{S_V \bar{H}}\,\frac{dL}{Ldt} + \frac{5\gamma\Omega b D_{bB}}{kT}\ . \tag{12}$$

The data can be plotted according to these equations to yield straight lines from which the boundary diffusivity of the anion can be calculated from the intercept and the volume diffusivity of the cation can be calculated from the slope. This form of the equations would be useful in the transition range, where the control of sintering is not clearly cationic or anionic. It must be pointed out that no data exist to test the accuracy of these equations under experimental conditions.

3 SPACE CHARGE AND IMPURITY EFFECTS

The defect concentration near vacancy sources and sinks, such as surfaces, grain boundaries, and edge dislocations, in ionic compounds has been shown to be different from the bulk concentrations because of the different energy of formation of the cation and anion vacancies or, for the case of Frenkel solids, the vacancies and the interstitials.[7-9] For instance, in sodium chloride, the cation-vacancy-formation energy at a vacancy source or sink is less than the anion-vacancy-formation energy. There would tend to be excess cation vacancies in pure NaCl, except that they are negatively charged; a potential is built up which repels cation vacancies from and attracts anion vacancies to the bulk. Thus, the neutrality condition is satisfied in the bulk, with equal numbers of cation and anion vacancies, while there is a higher concentration of cation vacancies near the vacancy sources and sinks. Cation self-diffusion would thus be enhanced in this near-surface region, while anion self-diffusion would be retarded.

If a divalent substitutional impurity is added to the NaCl, the bulk cation-vacancy concentration will increase and can be much higher than the surface concentration, while the bulk anion-vacancy concentration will decrease. Under these conditions, the space charge will have opposite sign, and the anion-vacancy concentration will be greater near the surface than in the bulk. The point-defect concentrations in the bulk will be more sensitive to the presence of aliovalent impurities than they will be in the near-surface regions, since their concentrations at the surface will be independent of impurity content.

The foregoing near-surface impurity effects are based upon the assumption that the impurity distribution is a function only of the electrical potential. The influence of impurities which segregate to the vacancy sources and sinks, because of strain energy or other causes, on transport near surfaces, grain boundaries, and dislocations has not been studied theoretically, and few experimental data are available.

The diffusivities of Eqs. (9) and (10) may be sensitive to the presence of impurities and deviations from stoichiometry. Knowledge of these effects on grain-boundary diffusion is quite limited indeed, while volume diffusion can be discussed at least in terms of ideal behavior. Volume diffusivities in single crystals of alkali and silver halides as influenced by aliovalent dopants is fairly well understood. The influence of dopants and deviations from stoichiometry of simple oxides is at least reasonably well understood in theoretical terms and some of the expectations are borne out experimentally.[19] However, the amount of data on such systems is still rather limited. Although the bulk values of the diffusion coefficient may be known, it must be remembered that the values near the grain boundaries and free surfaces may be altered by the presence of a space charge. The influence of impurities on sintering may not, therefore, be the same as that expected from the bulk behavior.

4 EXPERIMENTAL INVESTIGATIONS

A study of sintering of CoO was undertaken[11] to investigate the influence of deviations from stoichiometry on sintering of a material for which the phase diagram and defect equilibrium constants have been determined[20,21] and the bulk diffusivities of both the cations and the anions have been measured experimentally[22-24]. The initial sintering of uniform spherical particles prepared by dropping powder through a vertical oxygen-gas-fired MgO tube furnace followed volume diffusion kinetics for the first 0.5 percent linear shrinkage when plotted according to Eq. (1). The retardation beyond this range of sintering was caused by the development of facets on the surfaces of the spheres and a corresponding reduction in the driving force for sintering. However, the volume diffusivities calculated from the early portion are in agreement with those obtained from intermediate-stage-sintering results.

The intermediate stage of sintering was not plagued by the faceting problem since the curvature was measured explicitly. As in the initial stage, volume-diffusion kinetics were observed. The apparent diffusion coefficients calculated from these data were independent of oxygen partial pressure below about 10^{-3} atmosphere and increased with increasing P_{O_2} above this pressure at 1200°C.

The apparent volume-diffusion coefficients calculated from the initial- and intermediate-stage data are illustrated in Fig. 1, along with the values determined in air by direct methods. It is immediately apparent that the two sintering experiments gave comparable results and that the calculated diffusion coefficients agree with neither the cation nor the anion volume-diffusion coefficients. The possibility of misidentification of the sintering mechanism was ruled out because of the agreement between the initial- and intermediate-stage data and the fact that each of these followed the models for volume diffusion. If the data are forced to fit the grain-boundary-diffusion model, the intermediate- and initial-stage results are not in agreement. It was therefore proposed that the calculated apparent volume-diffusion coefficient represented that in the regions immediately adjacent to the grain boundary. An idealized model for the space charge in CoO predicts that the near-surface concentration of doubly ionized oxygen vacancies would be greater than the bulk concentration. This could account for some of the observations, but does not explain the apparent break in the activation-energy curve nor the higher activation energy for specimens sintered in atmospheres rich in carbon dioxide. (It was found consistently that the presence of significant amounts of carbon dioxide in the sintering atmosphere caused a decrease in the diffusion coefficient at lower temperatures and an increase at higher temperatures.) Explanations of these effects are not known at the present time.

As an interesting sidelight, it might be mentioned that the specimens which showed the highest overall densification rate during intermediate-stage sintering at 1200°C and various oxygen partial pressures resulted in the lowest calculated diffusion coefficient. This is because the grain

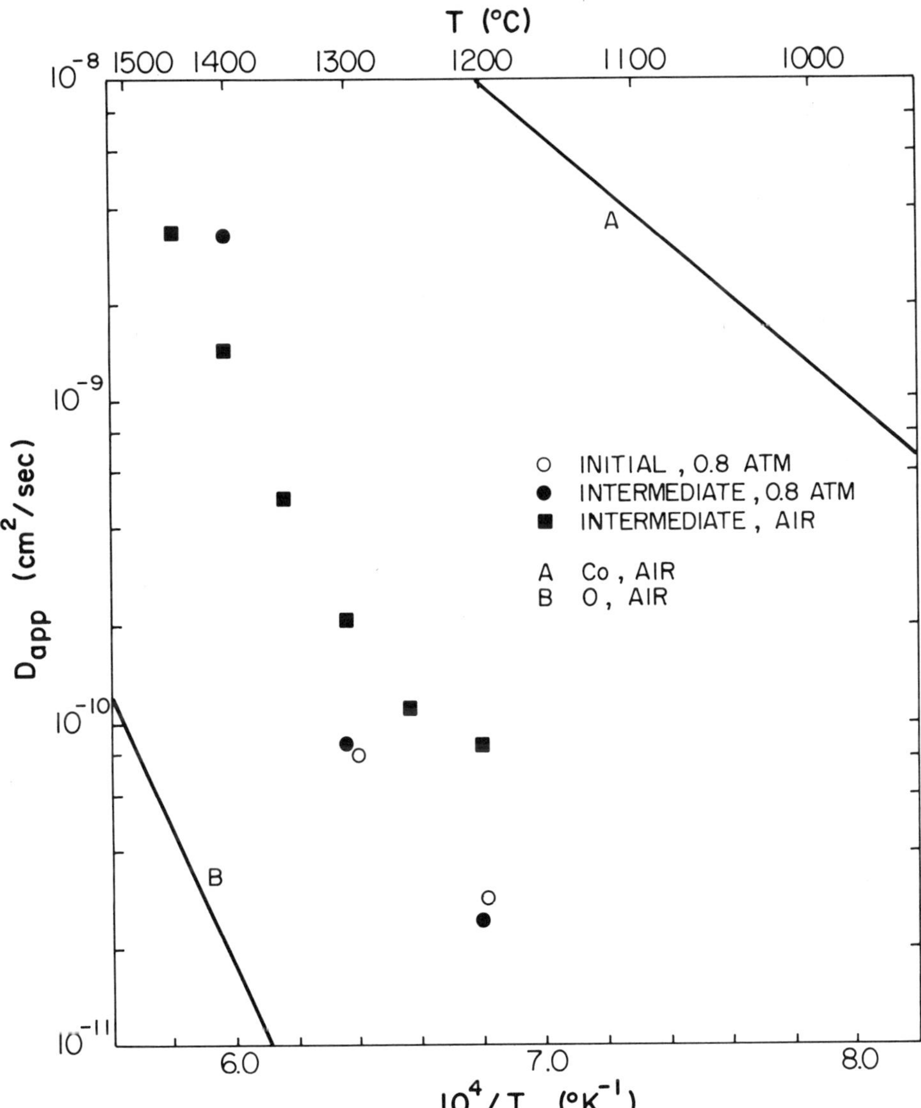

Fig. 1. Apparent volume-diffusion coefficients calculated from initial- and intermediate-stage sintering of CoO compared with oxygen and cobalt self-diffusion coefficients measured directly.

growth-rate constant was found to be directly proportional to the volume-diffusion coefficient calculated from the sintering data; the high densification rate was caused by a small grain size, which more than compensated for the low diffusion coefficient.

Aluminum oxide is probably the most extensively studied ceramic in examining the role of impurities on sintering.[25-38] While the general effects of various dopants are fairly well described, there is no general agreement about the mechanisms of sintering. Sintering of alumina has been ascribed to volume diffusion[29,30,33-38], grain-boundary diffusion[27-31], and grain-boundary diffusion changing to volume diffusion upon addition of TiO_2[31]. Unfortunately, in none of the experiments upon which these conclusions were based were sufficient data obtained to permit a conclusive identification of sintering mechanisms. In one study, necks between sintered spheres were measured on fractured surfaces at low shrinkages[33], but this would give a low estimate of the neck size since fracture would tend to occur at the smallest necks.

It has been observed experimentally that both divalent and quadrivalent cations enhance densification in the initial stage of sintering[29-31,33]. Unfortunately, it is not known how these dopants may be changing the surface-diffusion coefficients and thereby altering the sintering kinetics. Ignoring this for the time being and assuming that the mechanism is solely volume diffusion, the enhancement by both divalent and trivalent dopants indicates that the cation is rate controlling and that the cation defect structure is of the Frenkel type.[33] In other words, the incorporation of, for instance, TiO_2 as a substitutional impurity would give rise to cation vacancies. On the other hand, substitutional FeO would increase the concentration of cation interstitials.

If oxygen diffuses rapidly enough along the grain boundaries that aluminum diffusion is rate controlling[37,38], an increase in either aluminum vacancy or interstitial concentration would increase the aluminum self-diffusion coefficient and, presumably, the sintering rate.

The sintering data in the literature for nominally pure Al_2O_3, when interpreted as volume-diffusion kinetics, give calculated volume-diffusion coefficients which are, in some cases, more than an order of magnitude higher than that for aluminum in polycrystalline Al_2O_3. This may either indicate a misidentification of the sintering mechanism or space-charge effects near the grain boundary. Equation (9) can be utilized to examine the implications of sintering controlled by volume diffusion of aluminum. If M in Eq. (9) reduces to $\alpha V_v D_{vA}/\beta$, which it must for the previous condition to hold, the following must all be satisfied for sintering of spheres with an average diameter of 10^{-3} cm:

$$\frac{bD_{bO}}{D_{vAl}} \geqslant 2 \times 10^{-4} \tag{13}$$

$$\frac{bD_{bAl}}{D_{vAl}} \leqslant 2 \times 10^{-6} \tag{14}$$

$$\frac{bD_{bO}}{D_{vO}} \geqslant 4 \times 10^{-2} \tag{15}$$

Note that the particle size is on the right-hand side of each of these inequalities; each ratio must be an order of magnitude smaller for 10^{-4}-cm spheres. At this point, our lack of data on grain-boundary diffusion prevents further evaluation. However, the ratios in Eqs. (14) and (15) are significantly smaller and larger, respectively, than similar ratios in metals at similar fractions of the absolute melting temperature. This is probably reason enough to be cautious in accepting volume diffusion as the mechanism of sintering of alumina. Add to this the uncertainties in the experiments, and one must conclude that the sintering mechanisms of alumina are still in doubt. There are no ceramics for which the situation is any better.

There thus remain many unanswered questions on the influences of deviations from stoichiometry and dopants on the fundamental mass-transport phenomena during sintering of oxides. More work is required to understand and identify the sintering mechanisms, the rate-limiting species during sintering, and the appropriate defect model for the materials studied. Simplified theoretical approaches indicate many interesting features which can be compared with the sintering results. However, careful experiments are needed in which the instantaneous geometry of the compact is characterized as much as possible so that extraneous influences can be minimized in interpreting the results.

The densification of materials depends not only upon the diffusion coefficients, but also upon the diffusion-path length, driving force, and area across which diffusion can take place; these are related to the shrinkage rate through Eqs. (1) and (2), (7) and (8), or (11) and (12). Grain-growth inhibition by dopants has resulted in ultimate densification of a few oxides.[28,39,40] Pore elimination is possible if the pores remain on the grain boundaries and if they contain no insoluble gases.

The effect of dopants on grain-growth rates cannot yet be predicted theoretically, but rapid grain growth is often associated with rapid initial densification. For CoO it was observed that the grain-growth-rate constant was directly proportional to the volume diffusivity calculated from the intermediate-stage-sintering data.[11] This indicates that grain growth and densification are governed by the same process. Further work is required to more clearly describe the effects of dopants on grain growth and other aspects of microstructure evolution.

4 CONCLUSIONS

The sintering rate of oxides is controlled by the rate of diffusion of the slowest species, taking into account parallel volume and grain-boundary diffusion. The volume-diffusion coefficient appropriate to sintering is that near the grain boundary and will be influenced by any space-charge region that may be present. Dopants and nonstoichiometry can affect sintering through their influence on diffusion coefficients and on grain growth. The conclusions of most sintering experiments must be viewed with caution in light of the general lack of simultaneous measurements of densification rate and microstructure.

REFERENCES

1. Johnson, D. L., *J. Appl. Phys.*, **40**, 192 (1969).
2. Johnson, D. L., and Clarke, T. M., *Acta Met.*, **12**, 1173 (1964).
3. Seidel, B. R., and Johnson, D. L., *Phys. Sintering*, **3**, 143 (1971).
4. Johnson, D. L., *Modern Developments in Powder Metallurgy*, Vol. 4, H. H. Hausner (Ed.), Plenum Press, New York (1971), p. 189.
5. Johnson, D. L., *Powder Metallurgy for High Performance Applications*, J. J. Burke and V. Weiss (Eds.), Syracuse University Press, Syracuse, N. Y. (1972), p. 139.
6. Johnson, D. L., *J. Am. Ceram. Soc.*, **53**, 574 (1970).
7. Frenkel, J., *Kinetic Theory of Liquids*, Oxford, Clarendon Press, London (1946).
8. Lehovec, K., *J. Chem. Phys.*, **21**, 1123 (1953).
9a. Kliewer, K. L., and Koeller, J. S., *Phys. Rev.*, **140**, A1226 (1965).
b. Kliewer, K. L., *ibid.*, A1241.
c. Kliewer, K. L., *J. Phys. Chem. Solids.*, **27**, 705 (1966).
d. Kliewer, K. L., *ibid.*, p. 719.
10. Doi, A., Ph.D. Thesis, Northwestern University (1972).
11. Kumar, P., Ph.D. Thesis, Northwestern University (1972).
12. Smart, J. S., Ph.D. Thesis, Northwestern University (1968).
13. Cahn, J. W., *Trans. AIME*, **239**, 610 (1967).
14. DeHoff, R. T., *ibid.*, p. 617.
15. DeHoff, R. T., and Aigeltinger, E. H., *Perspectives in Powder Metallurgy*, Vol. 5, J. S. Hirschhorn (Ed.), Plenum Press, New York (1970).
16. Mendelson, M. I., *J. Am. Ceram. Soc.*, **52**, 443 (1969).
17. Smith, C. S., and Guttman, L., *Trans. AIME*, **197**, 81 (1953).
18. Gordon, R. S., *J. Am. Ceram. Soc.*, **56**, 147 (1973).
19. Kroeger, F. A., *The Chemistry of Imperfect Crystals*, North-Holland Publishing Co., Amsterdam (1964).
20. Fisher, B, and Tannhauser, D. S., *J. Electrochem. Soc.*, **111**, 1194 (1964).
21. Bransky, I., and Wimmer, J. M., *J. Phys. Chem. Solids*, **33**, 801 (1972).
22. Carter, R. E., and Richardson, R. D., *J. Metals*, **6**, 1244 (1954).
23. Chen, W. K., and Peterson, N. L., *Radiation Effects*, **40**, 141 (1970).
24. Chen, W. K., and Jackson, R. A., *J. Phys. Chem. Solids*, **30**, 1309 (1969).
25. Smothers, W. J., and Reynolds, H. J., *J. Am. Ceram. Soc.*, **37**, 588 (1954).
26. Jones, J. T., Maitra, P. K., and Cutler, I. B., *J. Am. Ceram. Soc.*, **41**, 353 (1958).
27. Johnson, D. L., and Cutler, I. B., *J. Am. Ceram. Soc.*, **46**, 545 (1963).
28. Jorgensen, P. J., and Westbrook, J. H., *J. Am. Ceram. Soc.*, **47**, 332 (1964).
29. Keski, J. R., and Cutler, I. B., *J. Am. Ceram. Soc.*, **48**, 653 (1965).

30. Keski, J. R., and Cutler, I. B., *J. Am. Ceram. Soc.*, **51**, 440 (1968).
31. Bagley, R. D., Cutler, E. B., and Johnson, D. L., *J. Am. Ceram. Soc.*, **53**, 136 (1970).
32. Brook, R. J., *J. Am. Ceram. Soc.*, **55**, 114 (1972).
33. Rao, W. R., Ph.D. Thesis, University of Utah (1972).
34. Kuczynski, G. C., Abernathy, L., and Allen, *Plansee Proceedings*, Pergamon Press, London (1958), p. 1.
35. Coble, R. L., *J. Am. Ceram. Soc.*, **41**, 55 (1958).
36. Coble, R. L., *J. Appl. Phys.*, **32**, 793 (1961).
37. Paladino, A. E., and Coble, R. L., *J. Am. Ceram. Soc.*, **46**, 133 (1963).
38. Coble, R. L., and Gupta, T. K., *Sintering and Related Phenomena*, G. C. Kuczynski, N. A. Hooton, and C. F. Gibbon (Eds.), Gordon and Breach, New York (1967), p. 423.
39. Jorgensen, P. J., and Anderson, R. C., *J. Am. Ceram. Soc.*, **50**, 553 (1967).
40. Greskovich, C., O'Clair, C. R., and Curran, M. J., *J. Amer. Ceram. Soc.*, **55**, 324 (1972).

DISCUSSION on Paper by D. Lynn Johnson

KRÖGER: You mentioned the fact that space charges in regions immediately adjacent to surfaces and subgrain boundaries will give rise to increased diffusion in these regions and that in the analysis these effects would show up as increased bulk diffusion.

It seems to me that the contribution by the most important part of the subinterface area (i.e., the area close to a surface or interface) would appear instead as increased surface or interface diffusion. The reason is that due to the variation of D with distance from the surface or interface, there will be a channelling of the diffusion parallel to the surface or interface. For a proper description, the diffusion profile would have to be incorporated in the model.

JOHNSON: I agree; the model is based upon D being uniform with distance from the boundary; if it isn't, then we have to worry about it. If, however, the space-charge region is on the order of aR or greater in extent, one would observe volume-diffusion kinetics with an enhanced D_v.

STUBICAN: How close were your particles to being spherical in your sintering experiments with CoO?

JOHNSON: The initial-stage work was done on spherical particles with a narrow size distribution. The intermediate stage does not depend upon having spherical particles.

STUBICAN: For the theory of sintering, surface diffusion coefficients are very important. What are the chances that they will be determined, at least for the most used oxides.?

JOHNSON: Reliable direct measurements are difficult because of the strong influence of trace amounts of impurities, which can segregate or adsorb on the surface.

HEUER: I would guess that because $D_{v_{Al}} > D_{v_O}$ in Al_2O_3, the space charge is such as to enhance D_{b_O} relative to D_{v_O} and depress $D_{b_{Al}}$ relative to $D_{v_{Al}}$, exactly in the way your calculations indicated.

JOHNSON: The effect of space charge on the grain-boundary diffusivities has not been discussed. The space charge influences diffusivities *near* the surface and grain boundary. Whether a diffusivity is enhanced or depressed in the region depends upon the defect concentration, which in turn depends upon the formation energies, temperature, and impurities. I am not sure about Al_2O_3.

SELTZER: The results for CoO showed that diffusivities obtained from sintering kinetics were lower than volume diffusivities for cobalt and higher than oxygen volume diffusivities. This would suggest that oxygen volume diffusion plays a major role in the sintering process. In addition, it was shown that sintering rates increase with P_{O_2} in the range $10^{-3} < P_{O_2} < 1$ atm. Could it be inferred from these data that the sintering is controlled by volume diffusion of interstitial oxygen?

JOHNSON: The only thing that I can say is that I believe the results are valid; I infer this because we have had such close agreement with the published tracer diffusion coefficients in the case of metals. Because we lack good diffusion coefficients for boundary diffusion in the oxides, I really cannot say we have a good set of data that agree with the literature. But I do not believe that we would be so many orders of magnitude in error as to miss boundary diffusion, on the basis of our results with metals. It is true that oxygen-interstitial concentration would increase with increasing P_{O_2}, but I am not sure one can answer your question in the affirmative.

FINAL-STAGE SINTERING AND GRAIN GROWTH IN OXIDES

Paul J. Jorgensen

Stanford Research Institute
Menlo Park, California

ABSTRACT

The atomic transport processes that result in densification during the sintering of oxides are controlled primarily by volume diffusion or by grain-boundary diffusion. Sintering can be described phenomenologically in terms of diffusion of atomic species from internal surfaces (i.e., grain boundaries) to pores. The flux of atoms from the grain boundaries is a function of the appropriate rate-limiting diffusion coefficient and the chemical-potential gradient. The purpose of this paper is to show that changes in the chemical-potential gradient can account for the densification of oxides that have been sintered to theoretical density.

The densification of Y_2O_3 doped with ThO_2 is described since the addition of 5 mole percent or more of ThO_2 to Y_2O_3 allows the densification to proceed to theoretical density. The requirement of 5 mole percent dopant for successful densification casts doubt on a mechanism entailing the concentration of point defects, and it has been shown by indirect experimental techniques that the thoria solute is segregated at the yttria grain boundaries.

Grain growth occurs during densification, and this geometric change in the microstructure causes a decrease in the chemical-potential gradient, which is the driving force for sintering. Grain-growth kinetics of Y_2O_3 at 2000°C follow the theoretical grain-growth law, $D^2 - D_o^2 = kt$ (where D_o is the initial grain size, D is the grain size after time t, and k is a constant), and show a decrease in the grain-boundary mobility as the concentration of the thoria in solid solution increases. When the amount of thoria exceeds the solid-solubility limit, the grain-growth kinetics follow a cubic-grain-growth law and can be described in terms of a grain-growth process that is controlled by the coalescence of the second phase.

If the assumption is made that the velocity of the average grain boundary is proportional to dD/dt, where D is the average diameter of the yttria grains, the effect of the thoria solute on the grain growth of yttria can be described in terms of an impurity drag as described by Cahn for low-velocity grain growth with a low driving force. Thus the inhibition of the grain growth or the impurity drag of the ThO_2 solute in the yttria grain boundaries causes the chemical-potential gradient in doped samples to be greater than in undoped samples, and this inhibition of the grain growth allows the pores to remain on grain boundaries and allows the densification to proceed to completion, i.e., to theoretical density.

1 INTRODUCTION

The atomic transport processes that result in densification during the final-stage sintering of oxides are primarily controlled by volume diffusion or by grain-boundary diffusion. Sintering can be described phenomenologically in terms of diffusion of atomic species from internal surfaces (i.e., grain boundaries) to pores.[1-5] The diffusion flux is a function of the appropriate rate-limiting diffusion coefficient (normally the diffusion coefficient of the slowest diffusing species) and a function of the chemical-potential gradient. An oxide may sinter to theoretical density if a relatively high diffusion flux between the grain boundaries and the pores can be maintained, and this is often accomplished by adding certain solutes to the oxides. For example, an MgO solute is added to Al_2O_3 in order to sinter to theoretical density, ThO_2 or ZrO_2 solutes are added to Y_2O_3, and Y_2O_3 or CaO solutes are added to ThO_2.

Several mechanisms can be invoked to explain the effect of the added solutes on the sintering kinetics. For example, Reijnen[6] attributes increased densification to a favorable modification of the defect structure in which the concentration of the slowest diffusing species is increased. Other explanations involve the inhibition of grain growth caused by solute segregation at the grain boundaries[7,8] or by the presence of a second phase[9]. Finally, the formation of a liquid phase could also explain the increased diffusion kinetics. Each of these mechanisms probably occurs during the sintering of various oxides under certain circumstances.

In this paper we will be concerned only with chemical additives that are in solid solution, thereby eliminating consideration of the effects of second phases or liquid phases, and we will concentrate on the Y_2O_3-ThO_2 system since both oxides have been sintered to theoretical density, with the other oxide acting as the added solute.[8,10] The amounts of solute required for sintering either yttria or thoria to theoretical density are well below the solid solubility limits.

Subbarao et al.[11] studied the defect structure and electrical conductivity of the yttria-thoria system and concluded from density and lattice-parameter measurements that Y^{+3} substitutes for Th^{+4} in ThO_2, and that electroneutrality is preserved by the formation of oxygen-ion vacancies. Substitution of Th^{+4} for Y^{+3} in the Y_2O_3 lattice could result in cation vacancies or in oxygen-ion interstitials. The beneficial action of the ThO_2 solute on the sintering of Y_2O_3 obscured the distinction between yttrium vacancies and oxygen interstitials, but Subbarao concluded that oxygen-ion interstitials are more probable.

In both thorium oxide and yttrium oxide the cation diffusion coefficients are less than the anion diffusion coefficients.[12-17] For example, oxygen diffusion coefficients in thoria are 10^7 times greater than the thorium diffusion coefficients.[18] The anion and cation diffusion coefficients in Y_2O_3 are not as well known and therefore these data are shown in Fig. 1 as a function of reciprocal temperature.

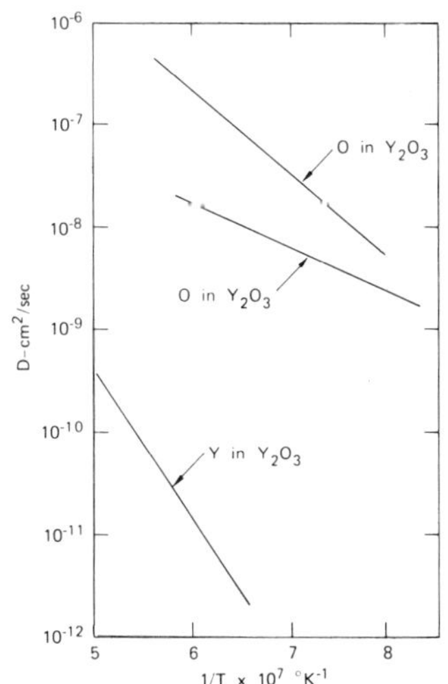

Fig. 1. Cation and anion diffusion coefficients in Y_2O_3 as a function of reciprocal temperature.

An increase in the concentration of oxygen-ion vacancies as a result of adding Y_2O_3 solute to ThO_2 would not increase the diffusion flux resulting in densification since the flux should be controlled by the slowest diffusing species. In fact, the diffusion flux could be decreased owing to the presence of a Shottky equilibrium. Thus, the improved sinterability of ThO_2 with the addition of Y_2O_3 as a solute cannot be explained solely on the basis of an increased diffusion flux due to the formation of a favorable defect structure.

The addition of ThO_2 solutes to Y_2O_3 could increase the diffusion flux during the sintering of Y_2O_3 if yttrium-ion vacancies are formed, and therefore the majority of this paper will be concerned with the final-stage sintering and grain growth of Y_2O_3 doped with various concentrations of ThO_2 solute in order to develop the concept that solutes in oxides can segregate to grain boundaries and produce an increase in the relative chemical-potential gradient by inhibiting the grain growth that occurs during the densification process.

During the intermediate stages of sintering, pores are formed at the grain boundaries of the polycrystalline sample, and grain growth occurs. As the grain size increases, the average distance between pores increases and the average diffusion distance between the pores and the grain boundaries increases. If the grain growth is inhibited, shorter distances predominate and a higher diffusion flux is maintained by a relative increase in the chemical-potential gradient. When the grain growth is not inhibited, as in the sintering of relatively pure oxide powders, pores become entrapped within the grains by grain boundaries breaking away from the pores. This phenomenon is called discontinuous grain growth, and the resulting microstructure in the case of Y_2O_3 is shown in Fig. 2.

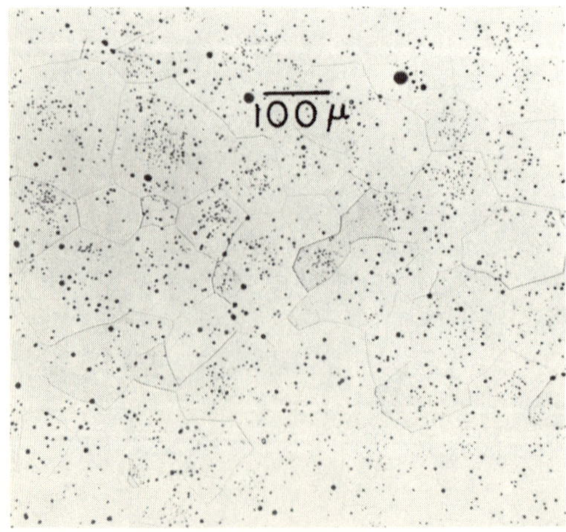

Fig. 2. Microstructure of sintered Y_2O_3 exhibiting discontinuous grain growth.

When 10 mole percent ThO_2 is added to the Y_2O_3 before sintering, complete pore removal or theoretical density is achieved. A typical microstructure of the theoretically dense sintered oxide is shown in Fig. 3. The microstructure of the theoretically dense thoria-doped Y_2O_3 is single phase and can be ground into a powder finer than the original grain size and again sintered to theoretical density. This indicates that the achievement of theoretical density does not occur by a transient phenomenon due to the methods of adding the ThO_2 solute.

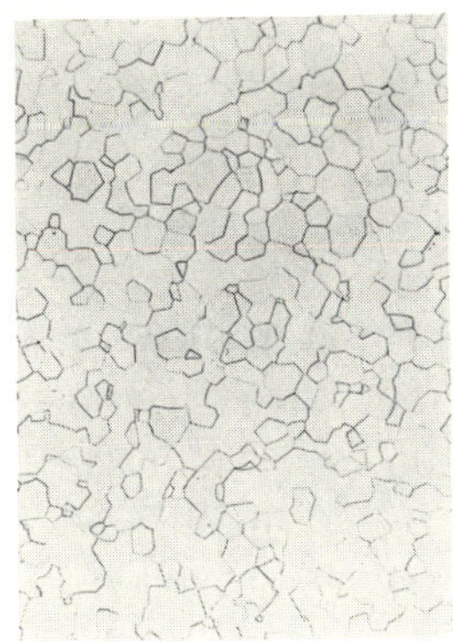

Fig. 3. Microstructure of Y_2O_3 containing 10 mole percent ThO_2.

Jorgensen and Anderson[8] provided indirect evidence that the ThO_2 solute added to Y_2O_3 segregates to the grain boundaries. Since then, more direct techniques such as Auger spectroscopy of grain-boundary fracture surfaces have failed to show conclusively an increase in the thoria concentration at the grain boundaries. Therefore, some of the indirect evidence, such as microhardness studies and surface-activity measurements presented previously, is reviewed, and grain-boundary-mobility studies have been extended to provide additional evidence for solute segregation at the grain boundaries in this system.

2 EXPERIMENTAL PROCEDURE

The samples used in this study were prepared by combining 99.99 percent yttria powder with calculated amounts of thorium nitrate. Since thorium nitrate is water soluble, a water slurry was made of the powders and mixing was accomplished by continuously stirring with a magnetic stirrer during drying to ensure uniform distribution of the solute in the yttria. After drying, the powders were calcined at 830°C in air for 5 hours, hydrostatically pressed into cylinders at 50,000 psi, cut into specimens about 1 cm in diameter and 2 mm thick, and fired in a hydrogen atmosphere saturated with water vapor at 0°C. A molybdenum resistance furnace with molybdenum radiation shields was used to achieve the 2000°C temperature used throughout this study. All samples were quenched from the sintering temperature to room temperature in approximately 30 seconds.

Hardness was measured with a Kentron microhardness tester using a 10-gram load; a 5-gram load was used for the hardness profiles taken across the grain boundaries. Each reported microhardness value was the average of five indentations.

The hardness specimens were ground to expose a section near the center of the specimen. The specimens were then polished and given a very light etch in boiling 20% HCl to reveal the microstructure. Hardness profiles across grain boundaries were recorded as hardness versus the perpendicular distance from the grain boundary to the center of the indentation.

Jorgensen and Westbrook[7] found that the adsorption of water on the surface of polished oxide specimens greatly affected the hardness. Therefore, after polishing, the specimens were heated for 1 hour in dry argon at 300°C and were then transferred immediately from the furnace into a container of toluene. All hardness indentations were made under toluene, and the widths of the impressions were measured with an oil-immersion lens.

The surface-segregation experiments were conducted by placing an alpha counter 2.5 cm from the surface of a thoria-doped yttria specimen. The alpha counter was positioned at this distance to measure a surface layer of minimum thickness. The mean range in air of an alpha particle emitted from ^{232}Th is 2.53 cm. The signal from the alpha counter was fed into an electronic counter and recorded.

The specimens were prepared for counting by annealing at 2000°C and then quenching to room temperature. They were then polished on one side, and the activity of the as-quenched surface was compared with that of the polished surface. After the as-quenched surface was counted, the specimen was then etched and counted in an attempt to determine the thickness of the proposed segregated layer.

Lattice-parameter measurements were made using a Debye-Scherrer technique combined with a precision computer program to determine the solid solubility limit for ThO_2 in Y_2O_3 at 1800, 1900, and 2000°C.

The average grain-boundary velocities for various concentrations of ThO_2 solute were determined from grain-growth data. The grain size was measured by counting the number of grain boundaries intersected by random straight lines of known length. The grain sizes were assumed to be 1.5 times the average distance between the intercepts.[19]

3 RESULTS AND DISCUSSION

The purpose of this paper is to show that the addition of a solute to an oxide can inhibit the grain-growth kinetics and thereby maintain a relatively higher chemical-potential gradient between the pores and the grain boundaries during sintering. The proposed mechanism by which the grain growth is inhibited is segregation of the solute to the grain boundaries where the solute then exerts a drag on the grain-boundary motion.

The amount of thoria solute required to successfully sinter yttria to theoretical density at 2000°C is approximately 5.0 mole percent. A thoria-solute concentration of 2.0 mole percent does not produce a theoretically dense structure, whereas the addition of 7 to 10 mole percent produces optimum optical properties. These amounts of thoria solute are well below the solid solubility limit at the normal sintering temperature of 2000°C, as shown in Fig. 4. The solid solubility limits for 1800°C and 1900°C are also shown.

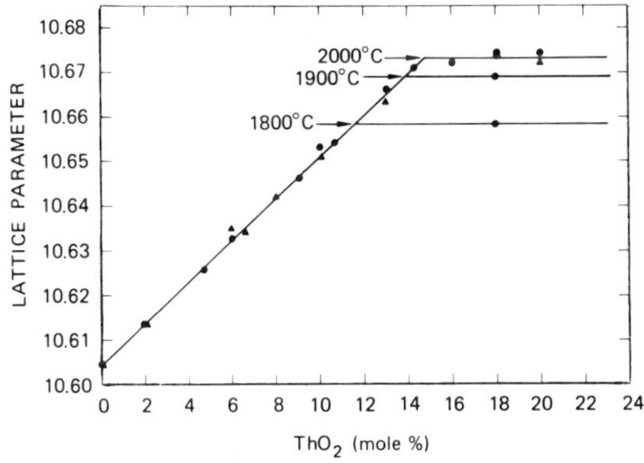

Fig. 4. Lattice parameters of Y_2O_3 as a function of ThO_2 concentration.

Since the amount of thoria solute required to sinter yttria to theoretical density is within a single-phase field of the thoria-yttria phase diagram, the probability of a second-phase mechanism controlling the grain-growth kinetics is unlikely.

Direct measurement of an increase in the thoria concentration at the grain boundaries of yttria to provide proof for a solute-segregation mechanism have been unsuccessful. The direct techniques that have been employed include electron-beam-microprobe measurements, autoradiographic techniques, and Auger spectroscopy of intergranular fracture surfaces. Why the latter technique failed to reveal an increase in concentration at the grain boundaries is not understood, assuming that an increased concentration of thoria existed at the grain boundaries.

Indirect techniques such as microhardness measurements, surface-activity measurements and grain-growth studies as a function of solute concentration provide data that are most easily explained on the basis of a solute-segregation mechanism.

3.1 Microhardness Measurements

Westbrook and Aust[20] measured the microhardness of grain boundaries in dilute high-purity alloys of lead and compared the measured values with the microhardness of regions remote from a grain boundary. They observed that the hardening of the boundary varied systematically with amount of solute and the structure of the boundary and, further, that these results correlated with grain-boundary-mobility studies.[21,22] Microhardness studies were also used to infer solute segregation in Al_2O_3[7] and in Y_2O_3[8].

The grain-boundary hardness of undoped Y_2O_3 is identical with the hardness measured in the center of the grains, as shown in Fig. 5. These microhardness measurements were made on yttria samples, quenched from 2000°C, containing various amounts of thoria. As the concentration of the thoria solute increases, the microhardness of the doped yttria structure increases until a solubility limit is reached. This solubility should correspond to the 2000°C solubility limit shown in Fig. 4. The inherent scatter of the microhardness measurements (error limits are shown on the ordinate of Fig. 5) coupled with the low slope of the hardness-versus-composition data make accurate solubility-limit-measurements difficult. However, it can be concluded from these data that solution hardening of the yttria lattice occurs as the concentration of the thoria solute increases to the solubility limit.

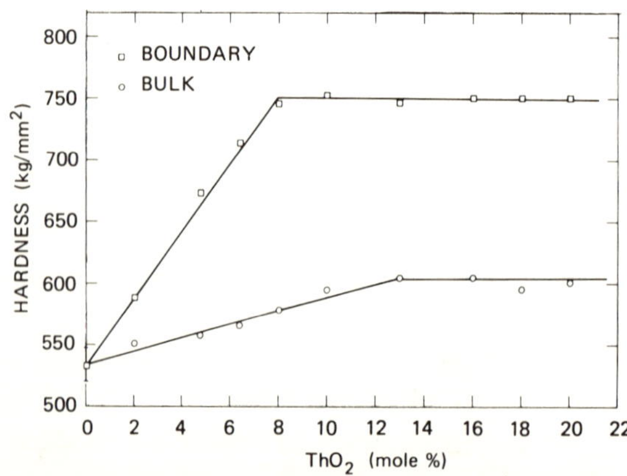

Fig. 5. Grain-boundary and bulk or lattice hardness of Y_2O_3 as a function of ThO_2.

The grain-boundary hardness also increases to a plateau value as thoria is added to yttria. The increase in grain-boundary hardness is greater than the increase in hardness of the yttria lattice and, therefore, it is reasonable to postulate that the concentration of thoria in the grain boundaries is greater than in the corresponding yttria lattice. The break in the curve for grain boundary versus thoria concentration corresponds to an overall concentration of 8.0 mole percent; however, the actual concentration in the grain boundary at this limit is much greater than the lattice solubility limit because of the greater initial slope of the grain-boundary-hardness curve in Fig. 5. If we assume that the initial hardness-versus-concentration slope for the yttria lattice can be applied to the grain-boundary region, then, by extrapolating these data to the grain boundary hardness plateau, we would calculate a grain-boundary composition of approximately 39 mole percent. If we also assume that the grain-boundary volume is negligible with respect to the sample volume, the thoria-solute concentration in the grain boundary is approximately a factor of five greater than that in the lattice.

The width of the proposed grain-boundary segregation is difficult to measure with a microhardness indenter, because the size of the average indentation in Y_2O_3 is between 4 and 5 microns when a 5-gram load is used on the indenter and because hardness measurements can be influenced by obstacles to plastic deformation beyond the actual indentation. Therefore a hardness profile across a grain boundary is not identical to the concentration profile, but it indicates that the concentration increases at the boundary. A hardness profile across the boundary is shown in Fig. 6. The hardness values do not correspond to the values shown in Fig. 5 because of the smaller load (5 grams), which allowed the indentations to remain as small as possible.

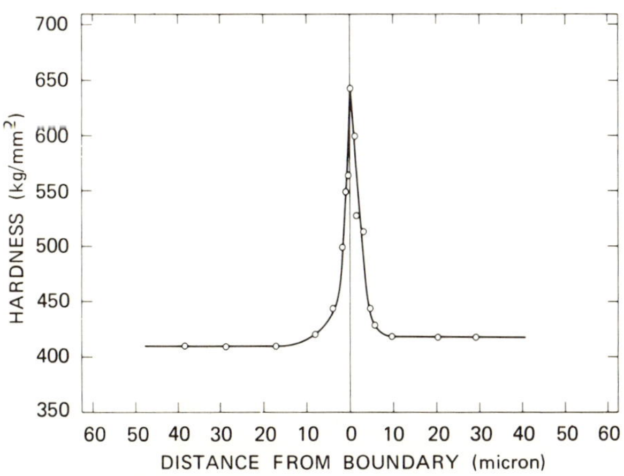

Fig. 6. Hardness profile across a grain boundary of Y_2O_3 containing 10 mole percent ThO_2.

3.2 Surface-Activity Measurements

The approximate width of the enriched or segregated region can be obtained by combining etching and surface-activity measurements, if it is assumed that a free surface is similar to one-half of a grain boundary. Therefore, the surface activity of a sample containing 10 mole percent thoria in yttria was heat treated at 2000°C and then quenched. The surface activity was measured as described above. The experimental arrangement is shown in Fig. 7. The surface activity of the free surface was compared with the activity of a freshly polished surface, and it was found that the quenched surface activity in counts per minute was 17.91 ± 0.02 and the polished surface activity was 17.06 ± 0.55. The plus or minus values correspond to three sigma values, i.e., 99 percent of the measured counting rates fell between these limits. The difference between the values indicates a segregation of thoria to the surface; however, the value of the difference is expected to be low because the measurement counted a surface that extended into the bulk of the sample. Etching of the quenched surface caused the activity of that surface to decrease to the value of the polished surface. The thickness of the removed oxide layers was determined by assuming a linear rate of chemical attack with time and by weighing the specimen before and after etching. The removal of the first layer (85 Å thick), calculated by interpolating the weight and time of the etching data, completely removed the excess solute concentration. This experiment therefore provides an upper limit to the width of the segregated-solute region, since segregation at a free surface may be greater than that at a grain boundary.

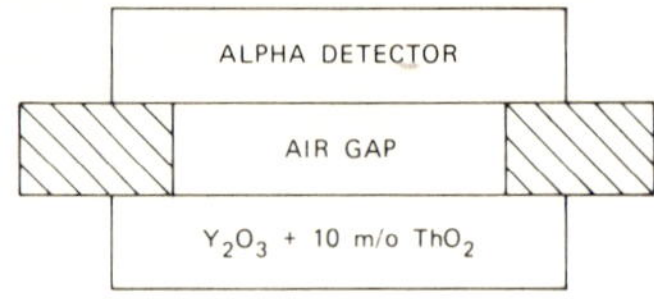

Fig. 7. Schematic of the experimental arrangement for measurement of surface activity of ThO_2-doped Y_2O_3.

3.3 Grain-Growth Studies

If a solute segregates to the grain boundary, we expect a decrease in the grain-boundary energy and also a decrease in the grain-boundary mobility. Cahn[23] has examined the effect of a solute atmosphere on the velocity of a grain boundary, and concluded that slowly diffusing solutes exert a greater drag on the grain boundary at low grain-boundary velocities. It is easier to study the effect of solute concentration on grain-boundary velocities by measuring the grain-growth kinetics and calculating a grain-boundary velocity for a constant grain size than it is to measure the grain-boundary energies as a function of solute concentration. Therefore, the grain-growth kinetics of yttria doped with various amounts of thoria solute were studied at 2000°C, and these data are shown in Fig. 8. The lines drawn through the data points were determined by a least-squares fit of the data. Since a change in slope occurs in the grain-growth data at the solid solubility limit, the calculated slopes or grain-growth exponents are listed as a function of composition in Table I.

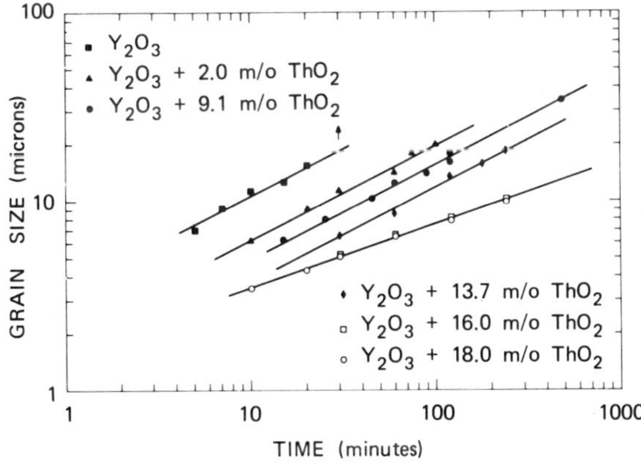

Fig. 8. Grain size as a function of time for Y_2O_3 samples containing various amounts of ThO_2.

Table 1. Calculated Grain-Growth Exponents as a Function of Composition

Composition	Grain-Growth Exponent
Y_2O_3	0.54
Y_2O_3 + 2.0 mole % ThO_2	0.50
Y_2O_3 + 9.1 mole % ThO_2	0.49
Y_2O_3 + 13.7 mole % ThO_2	0.50
Y_2O_3 + 16.0 mole % ThO_2	0.33
Y_2O_3 + 18.0 mole % ThO_2	0.33

The driving force for normal grain growth is the reduction of the free energy of the system in which the total grain-boundary area is reduced. Burke[24,25] has outlined the assumptions used in deriving a kinetic expression for normal grain growth. These include the assumptions that the grain-boundary mobility and grain-boundary energy are independent of grain orientation or grain size and that the driving force for grain growth is inversely proportional to the grain diameter. Using these assumptions, the following expression can be derived for normal grain growth:

$$D^2 - D_o^2 = A\mu\sigma t \quad , \tag{1}$$

where D is the grain size at time t, the initial grain size is equal to D_o, A is a geometric constant, σ is the grain-boundary energy, and μ is the grain-boundary mobility. If $D >> D_o$, which is the case in the present study, then a plot of logarithm D versus logarithm t (such as Fig. 8) should yield a straight line with a slope of 0.50. The grain-growth kinetics obtained within the solid solubility region can be described by Eq. (1), as verified by the exponents listed in Table I.

The grain-growth kinetics obtained for compositions in the two-phase region have grain-growth exponents equal to 0.33, and this behavior can be explained on the basis of the grain growth being controlled by the coalescence of a soluble second phase.[26] In this type of grain growth, the second-phase particles, d, grow according to the following law[27,28]:

$$d^3 = kt \quad , \tag{2}$$

where k is a proportionality constant. An yttria sample doped with 16 mole percent ThO_2 and containing large second-phase grains was prepared to test the applicability of Eq. (2). Plotting logarithm d versus logarithm t gave a slope of 0.33, thus experimentally confirming Eq. (2). Second-phase particles impede grain-boundary movement because new grain-boundary area must be formed equal to the cross-sectional area of the second-phase particles as the boundary moves away from the particle. Zener[29] showed that grain growth stops according to the following relationship:

$$f = \frac{4d}{3D} \quad , \tag{3}$$

where f is the volume fraction of the second-phase particles. When the Zener relationship is valid, and when the second-phase particles grow by coalescence according to Eq. (2), we expect the primary grains to grow according to the cube root of time, as shown in Fig. 8. To verify that this explanation is correct, we can calculate the expected second-phase particle size at 2000°C in an yttria sample doped with 16.0 mole percent ThO_2, using the solubility limit shown in Fig. 4, and compare this value with an experimentally measured value. The calculated value was 1500 Å and X-ray line-broadening measurements gave a second-phase particle size of 1500 ± 300 Å. Hence the Zener relationship is valid for grain growth of yttria containing a ThO_2 second phase, and the grain-growth kinetics can be adequately explained on this basis.

The grain-growth kinetics in the single-phase region should provide a means of determining whether solute segregation to the grain boundaries occurs in the $Y_2O_3 - ThO_2$ system. If we assume that the velocity of the average grain boundary is equal to dD/dt, we can obtain values of the grain-growth constants, $A\mu\sigma$, from the data in Fig. 8, and knowing the grain size, we can calculate the grain-boundary velocity. The calculated grain-boundary velocities are in the range of 10^{-7} to 10^{-6} cm/sec, and, therefore, low velocity behavior is expected with regard to Cahn's impurity-drag theory[23], i.e.,

$$\frac{dD}{dt} = V = \frac{P}{\lambda + \alpha C_o} , \tag{4}$$

where V is the grain-boundary velocity, P is driving force for grain-boundary motion, λ is an intrinsic drag coefficient, α is a constant containing the reciprocal of the mobility multiplied by a summation of the adsorption energies of the solute on the grain boundary, and C_o is the solute concentration.

A plot of dD/dt versus the solute concentration is shown in Fig. 9. The line drawn through the data was calculated using Eq. (4), and was forced through the data at 0 and 13.7 percent. The general shape of the curve is consistent with the theory of impurity drag and provides further evidence that thoria segregates to the grain boundaries of yttria.

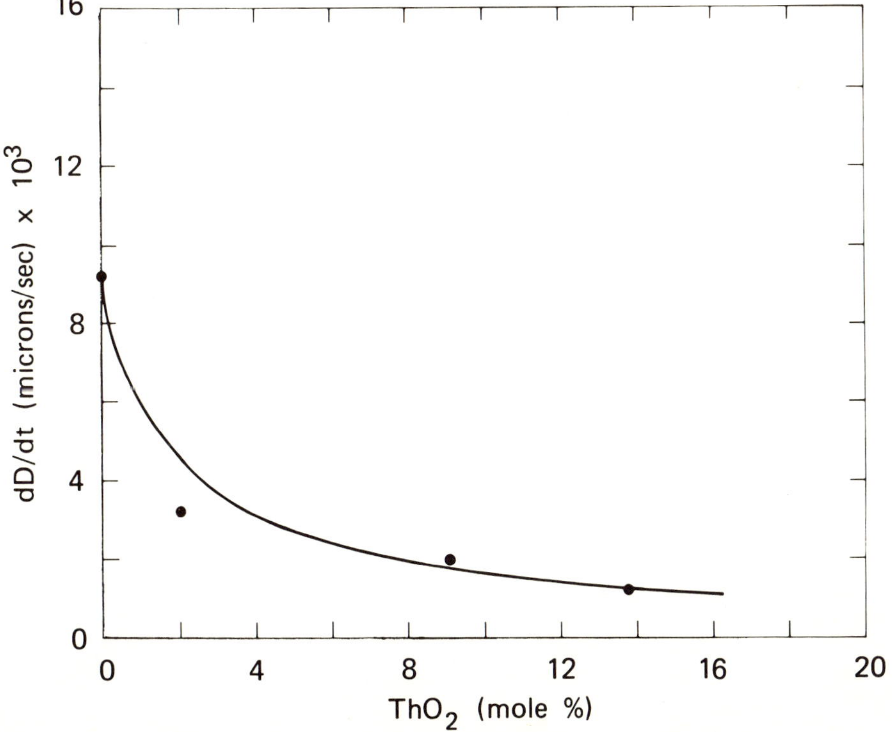

Fig. 9. Grain-boundary velocity at 2000°C as a function of thoria solute concentration.

4 CONCLUSIONS

The grain growth of yttria is inhibited by the addition of a thoria solute and can be described by theoretical grain-growth laws. The rate of grain growth in the single-phase solid-solution region is probably controlled by the thoria solute exerting a drag on the grain-boundary velocity. Indirect evidence for solute segregation to the grain boundaries of yttria is provided by microhardness and surface-activity measurements. The fact that grain growth occurs during the sintering of oxides to theoretical density and because the grain growth of yttria is inhibited by the addition of thoria, we conclude that a higher diffusion flux from the grain boundaries to the pores is maintained when thoria is present during sintering primarily by maintaining a higher chemical-potential gradient.

REFERENCES

1. Alexander, B. H., and Balluffi, R. W., "Mechanism of Sintering of Copper", *Acta Met.*, **5** (11), 666-77 (1957).
2. Burke, J. E., *J. Am. Ceram. Soc.*, **40** (3), 80-85 (1957).
3. Seigle, L., *Kinetics of High-Temperature Processes*, W. D. Kingery (Ed.), Technology Press of Massachusetts Institute of Technology, Cambridge, and John Wiley & Sons, Inc., New York (1959), pp. 172-78.
4. Van Buren, H. G., and Hornstra, J., *Reactivity of Solids*, J. H. de Boer, W. G. Burgers, E. W. Gorter, J.P.F. Huesse, and G.C.A. Schuit (Eds.), Elsevier Publishing Co., Princeton, N.J. (1961), p. 112.
5. Coble, R. L., and Burke, J. E., *Progress in Ceramic Science*, Vol. 3, J. E. Burke (Ed.), Pergamon Press, New York (1964), pp. 197-251.
6. Reijnen, P.J.L., *Problems in Nonstoichiometry*, A. Rabenau (Ed.), North Holland Publishing Company, Amsterdam (1970), pp. 219-238.
7. Jorgensen, P. J., and Westbrook, J. H., *J. Am. Ceram. Soc.*, **47**, 332 (1964).

Jorgensen, P. J., and Anderson, R. C., *J. Am. Ceram. Soc.*, **50**, 553 (1967).

9. Jorgensen, P. J., and Schmidt, W. G., *J. Am. Ceram. Soc.*, **53**, 24 (1970).
10. Anderson, R. C., U. S. Patent, 3,574,645 (April 13, 1971).
11. Subbarao, E. C., Sutter, P. H., and Hrizo, J., *J. Am. Ceram. Soc.*, **48** 443 (1965).
12. Edwards, H. S., Rosenberg, A. F., and Bittel, J. T., U. S. Air Force Final Report, Contract No. AF33(657)-8470, July 1963, 146 pp. (NAS Document #N63-20007).
13. Harkins, R. J., and Alcock, C. B., *J. Nucl. Mater.*, **26** 112 (1968).
14. King, A. D., *J. Nucl. Mater.*, **38**, 347 (1971).
15. Berard, M. F., Wirkus, C. D., and Wilder, D. R., *J. Am. Ceram. Soc.*, **51**, 643 (1968).
16. Wirkus, C. D., Berard, M. F., and Wilder, D. R., *J. Am. Ceram. Soc.*, **50**, 113 (1967).
17. Berard, M. F., and Wilder, D. R., *J. Am. Ceram. Soc.*, **52**, 85 (1969).
18. Etsell, T. H., *Z. Naturforsch.*, **27A**, 1138 (1972).
19. Fullman, R. L., *Trans. AIME*, **197**, 447 (1953).
20. Westbrook, J. H., and Aust, K. T., *Acta Met.*, **11**, 1161 (1963).
21. Aust, K. T., and Rutter, J. W., *Trans. AIME*, **215**, 119 (1969).
22. Rutter, J. W., and Aust, K. T., *Trans. AIME*, **218**, 682 (1960).
23. Cahn, J. W., *Acta Met.*, **10**, 789 (1962).
24. Burke, J. E., *Ceramic Microstructures*, Proc. 3rd Int. Mater. Symp., Berkeley, Calif., June 1966, p. 681.
25. Burke, J. E., and Turnbull, D., *Progress in Metals Physics*, Vol. III, Bruce Chalmers (Ed.), Pergamon Press, London (1952), pp. 220-92.

26. Hillert, M., *Acta Met.*, **13**, 227 (1965).
27. Greenwood, G. W., *Acta Met.*, **4**, 243 (1956).
28. Lifschitz, I. M., and Slyozov, V. V., *Zh. Eksp. Teor. Fiz.*, **35**, 479 (1958).
29. Zener, C., personal communication to C. S. Smith, *Trans. AIME*, **175**, 15 (1949).

DISCUSSION on Paper by Paul J. Jorgensen

CARO: Solubility of ThO_2 in A type La_2O_3 occurs as coherent intergrowth of pure phase microdomains ([111] cubic and c hexagonal axis common), which might be ordered, giving "compounds" on X-ray patterns. Similarly, solubility of Y_2O_3 into A type Ln_2O_3 oxides occurs as microdomain insertions of Y_2O_3 with a well-defined epitaxial relationship. I suggest that the same phenomenon occurs in the Y_2O_3 – ThO_2 system; consequently, extended defects (i.e., oriented domains of pure ThO_2 or Y_2O_3) may be operative in the sintering process on both sides of the binary system.

HAUFFE: Your results are in agreement with the lattice-defect model. When ThO_2 is added to Y_2O_3, the oxygen-vacancy concentration is decreased and so is the diffusivity. In agreement with this model, you found a decrease in the grain size and an increase in the hardness. By doping with CaO you should observe a decrease in the hardness and an increase in the average grain size because the concentration of the mobile oxygen vacancies is increased. Have you carried out such experiments?

JORGENSEN: Let me switch systems to answer your question. In aluminum oxide, we think of it in terms of segregation coefficients and ask whether the solubility is greater in the liquid or in the solid. Therefore, for those systems in which K is less than 1 we get an increase in hardness and for those systems in which K is greater than 1 we get a decrease in hardness. For example, the addition of magnesium oxide to alumina increases the bulk hardness and the grain-boundary hardness, whereas for chromium oxide additions to alumina we get a decrease in the hardness of the grain boundary.

HEUER: With regard to this microhardness of Al_2O_3 doped with chromium, our work on plastic deformation of sapphire and ruby suggests that the bulk hardness should increase with chromium doping.

JORGENSEN: Let me clarify that a little. What I'm talking about here is segregation to the grain boundaries, and no matter whether the bulk hardness is increased or decreased, the segregation is decreasing the grain-boundary hardness. You may actually have a very slight increase in bulk hardness when comparing data for nondoped alumina with chromium-doped alumina, although you are essentially within the scatter band. But, you can see a dramatic decrease in the grain-boundary hardness which we can relate to segregation.

HEUER: The Auger work of Marcus and Fine showing no segregation of MgO in Al_2O_3 is inconsistent with the picture you have given us. How do you explain that?

JORGENSEN: Let me briefly describe the study made by Marcus and Fine [*J. Am. Ceram. Soc.*, **55**, 568 (1972)]. They analyzed the composition of grain boundaries exposed by intergranular fracture in Al_2O_3 doped with MgO. They found no MgO in the grain boundaries; in fact, they found no MgO at all. However, Tong and Williams [*J. Am. Ceram. Soc.*, **53**, 58 (1970)] performed a similar experiment several years ago using a solid source mass spectrometer. Although this technique required that they analyze layers up to 3000 Å deep at the grain boundaries, they observed grain-boundary concentrations up to 80 times greater than the bulk concentrations measured within the grains themselves after sectioning. In addition, Berrin and Sundahl [private comm. to D. Kingery, *J. Am. Ceram. Soc.*, **57**, 74 (1974)] state that at the voltage level of 300 kv used by Marcus and Fine, the Auger spectrometer cannot detect magnesium. Berrin and Sundahl repeated the experiment using 500 kv, observed magnesium, and picked up a segregation effect. These results are in agreement with those I have presented here.

SHEWMON: In the equations used in your presentation you have omitted any discussion of the role of the as-sintered pores on grain-boundary motion. Must the grain boundaries stay on pores to complete sintering or is it only the mean grain-boundary density that determines the sintering rate?

JORGENSEN: I feel that a high grain-boundary density alone is not sufficient to remove porosity. The pores must stay on the boundaries. All that is required is that the boundaries are slowed up enough so that the pores, which are mobile, can keep up with them. During grain growth, grain boundaries continue to sweep through the bulk, so that after sufficient grain growth, all the pores could be removed. Once the boundaries leave pores behind, the flux to the pores falls off so rapidly with distance that these pores do not disappear in reasonable times.

ILSCHNER: Decrease of migration mobility of the grain boundary is just one consequence of grain-boundary segregation relevant to final-stage sintering. In addition, there should be an effect of the doping on interface energies. Since you assume segregation, you imply a decrease in the grain-boundary energy. This might make it easier for the boundary to break free from the pore. A decrease in pore surface energy, moreover, should reduce the driving force for pore shrinking, slowing down the sintering rate. Would you care to comment on this issue?

JORGENSEN: We expect solute segregation to the solid-vapor interface as you suggest, and this will decrease the sintering rate. The magnitude of the decrease has not been measured, however we expect that the change in rate will be small relative to the effect of solute segregation on the boundary mobility because the sintering rate is inversely proportional to the cube of the grain size.

APPROACHES TO DESCRIBING THE KINETIC BEHAVIOR OF DISTRIBUTIONS OF SECOND-PHASE INCLUSIONS

Alan J. Markworth

Battelle
Columbus Laboratories
Columbus, Ohio

ABSTRACT

Some useful mathematical approaches to treating the kinetic behavior of systems of inclusions in materials are discussed. Such systems are described in terms of a continuous size-distribution function which satisfies a continuity relation that is analogous to the familiar Boltzmann integrodifferential equation. Specific problems to which these approaches are applied include the growth of a second phase and the early stages of coarsening of gas-filled pores. It is shown, for these situations, that certain important properties of the size distribution can be quantitatively described without requiring detailed knowledge of the "shape" of the distribution in size space. In addition, the use of similarity solutions to describe coarsening kinetics at asymptotic times is explored and found to offer significant insights into the nature of this process.

1 INTRODUCTION

A second-phase species which is evolving from a supersaturated host phase must be described in terms of geometrical properties as well as physical properties. In particular, if the second phase consists of a system of discrete, spatially distributed inclusions, the size and shape of each of these inclusions are factors which determine such gross properties as the average volume and interfacial area of second phase per unit volume. If the shapes of all the inclusions are geometrically similar, then their size distribution can be described in terms of a single size-characterizing parameter.

Materials containing second-phase dispersions are used in many applications, some of which are listed in Table I.[1] The construction of physically realistic models to describe the evolution of the second phase constitutes an essential component of materials development. For example, the utility of a metal that is strengthened by a fine dispersion of oxide particles is limited by the tendency of the particle distribution to coarsen via an Ostwald ripening process in which larger oxide particles grow at the expense of smaller ones; quantitative models have been developed to predict the coarsening kinetics of such systems.[2,3]

Unfortunately, the evolution in time of a real multiphase material is immensely complex, so that quantitative physical models are usually based upon a rather restrictive set of simplifying assumptions in order to bring the mathematical descriptions within the realm of tractability. For example, one very simple model for describing the sigmoidal growth with time of a second-phase species from supersaturated solution is represented by the familiar Johnson-Mehl-Avrami equation:

$$f = 1 - \exp[-(t/\tau)^n] \quad , \tag{1}$$

where f represents the fraction of material transformed at time t, with n and τ being appropriate characteristic parameters. It has been shown by Ham[4], however, that Eq. (1) represents a good approximation to a "real" situation only for relatively short times, and that earlier work, which had related precipitate shape to the value of n, was incorrect.

In the work presented here, certain useful approaches are developed for use in modeling the temporal evolution of a two-phase system, the second phase being assumed to consist of a collection of spherical inclusions (e.g., solid particles or pores) distributed both in size and in coordinate space within the host phase. These approaches are applied to the specific cases of the growth of a system of inclusions, the early stages of coarsening of gas-filled pores, and asymptotic coarsening kinetics.

Table I. Potentially Attractive Multiphase Systems[1]

Property	System Description
Dispersion-hardening — creep strength	Fine uniform dispersion of inert particles usually in metallic matrix, e.g., $Ni\text{-}ThO_2$.
Hardness and wear resistance	Large volume fraction of relatively large hard refractory phase in ductile matrix, e.g., WC-Co.
Superplastic properties	Intimate fine-grain mixture of phases produces stable superplastic material, e.g., Al-Zn, stainless steel ($\alpha + \delta$ Fe).
Lubrication and machinability	Distribution of a soft phase in a harder matrix, e.g., Pb in Cu or brass.
Superconducting properties	Dispersions of second phase in superconductors pin flux lines and thus raise effective transition temperature and critical fields, e.g., Ω phase in Ti and Ti alloys.
Galvanomagnetic properties	Dispersions of second phases in a semiconductor matrix are expected to alter galvanomagnetic properties.
Magnetoresistance, thermomagnetic, and thermoelectric properties	Continuous magnetic conducting phases present in a high-mobility semiconductor matrix can lead to large increases in magnetoresistance (Corbino effect). Conducting inclusions in thermoelectrics may improve thermal conductivity of such materials, e.g., PbTe. Similar inclusions in thermomagnetics could alter the properties of these materials.
Magneto-electro-mechanical interactions	Coupling between magnetostrictive phases in a piezoelectric matrix may offer potential as a new means for energy conversion, e.g., Ni or Ni alloys in barium titanate.
Magnetodielectrics	Dielectric materials having conductive and/or magnetic inclusions. Have application as a radar absorbing material (RAM). Performance is critically dependent on configuration of the inclusions, e.g., Permaloy in Al_2O_3.
Electromagnetic properties	The presence of metallic inclusions in ferrite matrices affects magnetic response as a result of an electronic interaction between the phases, e.g., Co or Ni in Ni or Co ferrite.
Semiconductor properties	Mixtures of immiscible semiconductors could lead to interesting combinations of semiconducting properties within one material, e.g., GaAs/CdS.

2 EVOLUTION OF THE SIZE DISTRIBUTION AND ITS PROPERTIES

2.1 The Continuity Equation and Moments of the Distribution

We shall describe the distribution of inclusions in terms of a continuous function f(R,**r**,t), defined such that f(R,**r**,t)dRd**r** is the number of inclusions, at time t, having radii within incremental range dR measured about R and lying within volume element d**r** measured at position vector **r** relative to the origin of spatial coordinates. A continuity equation for the distribution function can be derived by considering an incremental four-dimensional volume element (three coordinate-space dimensions, one radius-space dimension) and requiring that the net rate at which inclusions are added to this element equal the net rate of "flow" into the element plus the net rate of "creation" within the element. One thus obtains

$$\frac{\partial f}{\partial t} + \mathrm{div}(\mathbf{v}f) + \frac{\partial}{\partial R}(\nu f) = \sum_i \psi_i \quad , \tag{2}$$

where **v** and ν represent the inclusion velocity in coordinate space and radius space, respectively, and the functions ψ_i are various source functions describing the creation and annihilation of inclusions such as by nucleation and collision events. Clearly, Eq. (2) is analogous to the familiar Boltzmann integrodifferential equation describing the distribution function for molecules of a gas.[5] We assume, in our analysis, that **v** and ν are functions only of R, **r**, and t. It should be borne in mind, however, that different inclusions having the same size and environment, may actually grow at different rates owing to differing "inherent growth activity" related to interfacial properties.[6]

The configuration of a hypothetical distribution function at two different times is illustrated in Fig. 1. For this case, the function is becoming "narrower" in radius space as time increases. This particular behavior is qualitatively characteristic of the diffusion-controlled growth of a disperse system of second-phase particles[7,8] (discussed below in more detail), although other shape changes which occur in this growth process, such as changes in skewness, are not illustrated in the figure.

Clearly, many of the properties of a second-phase system that are of physical interest (e.g., inclusion concentration, mean radius, volume fraction) are related to the distribution function in terms of its algebraic moments, the ith moment M_i being defined as

$$M_i(\mathbf{r},t) \equiv \int_0^{\infty} R^i f(R,\mathbf{r},t) dR \quad . \tag{3}$$

Consequently, we shall be largely concerned, in this work, with the temporal behavior of selected size-distribution moments. Also, it should be noted that Eq. (2) can be transformed into a corresponding expression in terms of moments by multiplying through by a factor R^j and integrating over all radii.[9]

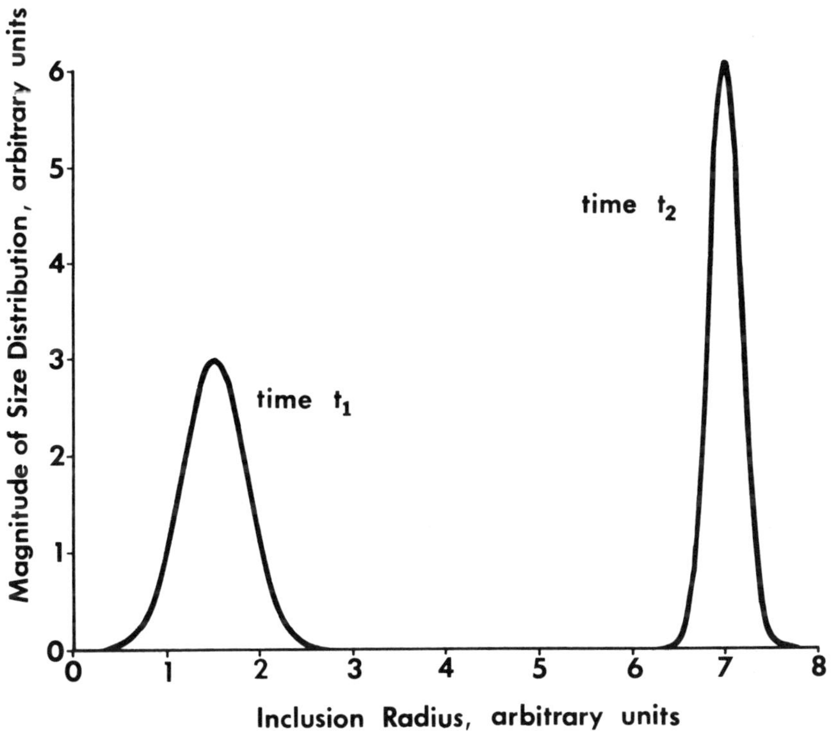

Fig 1 Variation with time of a hypothetical distribution of inclusions; the distribution becomes "narrower" with increasing time ($t_2 > t_1$); the "area" enclosed between the distribution and the size-axis remains constant, indicative of invariant inclusion concentration.

2.2 Temporal Variation of Moments

We shall confine further considerations to the situation for which all the inclusions are stationary relative to their surroundings ($\mathbf{v}$ = 0), no sources exist (ψ_i = 0 for all i), and the distribution function is, on the average, independent of spatial coordinates, in which case Eq. (2) reduces to

$$\frac{\partial f}{\partial t} + \frac{\partial}{\partial R}(\nu f) = 0 \quad . \tag{4}$$

From our definition of the distribution function, it is clear that the zeroth algebraic moment, M_0, is simply the average concentration of inclusions, which we assume here to be invariant in time. Also, we find it convenient to define a new function h equal to f/M_0, so that hdR is simply the net *fraction* of inclusions, at time t, having radii within range dR measured about R. We define the expectation value of a function g(R) with respect to h, as follows:

$$\langle g(R) \rangle \equiv \int_0^\infty g(R)h(R,t)dR \quad . \tag{5}$$

Furthermore, we assume that g can be expanded in a Taylor series over the entire range of R for which the function h may differ appreciably from zero; then, expanding about <R> and taking expectation values, we obtain

$$\langle g(R) \rangle = g(\langle R \rangle) + \sum_{j=2}^{\infty} \frac{1}{j!} g^{(j)}(\langle R \rangle)\mu_j \quad , \tag{6}$$

where $g^{(j)}(\langle R \rangle)$ is the jth derivative of the function g evaluated at <R> and μ_j is the jth central moment of the function h, defined by

$$\mu_j \equiv \langle (R - \langle R \rangle)^j \rangle \quad .$$

The importance of Eq. (6) lies in the fact that the expectation value of g is expressed entirely in terms of integral properties (i.e., moments) of the distribution function. In particular, if g varies sufficiently slowly over R, then

$$\langle g(R) \rangle \approx g(\langle R \rangle) + \frac{1}{2} g^{(2)}(\langle R \rangle)\mu_2 \quad , \tag{7}$$

where μ_2 is simply the variance of the distribution function, i.e.,

$$\mu_2 = \langle (R - \langle R \rangle)^2 \rangle = \langle R^2 \rangle - \langle R \rangle^2 \quad , \tag{8}$$

the square root of the variance, $\Delta \equiv (\mu_2)^{1/2}$, being the standard deviation, a measure of the "width" of the distribution in radius space. Equation (7), or more generally, Eq. (6), has been applied in various types of problems.[8,10-12]

Another useful relation can be obtained by differentiating Eq. (5) with respect to time, i.e.,

$$\frac{d}{dt} \langle g(R) \rangle = \int_0^\infty g(R) \frac{\partial}{\partial t} h(R,t) dR \quad .$$

From Eq. (4) we find that

$$\frac{d}{dt} \langle g(R) \rangle = - \int_0^\infty g(R) \frac{\partial}{\partial R} (\nu h) dR \quad . \tag{9}$$

Equation (9) can be integrated by parts, and assuming the product $g\nu h$ vanishes at both $R = 0$ and $R \to \infty$, we obtain

$$\frac{d}{dt} \langle g(R) \rangle = \langle \nu g^{(1)}(R) \rangle \quad . \tag{10}$$

The use to which these results [in particular, Eqs. (7) and (10)] can be put is illustrated below.

3 INCLUSION GROWTH

We first consider the growth, from solution, of a distribution of inclusions. To generalize the analysis, we shall work within an R^n size space, where n is an integer greater than or equal to unity, and to simplify notation, we define a new size variable $\xi \equiv R^n$. The distribution function $H(\xi,t)$ is defined such that $Hd\xi$ is the fraction of inclusions, at time t, within size range $d\xi$ measured about ξ, and the continuity relation satisfied by H is

$$\frac{\partial H}{\partial t} + \frac{\partial}{\partial \xi} (\nu_n H) = 0 \quad ,$$

where ν_n, the growth velocity in ξ space, is related to ν (the growth velocity in R space, i.e., $\nu = \nu_1$) such that

$$\nu_n = n\nu \xi^{\frac{n-1}{n}} \quad . \tag{11}$$

Likewise, the function H is related to h as follows:

$$H(\xi,t) = h(\xi^{1/n},t) n^{-1} \xi^{\frac{1-n}{n}} \quad ,$$

and the variance $\mu_{2,n}$ in ξ space is [after Eq. (8), noting that $\mu_{2,1}$ is simply what we had referred to above as μ_2],

$$\mu_{2,n} = \langle\xi^2\rangle - \langle\xi\rangle^2 \quad , \tag{12}$$

where the expectation values are now calculated with respect to the function H.

We can cover a rather broad class of problems by assuming

$$\nu = \theta(t)/R^k \quad ,$$

where θ is a positive, but otherwise unspecified, function of time and k is an integer greater than or equal to zero. Hence, from Eq. (11),

$$\nu_n = n\theta(t)\xi^{\frac{n-k-1}{n}} \quad . \tag{13}$$

One can use Eq. (10), generalized to ξ space, together with Eq. (13), to obtain

$$\frac{d}{dt}\langle\xi^\ell\rangle = n\ell\theta(t)\left\langle \xi^{\frac{n\ell-k-1}{n}} \right\rangle \quad , \tag{14}$$

where we take ℓ to be an integer greater than or equal to zero. Let us now use Eq. (14) to examine the behavior of the variance of the size distribution. To this end, we differentiate Eq. (12) with respect to time and combine the result with Eq. (14) to obtain

$$\frac{d\mu_{2,n}}{dt} = 2n\theta(t)\left(\left\langle \xi^{\frac{2n-k-1}{n}} \right\rangle - \langle\xi\rangle \left\langle \xi^{\frac{n-k-1}{n}} \right\rangle\right) \quad . \tag{15}$$

We assume that the size distribution is sufficiently "narrow" in ξ space to ensure the applicability of Eq. (7) generalized to our R^n size space, in which case we obtain an approximation for Eq. (15), i.e.,

$$\frac{d\mu_{2,n}}{dt} \approx 2(n-k-1)\theta(t)\langle\xi\rangle^{-\left(\frac{k+1}{n}\right)} \mu_{2,n} \quad . \tag{16}$$

The evolution of the standard deviation in ξ space, $\Delta_n \equiv (\mu_{2,n})^{1/2}$, can be evaluated by integrating Eq. (16), i.e.,

$$\Delta_n(t) \approx \Delta_n(t_o)\exp\left[(n\text{-}k\text{-}1)\int_{t_o}^{t}\theta(t)<\xi>^{-\left(\frac{k+1}{n}\right)}\,dt\right] \quad , \tag{17}$$

where t_o is some initial time. Now, one can use Eqs. (10) and (13) to obtain

$$\frac{d<\xi>}{dt} = n\theta(t)\left\langle \xi^{\frac{n\text{-}k\text{-}1}{n}}\right\rangle \quad . \tag{18}$$

To lowest order, Eq. (18) becomes

$$\frac{d<\xi>}{dt} \approx n\theta(t)<\xi>^{\frac{n\text{-}k\text{-}1}{n}} \quad ,$$

in which case

$$\theta(t)<\xi>^{-\left(\frac{k+1}{n}\right)}\,dt \approx d<\xi>/(n<\xi>) \quad . \tag{19}$$

Equation (19) can be substituted into Eq. (17), and the resultant integration can be carried out explicitly to yield

$$\frac{\Delta_n(t)}{\Delta_n(t_o)} \approx \left(\frac{<\xi>}{<\xi>_o}\right)^{\frac{n\text{-}k\text{-}1}{n}} \quad , \tag{20}$$

where $<\xi>_o$ is the value of $<\xi>$ at time t_o.

We see, from Eq. (20), that the variation, with time, of the standard deviation of a relatively narrow size distribution can be related quite simply to the variation of the mean size. We note that, since $<\xi>$ increases with increasing time, the distribution widens in ξ space if $n > k+1$, it narrows if $n < k+1$, and Δ_n is constant if $n = k+1$. The last result, in fact, follows immediately from Eq. (13), since ν_n is seen to be independent of ξ for $n = k+1$, in which case the distribution can undergo only pure translation in ξ space, with no simultaneous change of "shape". Some special cases, which have been treated elsewhere, include $k = 0$, $n = 1$ corresponding to the gas-diffusion-controlled growth of pores containing an ideal gas[13] and $k = 1$, $n = 1$ corresponding to the diffusion-controlled growth of precipitate particles[8]. For the latter case, the standard deviation in R space is seen to be inversely proportional to the mean radius, which corresponds to the well-known homogenization phenomenon that has long been known to exist for diffusion-controlled growth.[7,8]

Let us note three features of this analysis:

(1) Equation (20) contains no explicit dependence upon the function $\theta(t)$, and in this sense, is independent of *how fast* the growth process proceeds. In other words, the relative change in the standard deviation is determined entirely by the relative change in mean size.

(2) Nowhere in the analysis leading to Eq. (20) was any assumption made regarding details of the *shape* of the size distribution. The result is shape-independent except for the requirement that the distribution be relatively "narrow" in ξ space.

(3) Likewise, nowhere was it necessary to require that the function $H(\xi,t)$ represent the total size distribution. Equation (20) is valid for any subset of the distribution as long as (a) the same subset is considered for all times and (b) the same growth law, Eq. (13), applies to every inclusion within the subset. Whatever may be going on *outside* the subset (e.g., nucleation of new inclusions, impingement of adjacent inclusions, etc.) has no bearing on the procedure described here.

4 EARLY STAGES OF THE COARSENING OF GAS-FILLED PORES

Consider a solid containing a disperse distribution of intragranular gas-filled pores. We assume that the quasistationary diffusion of gas atoms within the solid is the kinetic process which controls the rate at which pores change in size. Then, taking the solid to be stress-free, the gas pressure within a pore of radius R is maintained virtually equal to the surface-tension restraint, $2\gamma/R$, where γ is the surface tension of the solid. In addition, we assume that the equilibrium solubility of the gas in the solid can be described by "Sievert's law", i.e., that the equilibrium concentration of gas in solution at the surface of each pore is equal to $K(2\gamma/R)^m$, where K is a temperature-dependent solubility parameter and m is a dimensionless parameter the value of which depends upon the nature of the gas; for example, m = 1 for an inert gas and m = 1/2 for a diatomic gas which must dissociate before undergoing dissolution from the gaseous phase into the solid. Under these conditions, one can show[14] that the growth velocity of a pore in R space is given by

$$\nu = \frac{1}{R_c^m} - \frac{1}{R^m} \quad , \tag{21}$$

where R_c is a time-dependent "critical radius" such that pores larger than or smaller than R_c are, respectively, growing or shrinking, and pores equal in size to R_c are instantaneously at rest in R space. Physical parameters that would be expected to appear in a coefficient on the right-hand side of Eq. (21) have been "absorbed" into the time scale.

Now, Eq. (8) can be differentiated with respect to time and the result combined with Eqs. (10) and (21) to yield, for the time rate of change of the variance,

$$\frac{d\mu_2}{d\tau} = 2\langle R\rangle \left\langle \frac{1}{R^m} \right\rangle - 2\langle R^{1-m}\rangle \quad , \tag{22}$$

where τ is the unit of time consistent with Eq. (21) and the expectation values are now assumed to cover the entire size distribution. Applying Eq. (7) to Eq. (22),

$$\frac{d\mu_2}{d\tau} \approx \frac{2m\mu_2}{\langle R\rangle^{m+1}} \tag{23}$$

from which we find, again setting $\Delta = (\mu_2)^{1/2}$,

$$\Delta(\tau) \approx \Delta(\tau_o) \exp \left(m \int_{\tau_o}^{\tau} \frac{d\tau}{\langle R\rangle^{m+1}} \right) \quad . \tag{24}$$

From Eqs. (10) and (21) we find

$$\frac{d\langle R\rangle}{d\tau} = \frac{1}{R_c^{\ m}} - \left\langle \frac{1}{R^m} \right\rangle \quad . \tag{25}$$

Now, the primary process that occurs during coarsening is the net transfer of gas atoms from smaller pores to larger ones, in which case $dM_2/d\tau \approx 0$. Hence, combining Eqs. (3), (10), and (21), it follows that

$$R_c^m = \langle R\rangle / \langle R^{1-m}\rangle \quad ,$$

in which case Eq. (25) becomes

$$\frac{d\langle R\rangle}{d\tau} = \frac{\langle R^{1-m}\rangle}{\langle R\rangle} - \left\langle \frac{1}{R^m} \right\rangle \quad . \tag{26}$$

Applying Eq. (7) to Eq. (26), we obtain

$$\frac{d\langle R\rangle}{d\tau} \approx -\frac{m\mu_2}{\langle R\rangle^{m+2}} \quad . \tag{27}$$

Combining Eqs. (23) and (27), we find that

$$d\langle R\rangle/d\Delta \approx -\Delta/\langle R\rangle \quad ,$$

so that if $\Delta << \langle R\rangle$, then a change in standard deviation will result in a change in mean radius, the magnitude of which is much smaller by comparison. Consequently, we can, to a good approximation, take $\langle R\rangle$ to be constant in Eq. (24) and equal to $\langle R\rangle_o$, its value at time τ_o, in which case we finally obtain

$$\Delta(\tau) \approx \Delta(\tau_o)\exp\left[\frac{m(\tau-\tau_o)}{\langle R\rangle_o^{m+1}}\right] \quad , \tag{28}$$

so that the standard deviation increases approximately in an exponential manner with time, the "time constant" being $\langle R\rangle_o^{m+1}/m$.

It is important to observe that the analysis leading to Eq. (28) is independent of the shape of the size distribution except for the "narrowness" assumption. Consequently, in the early stages of pore coarsening, before the distribution has widened appreciably, the standard deviation increases as predicted by Eq. (28) for any size distribution, as long as the same growth law, Eq. (21), is applicable to each pore within the distribution.

5 SIMILARITY SOLUTIONS AND OSTWALD-RIPENING KINETICS

The coarsening or Ostwald ripening of a system of inclusions consists of the growth of the larger inclusions at the expense of the smaller ones, so that the concentration of inclusions decreases as ripening proceeds. (We are now relaxing our earlier assumption of constant inclusion concentration.) A broad class of ripening phenomena can be described in terms of a growth-velocity function having the general form

$$\nu = \frac{1}{R^n}\left(\frac{1}{R_c} - \frac{1}{R}\right) \quad , \tag{29}$$

where n is an integer greater than or equal to zero and R_c is a time-dependent critical radius having the same general significance as discussed above. Physical parameters normally appearing in the growth-velocity function are again considered here to be "absorbed" in the time scale. Equation (29) does not cover all possible Ostwald-ripening phenomena [$m = 1/2$ in Eq. (21), being one example of a ripening process not included], but its generality is sufficient to demonstrate the procedures we shall present here. Now, in addition to Eq. (29), a condition which we require to be satisfied is that one of the algebraic moments of the size distribution is approximately invariant in time, i.e.,

$$M_j = \text{constant} \quad ,$$

where j is an integer greater than zero; for example, for a system of solid particles, the total particle volume is conserved during ripening, so that M_3 is constant in that case.

A great deal of physical insight into the kinetics of the process can be achieved by assuming that the distribution function asymptotically approaches a form which can be described in terms of a similarity solution, i.e.,

$$f(R,t) = t^p g(Rt^q) \quad , \tag{30}$$

where p and q are numerical constants dependent upon n and j. The validity of this assumption is found in succeeding analysis.

Equations (3) and (30) can be combined to show that

$$M_k(t) = \alpha_k t^{p-q(k+1)} \quad , \tag{31}$$

where

$$\alpha_k \equiv \int_0^\infty u^k g(u)du \quad .$$

Requiring that M_j be constant in time, we obtain

$$p = q(j + 1) \ . \tag{32}$$

Also, one can show, using Eqs. (3) and (4), that for the kth algebraic moment,

$$\frac{dM_k}{dt} = k \int_0^\infty R^{k-1} \nu f(R,t)dR \quad , \tag{33}$$

and substituting Eq. (29) into Eq. (33) and setting k = j, we find that

$$q = -(n + 2)^{-1} \quad . \tag{34}$$

Next, substituting Eqs. (32) and (34) into Eq. (31),

$$M_k(t) = \alpha_k t^{\frac{k-j}{n+2}} \quad . \tag{35}$$

It also readily follows, from Eqs. (29), (33), and (35), that

$$R_c = \left(\frac{\alpha_{j-n-1}}{\alpha_{j-n-2}} \right) t^{\frac{1}{n+2}} \quad ,$$

and

$$\left(\frac{k-j}{n+2} \right) \alpha_k = k \left(\frac{\alpha_{j-n-2}\alpha_{k-n-1}}{\alpha_{j-n-1}} - \alpha_{k-n-2} \right) \quad , \tag{36}$$

assuming that all the α's in Eq. (36) are indeed finite. Equation (36) represents a useful recursion relation for the α coefficients.

Consequently, we have obtained a great deal of useful information about ripening kinetics with no detailed description of the functional form of the distribution function itself. That description can be obtained by substituting the similarity form for the distribution function into the continuity equation [i.e., Eq. (4)] and solving the resultant differential equation for g(u), where $u \equiv Rt^q$. One does indeed find that the only variable appearing in that differential equation is u, so that our assumption of the similarity form for f(R,t) is justified. Some further uses of similarity solutions, as applied to studies of the diffusion-controlled and interface-controlled ripening of precipitate systems, have been described elsewhere.[15]

6 CONCLUSION

The evolution of distributions of second-phase inclusions in solids is an essential consideration in the behavior of many types of materials. Unfortunately, such problems are generally exceedingly complex, from a mathematical point of view, so that appropriate approximation methods are required in order to achieve a quantitative description of size-distribution kinetics. We have demonstrated here a number of such procedures that are useful in a broad class of problems. One of these procedures is applicable to distribution functions that are narrow in size space, in which case details of the shape of the distributions become unimportant. This approach can be used with problems involving the growth of inclusions as well as the early stages of coarsening. Another procedure involves the use of similarity solutions to describe coarsening kinetics at asymptotic times; a number of important properties of the size distribution, including the time dependence of its moments, can thereby be deduced without knowledge of the functional form of the distribution function itself. Thus, in general, many quantitative deductions can be made regarding the temporal behavior of properties of size distributions of inclusions without detailed knowledge of the actual size distribution. Such information is most useful, both for the analysis of experimental data and for application to the development of more complex models of precipitate evolution.

ACKNOWLEDGMENT

This work was supported, in part, by Battelle Memorial Institute.

REFERENCES

1. Gelles, S. H., Duga, J. J., and Beer, A. C., Private Communication, Battelle-Columbus.
2. Serebryakov, A. V., *Soviet Phys.-Solid State,* **11**, 778 (1969).
3. Fischmeister, H. F., Navara, E., and Easterling, K. E., *Metal Sci. J.,* **6**, 211 (1972).
4. Ham, F. S., *J. Phys. Chem. Solids,* **6**, 335 (1958).
5. Present, R. D., *Kinetic Theory of Gases,* McGraw-Hill Book Company, Inc., New York (1958), pp. 224 ff.
6. Randolph, A. D., and Larson, M. A., *Theory of Particulate Processes,* Academic Press, New York (1971), p. 43.
7. Reiss, H., *J. Chem. Phys.,* **19**, 482 (1951).
8. Markworth, A. J., *J. Colloid Interface Sci.,* **38**, 636 (1972).
9. Randolph, A. D., and Larson, M. A., *loc. cit.,* pp. 53 ff.
10. Markworth, A. J., and Baroody, E. M., *Metal Sci. J.,* **5**, 55 (1971).
11. Markworth, A. J., *Ber. Bunsenges. Phys. Chem.,* **75**, 533 (1971).
12. Luss, D., *J. Catal.,* **23**, 119 (1971).
13. Markworth, A. J., *J. Mater. Sci.,* **7**, 1225 (1972).
14. Markworth, A. J., *Metall. Trans.,* **4**, 2651 (1973).
15. Markworth, A. J., *Metallography,* **3**, 197 (1970).

DISCUSSION on Paper by Alan J. Markworth

HONIG: Can you refer this audience to a single source or compendium where your approximation procedures are derived?

MARKWORTH: The approximation procedures, together with applications to specific problems, are discussed in Refs. 8, 10-12, 14 and 15 of my paper.

CARO: Is it possible to take into account in your model, inclusions which have well-defined crystallographic relationships (epitaxial) to the host crystals? Growth of a second phase within a host strongly depends on the crystallographic orientation of the primary nuclei with respect to the neighboring crystals.

MARKWORTH: In order to apply the procedures I have discussed here, it is necessary to know, at least in approximate form, the functional relationship describing the growth rate ν If this is known for the case you describe, then these techniques can be applied.

O'KEEFFE: There are analytical solutions, for example the Lifshits-Wagner distribution function, to certain ripening problems. Does your treatment include these as special cases?

MARKWORTH: That is correct. One of the purposes of the analysis I presented here was to demonstrate the fact that a number of important features are common to different types of ripening phenomena and that a generalized or unified approach can be used to illustrate these features.

SCHMALZRIED: Can you comment on the possibilities of calculating average values of properties of heterogeneous materials using the moments of different order instead of complete distribution functions? For example, what do you need to know about these moments in order to calculate the average diffusion coefficient of a two-phase mixture of two isotropic materials?

MARKWORTH: For those situations in which one can derive approximate, closed-form expressions for various time-dependent moments of the size-distribution functions, he finds that these expressions contain algebraic combinations of various physical and experimental parameters that are pertinent to the process involved. His experimental data for a particular moment could, in principle, be used to evaluate the magnitude of the particular *combinations* of parameters characteristic of that moment.

COHEN: Your analysis is based on cutting off a Taylor series beyond the second term. This means that you are determining the variance but not higher moments. In many processes, distributions are not symmetric (e.g., Ostwald ripening), and for many properties the presence of larger amounts of smaller or larger sizes is quite important. Therefore, it would be useful to have procedures for higher moments.

MARKWORTH: Indeed, the size distributions characterizing second-phase inclusions evolving within a host medium are generally asymmetric. What I have shown here is that, as long as the distributions are relatively "narrow", some quantitative deductions can be made regarding certain of their properties (e.g., the variance) without requiring consideration of higher order properties such as skewness. Of course, the calculations presented here could be extended to descriptions of these higher order properties as well.

PRECIPITATION HARDENING IN CERAMICS

*A. H. Heuer and G. K. Bansal**

Department of Metallurgy and Materials Science
Case Western Reserve University
Cleveland, Ohio

ABSTRACT

The science of "heat treatment", i.e., the control of microstructure by use of subsolidus reactions involving changes of composition or symmetry or both, and attendant improvement of properties, has received less attention in ceramic systems than in metallic systems. (Typically, little microstructural control is available in ceramics *after* they have been sintered.) However, modification of properties through controlled heat treatment has much potential in ceramics; this paper summarizes results of one such study – precipitation hardening in spinel.

*Now with Battelle Memorial Institute, Columbus Laboratories, Columbus, Ohio.

1 INTRODUCTION

Extensive literature exists on the precipitation processes that occur in nonstoichiometric (alumina-rich) magnesium aluminate spinels[1-3], as well as on the effects of these precipitates on mechanical properties[4-7]. When this work began, however, there was no systematic microstructural/kinetic study of the precipitation reactions, nor had there been concurrent correlation with mechanical behavior. The results of our work are described in full elsewhere[8,9]; hence only a brief résumé is presented here.

2 EXPERIMENTAL PROCEDURES

Nonstoichiometric crystals grown by the Verneuil process, with a composition of $\sim MgO{:}3.5\ Al_2O_3$, were used for all studies. Outwater-Jerry double torsion specimens for determining crack velocity/stress intensity curves[10], slender beams for bend-strength determination, and polished {100} faces for microhardness measurements were used, either in the as-grown condition or after suitable aging. For convenience, a single aging temperature – 850°C – was employed, with aging times varying from 5 minutes to 10 days. The precipitation reactions are described first.

3 RESULTS AND DISCUSSION

It has been mentioned that the starting materials were as-grown supersaturated Verneuil crystals. They are free of visible precipitates but contain dislocations (typically 10^6 to 10^7 dislocation lines per cubic centimeter) and low-angle boundaries. Inasmuch as the precipitates usually form heterogeneously on dislocations and low-angle boundaries[3], previous transmission-electron-microscopy studies[3] had been performed on crystals deformed at elevated temperatures prior to aging; this enhanced the precipitation kinetics. For the present study, however, greater control over the microstructural evolution was obtained by allowing the precipitation reactions to proceed more slowly in undeformed crystals at a relatively low aging temperature (850°C). Under these conditions, microstructural changes were apparent after 5 to 15 minutes of annealing. Initially, the featureless background of as-received crystals became mottled (see Fig. 1) and diffraction halos were observed around every matrix diffraction spot in selected area diffraction patterns. These features are interpreted as indicating the presence of quasi-spherical "Guinier-Preston (GP) zones", the halos being the electron-diffraction equivalent of the diffuse X-ray scattering observed by Saalfeld and Jagodzinski in their X-ray studies.[1]

Fig. 1. "Mottled" substructure produced in nonstoichiometric spinel by annealing for 10 minutes at 850 °C. The inset shows a portion of the selected area diffraction pattern; a halo surrounds every diffraction spot.

Fig. 2. Microstructure after aging for 1 hour at 850 °C. The inset shows reflections from both the matrix and the particles present in the main figure.

Further aging (up to 1 hour) led to the coarsening of some of these zones at the expense of others, to give rise to particles roughly 100 Å in diameter (Fig. 2). At this stage of the aging, a new set of reflections appeared in the electron-diffraction patterns, indicating that the microstructure is analogous to that found in metallic systems containing "GP precipitates".

Further annealing led to the formation of metastable, intermediate, semicoherent precipitates. In agreement with Lewis' results[3], two distinct phases were found, both having monoclinic symmetry but different morphologies. In Lewis' study, both phases formed preferentially on dislocations and low-angle boundaries at roughly the same rate. In the present work (in which no attempt was made to increase the dislocation density prior to aging), the accicular type I intermediate phase was detected first after 4 hours of annealing. Since this phase appears first, its nucleation barrier must be smaller than either the type II precipitates (described below) or the alpha-alumina equilibrium phase. As annealing proceeds, these precipitates grow by depleting the matrix, which consists of the GP zones and GP precipitates formed initially. The first resolvable type II phase formed after 15 hours of annealing. It had a lathe-like morphology and different orientation relationships from those for the type I particles. (The orientation relationships for both intermediate phases are shown in Table I.) The alpha-alumina particles were first detected in samples annealed between 37 and 41 hours. Upon further annealing, these particles

Table I. Orientation Relationships for Intermediate Precipitates in Alumina-Rich Nonstoichiometric Spinel

Type I	(001) ppt	‖	{311} matrix
	[010] ppt	‖	<110> matrix
Type II	[010] ppt	‖	<100> matrix
	[001] ppt	‖	<110> matrix

grew at the expense of intermediate phases; these three types of precipitates are all visible in Fig. 3. After 10 days of annealing, some of the type I particles were still present with the coarse alpha-alumina particles (Fig. 4). Annealing at higher temperatures led to the development of alpha-alumina particles much more rapidly; Fig. 5, for example, shows the structure after 25 hours aging at 1050°C.

The Knoop microhardness (with 300-g load) and bend strength as a function of aging time are shown in Fig. 6 while the K_{Ic} (critical-stress-intensity factor for fast fracture) values are given in Table II. The two peaks in Fig. 6 can be correlated with the maximum development of "GP

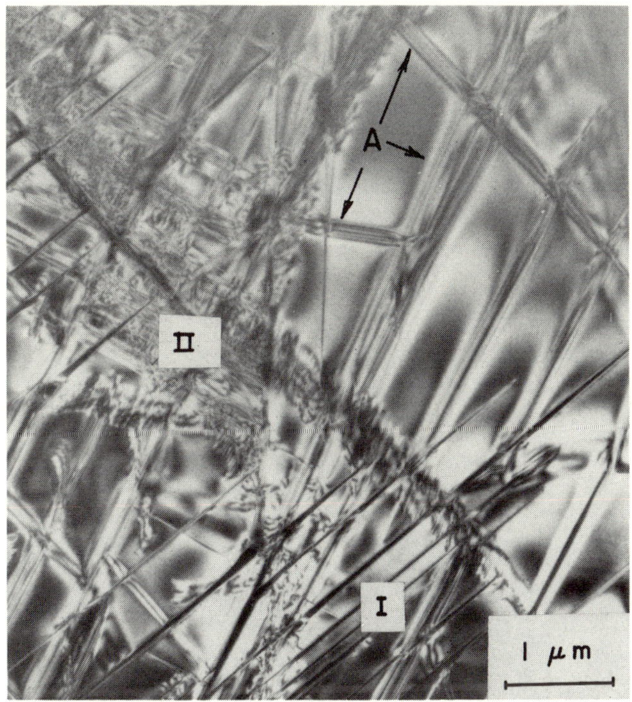

Fig. 3. Types I and II intermediate phases and the equilibrium α-Al_2O_3 phase (A) after aging 81 hours at 850°C.

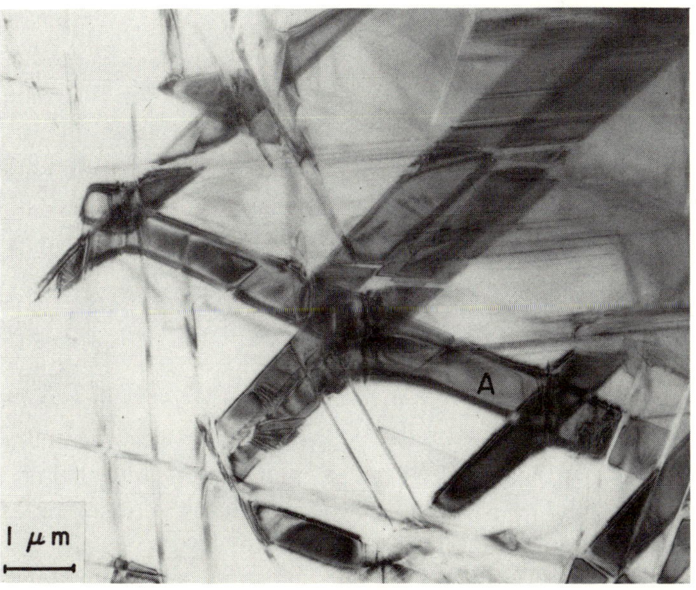

Fig. 4. Coarse platelets of α-Al_2O_3 (A) and some remaining type I particles (sample aged 10 days at 850°C).

zones" and the maximum development of the intermediate phases, respectively. The aging at times beyond the second peak coincides with the initial precipitation of alpha-alumina.

Table II. Critical-Stress-Intensity Factors for Alumina-Rich Nonstoichiometric Spinel

	K_{Ic}, 10^6 $N/m^{3/2}$
As-received supersaturated solid solution	1.5
Aged 15 minutes at 850° C	2.6
Aged 25 hours at 850° C	2.3

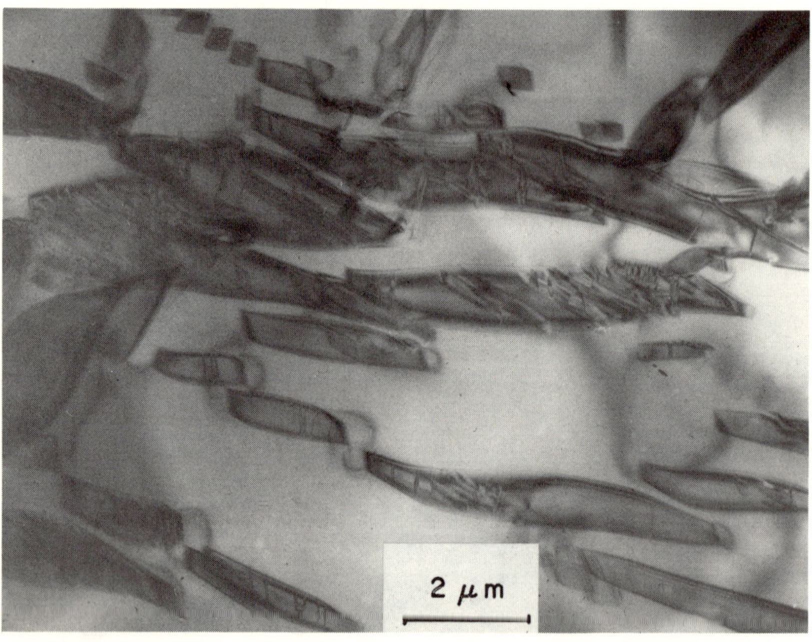

Fig. 5. Random orientations of α-Al_2O_3 plates in sample aged 25 hours at 1050°C.

While the hardness data can be understood by analogy with age-hardenable metallic alloys, i.e., the hardness increase arises from dislocation-precipitate interactions, the same is not true of the other properties. To begin with, fracture in spinel at room temperature is believed to be a purely brittle phenomenon. It is also believed that the GP zones are coherent with the matrix, the intermediate phases are semicoherent, and the equilibrium phase is incoherent. Following a model originally suggested by Lange[11] and further developed by Evans[12], the interaction of these coherent or semicoherent particles with propagating cracks is believed to be responsible for the improved bend strength and higher K_{Ic}. In particular, the best improvement is shown by the samples aged to the stage of GP zones, while aging to the intermediate stage is almost good. It is particularly significant that the particles are sufficiently small at the GP-zone stage so that samples do not lose their transparency.

In summary, precipitation hardening is shown to be an important strengthening mechanism for ceramic materials which exhibit brittle fracture. For such a hardening mechanism to be effective, it is important that the interface between the particle and the matrix be fully or partially coherent. Incoherent particles do not appear to provide any appreciable strengthening.

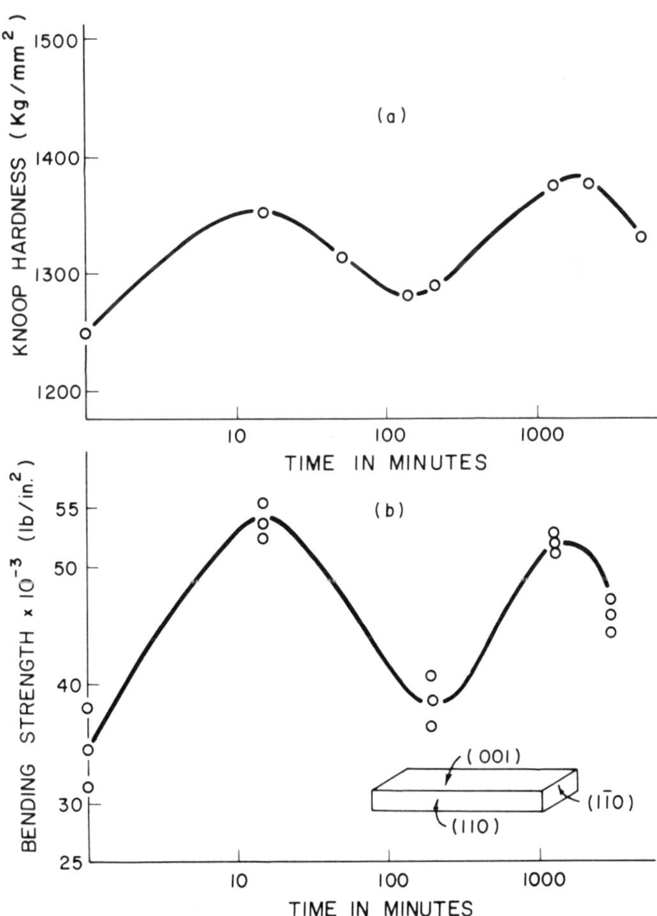

Fig. 6. Variation in Knoop hardness and bend strength with aging time at 850°C.

REFERENCES

1. Von Saalfeld, H., and Jagodzinski, H., *Z. Krist.*, **109**, 87 (1957).
2. Jagodzinski, H., *Z. Krist.*, **109**, 388 (1957).
3. Lewis, M. H., *Phil. Mag.*, **20**, 985 (1969).
4. Eppler, W. F., *Z. Angew. Mineral.*, **4**, 345 (1943).
5. Mangin, A., and Forestier, H., *Compt. Rend.*, **242**, 1893 (1956).
6. Grabmaier, J. G., and Falckenberg, R., *J. Am. Ceram. Soc.*, **52**, 648 (1969).
7. Falckenberg, R., *Mater. Res. Bull.*, **8**, 171 (1973).
8. Bansal, G. K., and Heuer, A. H., *Proceedings of the Conference on the Fracture Mechanics of Ceramics,* R. C. Bradt et al. (Eds.) (1973), p. 677.
9. Bansal, G. K., and Heuer, A. H., to be published in *Phil. Mag.* (1974).
10. Evans, A. G., *J. Mater. Sci.*, **7**, 1137 (1972).
11. Lange, F. F., *Phil. Mag.*, **22**, 983 (1970).
12. Evans, A. G., *Phil. Mag.*, **26**, 1327 (1972).

DISCUSSION on Paper by A. H. Heuer and G. K. Bansal

STUBICAN: This is to support your findings that during the precipitation process in the system $MgAl_2O_4 + Al_2O_3$, two metastable phases may form. One metastable phase with the composition close to the pure Al_2O_3 is well established. However, there is another metastable phase which can be obtained only by "topotaxial reaction" between the single crystal of $MgAl_2O_4$ and Al_2O_3. The composition of this phase was determined as $3Al_2O_3{\cdot}MgO$. It is quite possible that this is the phase which precipitates first in the system $MgAl_2O_4 + Al_2O_3$.

HEUER: We have done most of our work on nonstoichiometric spinels with approximate compositions $3.5Al_2O_3{\cdot}MgO$. Since the stable precipitate is Al_2O_3 containing a very small amount of MgO (~0.1 percent), the intermediate precipitates probably have a composition between $3.5Al_2O_3{\cdot}MgO$ and Al_2O_3, i.e., more aluminous than the phase you suggest. Is the symmetry of the phase produced by this "topotaxial reaction" known?

CLAUER: In your system you have suggested that the increase in K_{Ic} was caused by an interaction with the GP zones. This interaction must be due to some factor such as a strain field or a local increase of interfacial energy associated with the GP zones. In metals, some systems such as Al-Zn alloys have spherical GP zones with no coherency strain field. To test your contention, one could duplicate your experiments in an oxide system which contains GP zones with either a small or no associated strain field. Is it known that such systems exist in oxides?

HEUER: No. Unfortunately, very little is known about GP zones in oxide systems.

COHEN: You mentioned that the GP zones were spherical. It is a small point perhaps, but it is unlikely that such small regions are spherical; they must be faceted to fit into the lattice.

Concerning the study of fracture mechanics, can you tell whether the effect of GP zones and precipitates is different? Is the effect of one or the other on the surface energy or on crack growth known? You cannot tell this from crack velocity vs K_{Ic}.

HEUER: We have done scanning electron microscopy on fracture surfaces and they are consistently very rough – both for the case of GP zones and where we have small particles. In the overaged condition, the fracture surfaces are much smoother. I think the effect is due to interaction between the precipitates and a propagating crack. When the precipitate-matrix interface is coherent or semicoherent, the crack apparently either has to "bow" around the particle or to nucleate a secondary crack. In other words, this analogy is to interactions between dislocations and particles – we imagine the crack to have a pseudo "line tension".

RAPP: Several engineering applications of high-temperature electrochemical processes, particularly the electrochemical deoxidation of induction-stirred liquid metals [K. Oberg, et al., *Metall. Trans.*, **4**, 75 (1973)], are made unfavorable or are not yet feasible because of the extreme sensitivity of the fully or partially stabilized zirconia electrolytes to thermal shock. Your remarks concerning the study and use of thermal treatments to increase the strength, and therefore reduce the thermal-shock sensitivity of ZrO_2-CaO or ZrO_2-Y_2O_3 compositions, hopefully represent the initiation of some long overdue research in this critical area. To help us in our own electrochemical research, can you yet recommend a preliminary heat treatment of commercial zirconia ware which will reduce its sensitivity to thermal shock in a subsequent application at 1100 to 1600°C?

HEUER: Not yet. This is an area in which I have just started working with a new graduate student.

DEVELOPMENT OF SUBSTRUCTURES DURING CREEP AND EFFECT OF IONIC DEFECTS AND PRECIPITATES ON THE STRENGTH OF MAGNESIUM OXIDE

B. Ilschner and B. Reppich

University of Erlangen-Nürnberg
Erlangen, Germany

ABSTRACT

Compression tests with MgO single crystals, doped with Fe^{3+} to different levels, have been performed in various temperature ranges. In addition, microstructural aspects, and in particular the dislocation arrangement, have been studied by both optical and transmission electron microscopy. At high temperature, development of a steady-state creep process and a corresponding substructure is observed and interpreted by diffusion-controlled dislocation reactions similar to those in metals. At low temperatures, solid solution hardening due to point-defect complexes (dipole hardening, induced Snoek effect) has been found. At higher Fe^{3+} levels, spinel-phase precipitation causes particle hardening, the coherent spinel particles being cut by the dislocations in pairs.

1. INTRODUCTION

In ionic crystals such as alkali halides or oxides, point defects control the deformation behavior in two ways: (a) vacancy diffusion processes determine the annealing rate of dislocations generated by processes related to work hardening; (b) elastic interaction between point defects and dislocations determine the flow stress. Obviously, (a) is responsible for high-temperature plasticity, while (b) dominates at low temperatures, where diffusive flow rates are too small to contribute markedly to the deformation process.

Both effects have been studied in the present work, based on earlier investigations into the deformation behavior of $Fe_{1-y}O$[1,2] and of pure and doped alkali halides[3,4]. While pure wustite suffers from the well-known disadvantage of very high degrees of disorder, vacancy concentrations in magnesiowustite may be controlled both by the doping level and by the oxygen partial pressure. In the present investigation, a high Fe^{3+}/Fe^{2+} ratio was aimed for. The formation of Fe^{3+}-vacancy complexes due to attractive interaction of these defects appears to be an essential feature of this material. Models based on the assumption of a random distribution of point defects and dislocations are now considered to be insufficient to interpret the high-temperature mechanical behavior of metals. The same holds for ionic crystals and, therefore, particular attention has been directed toward the nonrandom dislocation arrangement in deformed MgO, characterized by a cell structure.

At low temperatures, deformation is carried by dislocations, representing linear elastic as well as electrostatic discontinuities in the lattice. These properties of dislocation lines cause elastic interactions with point defects, and in particular with defect "molecules" or complexes with asymmetrical stress (or potential) fields. Increasing the temperature may result in a partial dissociation of these complexes, so that measurements of the flow stress as a function of T were expected to yield relevant information (see Section 3.3).

A third mode of influence on mechanical strength of MgO by Fe^{3+} additions is through supersaturation and precipitation. Experiments designed to contribute to the understanding of this problem are dealt with in Section 3.4.

2 HIGH-TEMPERATURE PHENOMENA

2.1 Kinetics of Creep as Based on the Development of Substructures

Two types of elementary processes contribute to deformation at high temperature: (a) hardening, associated with the generation of new dislocation length, and (b) recovery, associated with the annihilation of dislocations. If the strain rate is denoted by $\dot{\epsilon}$ and the initial dislocation density by ρ_o, the dislocation density at any strain ϵ is given by

$$\rho(\epsilon) = \rho_o + \int_0^{\epsilon} (\dot{\rho}_+ - \dot{\rho}_-)\, \dot{\epsilon}^{-1}\, d\epsilon \quad . \tag{1}$$

Here, $\dot{\rho}_+$ and $\dot{\rho}_-$ are the dislocation-density changes due to the two above-mentioned processes. Both of them are controlled by diffusional processes ("climb"), giving additional freedom of motion to the dislocation lines. Further, both of them depend on external and internal stresses in a way discussed elsewhere in the literature.[5] For the present investigation, an essential consequence is that $\rho(\epsilon)$ will be constant if $\dot{\rho}_+ = \dot{\rho}_-$. Since the deformation rate

$$\dot{\epsilon} = \rho b v \quad , \tag{2}$$

$\rho(\epsilon)$ where b is the Burgers vector and v is the dislocation velocity, yields $\dot{\epsilon}$ = const under constant stress σ. This condition is termed steady state. Clearly, a dynamic equilibrium is established in the steady state between generation and annihilation processes for dislocations, so that methods of chemical reaction kinetics may be principally applied to creep kinetics.

The presentation of the problem by Eqs. (1) and (2), based on the notion of a random dislocation density ρ, is, however, an oversimplification. Recent investigations of the high-temperature deformation behavior of metals and nonmetals have shown that

(a) The dislocation arrangement, at least in the steady-state condition, is nonrandom, but characterized by cell walls
(b) These cell walls are not just low-energy "deposits" for inactive dislocations but, rather, essential sites for the annihilation process leading to recovery.

With these observations in mind, one is led to an approach characterized by three features[6]:

(1) Dislocation generation is primarily a "volume reaction" with the effective stress as driving force.
(2) Dislocation annihilation is essentially a "phase-boundary reaction", with the internal stress (due to elastic interaction of dislocations) as driving force.
(3) Cell walls nucleate and grow from random dislocation arrangements and adjust their distances according to a dynamic equilibrium between reactions (1) and (2).

From reactions (2) and (3) it follows that the mean cell-wall area (S_v) per unit volume is essential and should be a function of stress as soon as steady state is obtained: high stress produces high dislocation-generation rates and, thus, increases the average level of ρ within the cell walls. To counterbalance $\dot{\rho}_+$, a high cell-wall surface area is necessary in view of the phase-boundary-type reaction (2).

Experimental evidence for this kinetic equilibrium is demonstrated below.

2.2 Experiments and Results

Magnesium oxide crystals with an impurity level below 100 ppm as well as others doped with up to 1 wt % iron were purchased from a commercial source*. All samples were solution treated and annealed at 1773°K for 1 day in air and then quenched to room temperature with a cooling rate of 50 deg/min. Creep tests were conducted at temperatures between 1673 and 2073°K in argon atmosphere, in a compression apparatus described earlier.[7] The dislocation structure of the deformed crystals was investigated both by optical microscopy after producing etch pits, and by tranmission electron microscopy (TEM), using thin foils and 100 kv. Foil preparation and determination of foil thickness have been described elsewhere.[8,9]

Figure 1 demonstrates the actual situation in a MgO crystal deformed in compression at 1973°K. In view of the high-temperature anneal mentioned above, the initial dislocation density is quite low. Correspondingly, the internal stress level is low, so that the effective stress acting on the dislocations is close to the applied stress, and the initial creep rate is very high. As is shown by Fig. 1b, however, the high deformation rate leads to the generation of a large amount of new dislocation-line length, ρ reaching a peak value in the middle of the transient region. The increased dislocation density increases the internal stress level, reduces the effective stress, and, consequently, causes a decrease in creep rate (Fig. 1a).

At the same time, the increased density of moving dislocations within the crystal leads to more and more dislocation reactions, giving rise to configurations which may be considered as nuclei for cell-wall structures. After some initial strain, a clear cell structure such as in Fig. 2 may be observed. The growing perfection and surface area of these interfaces allows for a higher annihilation rate according to Mode (2) in the preceding section. For this reason, the dislocation density (Fig. 1b) drops again. Eventually, a dynamic equilibrium between dislocation generation within the cells and annihilation at cell walls is established, as is shown by

*Spicer Ltd., England.

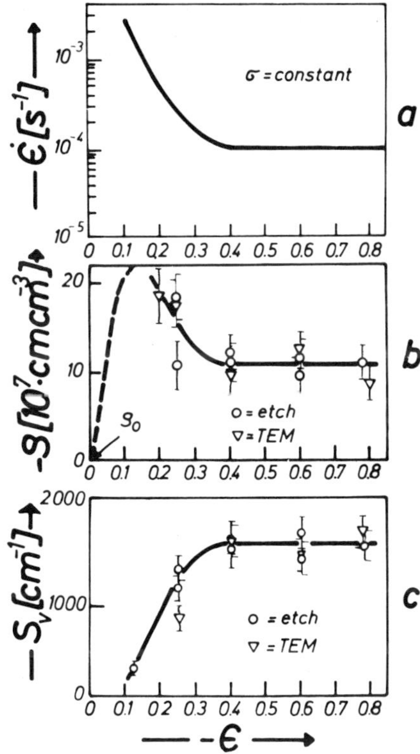

Fig. 1. Development of macroscopic and microscopic creep parameters during high-temperature creep of MgO single crystals, T = 1973°K, σ = 60.8 MN m⁻²: (a) creep rate ($\dot{\epsilon}$), (b) density (ρ) of dislocations not associated with cell walls, (c) cell wall area per volume (S_V).

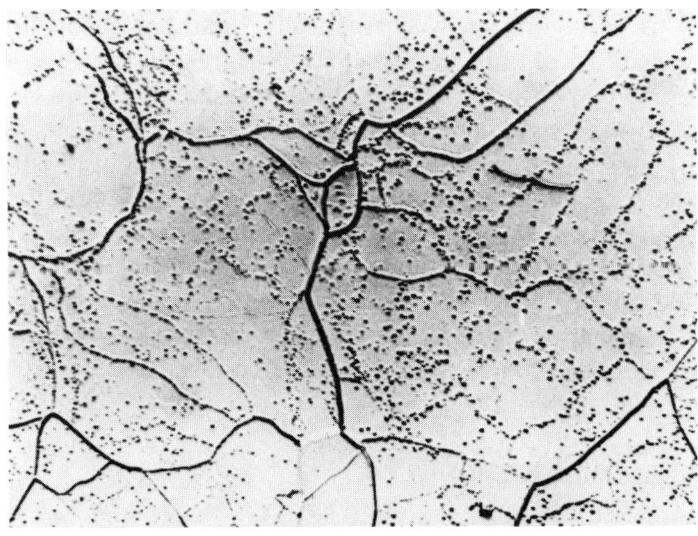

Fig. 2. Dislocation arrangement in {100} plane with cell structure in MgO single crystals crept at 1973°K in compression (steady-state condition).

the data points on the right-hand side of Fig. 1. It may be mentioned in passing that the creep rate $\dot{\epsilon}$ in these experiments was strongly dependent on stress. If one choses a power-law representation,

$$\dot{\epsilon} \sim \sigma^n \quad , \quad 4 < n < 100 \quad , \tag{3}$$

the high values – actually indicating exponential functions – apply for higher stresses.

The dependence of dislocation density and cell-wall-area size on stress has also been measured, Figs. 3a and 3b. Since $\rho \sim \sigma^{1..2}$ and $S_v \sim \sigma^{0.5}$, it follows that $S_v \sim \rho^{0.5}$ (Fig. 3c). This result is essential from the viewpoint of reaction kinetics, in that it reflects the balance between bulk reactions (ρ) and phase-boundary reactions (S_v), both of them controlled by the stress as a driving force.

2.3 Discussion

Lindroos[10] has discussed in considerable detail cell-wall formation in metals like aluminum. Here an example is given of this process in oxide crystals. Figure 4 shows a TEM photo of one cell-wall section in different stages of perfection. First, primary dislocations arrange themselves so as to form a simple, loosely connected structure (Fig. 4a). Dislocations (D) from another slip system become active after some strain hardening in the first one; they are not able to cut or cross the layer (a), so they form a cross grid (b). This configuration is energetically disadvantageous; free energy may be gained by rearrangements to configuration (c), making use of favorable Burgers vectors. However, "conversion" of (b) to (c) requires diffusion-controlled steps and, thus, some time or equivalent strain; note that the driving force for this rearrangement within the wall is low. If the load or stress is changed during an experiment, the distribution of cell sizes will be adjusted to the new equilibrium between (bulk) generation and (interface) annihilation processes. This adjustment occurs by knitting processes like in Fig. 4, or by cell-wall dissolution. The latter process is a reversal of cell-wall formation and involves opening of meshes and emission of individual dislocations into the adjacent lattice.

3 LOW-TEMPERATURE PHENOMENA

3.1 Point-Defect Complexes

An ionic crystal formed from divalent cations like MgO and containing trivalent impurity ions like Fe^{3+} presents a problem due to electro-

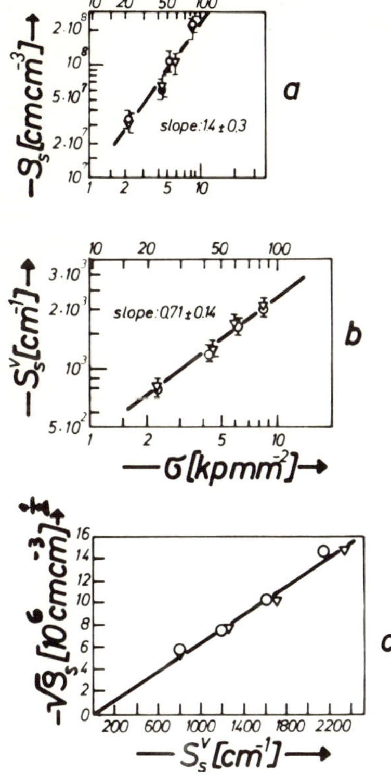

Fig. 3. Stress dependence of macroscopic and structural-creep parameters in the steady state as an indication for dynamic equilibrium (the index "s" stands for steady state): (a) dislocation density (ρ_s) not associated with cell walls, (b) cell wall area per volume (S_V^s), (c) relationship between ρ_s and S_V^s; T = 1973°K.

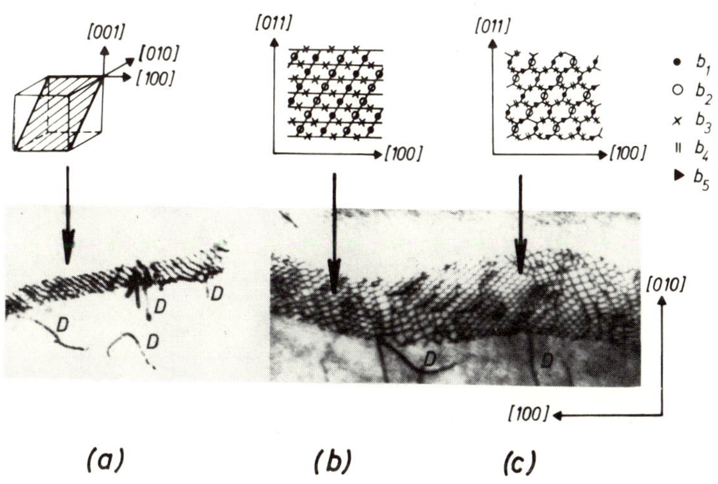

Fig. 4. Microstructural details of dislocation-network formation in the dislocation boundaries (see text); deformation conditions: T = 1973°K, σ = 83.3 MN m^{-2}, ε = 0.6.

static interaction. The substitution of the trivalent impurity requires formation of cation vacancies, which in turn represent a local deficiency of two positive elementary charges – equivalent to a double negative charge. The impurity, on the other hand, acts like a single positive charge in the otherwise neutral lattice. These charges will be subject to electrostatic interaction.

A homogeneous distribution of impurity ions and vacancies will therefore be unstable from an energetical point of view, though dissipation by entropy influences will counterbalance this association tendency at high temperatures.

An analysis of possible crystallographic configurations of such associated complexes in the MgO structure reveals a variety, schematically demonstrated in Fig. 5.[11] Elastic and electrostatic relaxation effects are not shown. A first-order approximation, taking into account only the sum of Coulomb interactions of the unrelaxed point defects, gives a sequence of energetic stability. Type Ia (Fig. 5) turns out to be most favorable from a Coulomb point of view, while IIa is very close to this energy level. If relaxation effects were taken into account, the sequence Ia – IIa might reverse.

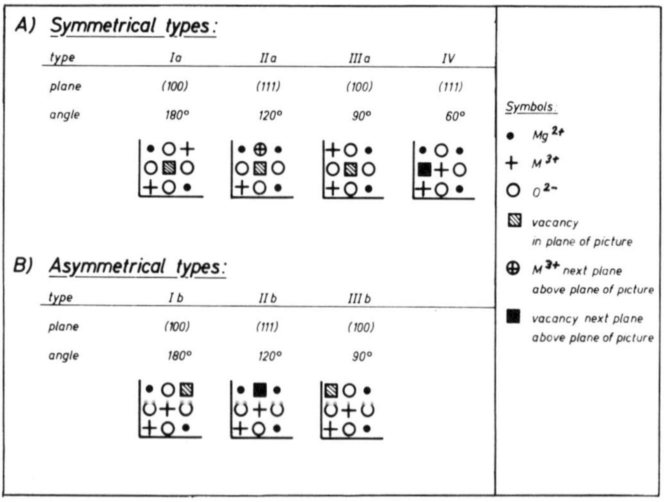

A) Symmetrical types:

type	Ia	IIa	IIIa	IV
plane	(100)	(111)	(100)	(111)
angle	180°	120°	90°	60°

B) Asymmetrical types:

type	Ib	IIb	IIIb
plane	(100)	(111)	(100)
angle	180°	120°	90°

Symbols:
● Mg^{2+}
+ M^{3+}
O O^{2-}
▧ vacancy in plane of picture
⊕ M^{3+} next plane above plane of picture
■ vacancy next plane above plane of picture

Fig. 5. Complex configurations $[M^{3+}VM^{3+}]$ in the MgO lattice; distance unit is $(a/2)\sqrt{2}$ in <110> direction.

These considerations are backed up by experimental evidence well known from the literature.[12,13] They form the basis of the experimental results reported below. Complexes with nonspherical symmetries ("dipoles") have a clear tendency for strong elastic interaction with dislocation lines, which have also a linear dimension. This holds for a parallel configuration of the dipole axis and the dislocation-line vector. If the temperature is not too low and the mean dislocation velocity not too high, the dipole axis may adjust to the dislocation line, by thermally activated interchange of the complex configurations in Fig. 5. This will be termed "induced Snoek effect" (see below). Of course, high temperature will lead to dissociation of these complexes and leave only isolated point defects with a weaker interaction tendency.

3.2 Experiments

Crystals such as described in Section 2.2 have been used. The concentration of [Fe^{3+} V Fe^{3+}] complexes was estimated using a method by Brynestad and Flood.[14] As an example: MgO doped with 2300 wt ppm iron was estimated to contain approximately 310 mole ppm of (Fe^2V) complexes. Stress-strain curves of <100>-oriented single crystals with cleaved {100} faces were measured in compression at temperatures between room temperature and 1873 K. Testing was performed in an Instron machine equipped with a high-temperature compression jig[15], Al_2O_3 rams, and sapphire-crystal spacers. The deformation rate was $3 \cdot 10^{-4}$ sec^{-1}.

3.3 Solid-Solution Hardening

The critical resolved shear stress (τ_o) of pure as well as Fe-doped MgO single crystals between room temperature and 1873 K for two deformation rates ($\dot{\epsilon}$) is illustrated in Fig. 6. Three distinct ranges of the temperature dependence of τ_o are observed:

Region I: In the low-temperature region T < 770 °K, τ_o decreases markedly with increasing temperature. Further, τ_o is dependent on strain rate for pure as well as doped material. At room temperature, a concentration dependence is found of the form[16] $\tau_o \propto C_D{}^{1/2}$, where C_D is the dipole concentration.

Region II: In this "plateau", T > 770°K, τ_o is independent of T and $\dot{\epsilon}$ for pure MgO for all temperatures investigated experimentally (1873°K), and for MgO doped with 2300 wt ppm Fe up to 1570°K. The concentration dependence can be described by[16] $\tau_o \sim C_D$.

Region III: For Fe-doped material, a high-temperature decrease of τ_o is observed for T > 1570 °K. This region has not yet been studied in detail with regard to the influence of $\dot{\epsilon}$ and iron concentration.

Our interpretation of these observations rests on the assumption, well accepted in the literature[17-20], that the solid-solution hardening in Fe-doped MgO is attributed to the ferric and not to the ferrous ion. On the other hand, it is not yet clear in what form the Fe^{3+} acts as a dislocation obstacle. In previous studies, electrostatic[18] and elastic interaction[21] of dislocations with nonassociated Fe^{3+} ions as well as elastic interaction with Fe^{3+} charge-compensating vacancy dipoles[17,18,20] (Fleischer interaction) have been discussed.

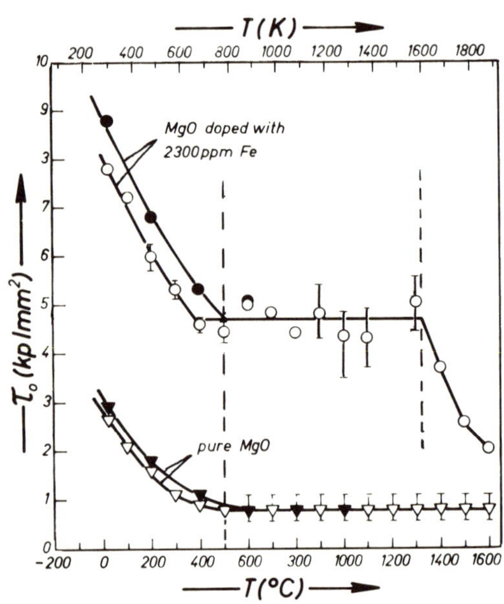

Fig. 6. Experimental dependence of the critical resolved shear stress (τ_0) of pure and 2300 wt ppm iron-doped MgO single crystals; ▽○, $\dot{\epsilon} = 3 \cdot 10^{-4} s^{-1}$; ▼●, $\dot{\epsilon} = 6 \cdot 10^{-3} s^{-1}$.

For monovalent ionic crystals, for example NaCl and LiF doped with divalent cations, the solid-solution hardening is quantitatively in agreement with the model of "tetragonal" hardening.[4,22] W. Frank[23,24] has developed a theory extending the ideas of Fleischer[25,26] and of Schoeck and Seeger[27].

Accordingly, the following regions/mechanisms can be distinguished:

I. The low-temperature decrease of τ_0 is determined by thermally activated passing of dislocations over obstacles set up by the tetragonal stress fields of elastic impurity cation-vacancy dipoles close to the slip plane. The critical resolved shear stress is obtained as[23] (the index "F" means Fleischer interaction):

$$\tau_0 = \tau_G + \tau_F \tag{4a}$$

$$\tau_F = A \frac{G}{d} \Delta\epsilon^* \sqrt{C_D} \, (1 - \sqrt{\Phi T})^2 \quad . \tag{4b}$$

τ_G is the contribution to τ_o due to long-range dislocation-dislocation interaction, G is the shear modulus, and d is of the order of 0.5 b, where b is the Burgers vector. $\Delta\epsilon^*$ is the "dipole strength" and characterizes the tetragonal stress field of the dipoles. The function Φ contains the $\dot{\epsilon}$ dependence and A is a numerical factor.

II. In the temperature-independent "plateau" at higher temperatures, τ_o is controlled by the induced Snoek effect of dipoles; in the stress field of moving dislocations, τ_o is[24]

$$\tau_o = \tau_G + \tau_S \tag{5a}$$

$$\tau_S = 1.48\ G\ \Delta\epsilon^*\ C_D \quad . \tag{5b}$$

The index "S" stands for Snoek interaction. Comparison with the experimental data as presented above gives satisfactory agreement with Frank's theory.[16] The concentration dependence at room temperature follows Eq. (4). The predicted $\dot{\epsilon}$ dependence is experimentally observed. A plot of $\sqrt{\tau_o - \tau_G}$ versus $\sqrt{T}$ according to Eq. (4b) for $T \leqslant 773°K$ is strictly linear. In the temperature-independent plateau, the linear concentration dependence according to Eq. (5) is found. A more detailed evaluation[16] of measured T, $\dot{\epsilon}$, and concentration dependencies for both regions gives a value for the dipole strength $\Delta\epsilon^*$ similar to those for alkali halides.[4,22] Therefore, it may be concluded that solid solution hardening in Fe-doped MgO at temperatures up to 1570°K is attributed to elastic dipoles consisting of symmetric (Fe^{3+} V Fe^{3+}) complexes.

The measured decrease of τ_o for MgO doped with 2300 ppm Fe above 1570°K (Fig. 6) may be explained by elastic and electrostatic interaction of nonassociated Fe^{3+} ions which are formed by thermal dissociation of dipoles.[4,18,21]

3.4 Precipitation Hardening

Magnesioferrite particles precipitate after isothermal aging as regular, coherent, stress-free octahedrons with spinel lattice (Fig. 7). The particle size distribution in the matrix up to the longest aging times (856 hours at 1070°K) is well described by a symmetrical distribution curve. At all particle sizes resolved by the electron microscope ($a > 50$ Å), the volume fraction of the precipitated phase has a constant value of $f = 0.006 \pm 0.0015$.

The dependence of room-temperature critical resolved shear stress (τ_o) on particle size is shown by Fig. 8. Starting from the value for "pure" MgO as a reference value, τ_0 reaches a broad maximum between 50 and 100 Å, followed by a marked decrease for a > 100 Å.

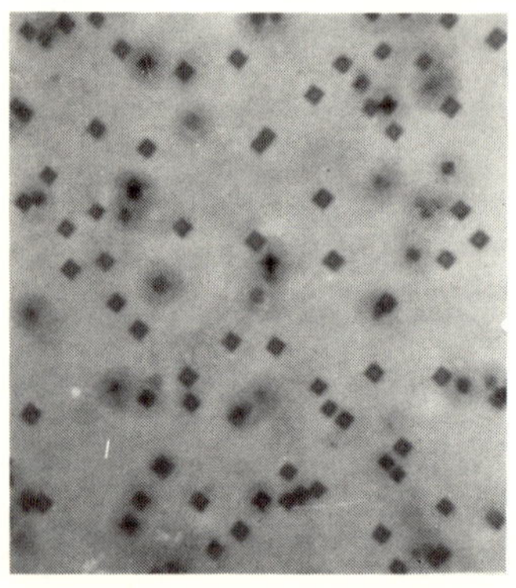

Fig. 7. Octahedral shaped magnesioferrite precipitates from supersaturated iron-doped MgO (TEM, 40,000X).

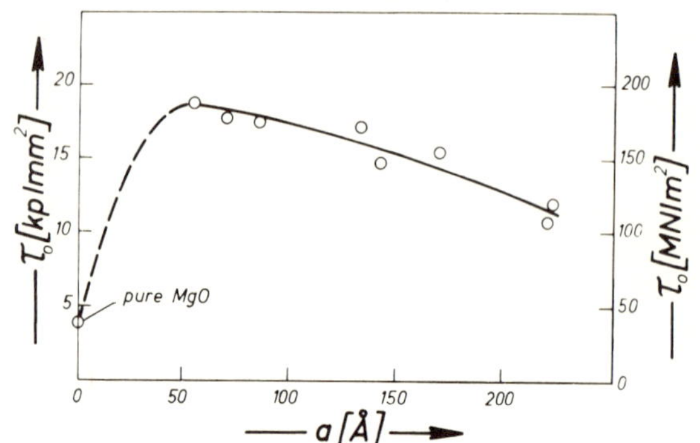

Fig. 8. Effect of particle size on the room temperature shear stress τ_0 in MgO-spinel system.

Particle strengthening is caused by interaction between particles and dislocations. In principle, two kinds of interactions are possible:

(a) Orowan bypassing is expected for "hard" particles, e.g., large coherent particles with strong coherency strain field, semicoherent, and incoherent particles. The Orowan stress is proportional, $1/a$, and could explain the observed decrease of τ_0 for a > 100 Å. On the other hand, a theoretical estimate of the Orowan stress leads to values twice as large as those measured experimentally.[28]

(b) Cutting or shearing is expected for coherent particles without or with small coherency strain field. The cutting stress is proportional, $\sqrt{a}$, and cannot explain the maximum and the decrease of τ_0 in Fig. 8.

Fine[28] and Wicks and Lewis[20] have given an interesting interpretation of the precipitation-dislocation interaction in MgO. According to these authors, strengthening is attributed to stacking faults in spinel particles when a matrix dislocation cuts particles. However, a rough calculation based on this idea[28] leads to a size-independent τ_0, in contrast to the observation. In a later study, Kruse and Fine[29] explain the τ_0 decrease by chemical hardening, that is, creation of particle matrix surface during cutting. A precise calculation[9], however, shows that the decrease of τ_0 for $a > 150$ Å cannot be understood by this mechanism alone.

Our interpretation is based on the model proposed by Gleiter and Hornbogen[30] for metallic systems with coherent, stress-free, ordered particles. These authors point out that in such systems, particles are cut by coupled pairs of dislocations. The leading dislocation of the pair creates disorder in the form of an antiphase domain boundary (APB) on the slip plane within the particle. The second trailing dislocation removes the APB created by the first one. The APB thus causes the pair coupling, which increases with increasing particle size. Consequently, in large particles, both dislocations of the pair will lie in the same particle. This reduces the required cutting stress considerably, so that Orowan bowing does not take place. In this way, particles of arbitrary size can be sheared, and τ_0 decreases with increasing particle size.

Our TEM observations suggest the application of this concept to the present system. As illustrated in Fig. 9, the magnesioferrite particles are in fact cut by dislocation pairs. The dislocation configuration indicated by arrows in Fig. 9 agrees precisely with that calculated by Gleiter and Hornbogen for the case of strong pair coupling. Further, fitting of the experimental data of Fig. 8, based on this model and, in addition, chemical hardening, gives a value for the APB energy on {110} planes of 855 erg/cm^2 and a particle-matrix surface energy of the same order of magnitude.[9]

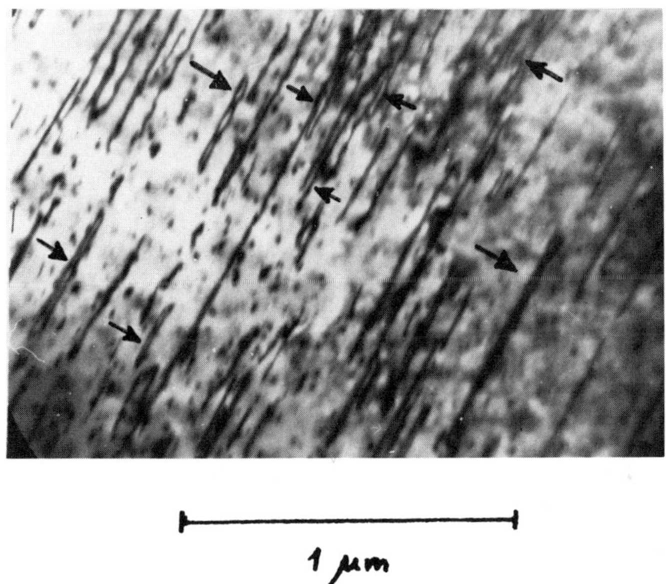

Fig. 9. Cutting of spinel precipitates in MgO by dislocation pairs (TEM); particle size a = 150 Å.

REFERENCES

1. Ilschner, B., Reppich, B., and Riecke, E., *Disc. Faraday Soc.*, **38**, 243 (1964).
2. Reppich, B., *Phys. Status Solidi,* **20**, 69 (1967).
3. Reppich, B., and Streb, G., *Phys. Status Solidi,* (a) **15**, 77 (1973); **16**, 493 (1973).
4. Reppich, B., *Acta Met.*, **20**, 557 (1972).
5. Ilschner, B., *Hochtemperatur-Plastizität*, Springer, Heidelberg (1973).
6. Ilschner, B., *Constitutive Equations in Deformation*, A. S. Argon (Ed.), to be published.
7. Hübner, W., Doctoral Thesis, University Erlangen-Nürnberg, 1971.
8. Hüther, W., and Reppich, B., *Phil. Mag.*, **28**, 386 (1973).
9. Knoch, H., and Reppich, B., in preparation.
10. Lindroos, V. K., and Miek-Oja, H. M., *Phil. Mag.*, **16**, 593 (1967); **17**, 119 (1968).
11. Ilschner, B., "Festkörperprobleme X", *Advances in Solid State Physics,* O. Madelung (Ed.), Pergamon, Vieweg (1970), p. 415.
12. Wertz, J. F., Orton, J. W., and Auzins, P., *J. Appl. Phys.*, Supplement to **33**, 322 (1962).
13. Schawlow, A. L., *J. Appl. Phys.*, Supplement to **33**, 395 (1962).
14. Brynestad, J., and Flood, H., Z. Elektrochem., **62**, 953 (1958).
15. Reppich, B., Blum, W., and Ilschner, B., *Ber. Deut. Keram. Ges.*, **44**, 41 (1967).
16. Reppich, B., in preparation.
17. Davidge, R. W., *J. Mater. Sci.*, **2**, 339 (1967).
18. Moon, R. L., and Pratt, P. L., *Proc. Brit. Ceram. Soc.,* No. 15, 203 (1970).
19. Srinivasan, M., and Stoebe, T. G., *J. Appl. Phys.*, **41** (9), 3726 (1970).
20. Wicks, B. J., and Lewis, M. H., *Phys. Status Solidi,* (a) **6**, 281 (1970).
21. Miles, G. D., Clarke, F.J.P., Henderson, B., and King, R. D., *Proc. Brit. Ceram. Soc.*, **6**, 325 (1966).
22. Frank, W., *Phys. Status Solidi,* **29**, 391 (1968).
23. Frank, W., *Z. Naturf.*, **22a**, 365 (1966).
24. Frank, W., *Z. Naturf.*, **22a**, 377 (1966).
25. Fleischer, R. L., *Acta Met.*, **10**, 835 (1962).
26. Fleischer, R. L., *J. Appl. Phys.*, **33**, 3504 (1962).
27. Schoeck, G., and Seeger, A., *Acta Met.*, **7**, 469 (1959).
28. Fine, M. E., *Trans. JIM,* **9** (Supplement), 527 (1968).
29. Kruse, E. W., III, and Fine, M. E., *J. Am. Ceram. Soc.*, **55**, 32 (1972).
30. Gleiter, H., and Hornbogen, E., *Phys. Status Solidi,* **12**, 235, 251 (1965).

DISCUSSION on Paper by B. Ilschner and B. Reppich

HEUER: I think it can be somewhat dangerous to argue for a single trend on the yield stress or creep rate as a function of nonstoichiometry. For example, in nonstoichiometric alumina-rich Mg-Al spinel, there is a strong lowering of the yield stress as the nonstoichiometry increases. A result, I think, of the change of stacking-fault energy with composition. In UO_{2+x}, a similar effect is found, but here this is understandable in terms of the defect equilibria, following Lidiard's analysis. Finally, in iron-doped MgO, a quite different trend may be observed, because of solid-solution hardening due to Fe^{3+}.

ILSCHNER: In case I have given the impression that I claimed generality for all oxide materials concerning our observations on solid solution hardening, I have certainly been misunderstood. For the NaCl structure, however, disorder as well as dislocation of associated defects with dislocations should be anticipated, and should outweigh other effects.

HUGGINS: When considering the mechanical behavior of oxides at lower temperatures, can one say anything about the relative influence of the elastic interaction between dislocations and lattice defects upon the number of mobile dislocations, on the one hand, and dislocation mobility, on the other?

ILSCHNER: I think that both are influenced by the same tendency, but to a different degree. The dislocation mobility is certainly affected to a high degree, since V depends on a high power of the effective stress. Further, the rate of generation of new mobile dislocation loops will be reduced for a given applied stress level, thus lowering their number.

WERTZ: How were the vacancy concentrations derived for your plots? If these were taken as half the concentration of Fe^{3+} ions, I would have regarded the 100 ppm Fe^{3+} concentrations in your "undoped" crystals as already heavily doped (and therefore one would expect high concentrations of doubly associated vacancies, i.e., two ions with one vacancy).

ILSCHNER: The vacancy concentration was taken to be half the ferric-ion concentration, and at high temperatures this is a fairly good approximation. I agree, though, that this procedure should not be taken too literally.

CLAUER: We have also studied the formation of the dislocation-creep substructure in MgO single crystals but used tension at 1400°C rather than compression. Our results are in very good agreement with your observations up to 0.10 creep strain, at which point the tensile creep specimens usually fractured. That is, our results showed increasing dislocation density with increasing creep strain and no subboundary formation up to 0.10 strain. However, when we prestrained the specimens at 1800°C to 0.10 tensile strain to introduce subboundaries, we found that this subgrain structure largely dissolved during subsequent creep at 1400°C. This is probably because in our studies the primary slip systems at 1400°C had orthogonal Burgers vectors and so were not effectively reacting with each other. Since reacting dislocations must be necessary to build up a "stable" subgrain boundary, the subboundaries slowly evaporate at 1400°C. The dislocations having oblique Burgers vectors, which presumably stabilize the subboundaries at 1800°C, slowly left the crystal by recovery processes at 1400°C and were not replaced.

ILSCHNER: In our investigations, which were performed in the temperature range 1300 to 1700°C, we could not observe any significant change with respect to the dislocation reactions involved in the buildup of subgrain boundaries. Therefore, we rather suppose that the dissolution of subboundaries, which you describe in your comment, is due to the transition between two different substructures. Probably, at 1400°C a stress different from that at 1800°C was applied, so that the substructure had to adjust to the new equilibrium cell configuration; this requires a certain amount of strain.

WELCH: What is your opinion about the relative importance of elastic and electrostatic interactions between point defects and dislocations in materials such as MgO? Does such a distinction between these sources of interaction energy make any difference in the interpretation of your results?

ILSCHNER: Well, I cannot give you a clear answer as to which case dominates here. It depends, of course, on the ionic size, and what we did was estimate the elastic interaction. With the elastic interaction alone we could arrive at an interaction constant consistent with the experimental data. I would point out that there are some sizable sources of error in this approach, but one can interpret the data with the elastic interaction alone within a factor of two or so.

COHEN: Elastic interactions of defects and dislocations at low temperatures and the role of vacancies at high temperatures in determining mechanical properties should be highly dependent on the state of aggregation of defects. Have you attempted to interpret the anomalous results for Fe_xO in terms of the clusters now known to exist in this material?

ILSCHNER: No, but it should be done.

KOFSTAD: The deformation rate of metal-deficient oxides increases with increasing cation-vacancy concentration. This behavior is unexpected if one assumes that the deformation rate is controlled by the diffusion of oxygen defects and that the oxygen defects are oxygen vacancies. The results could be explained, however, if one assumes that the oxygen defects are of oxygen-interstitial type. Oxygen self-diffusion in MgO has been shown to increase with increasing partial pressure of oxygen. This may also indicate that the predominant anion defects are oxygen interstitials.

ILSCHNER: There are other possibilities besides oxygen interstitials. For example, I feel that in a material containing impurities the jog density on edge dislocations will increase. This provides more sites for vacancy generation and absorption and may increase the climb mobility of dislocations with increasing impurity content.

CLAUER: We have carried out creep studies in CoO as a function of p_{O_2} and have observed that at constant temperature and stress, the creep rate increases with increasing p_{O_2}. For p_{O_2} approaching 1 atm, the creep rate increases approximately as $p_{O_2}{}^{1/2}$. As mentioned by Kofstad, if only simple, unassociated point defects are considered, then this result is consistent with the diffusion of oxygen interstitials in this regime.

O'KEEFFE: Further to Dr. Kofstad's comment, I would like to emphasize that in MgO as in other rock-salt-structured oxides (such as NiO), the evidence, such as it is, mainly shows oxygen diffusion coefficients increasing with oxygen activity and hence with increasing cation-vacancy concentrations and cation diffusivities. The most attractive conclusion is that oxygen diffusion occurs via interstitials. There is, after all, abundant evidence that sulfur diffusion occurs by an interstitial mechanism in close-packed sulfides, so one should not be too alarmed at the presence of oxygen interstitials in close-packed oxides.

ILSCHNER: I do not want to exclude oxygen diffusion via interstitials as a solution to the question of partial-pressure dependence, which is unexpected on the basis of an oxygen vacancy philosophy. Life would be so much easier, if conclusive evidence for the oxygen-interstitial diffusion could be presented by someone. Of course, in this case, strong interaction of these interstitials with dislocation lines in the sense of solid-solution hardening is expected at low temperatures. The effect is working in the right sense, compared with experiment; still, the number of oxygen interstitials may be quite small compared with the number of (associated) cation vacancies, so that the total effect is controlled mainly by the latter.

THE INFLUENCE OF STOICHIOMETRIC DEFECTS ON THE CREEP OF OXIDES WITH THE FLUORITE STRUCTURE

M. S. Seltzer, B. A. Wilcox, and A. H. Clauer

Battelle
Columbus Laboratories
Columbus, Ohio

ABSTRACT

Studies have been conducted at this laboratory and elsewhere to establish the effects of varying the oxygen/metal ratio on the steady-state creep behavior of oxides that crystallize in the fluorite-cubic structure, including UO_2, PuO_2, CeO_2, and mixed oxides of these compounds. Creep activation energies have been found to vary with stoichiometry, the maximum values being measured at or near the stoichiometric composition. The observed changes in activation energy with O/M level occur both in the power-law stress regime and when a diffusional creep mechanism is operative. The absolute values for the creep activation energies differ for these two cases however. The manner in which creep rates vary with O/M ratio suggests that the creep process is controlled by diffusion of cations via cation vacancies for hyperstoichiometric compositions and by diffusion of cation interstitials for hypostoichiometric specimens.

1 INTRODUCTION

The impetus for the extensive studies of high-temperature creep behavior of fluorite-structure oxides came from the need for accurate creep data for use in various computer codes that predict the swelling characteristics of nuclear fuels. Since UO_2 and mixed oxides of UO_2 and PuO_2 have homogeneity ranges encompassing tens of atomic percent at temperatures above 1000°C, it is particularly important to ascertain the influence of deviations from stoichiometry on the creep strength of these compounds. Among the fluorite-structure oxides whose creep properties have been investigated in recent years are the nuclear fuels UO_2, PuO_2, and ThO_2, and mixed oxides of these compounds, as well as CeO_2, which is often used as a prototype for the highly toxic PuO_2. In this paper the salient results of the studies performed to date are summarized, with emphasis on experimental data that provide a better understanding of the influence of deviations from the stoichiometric compositions on steady-state creep rates, $\dot{\epsilon}$.

2 EXPERIMENTAL RESULTS

UO_{2+x}

At temperatures below 1600°C, uranium dioxide exists only as a hyperstoichiometric compound, UO_{2+x}. Experiments on this oxide have therefore been limited to oxygen-excess compositions in the region of $UO_{2.000}$ to $UO_{2.16}$. Early bend-creep studies were performed by Scott, Hall, and Williams[1] and by Armstrong and Irvine[2], while more recent work has been conducted by Seltzer, Clauer, and Wilcox[3,4]. Armstrong and Irvine carried out bending creep experiments on polycrystalline material having an average grain size of 6 μm. Experiments were conducted in the viscous creep region, and creep rates were found to increase linearly with oxygen excess X (see Fig. 1). For values of $0.02 < X < 0.08$, the creep activation energy, Q_c, was found to be 56 ± 5 kcal/mole, independent of X, while for X = 0.16, Q_c was 63 ± 8 kcal/mole. Scott et al.[1] reported Q_c values of 65, 72, and $\geqslant$ 95 kcal/mole for creep of $UO_{2.16}$, $UO_{2.06}$, and $UO_{2.00}$, respectively.

Seltzer, Clauer, and Wilcox performed an extensive series of experiments on single crystals[3] and on polycrystalline specimens[4]. Constant-stress compression tests were performed under flowing hydrogen or CO/CO_2 gas mixtures to control the specimen stoichiometry during the experiments. The oxygen partial pressures required to maintain desired O/U levels at various test temperatures were determined from the pressure-

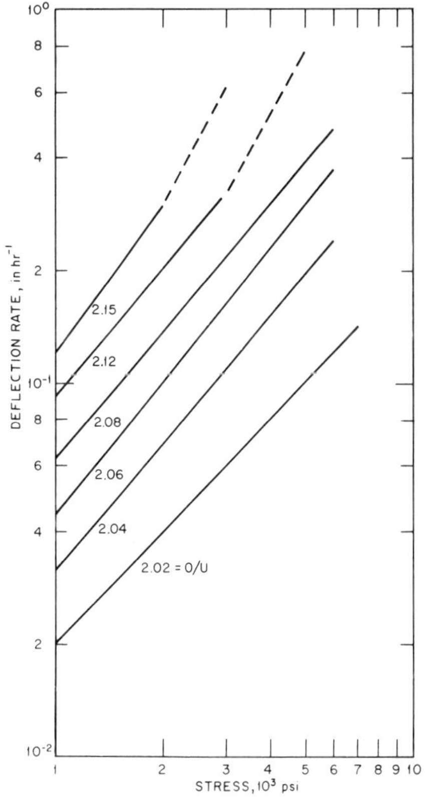

Fig. 1. Deflection rate as a function of stress for polycrystalline UO_{2+x} tested in bending at 1300°C.[2]

temperature-composition relationships derived by Perron.[5] The tests in hydrogen were designed to yield "stoichiometric" UO_2.

Results obtained for UO_{2+x} single crystals, tested as a function of stress in the composition range $0.0001 < X < 0.121$ at 1100°C, are presented in Fig. 2a. The creep rates are seen to increase strongly with the applied stress. Assuming the data can be fit by a power-law stress dependence of the form $\dot{\epsilon} \propto \sigma^n$, where σ is the applied stress and n is the stress exponent, then $n \approx 17$ for O/U levels $\geqslant 2.001$, while $n = 13$ for an O/U of 2.001 and $n = 7$ for stoichiometric UO_2. The results for tests conducted on specimens with oxygen contents greater than $UO_{2.001}$ can best be fit by assuming that the creep rates are proportional to the square of the oxygen excess, $\dot{\epsilon} \propto X^2$. At lower O/U ratios, the stoichiometry dependence cannot be so simply described because the stress dependence varies with composition.

Seltzer et al.[3] obtained more limited data at 1300°C and 1400°C for the stress dependence of nonstoichiometric single crystals (see Fig. 2b.) At high stresses the creep rates can be fit by a power-law dependence, with n increasing from 7 to 10 with increasing O/U ratio, but at low stresses, the creep rates for these single crystals decrease linearly with decreasing stress.

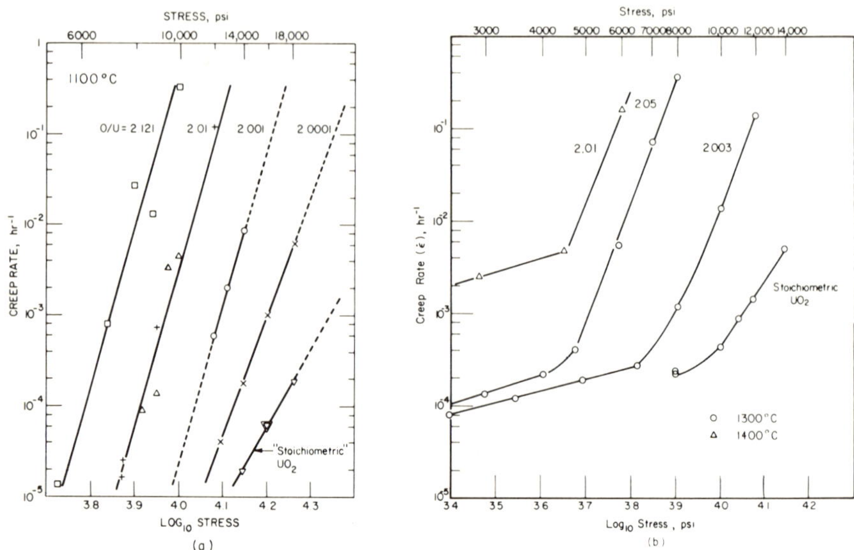

Fig. 2. Creep rate as a function of stress for UO_{2+x} single crystals tested in compression at 1100°C, 1300°C, and 1400°C.

In addition, the creep rates are nearly independent of composition in the linear, low-stress regime, in contrast to the power-law stress region where $\dot{\epsilon} \propto X^2$.

The temperature dependence for creep of UO_{2+x} single crystals was determined by Seltzer et al.[3] in the temperature range 1100 to 1400°C. Results obtained under an applied stress of 10,000 psi, where creep rates were fit by a power-law stress dependence are presented in Fig. 3, while data obtained at a stress of 3000 psi in the linear stress regime are plotted in Fig. 4. Examination of Fig. 3 reveals that creep rates increase with increasing O/U ratio and, furthermore, that the creep activation energies, Q_c, decrease with increasing oxygen excess, from 134 kcal/mole for $UO_{2.0001}$ to 57.5 kcal/mole for $UO_{2.062}$. These activation energies are plotted in Fig. 5 as a function of $\log_{10} X$.

The results included in Fig. 4, obtained at a stress where creep rates increase linearly with applied stress, are for tests performed in hydrogen to give stoichiometric UO_2 and in CO/CO_2 to give an O/U ratio of 2.01. The creep activation energies for these compositions were found to be 129.4 and 131.5 kcal/mole, respectively, which suggests that Q_c is independent of O/U ratio. In addition, the creep rates presented in Fig. 4 are seen to be essentially independent of composition, in marked contrast to those for creep at 10,000 psi where creep rates are several orders of magnitude greater for $UO_{2.01}$ as compared with those measured for stoichiometric UO_2.

Temperature, (C)
1400
1300
1200
1150
1100
2.062
2.01
2.003
2.001
10,000 psi
"Stoichiometric" UO2
Creep Rate (ε̇), hr⁻¹
0.60
0.62
0.64
0.66
0.68
0.70
0.72
1000/T, °(K)⁻¹

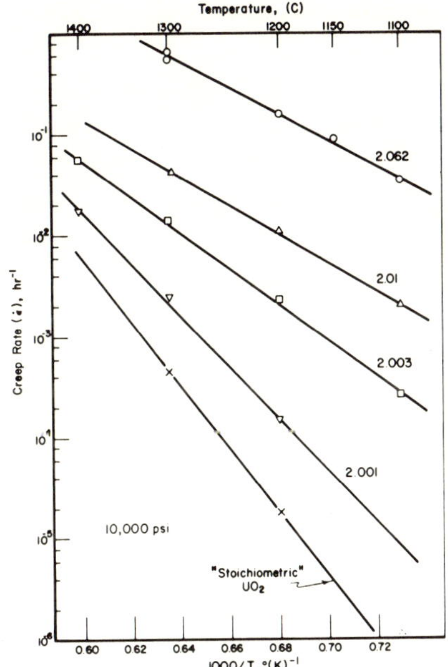

Fig. 3. Creep rate versus reciprocal of absolute temperature for UO_{2+x} single crystals tested at 10,000 psi.[3]

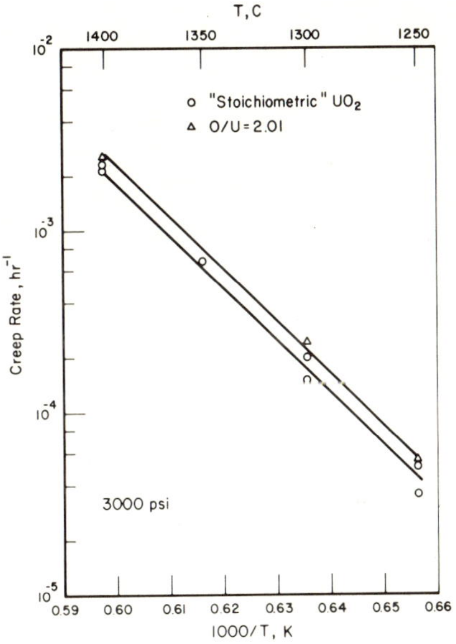

T, C
1400
1350
1300
1250
o "Stoichiometric" UO2
Δ O/U=2.01
3000 psi
Creep Rate, hr⁻¹
0.59
0.60
0.61
0.62
0.63
0.64
0.65
0.66
1000/T, K

Fig. 4. Creep rate versus reciprocal of absolute temperature for UO_{2+x} single crystals tested at 3000 psi, where $\dot{\epsilon} \propto \sigma$.[3]

1300°C
130
120
110
100
90
80
70
60
50
Qc, kcal/mole
ε̇ ∝ σ
ε̇ ∝ σⁿ (n = 4-17)
-4(2.0001)
-3(2.001)
-2(2.01)
-1(2.10)
LOG10 X

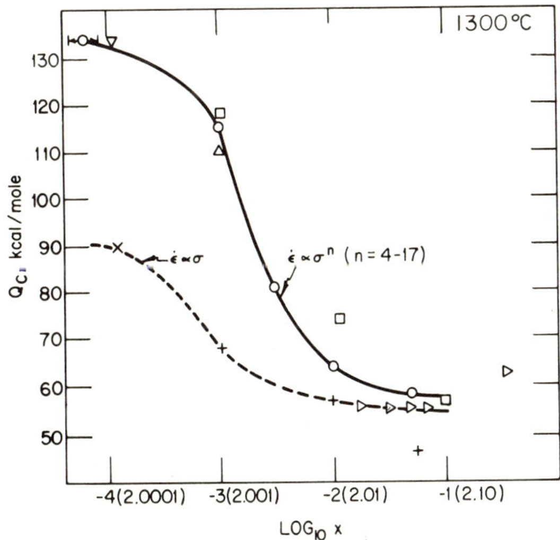

Fig. 5. Activation energies for creep of UO_{2+x} as a function of $\log_{10}X$: o single crystals[3]*; □ enriched polycrystals, 27 μm*[4]*; +, Δ depleted polycrystals, 6 μm*[4]*; X, ∇ polycrystals, 4 to 35 μm*[6]*; ▷ polycrystals, 6 μm, in bending*[2]*.*

Results obtained by Seltzer, Clauer, and Wilcox[4] for polycrystal UO_{2+x} with a grain size of 27 μm and a density of 97.8 ± 0.5 percent of theoretical at 1100°C and 1300°C, are presented in Figs. 6a and 6b as steady-state creep rate versus applied stress. For this grain size, at stresses above 3000 psi the creep rates can be fit by a power-law stress dependence where n = 7. At lower stresses, a transition to a lower stress dependence occurs, although a stress exponent of one was measured only at the highest temperature and O/U ratio studied, 1300°C and $UO_{2.10}$. In the high-stress region, creep rates increase with deviation from stoichiometry in the same manner as found for single crystals: $\dot{\epsilon} \propto X^m$ where $1.75 < m < 2$.

The creep activation-energy values calculated for O/U levels of 2.001, 2.01, and 2.10 from Figures 6a and 6b are 117, 75, and 54 kcal/mole, respectively. They are plotted as squares in Fig. 5, and are seen to agree well with the single-crystal activation energies, suggesting that the same creep mechanism is operative in both cases.

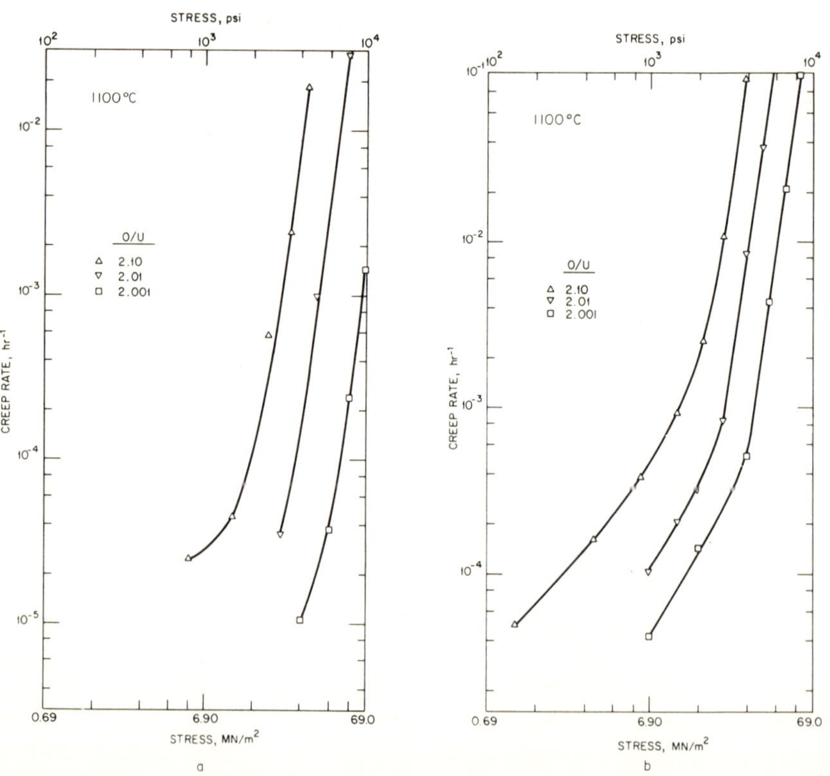

Fig. 6. Creep rate versus stress for enriched UO_{2+x} tested in compression at 1100°C and 1300°C; grain size = 27 μm.[4]

Seltzer et al. also performed creep tests on 95 percent dense specimens which had a grain size of 6 μm. These tests were conducted at 1100°C and 1300°C under stresses of 150 to 24,000 psi. Results are presented in Fig. 7 for specimens with O/U ratios of 2.001, 2.01, and 2.05. For these fine-grained samples, creep rates are seen to increase linearly with stress over most of the range of applied stresses, but the stress dependence increases at high stresses, where, if described by a power-law stress dependence, $n \geqslant 4$. In the region of linear stress dependence, the creep rates are found to increase linearly with X, as found previously by Armstrong and Irvine.[2] In the power-law stress-dependence region, however, creep rates increase with $X^{1.75}$. This is similar to the X^2 dependence found in the single-crystal studies.

Estimates of several creep activation energies can be obtained from the data in Fig. 7. For an O/U ratio of 2.001, $Q_c \approx 108$ kcal/mole in the high-stress region and 68.5 kcal/mole in the low-stress region. For O/U values of 2.01 and 2.05, Q_c values in the low-stress region are 55 kcal/mole and 45 kcal/mole, respectively. These activation energies are plotted as triangles (Δ) and pluses (+) in Fig. 5. Also included in this figure are Q_c values obtained by Bohaboy et al.[6] for polycrystalline UO_2 tested in hydrogen and by Armstrong and Irvine[2]. It was assumed that the value of 132 kcal/mole obtained for Q_c by Bohaboy et al. in the power-

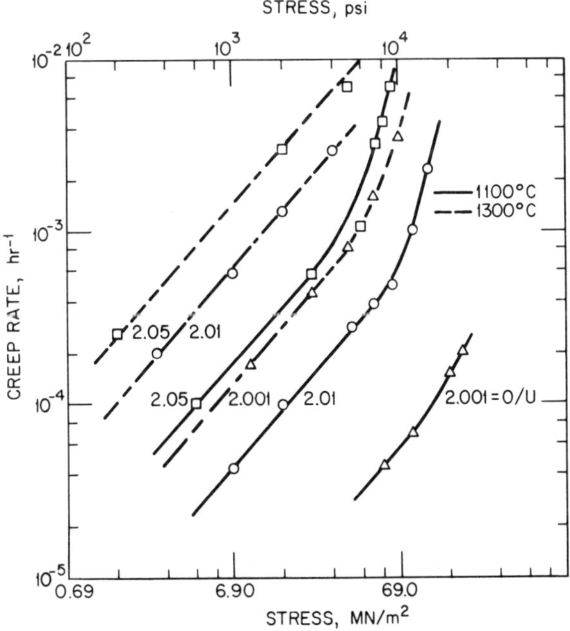

Fig. 7. Creep rate versus stress for depleted UO_{2+x} 6 μm grain size, tested in compression at 1100°C and 1300°C.[4]

law region corresponded to an O/U ratio of 2.0001. Their Q_c value of 90 kcal/mole for the linear stress region has been accordingly placed at the same O/U ratio. A multiple-regression analysis of all available polycrystalline UO_2 creep data was performed[7] to provide correlations which best describe UO_2 strain rates in terms of stoichiometry, temperature, stress, and grain size. In the low-stress region:

$$\dot{\epsilon} = \sigma G^{-2}\ \Delta^{-2.064}\ \exp\ (-4.866 + 102.5\Delta)\ \exp - \frac{Q(\Delta)}{RT}\ , \qquad (1)$$

where $Q(\Delta) = 50.02 + 43.63\ \exp\ (-863.3\Delta)$, while at high stresses:

$$\dot{\epsilon} = \sigma^{4.5} \Delta^{-2.911}\ \exp\ (-34.55)\ \exp - \frac{Q(\Delta)}{RT}\ , \qquad (2)$$

where $Q(\Delta) = 57.35 + 79.25\ \exp(-368.9\Delta)$, where Q is the creep activation energy in kcal/mole, Δ is the deviation from stoichiometry (Δ = X for UO_{2+x}), G is the grain diameter in μm, $\dot{\epsilon}$ is in hr^{-1}, σ is in psi, T is the absolute temperature, and R is the gas constant.

$(U,Pu)O_{2-x}$

In contrast to uranium dioxide, the mixed oxide $(U,Pu)O_2$ containing some 20 to 25 weight percent PuO_2 exists primarily as an oxygen-deficient hypostoichiometric material. Routbort and co-workers[8,9] have investigated the steady-state compressive creep of polycrystalline $(U,Pu)O_{2-x}$ as a function of O/M ratio in both the high- and low-stress creep regions. Tests were performed on mechanically mixed fuel pellets containing 25 weight percent PuO_2 in the temperature range of 1300 to 1700°C, at stresses of 1000 to 16,000 psi and at O/M ratios of 1.88 to 1.995. These workers used H_2/H_2O mixtures to obtain specimens of the desired composition. The specimens had grain sizes of 10 μm or 20 μm, with densities of 95 and 97 percent of theoretical, respectively.

Measurements were made primarily on grossly defective specimens in the range $0.01 < X < 0.12$ at temperatures above 1400°C. Under these conditions, creep rates varied only modestly with O/M ratio. At 1400°C, where the largest effect was observed, $\dot{\epsilon}$ increased by a factor of 10 as the O/M ratio decreased from 1.99 to 1.90. The creep activation energies were found to be influenced by O/M ratio, however, in both the lower- and high-stress regions. Activation energies obtained by Routbort and co-workers[9] for $(U,Pu)O_{2-x}$ are shown in Fig. 8 as a function of O/M ratio on a linear scale. Also included in this figure are the curves for UO_{2+x} obtained by Seltzer et al.[3,4]. As for the case of UO_{2+x}, the values of Q_c are higher in the high-stress region than in the viscous creep regime.

Maxima occur for apparent activation energies in both stress regions at an O/M ratio of about 1.99.

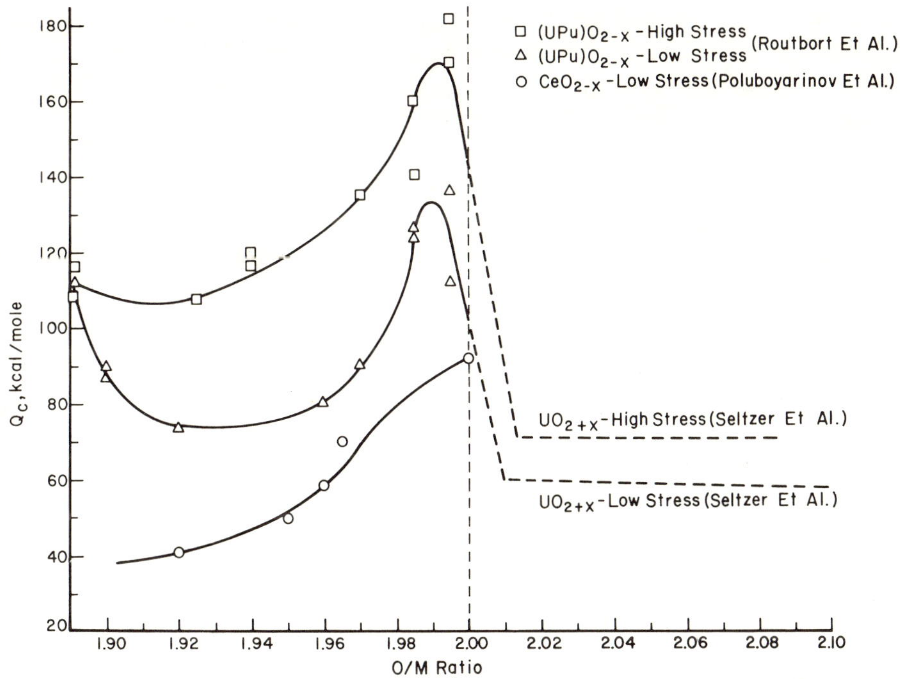

Fig. 8. Creep activation energies versus O/M ratio for (UPu)O_{2-x}, CeO_{2-x}, and UO_{2+x}.

CeO_{2-x}

Ceria, like (U,Pu)O_{2-x}, is capable of existing as a single-phase, hypostoichiometric oxide over a wide range of compositions ($CeO_{2.00}$ - $CeO_{1.72}$) at temperatures above 1000°C. Poluboyarinov and co-workers[10] have utilized a four-point bending technique to study the high-temperature creep behavior of CeO_{2-x}. For an applied stress of 13.65 kg/cm^2, the creep activation energies were found to decrease from 92.5 kcal/mole to 39 kcal/mole as the O/Ce ratio was decreased from 2.00 to 1.90 (see Fig. 8). As for the case of (U,Pu)O_{2-x}, creep rates were found to increase by about one order of magnitude as the O/M ratio was decreased from about 2.00 to 1.90. For this range of compositions, creep rates were proportional to some power m of the applied stress where $1.2 < m < 2.2$.

3 DISCUSSION

The experimental results presented in the previous section represent the most complete body of data available with regard to the influence of deviations from stoichiometry on the creep properties of a particular class of inorganic compounds. In theory, analysis of these data, in conjunction with results obtained from self-diffusion studies and from studies of other diffusion-controlled properties such as sintering and grain growth, should provide a satisfactory understanding of the atomic defect structure for the fluorite-structure oxides. As will be seen, a lack of certain supplementary data, together with some inconsistencies in the available information, does not permit such a definitive analysis to be made at this time.

The results included in Fig. 5 form a self-consistent picture of the effects of O/U ratio on the creep activation energies for UO_{2+x}. The data obtained by Seltzer et al.[4] for near-stoichiometric UO_{2+x} are in good agreement with the Q_c values obtained by Bohaboy et al.[6], while the Q_c data obtained by the former authors for UO_{2+x} in the linear stress region at large deviations from stoichiometry are in good agreement with the results of Armstrong and Irvine[2].

Furthermore, MacEwan and Hayashi[11] found that the activation energy for equiaxed grain growth in UO_{2+x} decreased from 109 kcal/mole to 65 kcal/mole as the O/U ratio was increased from near 2.00 to 2.02, while the rate of growth increased by four orders of magnitude if the O/U ratio was increased from near 2.00 to 2.23. These results are in good agreement with those obtained from measurements of steady-state creep rates in the high-stress region. In addition, Lay[12] has found that the initial sintering kinetics of UO_{2+x} were approximately proportional to the square of the oxygen excess in the range $2.03 < O/U < 2.16$ and that the activation energy for shrinkage was 55 kcal/mole for $O/U = 2.08$. This value is in accord with the creep activation energy for UO_{2+x} of the same composition, again measured in the high-stress creep regime.

Thus it appears that steady-state creep in the power-law stress region, initial sintering kinetics, and equiaxed grain growth of UO_{2+x} are all controlled by the same diffusion process. Since oxygen diffusivities in UO_{2+x} are five orders of magnitude greater than uranium diffusion rates in the temperature range 1000 to 1400°C[13], it is generally assumed that the various diffusion-controlled properties discussed above are limited by uranium diffusion. Several investigators[14-16] have in fact found that diffusivities for uranium in hyperstoichiometric urania are nearly proportional to the square of the oxygen excess. The X^2 dependence was predicted by Lidiard[17] for the case where uranium diffuses as U^{4+} via vacancies in the uranium sublattice. It might therefore be reasonable to assume that volume diffusion of uranium is the rate-controlling process for a number of diffusion-limited properties. The decrease in activation energies with increasing X could then be attributed to a decrease in the apparent energy of formation of uranium vacancies with increasing X, the activation energies of 50 to 60 kcal/mole found at O/U levels above 2.01 being assumed to represent the migration energy for uranium vacancies.

Although the various activation-energy dependencies on deviation from stoichiometry are self-consistent, the magnitudes and variations with O/U ratio of these energies cannot be correlated with all existing uranium self-diffusion activation energies. Lindner and Schmitz[18] measured the activation energies for diffusion of U^{233} in UO_2 under atmospheres of hydrogen and argon. They reported values of 104.6 kcal/mole and 52 kcal/mole for experiments performed in hydrogen and in argon, respectively. No O/U ratios were reported for these experiments, but if the reasonable assumption is made that the specimens annealed in hydrogen are assumed to be near-stoichiometric urania, then these diffusion activation energies may be in accord with the energies obtained from other physical-property measurements.

Hawkins and Alcock[14], on the other hand, found that uranium self-diffusion activation energies increased from 89 kcal/mole to 105 kcal/mole as the O/U ratio was increased from 2.01 to 2.10, in contrast to the activation energies for the other processes, described above, which decrease from 65 kcal/mole to 55 kcal/mole in this range of compositions. Marin and Contanin[19] also measured uranium self-diffusion activation energies in UO_{2+x} in the range $0.03 < x < 0.20$ and found the self-diffusion activation energy to be 85 kcal/mole, independent of O/U ratio.

No explanation can be offered for the differences in activation energies for creep, sintering, and grain growth on the one hand and uranium self-diffusion on the other. It should be mentioned that Matzke[16] has critized the data of Hawkins and Alcock[14] on the basis that the increase in uranium diffusion activation energies and in the preexponentials constants, D_o, with increasing O/U level are inconsistent with our current understanding of diffusion in highly defective oxides.

Assuming that the creep rates for UO_{2+x} in the high-stress region are limited by uranium volume diffusion, we must next ask what diffusion process controls creep behavior in the linear stress-dependency region. The lower Q_c values could be attributed to a grain-boundary-diffusion process. This implies that the observed viscous creep is due not to Nabarro-Herring diffusional creep, but rather to grain-boundary diffusion-controlled Coble creep. The predominance of one of these two processes can be established by determination of the grain-size dependence for creep rates. For Nabarro-Herring creep, $\dot{\epsilon}$, is proportional to the square of the reciprocal grain size, while for Coble creep the reciprocal cube dependence on grain size is predicted. Several investigators[6,20] have reported that creep rates for polycrystalline UO_{2+x} are proportional to $1/G^2$, implying that the creep activation energies in the low-stress region do correspond to uranium-volume-diffusion energies. If this is the case, we are left without an interpretation of the result that $\dot{\epsilon} \propto X$ in the low-stress region, rather than X^2 as predicted for uranium-volume-diffusion-controlled processes. In addition, some further interpretation is required for the higher Q_c values found in the power-law stress region. These energies could correspond to the uranium-volume-diffusion activation energy onto which is added an additional term such as the energy to form a jog, if creep is governed by the diffusion-controlled motion of jogged screw dislocations.

Data for the effects of deviation from stoichiometry for hypostoichiometric oxides are more limited as compared with those for the oxygen excess oxides. Thus, no self-diffusion measurements have been made on hypostoichiometric mixed oxides to compare with the available creep data. It does appear from the manner in which creep rates and activation energies for $(UPu)O_{2-x}$ and CeO_{2-x} vary with O/M ratio that steady-state creep in these materials is limited by the diffusion of metal interstitials, in the form of isolated point defects or in the form of some defect complex. Routbort et al.[8] have shown that the concentration of cation interstitials is approximately eight orders of magnitude greater than the concentration of cation vacancies for UO_{2-x} with an O/U ratio of 1.98.

The data in Fig. 8 imply that a change occurs in the rate-limiting process for steady-state creep from one involving cation vacancies to one involving cation interstitials as the O/M ratio is reduced, with the transition coming at a value slightly less than 2.00. The finding that creep activation energies for $(UPu)O_{2-x}$ vary with O/M ratio in both stress regimes, and that values for Q_c are higher in the power-law stress region than at lower stresses where viscous creep predominates, is exactly analogous to the situation for UO_{2+x}. From the facts that creep activation energies and the preexponential factors in the creep equations for high- and low-stress regions vary in approximately the same manner with the O/M ratio, Routbort et al.[8] concluded that the motion of the same defects controls creep behavior in both regions. These are assumed to be cation vacancies for O/M ratios $\geqslant$ 1.99 and cation interstitials, or complexes involving cation interstitials, for lower O/M ratios.

REFERENCES

1. Scott, R., Hall, A. R., and Williams, J., *J. Nucl. Mater.,* **1,** 39 (1959).
2. Armstrong, W. M., and Irvine, W. R., *J. Nucl. Mater.,* **9,** 121 (1963).
3. Seltzer, M. S., Clauer, A. H., and Wilcox, B. A., *J. Nucl. Mater.,* **44,** 43 (1972).
4. Seltzer, M. S., Clauer, A. H., and Wilcox, B. A., *J. Nucl. Mater.,* **44,** 331 (1972).

5. Perron, P. O., AECL-3072 (May, 1968).
6. Bohaboy, P. E., Asamoto, R. R., and Conti, A. E., GEAP 10054 (May, 1969).
7. Seltzer, M. S., Markworth, A. J., Clauer, A. H., and Wilcox, B. A., in BMI-1918, D. L. Keller (Ed.) (July, 1971).
8. Routbort, J. L., Javed, N. A., and Vogelwede, J. C., *J. Nucl. Mater.*, **44**, 247 (1972).
9. Routbort, J. L., and Vogelwede, J. C., *J. Am. Ceram. Soc.*, **56**, 330 (1973).
10. Poluboyarinov, D. N., Shapiro, E. Y., Bakunov, V. S., and Akopov, F. A., *Inorg. Mater.*, **2**, 336 (1966).
11. MacEwan, J. R., and Hayashi, J., *Proc. Brit. Ceram. Soc.*, **7**, 245 (1967).
12. Lay, K. W., *J. Am. Ceram. Soc.*, **54**, 18 (1971).
13. Belle, J., *J. Nucl. Mater.*, **30**, 3 (1969).
14. Hawkins, R. J., and Alcock, C. B., *J. Nucl. Mater.*, **26**, 112 (1968).
15. Marin, J. M., Michaud, H., and Contamin, P., *Compt. Rend.*, **264**, 1633 (1967).
16. Matzke, H. J., *J. Nucl. Mater.*, **30**, 26 (1969).
17. Lidiard, A. B., *J. Nucl. Mater.*, **19**, 106 (1966).
18. Lindner, R., and Schmitz, F., *Z. Naturforsch.*, **160**, 1373 (1961).
19. Marin, J. F., and Contamin, P., *J. Nucl. Mater.*, **30**, 16 (1969).
20. Armstrong, W. M., Irvine, W. R., and Martinson, R. H., *J. Nucl. Mater.*, **7**, 133 (1962).

DISCUSSION on Paper by M. S. Seltzer, B. A. Wilcox, and A. H. Clauer

CHIKALLA: What significance do you attach to the activation-energy maximum at O/M = 1.98, and is there any influence of plutonium in comparing Q_c for hyperstoichiometric UO_2 with Q_c for hyperstoichiometric (U, Pu)O_2?

SELTZER: The activation-energy maximum should not be expected to occur at the stoichiometric composition if there is a difference in the mobilities of cation vacancies and cation interstitials or in the defect complexes controlling creep on either side of O/M = 2.00.

Unfortunately, Q_c has been measured as a function of O/M only for hyperstoichiometric UO_{2+x} and hypostoichiometric (U, Pu)O_{2-x}, so that no direct comparison is possible. For O/M = 2.00, the activation energies appear to be nearly equal at 130 ± 10 kcal/mole.

HEUER: Is there any reason to discount Alcock's data for D_U? If not, it is a serious problem that the diffusion data do not support the trend of activation energy with nonstoichiometry that is found from the creep data.

SELTZER: As pointed out in the text, Alcock's data have been criticized as being inconsistent with our current understanding of diffusion in highly defective oxides. The general form of the creep activation-energy dependence on stoichiometry has been reported by several other investigators, the most recent being Burton and Reynolds [*Acta Met.*, 21, 1641 (1973)].

HEUER: Isn't there reason to be concerned with some of the creep experiments with enriched UO_2, where the absolute magnitude of the creep rate is much enhanced (presumably because of impurities added during the enrichment) compared with that of the usual specimens of UO_2 studied.

SELTZER: After correcting for grain size and stress, there is little difference between creep rates of enriched and depleted UO_2. These may certainly be due to impurities.

ILSCHNER: Creep-rate measurements on "commercial" UO_2-PuO_2 performed by H. Assmann (Ph.D. Thesis, Erlangen, 1972) appear to agree well with your results for comparable experimental conditions. In particular, effective activation energies, Q_{eff}, as determined from temperature-cycling data, were found to be 140 ± 30 kcal/mole (depending on sample structure, stoichiometry and pretreatment). However, we experienced considerable difficulty in relating this effective activation energy for creep to diffusion data, since Q_{eff} was influenced considerably by "cross terms" due to a temperature-dependent activation volume and stress-dependent Q_{eff}, which were difficult to evaluate.

LIDIARD: In discussing variations in the data on diffusion of uranium in UO_{2+x}, it should be remembered that there may be significant contributions from dislocation-enhanced and grain-boundary diffusion since the uranium point defects are minority defects. Indeed, Lundy has argued that these other mechanisms are dominant in many measurements, although, if this is so, it is not obvious how one explains the observed dependence of uranium self-diffusion on the degree of nonstoichiometry, X. However, Catlow's calculations on model UO_2 give the energy of formation of O^{2-} Frenkel defects as 5.3 ev and of Schottky defects as 10.3 ev, while the activation energy for U^{4+} vacancy jumps is found to be 5.1 ev. Thus, the activation energy Q for uranium self-diffusion in intrinsic UO_2 (i.e., pure, stoichiometric UO_2) should be

$$Q = 10.3 - 5.3 + 5.1 = 10.1 \text{ ev},$$

while in nonstoichiometric UO_{2+x}, Q should be 10.3 - 2 x 5.3 + 5.1 = 4.8 ev when D is proportional to X^2. Thus in nonstoichiometric UO_{2+x}, the measured self-diffusion coefficients would appear to be principally determined by the motion of U^{4+} vacancies, since both the magnitude of Q and the dependence of X are close to these predictions.

NICHOLS: In the low-stress region of creep in polycrystalline UO_2, you showed good agreement with grain-size dependence predicted by Nabarro-Herring creep. You also report reasonable agreement between your calculated activation energies and those for volume diffusion of uranium. How do your *quantitative* creep rates compare with Nabarro-Herring predictions using reported diffusivities of uranium?

SELTZER: The experimental creep rates are several orders of magnitude greater than the values predicted by the Nabarro-Herring equation. This discrepancy is common in diffusional creep studies and may be due to stoichiometry effects or to difficulties in measuring uranium diffusivities.

AGENDA DISCUSSION: CREEP AND SINTERING

*F. A. Nichols**

Bettis Atomic Power Laboratory

*P. E. Hart***

Battelle, Pacific Northwest Laboratories

The chairman opened the discussion with comments relative to our present state of knowledge concerning sintering processes and followed this with remarks on surface-layer effects on diffusion and threshold effects in diffusion creep processes.

Certainly, in the early stages of sintering, with diffusional distances of 1 μ or so, surface diffusion is controlling the neck-growth process. At later stages, when diffusional paths are several microns in length, *either* volume diffusion or vapor transport may control, e.g., for UO_2, volume diffusion should never be important, and vapor transport should control. As for

*Chairman

**Secretary

kinetics, the only one that has been analyzed accurately is surface diffusion where $x/a \propto t^n$, $5 \lesssim n \lesssim 8$, $\bar{n} \sim 6$. For volume diffusion and volume transport, we expect $\bar{n} < 6$, but an accurate analysis which follows actual changes in profile is not available.

With regard to densification (shrinkage), it was noted that, of course, this occurs simultaneously with neck growth but, in the early stages at least, one can probably make a correction for its contribution to neck growth and get reasonable agreement between measurement and predicted surface-diffusion control. This has been done in a few cases. At intermediate stages, separation is not so easy. But surely, gradients in curvature still exist in the neck, causing continued neck growth by surface diffusion, volume diffusion, and vapor transport as before, with the *neck surface* serving as both source and sink for defects. Thus, grain boundary or volume diffusion with the grain boundary alone acting as an atom source cannot produce sintering. Atoms must redistribute themselves in the neck by surface diffusion, volume diffusion, or vapor transport. Now if this occurs *primarily* by surface diffusion (i.e., rather early stages), it is probably much *faster* and hence not rate controlling. The models such as discussed by Johnson ignore this process and this is satisfactory as long as it is very rapid. But in intermediate stages it may not be negligible. Thus the models could be quite wrong. In addition, the models of necessity use much-oversimplified geometric assumptions which could certainly alter the results quantitatively and could significantly alter the time laws obeyed. This could lead to erroneous conclusions as to rate-controlling mechanisms. For example, the early models of Kuczynski erroneously predicted

$$x/a \propto t^7 \text{ for surface diffusion}$$

$$x/a \propto t^5 \text{ for volume diffusion.}$$

Since data typically show $x/a \propto \sim t^{5.5}$, he and many following him concluded that volume diffusion controlled neck growth. We now know this was quite wrong. Early neck growth is normally surface-diffusion controlled and probably *never* is surface-diffusion negligible. An exact formulation would require a complicated numerical solution which follows *exact* shape changes. This has not been done.

Commenting on a point raised earlier by Kröger as to how *thick* the boundary-diffusion layer is, the chairman pointed out that in *any* mass transport problem, we get *only* the product C_iD_i. For volume diffusion, we assume $C_i \propto 1$/molecular volume and calculate D_i. For boundary diffusion, we get a concentration/unit area times a boundary diffusivity. One must invoke independent evidence or models to choose the proper concentration of diffusing species. Since the mechanism for boundary (or surface) diffusion is not yet clarified, one often simply *defines* the concentration as the atomic concentration. The resulting value for D_b is subject to that rather arbitrary selection. An equivalent assumption is that the boundary-layer thickness δ is ~ one atomic distance, a. In utilizing such results, one must use the *same* choice for C_i (or δ). Most independent studies with metals indicate that the choice of $\delta \sim a$ is quantitatively reasonable. The answer is not at hand yet for ionic solids, but the existence of space-charge effects certainly raises the possibility that a much broader region neighboring a boundary could have a significantly *different* diffusivity from the bulk. It seems that it might be enhanced in some cases and depressed in others, depending upon the defect structure of the space-charge layer and the rate-controlling mechanism for bulk diffusion. The rate-controlling species may even differ from that in the bulk. But it would be much less confusing if this were thought of (as it is in fact) as a *volume* diffusivity differing from that in the bulk. The term boundary diffusivity is then reserved for the case of truly two-dimensional transport along the boundary surface (thickness $\delta \sim a$).

With regard to final-stage sintering (densification), the situation seems greatly simplified. The porosity is spheroidized and densification with no complex shape changes occurs. If gas of negligible solubility is entrapped in pores, densification ceases when

$$p = \frac{2\gamma}{r} \text{ in the pores,}$$

where

p = pressure
γ = surface energy per unit area
r = pore radius.

With finite solubility, gas can diffuse out and densification can in principle proceed to completion, i.e., theoretical density. This is a question of strong interest to ceramists: What are the requirements to obtain theoretical density? Why do we sometimes observe end-point densities? Entrapped insoluble gas is one obvious answer. If gas can be eliminated, are there other barriers to obtaining theoretical density? Jorgensen previously suggested one way, i.e., the boundaries move away from the pore and densification stops or becomes very slow. Dislocations are apparently not effective defect sources or sinks, under such conditions, since calculated rates are much slower than measured rates.

The chairman then discussed threshold stresses, σ_o, for diffusion-controlled creep which have been observed in several metals and in UO_2. Data are not accurate enough for one to know whether there is a true threshold stress or whether the data simply indicate a high stress dependence. One wonders if end-point densities are not related to this effect. Now, for creep as for any activated process, velocity is proportional to driving force when that force is low enough. Thus, one can question Seltzer's statement that no model predicts $\dot{\epsilon} \propto \sigma$ at low stresses for single crystals. There are several in fact. The question is: which one or ones are important for his UO_2? But, at very low stresses, the linear dependence on σ gives way, in several instances, to a threshold stress below which $\dot{\epsilon}$ becomes immeasurably low. What is the origin of this apparent threshold? Is this related to the end-point densities of sintering studies? What hinders the operation of the boundary as a source or sink? Is there any way to eliminate the σ_0, and if so, would this permit sintering to theoretical density? Does the presence of the pores on the boundary facilitate its operation as a sink? Is this why densification stops as soon as the boundaries leave the pores? If so, we should not see σ_0's in the creep of bodies with grain-boundary porosity. Do we have any information in this area? A related question concerns Ashby's proposal of a somewhat different diffusion creep mechanism. In his so-called diffusion-accommodated flow, which involves smaller diffusion distances, neighbor switching events occur. Ashby's model predicts that grains remain approximately equiaxed. What do our data show us? Do UO_2 grains show the same strain as the macroscopic strain? Or do they remain equiaxed?

Stringer commented that grain-boundary width, so far as diffusion is concerned, is not very well defined in oxides. In addition to the "space charge region" effect that has been mentioned, the diffusion properties are sensitive to impurities, which have a marked tendency to segregate to grain boundaries. The relatively heavily doped region *near* the grain boundary will then have a different diffusivity, and as a result, the apparent grain-boundary width in a diffusion experiment may well be strongly dependent on impurity content. He thinks that this has been shown for anion diffusion in alkali halides by Prof. Benard's group. The conventional model used to describe grain-boundary diffusion in mathematical terms involves defining a slab of width δ with a diffusivity D_b. It is often difficult to determine appropriate values for δ: in metals, values ranging from a few to several thousand Angstroms have been chosen by different investigations. It would seem difficult to choose a realistic value for oxides at the present time.

In response, Nichols stated that in dealing with diffusivities in or near an interface which differ from the corresponding values in the bulk material, one must keep in mind two basically different aspects of the phenomenon. The first aspect, exemplified by the Fisher model for grain-boundary diffusion or the Hart model for dislocation pipe diffusion, deals only with the assumption that in or near the interface, a component is transported at a different (usually larger) rate. This phenomenon can be treated as bulk (volume) diffusion through a composite material where one assumes for simplicity that the second "component", i.e., the material near the interface, has a different fixed diffusivity. The important distinction here is that the interface serves as a "short-circuit" diffusion path but the *movement* of the interface is not involved. One can assume different effective "widths" of the short-circuit path and in so doing obtain from any given experimental results correspondingly differing diffusivities for the region. These two quantities always appear as a product in any experimental results; one needs to invoke additional independent information in order to separate them. Although we sometimes forget the fact, this is true for any diffusional transport, including volumetric diffusion. In bulk diffusion, we obtain the equivalent product of concentration and diffusivity of the mobile species and usually we can evaluate the concentration by other means and so separate the two factors. This separation is not so straightforward for surface and line diffusion. On the other hand, in some problems such as diffusion creep or sintering, the interface is not simply a short-circuit diffusion path but serves as a source or sink for defects and thereby undergoes net normal movement. In this case, the flux need not be simply parallel to (along) the interface but may also be *normal* to it. The former is interfacial diffusion, the latter is volume diffusion and the distinction must be made. One cannot simply widen the "effective" width of the path and treat the entire effect mathematically as interfacial diffusion. This is because of the operation of the interface as a source or sink and the fact that the normal velocity of the interface is directly proportional to the flux entering (or leaving) normally but is proportional to the *divergence* of the flux along the interface. This basic difference gives rise to different kinetic dependencies for the two means of transport and is the origin of the rigorous scaling laws developed by Herring. An illustration is the dependence of diffusion creep on the inverse square of the grain size for volume diffusion and the inverse cube of the grain size for boundary diffusion. Since no mechanism for defect generation or annihilation exists away from the interface even if a region of enhanced diffusivity exists around the boundary of oxides due, say, to space-charge effects, the distinction must be maintained between boundary diffusion and volume diffusion for transport along and normal to the interface, respectively. Even if the two diffusivities happen to equal each other, the two components of flux produce their characteristic kinetic dependencies or scaling laws. The application of the scaling laws to experiments carried to equivalent shapes but with differing dimensions can be used to define *unambiguously* the rate-controlling process when one dominates. Now, if one is analyzing the volume diffusion kinetics of such a process in which a region of enhanced diffusivity exists around the interface, this region should be assigned its own diffusivity and the bulk region beyond it should be assigned its (bulk) value, with continuity of flux applied at the transition between the two regions. (Apparently, no such analyses have yet been done.) If only one diffusivity is assigned and the analysis carried through for volume diffusion, then one can get from experimental comparisons only an "average" diffusivity. If most of the volume flux is outside the "enhanced" region, this average value should be very near the bulk value. If the region is wide enough that most of the volumetric current is through this region, then the average value should be near the enhanced value; but, in either event, it must obey a volume-diffusion scaling law.

Stringer responded that what one gets from solving the diffusion equation with appropriate boundary conditions is indeed a concentration times a diffusion coefficient, but stated his belief that the information at hand does not permit distinguishing between an event that took place at a boundary (presumably having a very small width) and one taking place in a region near the grain boundary, which might therefore involve a much larger width and be less easy to relate to any specific physical picture.

Nichols reiterated that one must in fact decide (or assume) where defects are being created and where they are being destroyed in order to analyze a particular process involving mass transport. You establish this as a boundary condition, and then you solve the diffusion equation. Solution of the diffusion equation for along the grain boundary and normal to it will give you different kinetics.

In reply to Stringer's earlier comments on grain boundaries, Heuer said that he had looked at one of the "dirty" MgO bicrystals of Wuensch and Vasilos that showed enhanced grain-boundary diffusion of nickel ("clean" bicrystals do not show enhanced diffusivity) in the electron microscope and found many small particles, as shown in Figure 1. In response to Cohen's query as to whether some of the boundary-thickness effects could be due to particles being ball milled before sintering, Johnson stated that in the Bagley study to which he referred earlier, the powders had been calcined before sintering and were not subsequently ball milled.

Cohen asked if a term for plastic deformation should be included in the sintering equation in addition to grain-boundary and surface terms, and Johnson responded that in his opinion, during initial stages of sintering when stresses are high, there is a high probability of plastic flow. This would show quite different kinetics; there is little likelihood of confusing this with boundary- and volume-diffusion kinetics. Johnson went on to comment that he also had some preliminary data on intermediate-stage sintering of Y_2O_3 and on the same powder doped with 1 and 0.1 mole percent MgO and 5 and 10 mole percent ZrO_2 at 1600°C. Volume-diffusion kinetics were observed. Doping with MgO gives a monotonic increase in the apparent volume-diffusion coefficient, enhanced sintering rates, and spectacular grain growth. On the other hand, ZrO_2 causes an

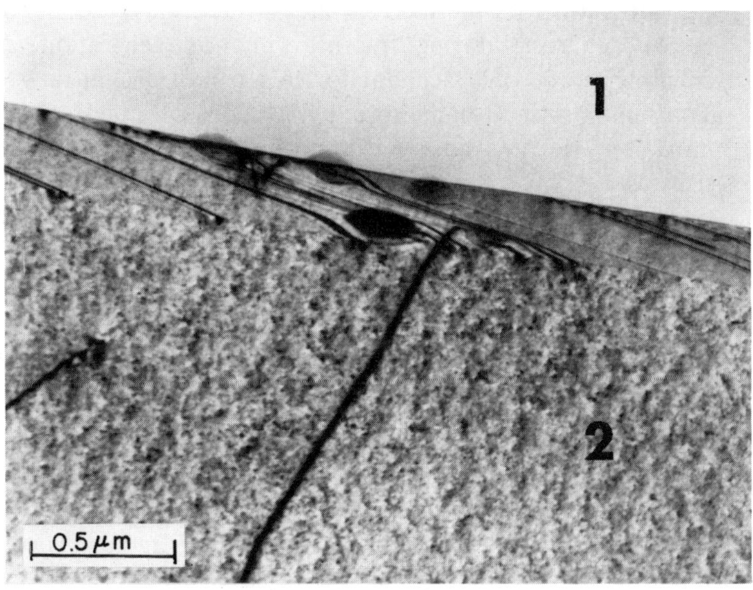

Fig. 1. Electron micrograph of MgO bicrystal diffusion specimen. (The contrast conditions are such that crystal 2 is much more strongly diffracting than crystal 1.) The particles present in the grain boundary are clearly visible, as well as some grain boundary dislocations. (The mottled structure in crystal 2 is a result of surface damage introduced during the ion thinning preparation of the electron microscope foil.)

initial decrease of the apparent volume-diffusion coefficient, followed by an increase at the higher concentrations. Yttria can be sintered to theoretical density, as Jorgensen discussed, even in this range of reduced diffusivity, because the grain growth is also reduced. The observed diffusion coefficient can be explained in terms of Frenkel disorder on the cation sublattice if the interstitials have a higher mobility than the vacancies. Magnesia would create yttrium interstitials, while zirconia would create yttrium vacancies and reduce the interstitial concentration. Zirconia would initially lower the diffusion coefficient through reduced interstitial concentration, but then increase it as the vacancy concentration becomes sufficiently large.

Shewmon expressed interest in what evidence Heuer has that makes him believe that precipitates influence crack propagation instead of a deformation process that may accompany the fracture. He questioned what the fracture process is and how much lattice deformation was involved. Heuer responded that, in his opinion, fracture in brittle solids like spinel and alumina does not involve plastic deformation. The following evidence may be cited:

(a) Fracture surface energies, particularly for single crystals, are of the order of magnitude (within a factor of 2 to 3) of our estimates of the thermodynamic surface energies. Several processes occurring during fracture (phonon generation, multiple cracking, etc.) could account for this size of discrepancy.

(b) Transmission-electron-microscopy studies of crack tips in alumina by Hockey and Wiederhorn of NBS showed that the stresses ahead of the crack could cause the formation of dislocations only at temperatures above 400°C. The similar mechanical behavior of spinel and alumina suggests that spinel would similarly act perfectly brittle.

(c) Calculations of the theoretical shear strength and theoretical cleavage strength are so similar (ratio of ~ 1) that there is no compelling theoretical reason to believe that plastic deformation must occur at the high stresses at crack tips.

Cohen stated that in a brittle material, once a crack is initiated it grows. Whether fracture is controlled by crack initiation or crack growth can be distinguished by the grain-size dependence. If fracture is controlled by crack initiation, there will be no grain-size dependence. If it is controlled by growth, there will be a grain-size dependence. To this comment, Heuer replied that there is a grain-size dependence in alumina. At small grain sizes the fracture stress increases at a rate somewhat less than the square root of the grain size, while at large grain sizes it goes as the grain size, due to flaws. The critical-stress-intensity factor is increased by precipitates. He further stated that he feels it is more appropriate to look at fracture from the standpoint of stress-intensity factors than from the standpoint of crack initiation and crack growth. Cohen injected his opinion that fracture mechanics is an interesting approach for design purposes, but that if you want to control cracking you have to understand how cracks are nucleated and how they propagate. Heuer saw a difficulty with that approach in that oxides exhibit subcritical crack growth which is associated with the environment.

Jorgensen was disturbed that Heuer assumed the thermodynamic surface energy of alumina to be 5,000 ergs/cm^2. He stated that Wiederhorn has actually found values of 7,000 ergs/cm^2 at room temperature and that we know from Kingery's work that it is 905 ergs/cm^2 at 1000°C. To this, Heuer countered that Kingery's values were measured on fairly-impure alumina and stated further that if you have plastic deformation you'll have $\gamma \sim 10^{+4}$; therefore a factor of two variation does not do anything. A γ of 5,000 implies little or no plastic deformation. Shewmon disagreed and pointed out that for "brittle" metals where there *is* deformed surface, γ falls in the 5,000 to 10,000 ergs/cm^2 range.

Returning to Stringer's question on the thickness of the subsurface region, Tallan commented that the thickness of the region is really tied to the Frenkel-Lehovec space-charge layer. In Lehovec's original work on sodium chloride, he calculated an effective thickness of 100 Å. That effective thickness is a function of the difference in the energy of formation; in some cases, we could get grain-boundary thicknesses well up in the μm range. Lidiard commented, in connection with discussion of the Frenkel-Lehovec space-charge thickness around dislocations and at surfaces, that the width is given by the Debye-Hückel screening length κ^{-1}, where

$$\kappa^2 = \frac{4 \pi \Sigma n_i q_i^2}{\epsilon_o kT}$$

where

n_i = concentration of defects of type i
q_i = effective defect charge
ϵ_o = dielectric constant.

He went on to point out that it is important to note that the width is independent of the details of the system such as the heat of formation of the complementary defects, the shape of the space-charge region, etc. Typically, with a concentration of defects of 10^{-6} mole fraction, these widths are on the order of several hundred Angstroms, whereas in doped crystals containing 10^{-4} mole-fraction defects, the width of the band is approximately 50 Å. In any given material, the width of the region varies inversely as the square root of the defect concentration.

Schmalzried enquired as to the number of defects in the space-charge region, and Lidiard responded that normally it is not very many compared with what you would expect to have because of impurities. Schmalzried commented that, typically, he estimated the concentration to be 10^{+16} to 10^{+17}. Therefore, he concluded that the space charge cannot strongly influence either the intrinsic defect concentration or diffusion due to defects or impurities. Lidiard felt that this is a fair conclusion. In cases he has examined in detail, space charge does not have an overwhelming effect on diffusion when compared with effects due to impurities.

Smeltzer cautioned that the utilization of the Debye screening distance for the electrical space charge to describe the grain-boundary width of a metal oxide at high temperatures met in a sintering environment is to be treated with reservation. A classification should at least be made of the oxides. Its application to insulating-type oxides such as alumina, BeO, and MgO may be valid since the low defect concentration generated by thermal energy lies in an appropriate range. On the other hand, its application to the divalent transition metal oxides such as FeO, CoO, and MnO is not correct since the point-defect concentrations would be too high. The nonstoichiometric phase fields of these oxides markedly broaden with increasing temperature. Simple application of the Debye theory would lead to a screening distance less than the interatomic distance.

Moreover, several transition metal oxides exhibit a transition from semiconductance to metallic conductance. A more realistic general parameter for describing the grain-boundary width could be based on crystallography and the type of boundaries determined by ledges and dislocation-array structures in and within the region of the boundary. To this, Heuer responded that he has looked at several different polycrystalline ceramics in the electron microscope and believes that, to a good approximation, oxide grain boundaries are quite similar to those in metals. Of course, for ionic crystals there may be some high angle boundaries that are very energetic where ions of like charge are in proximity. The work on MgO by P. Chaudhari et al., of IBM supports this. These researchers found in bicrystals produced by burning magnesium a preference for the coincidence orientations of fcc crystals, except that one or two orientations were not present.

Farnsworth then turned the discussion to the subject of threshold-stress limits for creep, and Seltzer stated that it has been seen for Fe_2O_3; however, on UO_2 down to 300 psi he found none. Nichols commented that the concept of a threshold stress has been getting a lot of activity by the creep people. Everything was on metals until the paper on UO_2. He further stated that he does not believe there is any contradiction between their work and the work of Seltzer. Seltzer's work probably did not cover the range of variables for which a threshold stress has been seen in UO_2. Clauer responded that their results were on fine-grain urania, approximately 6-μm grain diameter, which showed no sign of a threshold stress, although their lowest stress levels were below Burton's in the same temperature range. The most obvious difference between the two sets of experiments is that the threshold stress was observed in stoichiometric UO_2 during creep in a hydrogen atmosphere, whereas the threshold effect was absent down through the lowest stress levels of about 300 psi for creep in nonstoichiometric UO_2. The stoichiometry may influence the threshold stress. Nichols reiterated that the threshold stress is a documented effect in metals. It was first seen in metals where there were particles in the grain boundaries; since then it has been seen in cases where no particles are present in the grain boundaries. Probably, people did not see it previously because they did not go low enough in stress and low enough in temperature. Stoichiometry effects may come in as well, as Dr. Clauer mentioned.

Stringer asked if anyone would comment on why plasticity increases with departure from stoichiometry, as this has certain implications in the oxidation of metals. Seltzer replied that in creep studies of both n- and p-type lead sulfide, minimum creep rate and a maximum activation energy were found for stoichiometric lead sulfide, as is the case for uranium carbide and CoO.

After some discussion on the apparently high Al^{+3} diffusion rates one deduces from creep of Al_2O_3, Niesz offered a possible explanation. In some of the work on UO_2 at Battelle, high creep rates were found for isolated samples that appeared by standard microstructural and chemical analysis to be the same. These samples never really showed true steady-state creep; rather, they pulled apart at the grain boundaries with little or no steady-state creep. This problem was traced to a UO_2-SiO_2 eutectic that penetrated the boundary-line intersections. The total silicon concentration was only about 100 ppm. These samples were sintered at temperatures which exceeded the UO_2-SiO_2 eutectic temperature. Most oxide creep data in the literature are for samples which show grain-boundary separation. The apparently high diffusion rate for the aluminum ion seen in creep may be a result of a similar factor. Niesz emphasized that his experience with UO_2 and the general low solubility of oxides, particularly alumina, for impurities (particularly SiO_2) suggest that grain-boundary separation observed in oxide creep may not be intrinsic to oxides.

In this regard, Heuer commented that he has used the electron microscope to look at the microstructure of fine-grained, polycrystalline alumina crept in bending to small strains (less than 3 percent). As shown in the figure, the grain boundaries looked quite sound. Small triple-point voids developed owing to nonconservative grain-boundary sliding. For samples crept to larger strains in compression, the grain boundary began to separate owing to growth of triple-point voids. In fact, the strain rate $\dot{\epsilon}$ is inversely proportional to the grain size in these latter studies, because of the predominance of nonconservative grain-boundary sliding in the creep process.

Returning to the field of sintering, Kofstad stated that in the growth of compact scales during the oxidation of metals, the rate-determining step for electronically conducting scales is the diffusion of ions. For ionically conducting scales, the growth rate is determined by the transport of electrons. Kofstad wondered if there are any aspects of sintering in ionically conducting compounds where electron transport could be a rate-determining factor. Possible cases could be defects having zero effective charge or oxygen being transported through the gas phase. To this suggestion, Johnson replied that transfer of oxygen through the vapor phase while the cation moves through the crystal could occur only at the particle surfaces during sintering, and this would not cause densification. Densification requires movement of matter from the grain boundaries to the pores. Parallel cation diffusion does not require electron transfer. However, Anderson felt that Dr. Kofstad's point was relevant to the situation in initial stages of sintering where there are parts of the system with high curvature. The shift of chemical potentials resulting from curvature implies a change in equilibrium pressure and a change in composition in the surface layer of a nonstoichiometric system. There will be some transfer of oxygen through the gas phase in this case. Johnson agreed that this could apply but pointed out that one must remember that the vapor transport of the oxygen could occur only on the surface of the particles. In the case of grain-boundary grooving in nickel oxide, it has been proposed that the nickel diffuses through the bulk of the oxide, while the transport of the oxygen is by vapor transport. Nickel oxide is a semiconductor so there is no difficulty in the transfer of charge. Welch wondered how great a limiting factor charge transfer would be in general, since he felt that transfer of electrons could be very rapid along the surface of the particles.

Goodenough asked whether, in the sintering of Al_2O_3, having trapped air in the pores, one can sinter to theoretical density or does one find that the oxygen partial pressure is not sufficient to put oxygen interstitials into the lattice? Jorgensen responded that if you sinter in oxygen, there is enough solubility of the oxygen in alumina to get rid of it. If one sinters in nitrogen or argon, however, pressures develop and one cannot get rid of all the porosity. If one sinters in air, one can densify a lot further than one can in nitrogen. Kröger, however, demurred from this rationale and stated that it was his opinion that interstitials had nothing to do with it and that the main point is the diffusion coefficient of oxygen, independent of interstitials. Goodenough then asked about the case of transition metal oxides, and Jorgensen replied that, as far as he knew, when you sinter in pure oxygen, you are not limited in sintering oxides.

Seltzer asked if there were a general rule that the covalently bonded materials such as silicon nitride cannot be sintered to a theoretical density, and Jorgensen responded that they do have a difficult time sintering covalent compounds. Heuer, however, pointed out that they are easy to hot press. Jorgensen then commented that in the case of hot pressing of silicon nitride, one puts a little bit of magnesium in and probably ends up with a bit of liquid phase, giving liquid-phase sintering.

Nichols, in response to an earlier comment by Niesz, returned the discussion to the creep of oxides and stated that one has to be very careful in interpreting steady-state creep, particularly at high strains. This can be very bad in bending but is not so serious in compression. Niesz replied that their experiments were in compression.

Ilschner then stated that considerable strains have to be built up before the steady-state condition can be obtained. If one is just interested in the relative effect of factors like temperature, oxygen partial pressure, or impurity levels on creep rate, it is sufficient to compare creep-rate data from within the transient region since the basic mechanisms involved are the same. Rates should be compared at approximately equal strains, however.

After some discussion on possible mechanisms for low-stress creep, Nichols stated that in Nabarro-Herring creep, the grain should show as much strain as the sample does. Ilschner did not agree but felt that the grain boundaries would migrate and that, under normal circumstances, the grain boundaries will reshape during creep. Nichols questioned whether this is not merely a rationale for not seeing grain elongation and pointed out that the lack of grain elongation during low-stress creep is one of Ashby's main arguments for his mechanism of diffusion-accommodated flow.

Part Five

DEFECT DEPENDENT PROCESSES II
(Oxidation, Solid State Reactions, Electrochemistry)

DIFFUSIONAL GROWTH OF OXIDE SCALES ON BINARY ALLOYS

W. W. Smeltzer

Department of Metallurgy and Materials Science
McMaster University
Hamilton, Ontario, Canada

ABSTRACT

A ternary diffusion model is invoked to describe the high-temperature oxidation of binary alloys when parabolic kinetics represent the growth of a single or multilayer scale and internal oxidation. Equations derived by Wagner for metal diffusion through p-type ternary oxide scales under chemical-activity gradients are shown to lead to a description of the scaling kinetics of several iron-, nickel-, and cobalt-base binary alloys. The consistency of this model is tested by evaluating the metal diffusion coefficients and the nonstoichiometry of the ternary oxides, and extrapolating these parameters to those of the oxides formed by parabolic scaling of the pure metals.

1 INTRODUCTION

It is appropriate to consider kinetic and thermodynamic interactions encountered in ternary systems when interpreting the high-temperature oxidation properties of binary alloys. The kinetics are usually controlled by diffusion since the reactants are separated by a superficial scale soon after the beginning of the reaction. In this situation, the free-energy change takes place over a large distance, and any microscopic volume element is substantially at equilibrium with itself and characterized by chemical potentials for each species. Wagner[1] has developed a model to describe scale growth using ternary-diffusion-flux equations, together with appropriate boundary conditions. He was concerned with methods for calculating the local distributions of the metals in the scale on an alloy exhibiting negligible oxygen solubility as a basis for calculating the scaling rate. Coates and Dalvi[2] and Whittle[3] have extended this analysis to account for oxygen solution and internal oxide precipitation in the alloy.

Results dealing with metal distributions in the scales determined by the electron-microprobe technique can be assessed using the diffusion equations. One seeks correlations between the parabolic oxidation kinetics of the pure metal and for the metal with increasing concentration of the alloying element, because relations of the metal diffusivities with the point defects and thermodynamic properties of the ternary oxides are obtained from the binary oxides. The alloys considered here are composed of metals ranging from chromium to nickel in the first transition series, and they form scales containing the metal conducting p-type monoxides of NaCl structure.

2 DIFFUSION FORMALISM

According to the phenomenological diffusion theory of Onsager[4,5], the flux-force relations are linear for a system near equilibrium. The flux J_i of component i in an isotropic ternary solid subjected to chemical gradients is

$$J_i = -\sum_{k=1}^{3} L_{ik} \frac{\partial u_k}{\partial x} = -\sum_{j=1}^{2} D_{ij} \frac{\partial c_j}{\partial x} , \tag{1}$$

where L_{ik} and D_{ij} are mobilities and diffusivities, respectively, and $\partial \mu_k / \partial x$ and $\partial c_j / \partial x$ are the chemical-potential and concentration gradients. Assuming that $L_{ii} \gg L_{ik}$ and that the mobility is related to the self-diffusion coefficient D_i by the Nernst-Einstein relation[6,7], Eq. (1) reduces to

$$J_i = -L_{ii} \frac{\partial \mu_i}{\partial x} = -c_i D_i \frac{\partial \ln a_i}{\partial x} , \qquad (2)$$

where a_i is the activity of component i.

It is convenient to use Eq. (2) for the reactant flux in a scale, since strong thermodynamic interactions generally prevail. The alternative formalism using diffusivities and concentrations is more suitable for considering the alloy if the expressions for the activity coefficients, γ, are known. Thus, for the ternary phase (binary alloy containing oxygen in solid solution),

$$D_{ii} = D_i\left(1 + c_i \frac{\partial \ln \gamma_i}{\partial c_i}\right);\ D_{ij} = D_i c_i \frac{\partial \ln \gamma_i}{\partial c_j}\ ;\ \ i, j = 1,2\ . \qquad (3)$$

If the alloy exhibits negligible oxygen solubility, metal diffusion can be interpreted utilizing the Darken equation for the interdiffusion coefficient.[8]

Since the concentration of one component in a ternary system is dependent under isothermal and isobaric conditions, the two independent concentrations occur as parametric solutions of the diffusion equations. Consequently, the unique diffusion or composition path in a ternary diffusion couple can be plotted on the ternary isotherm. Surveys describing construction of these paths and their use for interpreting alloy-oxidation phenomena were recently presented by Kirkaldy[9] and Dalvi and Coates[10]. Diffusion and composition paths have proven useful for elucidating the oxidation properties of binary copper, iron, nickel, and zirconium alloys.[9-11]

3 DIFFUSION MODEL

The alloy AB is regarded as one or two phase and to be given as a plate virtually infinitely thick. Since the solids are considered as isotropic but with different molar volumes, the volume change associated with oxidation is taken to cause one-dimensional plastic flow in the oxide normal to the alloy surface. Equations are therefore formulated for diffusion in the x direction only. The scale, which is treated as compact oxide exhibiting predominantly electronic conductance, thickens by the outward diffusion of metal cations. Small deviations from stoichiometry of the ternary oxide are presumed to occur by point defects and the binary oxides dissolved in the scale are denoted as AO and BO. The molar volumes of the ternary oxide solid solution (AB)O and the alloy are regarded as not being dependent on composition.

The model for describing growth of a scale is shown in Fig. 1. It is advantageous to use two diffusion reference frames: the u frame whose origin is at the original surface describes transport in the alloy; the x frame, which is attached to the scale/alloy interface, describes transport in the oxide. The alloy solute and oxygen compositions $N_B{}^o$, $c_O{}^o$ correspond to boundary values at u = ∞. In most cases $c_O{}^o \simeq 0$. The values of the solute metal composition ξ'' and the oxygen activity a_O'' at the outer oxide surface serve as remaining boundary parameters. (Single and double primes refer to the inner and outer scale interfaces throughout the text).

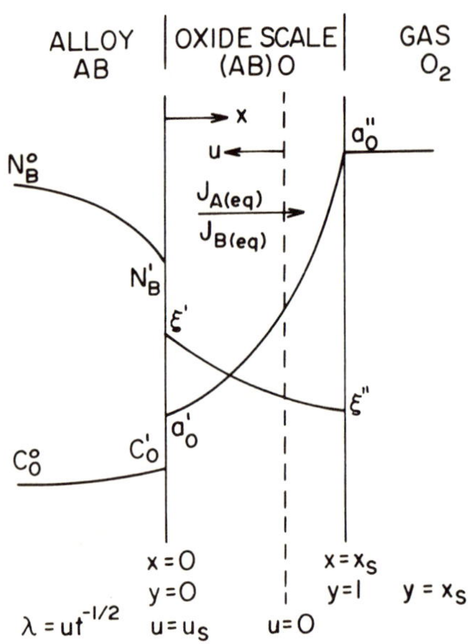

Fig. 1. Ternary diffusion model for growth of an oxide scale on a binary alloy with notations used in the text.

Wagner[1] expressed the flux of metals A and B relative to the oxygen sublattice of the oxide in equivalents per unit area by Eq. (2). Upon relating the metal activities to the oxide activities a_{AO} and a_{BO}, and the oxygen activity a_O, he reformulated this equation as

$$J_{A(eq)} = -D_A\left(\frac{1-\xi}{V_{eq}}\right)\left(\frac{\partial \ln a_{AO}}{\partial \xi}\frac{\partial \xi}{\partial x} - \frac{z_A}{z_O}\frac{\partial \ln a_O}{\partial x}\right) \qquad (4)$$

$$J_{B(eq)} = -D_B\,\frac{\xi}{V_{eq}}\left(\frac{\partial \ln a_{BO}}{\partial \xi}\frac{\partial \xi}{\partial x} - \frac{z_B}{z_O}\frac{\partial \ln a_O}{\partial x}\right)\,, \qquad (5)$$

where $(1-\xi)/V_{eq}$ and ξ/V_{eq} are the local cation concentrations, z is a valency, and $\partial \ln a_{AO}/\partial \xi$ and $\partial \ln a_{BO}/\partial \xi$ are known quantities characteristic of the quasi-binary mixed oxide (AB)O.

If a negligible amount of oxygen dissolves in the alloy, the metal flux is related to the scale-thickening rate by

$$\left(J_{A(eq)} + J_{B(eq)}\right) V_{eq} = \frac{dx_s}{dt} = \frac{k}{2\sqrt{t}} \quad , \tag{6}$$

where k is the parabolic oxidation-rate constant. By defining the fractional distance within a scale by the dimensionless parameter $y = x/x_s$, the parabolic scaling constant is given by the following first-order differential equation upon substituting Eqs. (4) and (5) into Eq. (6)

$$D_A(1-\xi)\left(-\frac{\partial \ell n a_{AO}}{\partial \xi}\frac{d\xi}{dy} + \frac{z_A}{z_O}\frac{d\ell n a_O}{dy}\right) + D_B\xi\left(-\frac{\partial \ell n a_{BO}}{\ell \xi}\frac{d\xi}{dy} + \frac{z_B}{z_O}\frac{d\ell n a_O}{dy}\right) = k \quad . \tag{7}$$

Applying the continuity equation to Eq. (7) for the solute metal, the second-order differential equation

$$yk\frac{d\xi}{dy} = \frac{-d}{dy}\left[D_B\xi\left(\frac{\partial \ell n a_{BO}}{\partial \xi}\frac{d\xi}{dy} - \frac{z_B}{z_O}\frac{d\ell n a_O}{dy}\right)\right] \tag{8}$$

is obtained to describe the metal profiles across the scale.

The distribution of the alloying element in the alloy (see Fig. 1) is a solution of the second-order differential equation

$$\frac{d}{d\lambda}\left(D_{AB}\frac{dN_B}{d\lambda}\right) + \frac{1}{2}\lambda\frac{dN_B}{d\lambda} = 0; \lambda = \frac{u}{\sqrt{t}} \quad , \tag{9}$$

where D_{AB} is the alloy interdiffusion coefficient. If diffusion in the alloy is much more rapid than in the scale, N_B' obviously approximates closely the alloy composition N_B^o. On the other hand, when diffusion in the alloy is the slowest step, the ratio of the metal fluxes in the scale at y = 0 equals the ratio of the metal equivalents in the alloy. Equations (4), (5), and (7) give

$$\xi_{alloy} = \frac{z_B N_B^o}{z_A N_A^o + z_B N_B^o} = \frac{1}{k}\left[D_B\xi\left(\frac{-\partial \ell n a_{BO}}{\partial \xi}\frac{d\xi}{dy} + \frac{z_B}{z_O}\frac{d\ell n a_O}{dy}\right)\right]_{y=0} \tag{10}$$

The mass balance for B at the scale/gas interface

$$J_{B(eq)} = \frac{\xi''}{V_{eq}} \left(\frac{dx_s}{dt} \right)_{x=x_s} \tag{11}$$

becomes, upon substituting Eqs. (5) and (6), the boundary expression

$$-\left[D_B \xi \left(\frac{\partial \ln a_{BO}}{\partial \xi} \frac{d\xi}{dy} - \frac{z_B}{z_O} \frac{d \ln a_O}{dy} \right) \right]_{y=1} = \xi'' k \quad . \tag{12}$$

A phenomenological diffusion model is therefore available for describing scale growth if the cation diffusivities in the ternary oxide, D_A and D_B, are known as functions of composition ξ and oxygen activity a_O. The oxidation constant k, ξ and a_O as a function of the scale dimensionless parameter y, and N_B as a function of the metal-distance time parameter λ can be calculated from the first-order differential equation [Eq. (7)] and the second-order differential equations [Eqs. (8) and (9)] using Eqs. (10) and (12) as auxiliary expressions.

4 TERNARY METAL OXIDE PROPERTIES

The previously deduced differential equations are solvable only by numerical methods. All of the required parameters, moreover, are not known for any specific system. It is necessary accordingly to establish the effect of cations exhibiting variable valence on the defect structures and the diffusivities in ternary transition metal oxides.

If association of cation vacancies and positive holes is neglected, the defect equations for pure AO and BO can be written as

$$1/2\ O_2 + \alpha A^{2+} = O^{2-} + V_M^{\alpha'} + \alpha A^{3+} \tag{13}$$

$$1/2\ O_2 + \alpha B^{2+} = O^{2-} + V_M^{\alpha'} + \alpha B^{3+} \quad . \tag{14}$$

$V_M^{\alpha'}$ is a cation vacancy with α negative charge relative to the lattice, and the metal and oxygen ions occupy normal lattice sites. Upon mixing these oxides to form a solid solution, substitution of equally charged cations sustains electrical neutrality. If the formation energy of B^{3+} from B^{2+} is much less than this energy for the A cation, the reaction in Eq. (14) will shift to the right and lead to a higher vacancy concentration. When the equilibrium constant K for this reaction is much larger, $X_{B^{2+}} > X_{B^{3+}}$ and the vacancies are in the doubly ionized state, the variation of vacancy concentration with metal composition and oxygen pressure is given by

$$X_V = (K/4)^{1/3}\, p_{O_2}{}^{1/6} \xi^{2/3} \tag{15}$$

Since the ternary oxide is treated as being composed of various proportions of AO and BO, the Gibbs-Duhem equation for an ideal or Henrian solution becomes

$$(1-\xi)d\ell n a_{AO} + \xi d\ell n a_{BO} = 0; \quad \frac{d\ell n a_{BO}}{d\ell n \xi} = 1 \ . \tag{16}$$

The self-diffusion coefficient D_M for cation M^{2+} is

$$D_M = D_V(M) X_V \, ; \, M = A, B \quad , \tag{17}$$

where $D_V(M)$ is the vacancy diffusivity. Thus, from Eqs. (15) and (17),

$$D_A = D_V(A)\,(K/4)^{1/3} p_{O_2}{}^{1/6}\ \xi^{2/3} \, ;$$

$$D_B = D_V(B)\,(K/4)^{1/3} p_{O_2}{}^{1/6}\ \xi^{2/3} \quad . \tag{18}$$

When interactions between the defects do occur, measurements of the self-diffusivities at different oxide compositions are needed, or a method is needed for determining the vacancy diffusivities from a more sophisticated model. The ternary oxides to be considered here are of the cubic rock-salt structure. It is reasonable that the ratio of the vacancy diffusivities defining the jump frequencies of the two cations into an adjacent vacancy be equated as a first approximation to this factor for the constituent binary oxides. Equation (17) gives

$$D_A \simeq D_A^o \frac{X_V}{X_V^{AO}} \, ; \, D_B = D_B^o \frac{X_V}{X_V^{BO}} \quad , \tag{19}$$

where D_A^o and D_B^o are the tracer diffusion coefficients in the pure oxides containing vacancy atom fractions $X_V{}^{AO}$ and $X_V{}^{BO}$. Hence,

$$D_B = pD_A \quad , \tag{20}$$

where the constant p is the ratio of the cation mobilities.

5 OXIDE GROWTH ON PURE METALS

Equation (2) describes reactant transport by ambipolar diffusion through an oxide layer of prevailing electronic conductance. It gives the classical expression derived by Wagner for the rational parabolic rate constant upon integrating the flux across the instantaneous layer thickness[7]:

$$k_r(\text{eq/cm sec}) = \bar{c}_{eq} \int_{a_O'}^{a_O''} \frac{z_A}{|z_O|} D_A \, d\ln a_O \tag{21}$$

This rate constant for wustite (FeO) growth on iron was evaluated by Himmel et al.[12] using Eq. (21) and determinations of the iron-tracer diffusion coefficient and oxygen activities for this oxide over its sub-stoichiometric field. As shown in Table 1, the calculated and experimental values are in close agreement. Results from other investigations on the growth of wustite on iron at temperatures higher than 700°C in oxygen[13,14], CO_2-CO[15], and H_2O-H_2[16] atmospheres are also in accord with these theoretical predictions.

*Table I. Measured and Calculated Parabolic-Oxidation-Rate Constants**

Temp., °C	Experimental	Calculated	K_p(exptl.)/K_p(calcd)
	$Fe + 1/2\ O_2 = FeO$[12]		
800	2.3×10^{-4}	2.3×10^{-4}	1.0
897	5.0×10^{-4}	4.8×10^{-4}	1.04
983	8.2×10^{-4}	7.7×10^{-4}	1.07
	$Co + 1/2\ O_2 = CoO$[17,18]		
1000	1.56×10^{-4}	1.65×10^{-4}	0.90
1148	3.05×10^{-4}	3.35×10^{-4}	0.88
1350	8.85×10^{-4}	8.26×10^{-4}	1.16

*Pressure: 1 atm.

Parabolic rate constants for the scaling of cobalt to cobaltous oxide (CoO) were estimated by Carter and Richardson using determinations of the cobalt-tracer diffusion coefficient.[17,18] The values of this diffusivity were proportional to $(p_{O_2})^{1/n}$, but 1/n varied from 0.27 to 0.35 in the range 1100 to 1300°C. Equation (21) was therefore integrated to obtain the following expression:

$$k_r = \bar{c}_{eq} \frac{z_{Co}\, n}{|z_O|^2} D^o_{Co} \left[(p_{O_2})^{1/n} - \left(p^D_{O_2}\right)^{1/n} \right], \tag{22}$$

where D^o_{Co} is the tracer diffusivity at 1-atm pressure and $p^D_{O_2}$ is the dissociation pressure of the oxide. The calculated and experimental values, Table 1, are approximately equal.

Khoi[19] has recently demonstrated that these principles apply to the oxidation of nickel at temperatures exceeding 900° C. This is strikingly illustrated by plotting the parabolic rate constants for nickel oxide (NiO) formation in Arrhenius form as shown in Fig. 2. The full lines represent the temperature coefficient of the rate constants estimated from nickel-tracer diffusion in nickel oxide crystals.[20-22] Values obtained by graphical integration of Eq. (21) were determined from the diffusivities obtained at different oxygen pressures.[21] The atom fraction of vacancies in nickel oxide is extremely small, and assuming that they are divalent, Eq. (21) reduces to $k_r = 3\ \overline{c}_{eq}\ D_{Ni}$. This formula was also used to calculate k_r. The plots demonstrate that the calculated values by both methods are consistent with those obtained experimentally above 900° C. The much larger experimentally determined parabolic oxidation constants at lower temperatures result from more rapid nickel diffusion along boundaries of the small oxide grains in the scales.[26-28]

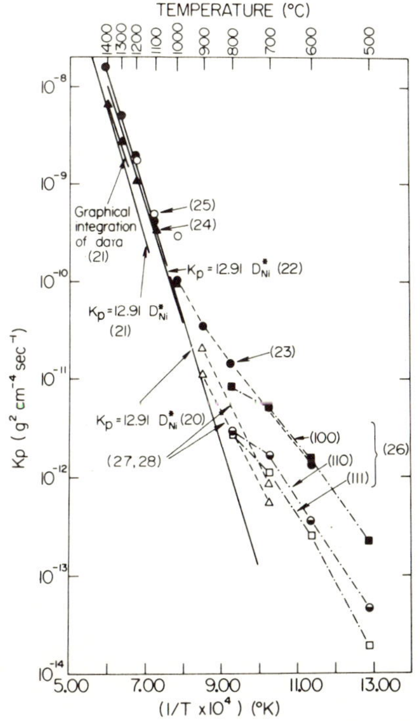

Fig. 2. The temperature coefficient of the parabolic oxidation constants for nickel oxide formation on polycrystalline and crystal faces of nickel (>99.995% pure).[19]

The agreement of the calculated and experimental parabolic oxidation constants for these transition metals is adequate for interpreting the influence of alloying elements on the scaling properties of the metals. Unfortunately the values for these constants and the metal diffusivities are not sufficiently accurate for distinguishing refinements that could be applied to account in more detail for the properties of the point defects in the oxides. It was assumed that a tracer diffusivity was directly related to the cation mobility and the vacancy concentration. This latter concentration was regarded as proportional to the oxide nonstoichiometry. Moreover, the mobility was assumed to be constant and the vacancies to exhibit ideal or Henrian solution properties. Continuing work is showing, however, that vacancies in these oxides may occur in several electrically charged states and interact to form complexes which may even in the case of wustite lead to secondary phases in the nonstoichiometric field. Accordingly, the explanation offered for oxidation of the metals based upon rather simplified assumptions for the defect properties of the oxides must be regarded as an approximation to the reaction mechanism.[29] These limitations also apply to the properties of the ternary oxides on the basis of those for the binary oxides since the influence of the above factors on diffusion has yet to be distinguished.[30,31]

6 OXIDE GROWTH ON BINARY ALLOYS

Sufficient appropriate results on the oxidation of a binary alloy were not available for analysis when Wagner introduced the model in 1969. He was able to predict, notwithstanding, features of the oxidation properties of Ni-Cr alloys. More recently, additional information has become available on the distribution of cations in the scales formed on Ni-Co[32,33], Ni-Fe[34,35], Fe-Mn[36], and Co-Fe[37] alloys which can be employed for more fully testing predictions from the model using compositional, diffusional, and thermodynamic properties of the binary and ternary oxides.

Ni-Cr Alloys: The scale formed on nickel containing small chromium additions is (NiCr)O in which the cation-vacancy concentration is determined by trivalent chromium substituting for divalent nickel according to the reaction

$$CrO_{1.5} = Cr^{3+}_{Ni} + 1/2\ V''_{Ni} + 3/2O^{2-}\ . \tag{23}$$

$NiCr_2O_4$ becomes stable with trace amounts of ~10 ppm of chromium in nickel, and the stability field of nickel oxide extends to an alloy chromium content <1 wt %.[10] Accordingly, it was assumed that the metal interface was substantially pure nickel owing to depletion of chromium by selective oxidation. For $N^{o}_{Cr} \ll 1$, $\xi \ll 1$, $Cr^{3+}_{Ni} = 2X_V = \xi$ and $d\ell n a_{CrO_{1.5}}/d\ell n\xi = 1.5$, Eqs. (7) and (8) reduce to

$$D^{o}_{Ni}\xi \frac{d\ell n a_O}{dy} = k \qquad (24)$$

$$\xi \frac{d^2\xi}{dy^2} + \left(\frac{d\xi}{dy}\right)^2 - \frac{y-p}{1-p}\left(\frac{d\xi}{dy}\right)_{y=1} \frac{d\xi}{dy} = 0 \quad , \qquad (25)$$

where $p = 3D^{o}_{Cr}/2D^{o}_{Ni}$. Using results for T = 1000° C, ξ was estimated from the above equations for different values of p. If $p < 1$, distinct chromium depletion occurs in the scale near its outer interface. Conversely, if $p > 1$, chromium accumulation occurs near this interface.

The first limiting case is in reasonable accordance with results presently available. The chromium diffusivity in nickel oxide, according to Eq. (23), is directly dependent on its concentration. Perkins and Rapp[38] have shown that this is nearly so for small chromium concentration; however, diffusion involves chromium associated with a vacancy as a two-member complex. The ratio of the chromium to nickel mobility is ~0.1. Thus, chromium is depleted in the scale over its outer region corresponding to the case $p < 1$. This conclusion can be only qualitatively confirmed from presently known results, since internal oxidation and inclusions of $NiCr_2O_4$ interfere with precise determination of the metal profiles in the (NiCr)O scale.[39]

Ni-Co Alloys: Metal distributions in the phases of these alloys oxidized in oxygen at 1000°C were determined by Wood and Ferguson.[32] Dalvi and Coates[33] carried out a comparison of the cobalt gradients in the (NiCo)O scales to those calculated by invoking known properties of the alloy-oxide composite system. In this case, cobalt was enriched towards the scale surface by diffusing against its own concentration gradient.

Cobalt and nickel oxides form a continuous solid solution which approximates closely the Henrian solution properties.[40] The vacancy concentration in (CoNi)O decreases nearly exponentially with increasing nickel content.[41] In view of these findings, Dalvi and Coates[33] suggested that the atom fraction of vacancies is given by

$$X_V = \left[X_V^{NiO}\right]^{1-\xi} \left[X_V^{CoO}\right]^{\xi} a_O^{1/\nu} \quad , \qquad (26)$$

where X_V^{NiO} and X_V^{CoO} are the vacancy fractions in the pure oxides and ξ is the equivalent fraction of CoO in (CoNi)O. The following relationships are therefore obtained upon substituting Eqs. (19), (20), (26), and literature results for 1000° C into Eq. (17):

$$D_{Co} = D^o_{Co} A^{(\xi-1)} a_O^{1/\nu} ; \; D_{Ni} = pD^o_{Co} A^{(\xi-1)} a_O^{1/\nu} \quad , \tag{27}$$

where $A = X_V^{CoO}/X_V^{NiO} = 6$ and $p = D^o_{Ni}/D^o_{Co} = 0.5$.[33]

These expressions were substituted into Eqs. (7) and (8) using Eqs. (10) and (12) for material balance at the scale interfaces as auxiliary equations. These equations were solved simultaneously using a computer to obtain ξ and a_O as a function of y using trial values of ν. The best result was obtained for $\nu = 2.5$, which corresponds to a $p_{O_2}^{1/5}$ dependence for the vacancy concentration (Henrian or ideal solution of divalent vacancies gives a $p_{O_2}^{1/6}$ dependence). The agreement of the calculated to the observed cobalt distribution in the scale shown in Fig. 3 appears to be satisfactory and consistent with the diffusion model. Thus, a relatively minor difference equal to a ratio of 2.0 in the metal mobilities, which agrees with recent diffusion measurements[42], causes a significant enrichment of the more mobile cobalt in the scale near its outer interface. The estimated oxygen-activity gradient is also shown in Fig. 3. In this case, regions of steeper oxygen-activity gradient resulted in a steeper gradient of cobalt due to its slightly more rapid mobility.

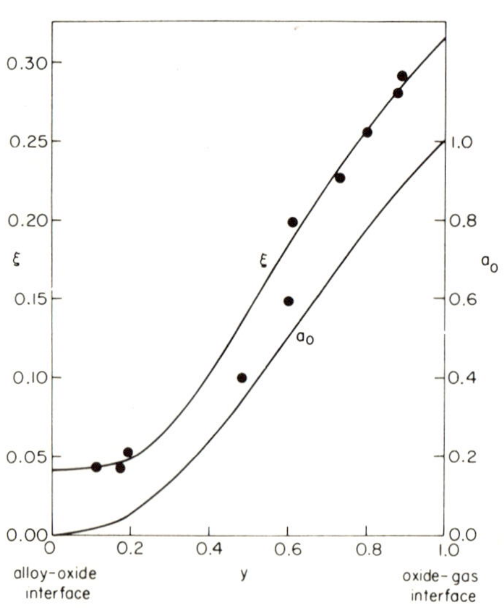

Fig. 3. Equivalent fraction of CoO, ξ, and the oxygen activity, a_O, as a function of normalized distance coordinate y in the (NiCo)O scale formed on a Ni-10.9% Co alloy oxidized at 1000°C in oxygen; solid line represents calculated values; experimental points are from ref. 32.

Fe-Mn Alloys: Parabolic kinetics for growth of (FeMn)O scales are obtained upon oxidizing these alloys in CO_2-CO atmospheres. The rate constants and compositions of scales formed on the alloys containing up to 60 percent manganese at 1000°C were determined by Mayer and Smeltzer.[36] The plots in Fig. 4 illustrate that the manganese concentrations in the scales, which decreased slightly with increasing thickness, were approximately the same as the alloy compositions. The relatively flat manganese profiles correspond to linear gradients.

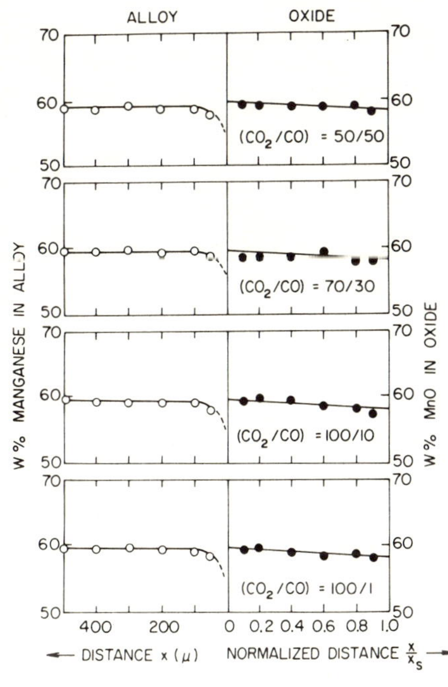

Fig. 4. Manganese-concentration profiles in the metal and (FeMn)O scale of an Fe-59% Mn alloy oxidized in CO_2-CO atmospheres at 1000°C; scale thickness is standardized to $y = x/x_s$.[36]

It is a characteristic property of the diffusion model that the scale and alloy approximate to the same solute composition if the cation diffusivities in the scale are much larger than the interdiffusion coefficient for the alloy. This factor is $\sim 10^4$ for this system. Since the metal gradients in the scale are linear, Eq. (7) for the metal flux and Eq. (12) for the material balance at the scale surface were solved to determine the relative mobilities of iron and manganese in (FeMn)O. This solid solution is practically ideal.[43] Substituting $d\ell n a_{MnO}/d\ell n\xi = 1$ and $b = d\xi/dy$ into Eqs. (7) and (12) gave the following simple equation for the self-diffusivities:

$$D_{Mn} = pD_{Fe} = \frac{k\xi''(p-1)(1-\xi'')}{b} \quad . \tag{28}$$

A value of $p = 0.99$ was obtained. This small mobility difference of the cations accounted for the very small manganese depletion across a scale. Knowing p, the iron self-diffusion coefficient was evaluated as a function of manganese content in (FeMn)O.

Since the value of the iron diffusivity in wustite is proportional to nonstoichiometry[12,44], Eq. (19) may be written as

$$D_{Fe} = D^{o}_{Fe} \frac{X_V}{X_V^{FeO}} = D^{o}_{Fe} \frac{(X-1)X^{o}}{(X^{o}-1)X} \quad , \tag{29}$$

where X^{o} and X are the ratios of the atom fractions of oxygen to metal in pure wustite and the ternary oxide. The nonstoichiometry of (FeMn)O was calculated and a comparison of these values to those determined by chemical analyses is shown in Fig. 5. They are in good agreement, which offers substantiation for the diffusion model adopted to describe scaling of these alloys.

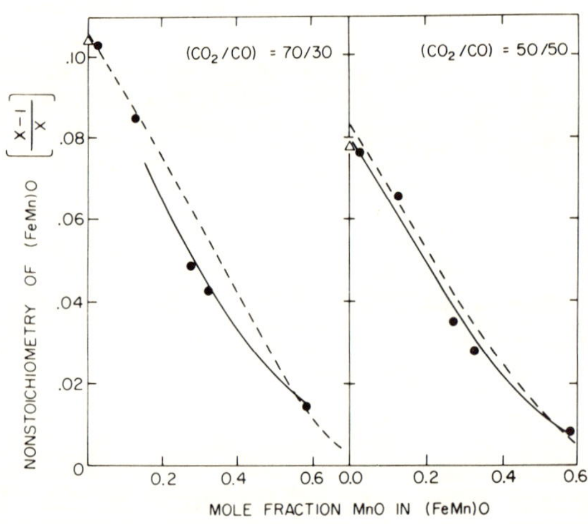

Fig. 5. Nonstoichiometry of the (FeMn)O ternary oxide at 1000°C[36]; dashed curve represents values obtained on equilibrating (FeMn)O in atmospheres undergoing the water-gas reaction.

Co-Fe Alloys: Cobaltous and ferrous oxides form a nearly ideal solid solution in all proportions. Aukrust and Muan[45] determined the composition of (CoFe)O at 1200° C subjected to oxygen pressures between 10^{-9} and 1 atm. The metal-tracer diffusion coefficients were determined by Crow[22] for this oxide containing very dilute iron concentrations. At 1200° C, D_{Fe}/D_{Co} = 1.6 at pressures to 10^{-3} atm and 0.7 at pressures greater than 10^{-1} atm. Metal diffusion in (CoFe)O is much more rapid than in the alloys.

We would expect in light of these findings that oxidation of the alloys would lead essentially to superficial scale formation. This was found by Mayer and Smeltzer.[37] Alloys containing up to 10 percent iron exposed at 1200° C to oxygen at 10^{-4} and 10^{-3} atm oxidized parabolically as a result of (CoFe)O diffusional scale growth. Because iron diffused more rapidly than cobalt, its concentration was larger with increasing outward distance in the scales. This feature is shown by the plots in Fig. 6.

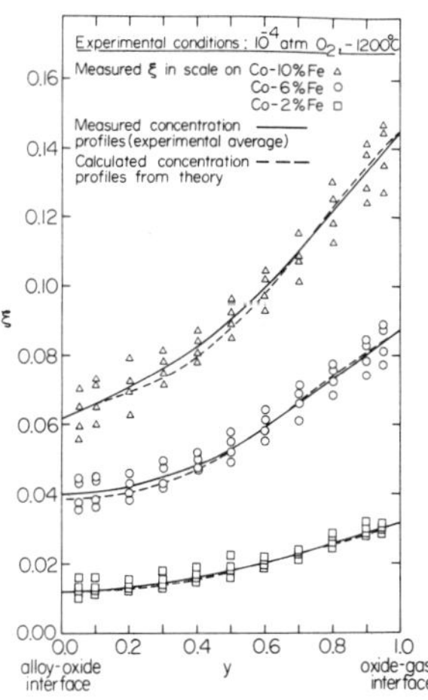

Fig. 6. Iron composition within the (CoFe)O scales formed on Co-Fe alloys containing 2, 6, and 10% Fe at 1200°C in oxygen at 10^{-4} atm.[37]

An attempt to interrelate the parabolic oxidation constants with the normalized iron gradients was confined to the scales formed at these low pressures in order that the ratio of the metal diffusivities remained constant (p = 1.6). Owing to the paucity of results on the defect structure of (CoFe)O, an empirical relationship was assumed for its non-stoichiometry as a function of iron composition and oxygen activity. It was obtained by postulating that the cation vacancies predominantly arose from iron being present in the oxide as the trivalent state [reaction in Eq. (14) displaced far to the right]. A plot

$$X_V = ba_O^n \xi \quad , \tag{30}$$

where b and n are empirical constants, was an adequate relation for iron contents encountered in the scales. The vacancy concentration was then related to the diffusivities by Eq. (17) and the metal compositions to oxygen activities by Eq. (16).

Equations (7) and (8) describing the metal flux and its divergence, respectively, were simultaneously numerically integrated, knowing boundary values at y = 0 from auxiliary Eq. (10) for material balance at the inner scale interface. The generated iron profiles are shown in Fig. 6. The calculated profiles correspond to those experimentally obtained in both magnitude and form. Since iron diffused slightly more rapidly than cobalt, it was most significantly enriched in the outer regions of the scales where the influence of the oxygen-activity gradient was most pronounced. This feature, nevertheless, is not general since subsequent work has shown that scales formed at oxygen pressures higher than 10^{-3} atm exhibit bell-shape profiles, presumably, as a result of a variation in the ratio of the metal diffusivities for (CoFe)O with oxygen activity.[46]

Ni-Fe Alloys: Parabolic oxidation kinetics of alloys containing up to 25 percent iron at 1000°C account for growth of superficial scales containing an inner (NiFe)O layer and an outer two-phase $NiFe_2O_4$-(NiFe)O layer, and for penetration of internal oxidation zones containing $NiFe_2O_4$ precipitates.[34,35] A model exhibiting typical metal profiles is given in Fig. 7. The iron concentration increases in the outer region of the (NiFe)O layer and it subsequently decreases slightly in the two-phase layer. Thus, iron diffuses against its own concentration gradient only in part of the scale.

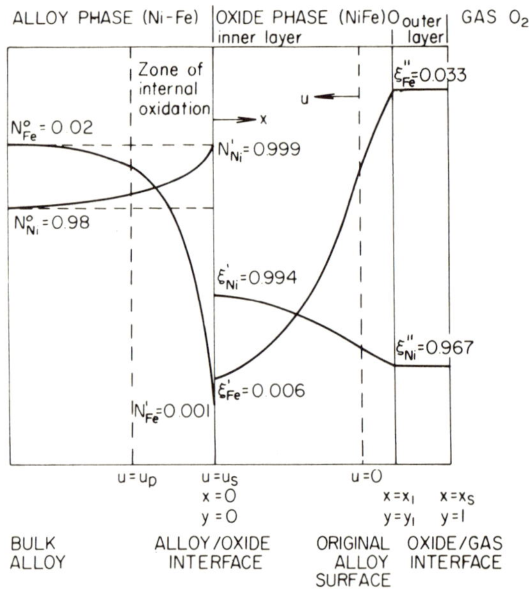

Fig. 7. Diffusion model for the oxidation of a Ni-2% Fe alloy oxidized in oxygen at 1000°C.[35]

One can interpret the oxidation mechanism despite these complicated features. Since information is available on the thermodynamics of the Fe-Ni-O system[47,48], the metal- and oxygen-activity gradients were estimated from the diffusion equations by modifying the boundary conditions to meet those for growth of a duplex scale. It was possible to obtain a tractable mathematical treatment for the (NiFe)O layers of the scales on the alloys containing up to 5 percent iron based on the equations for the metal flux and divergence, using an auxiliary equation for mass balance at the outer surface of the inner layers. This simplification came about because the amount of spinel was negligible in the outer layers of these scales.

Using the functional forms for D_{Ni} and D_{Fe} in (NiFe)O given by Eq. (18), Eqs. (7), (8), and (12) were solved for the iron- and oxygen-activity gradients in the (NiFe)O layers. These equations were initially solved for the case of the Ni-2% Fe alloy with trial values of $p = D^o_{Fe}/D^o_{Ni}$ to obtain ξ and a_O as a function of y. Since thermodynamic results for the boundary values were available, the equations gave k/D^o_{Ni}. Because k was experimentally known, D^o_{Ni} was calculated. The estimated and observed values of ξ are shown in Fig. 8. This analysis lacks closure because p and D^o_{Ni} were obtained using experimental boundary values. Consequently, the equations were subsequently solved with the same values of p and D^o_{Ni} as for the (NiFe)O layer on the Ni-4% Fe alloy. As shown in Fig. 8, correspondence was again found to the iron profiles. This agreement satisfied the assumption that the cation mobilities remained constant within this range of iron composition. The ratio of the iron-to-nickel mobilities by these calculations was 3.0, which is to be contrasted to a value of ~7 known with certainty to only an order of magnitude from the diffusivities and vacancy concentrations in the binary oxides.[35]

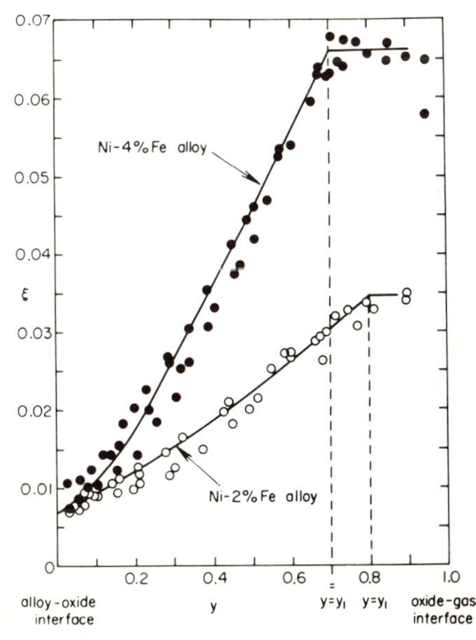

Fig. 8. Equivalent fraction of FeO, ξ, versus the normalized distance coordinate y in the (NiFe)O scale layers formed on the Ni-Fe alloys containing 2 and 4% Fe[35]; solid curve represents calculated values.

The consistency of these unusual metal profiles with thermodynamic requirements is illustrated by superimposing the composition path on the Fe-Ni-O isotherm. The effective diffusion parameter in an internal oxidation zone of an alloy is the metal interdiffusion coefficient if the variation of the oxygen solubility with respect to the alloying-element concentration is negligible.[49] A composition path based on solutions of the diffusion equations for the alloy-oxide composite system is shown in Fig. 9.[35] The path crossing tie lines in the spinel-nickel oxide field represents compositions in the oxide layer adjacent to the atmosphere, since nickel oxide cannot sustain a large iron concentration at high oxygen activity. A distinct (NiFe)O layer was maintained, however, on the alloy. Iron was enriched in this layer towards its outer interface to values determined by the combined effects of metal mobilities and oxygen activity. The composition path next crosses the spinel-nickel oxide-alloy field to intersect the spinel-alloy field. This latter field defines the oxide in the internal oxidation zone; the fraction of oxide follows the lever rule and the alloy and spinel compositions in the internal oxidation zone are given by the ends of the tie lines cut by the composition path. The final segment of the path is traversed in the field corresponding to the unoxidized alloy.

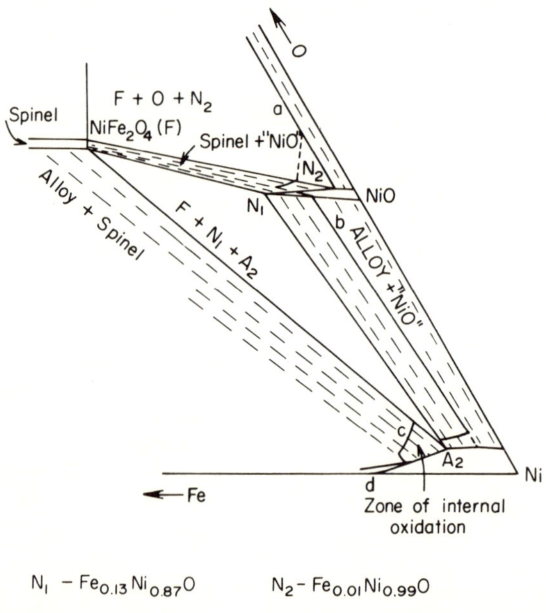

Fig. 9. A composition path. abcd, for the metal profiles in the superficial scale and internal oxidation zone shown in Fig. 7 superimposed on the Fe-Ni-O isotherm (not to scale)[35].

SUMMARY

A ternary diffusion model describes the high-temperature oxidation of several binary alloys composed of metals ranging from chromium to nickel in the first transition series. The effect of cations exhibiting variable valence on the defect concentration and the diffusivities in the ternary p-type cubic monoxides was considered. The ratio of the mobilities defining the relative migration rates of the two different cations via vacancies in a ternary oxide was taken as equal to this factor for the mobilities in the particular binary oxides. Oxidation of Ni-Co, Fe-Mn, and Co-Fe alloys led essentially to superficial scale formation and an average scale solute composition approximating that of the alloy, since the cation diffusivities of the ternary oxides were much larger than the interdiffusion coefficients of the alloys. Although Ni-Cr and Ni-Fe alloys exhibited more complicated oxidation behavior by formation of spinel oxides as precipitates in the alloys and scales, the ternary diffusion model, in conjunction with compositional, diffusional and thermodynamic results, gave a consistent description of the oxidation phenomena.

ACKNOWLEDGMENT

This paper was prepared under auspices of the National Research Council of Canada.

REFERENCES

1. Wagner, C., *Corrosion Sci.,* **9**: 91 (1969).
2. Coates, D. E., and Dalvi, A. D., *Oxidation of Metals,* **2**, 331 (1970).
3. Whittle, D. P., "High Temperature Corrosion of Aerospace Alloys", Proc. 34th AGARD Meeting (1972), p. 157.
4. Onsager, L., *Phys. Rev.,* **37**, 405 (1931); **38**, 2265 (1931).
5. Onsager, L., *Ann. N.Y. Acad. Sci.,* **46**, 241 (1945-46).
6. Darken, L. S., *Atom Movements,* ASM, Cleveland, Ohio (1951), p. 1.
7. Wagner, C., *ibid.,* p. 153.
8. Darken, L. S., *Trans. AIME,* **175**, 184 (1948).
9. Kirkaldy, J. S., *Oxidation of Metals and Alloys,* ASM, Cleveland, Ohio (1971), p. 101.
10. Dalvi, A. D., and Coates, D. E., *Oxidation of Metals,* **5**, 113 (1972).
11. Urbanic, V. F., and Smeltzer, W. W., *J. Electrochem. Soc.,* **119**, 1527 (1972).
12. Himmel, L., Mehl, R. F., and Birchenall, C. E., *Trans. AIME,* **197**, 827 (1953).
13. Smeltzer, W. W., *Acta Met.,* **8**, 377 (1960).
14. Goursat, A. G., and Smeltzer, W. W., *J. Electrochem. Soc.,* **120**, 390 (1973).
15. Pettit, F. S., and Wagner, J. B., Jr., *Acta Met.,* **12**, 35 (1964).
16. Turkdogan, E. T., McKewan, W. M., and Zwell, L., *J. Phys. Chem.,* **69**, 327 (1965).
17. Carter, R. E., and Richardson, F. D., *Trans. AIME,* **200**, 1244 (1954).
18. Carter, R. E., and Richardson, F. D., *Trans. AIME,* **203**, 336 (1955).

19. Khoi, N. N., "The Growth and Structure of Nickel Oxide Formed on the (100) (110) and (111) Crystal Faces of Nickel", Ph.D. Thesis, McMaster University (1972).
20. Lindner, R., and Akerstrom, A., *Disc. Faraday Soc.,* **23**, 133 (1957).
21. Volpe, M. L., and Reddy, J., *J. Chem. Phys.,* **53**, 1117 (1970).
22. Crow, W. B., Report ARL-70-0090, Project No. 7021, Office of Aerospace Research, USAF., Wright-Patterson Air Force Base, Ohio (1970).
23. Paidassi, J., and Berry, L., *Compt. Rend. Acad. Sci., Paris,* **262**, 1553 (1966).
24. Fueki, K., and Wagner, J. B., Jr., *J. Electrochem. Soc.,* **112**, 384 (1965).
25. Douglass, D. L., *Corrosion Sci.,* 8, 665 (1968).
26. Herchl, R., Khoi, N. N., Homma, T., and Smeltzer, W. W., *Oxidation of Metals,* **4**, 35 (1972).
27. Graham, M. J., Sproule, G. I., Caplan, D., and Cohen, M., *J. Electrochem. Soc.,* **119**, 883 (1972).
28. Caplan, D., Graham, M. J., and Cohen, M., *ibid.,* p. 1287.
29. Smeltzer, W. W., *Oxidation of Metals and Alloys,* ASM, Cleveland, Ohio (1971), p. 119.
30. Goodenough, J. B., *Prog. Solid State Chem.,* **5**, 145 (1971).
31. Kofstad, P., *Nonstoichiometry, Diffusion and Electrical Conductivity in Binary Metal Oxides,* Wiley-Interscience Publishers, New York (1972).
32. Wood, G. C., and Ferguson, J. M., *Nature,* **208**, 369 (1965).
33. Dalvi, A. D., and Coates, D. E., *Oxidation of Metals,* **3**, 203 (1971).
34. Dalvi, A. D., and Smeltzer, W. W., *J. Electrochem. Soc.,* **118**, 1978 (1971).
35. Dalvi, A. D., and Smeltzer, W. W., *J. Electrochem. Soc.,* **121**, 386 (1974).
36. Mayer, P., and Smeltzer, W. W., *J. Electrochem. Soc.,* **119**, 626 (1972).
37. Mayer, P., and Smeltzer, W. W., *J. Electrochem. Soc.,* **121**, 538 (1974).
38. Perkins, R. A., and Rapp, R. A., *Met. Trans.,* **4**, 193 (1973).
39. Wood, G. C., *Oxidation of Metals and Alloys,* ASM, Cleveland, Ohio (1971), p. 201.
40. Stiglich, J. J., Jr., Cohen, J. B., and Whitmore, D. H., *J. Am. Ceram. Soc.,* **56**, 119 (1973).
41. Zintl, G., *Z. Phys. Chem. N.F.,* **48**, 340 (1966).
42. Chen, W. K., and Peterson, N. L., *J. Phys. Chem. Solids,* **33**, 881 (1972).
43. Foster, P. K., and Welch, A.J.E., *Trans. Faraday Soc.,* **52**, 1636 (1956).
44. Hembrec, P., and Wagner, J. B., Jr., *Trans. AIME,* **245**, 1547 (1969).
45. Aukrust, E., and Muan, A., *Trans. AIME,* **227**, 1378 (1963); **230**, 1395 (1964).
46. Mayer, P., and Smeltzer, W. W. (to be published).
47. Dalvi, A. D., and Smeltzer, W. W., *J. Electrochem. Soc.,* **117**, 1431 (1970).
48. Davies, H., and Smeltzer, W. W., *ibid.,* **119**, 1362 (1972).
49. Kirkaldy, J. S., *Can. Met. Quart.,* **8**, 35 (1969).

THE IMPORTANCE OF SHORT-CIRCUIT AND RELATED TRANSPORT PROCESSES IN HIGH-TEMPERATURE OXIDATION

John Stringer

University of Liverpool
Liverpool, England

ABSTRACT

Possible modes of transport of the reactants through a growing oxide scale are reviewed. These include gaseous transport through pores or microcracks, diffusion through grain boundaries, and diffusion along dislocations. The experimental evidence for each mechanism is summarized. Finally, examples of some metals where short-circuit paths are believed to be important are presented.

1 INTRODUCTION

In simple theory, the growth of a continuous, adherent, compact oxide layer on the surface of a metal involves the bulk diffusion of the reactants across the scale to a site where reaction between them takes place. In general, the diffusion rate of one species greatly exceeds the rate of diffusion of the other, and one can therefore regard only one species as mobile. The migrating defect is usually ionized, so that the simultaneous transport of an ionic and an electronic defect is required. The *slower* moving of these effectively determines the rate: a field is established across the oxide until the flux of both charged components becomes equal. Wagner has developed expressions for the reaction rate, and these are described in the standard texts.[1,2] For simplicity, only the case where the ionic defect is slower moving will be considered (that is, the oxide is an electronic conductor).

It is immediately apparent that if the interface reactions are rapid and there is local electrical neutrality throughout the scale, the flux of material through the scale will be inversely proportional to its thickness, which corresponds to a parabolic rate law. In Wagner's theory[3], the parabolic rate constant for the growth of a scale of composition M_aX_b may be related to the diffusion coefficients D_M and D_X of the cations M^{m+} and the anions X^{r-} by the expression

$$K_p = \frac{c_x}{bk\,T} \int_{\mu_x^i}^{\mu_x^o} \left(\frac{m}{r} D_M + D_X \right) d\mu_x \quad , \tag{1}$$

where μ_x is the chemical potential of the anion (oxygen in an oxidation reaction) and the superscripts o and i indicate the metal/oxide and oxygen/oxide interfaces, respectively; c_x is the concentration of X. A similar expression can be written in terms of the cation activities.

Kofstad[2] points out that in using Eq. (1) to test the validity of Wagner's theory it is important to note that the diffusion coefficients must be determined under the proper experimental conditions. For p-conducting oxides, the self-diffusion coefficients must be determined in oxides in equilibrium with oxygen gas having the same partial pressure as that used in the direct experimental determination of the parabolic rate constant; while for n-conducting oxides, the oxide should be in equilibrium with the metal. Furthermore, if tracer diffusion coefficients are used, care must be taken to correct for the correlation coefficient. For an oxide with cation vacancies as the principal defect and for which the cation diffusion coefficient is much greater than that of the anion, Eq. (1) becomes

$$K_p = \frac{p+1}{a} \int_{[V_m{}^{p-}]^i}^{[V_m{}^{p-}]^o} D_V \, d\left[V_m{}^{p-} \right] \quad , \tag{2}$$

where p is the effective charge on the cation vacancy V_M and D_V is the diffusion coefficient of the vacancies. On this basis, comparisons have been made between the calculated and observed oxidation rate for a number of metals. For the growth of CoO on cobalt in 1 atm O_2, Carter and Richardson[4] obtained the results shown in Table I.

Table I. Calculated and Measured Rates of Oxidation of Cobalt at 1 Atmosphere[4]

	Rate Constant, K_p, $(g\ O_2)^2\ cm^{-4}\ sec^{-1}$	
Temperature, °C	Experimental	Calculated From the Tracer Diffusion Coefficient
1000	2.43×10^{-8}	2.72×10^{-8}
1148	9.3×10^{-8}	10.6×10^{-8}
1350	7.8×10^{-7}	6.7×10^{-7}

Kofstad[2] points out that the correlation factor has been neglected, and that the calculated results should be approximately 25 percent higher. If the diffusion coefficient does not vary appreciably with the vacancy concentration, Eq. (2) becomes

$$K_p = \frac{p+1}{a} D_V \left\{ \left[V_M{}^{p-}\right]^o - \left[V_m{}^{p-}\right]^i \right\} , \tag{3}$$

and for the oxidation of cobalt in 1 atm O_2, $\left[V_m{}^{p-}\right]^o >> \left[V_m{}^{p-}\right]^i$ and $a=1$; noting that $D_V \left[V_M{}^{p-}\right]^o = D^o_{Co}$, the diffusion coefficient of cobalt in CoO at the oxide/oxygen interface,

Eq. (3) becomes

$$K'_p = (p+1)\ D^o_{Co} \quad . \tag{4}$$

(The prime indicates that the rate constant is expressed in units of $cm^2\ sec^{-1}$). Kofstad[2] has used this expression with the data of Carter and Richardson to estimate the charge on the cation vacancy, and a similar calculation has been performed for copper vacancies in Cu_2O by Moore and Selikson[5]. It appears that the effective charge varies with temperature, but the variation does not seem to be great.

In general, the temperature dependences of both parabolic rate constants and diffusion coefficients follow an Arrhenius expression, and it follows from Eq. (4) that the activation energies should be equal and the preexponential frequency factors should differ by a numerical factor generally less than five. This is an easier test of the validity of the theory than is an attempt to calculate the reaction rates because of the experimental scatter in the determination of both rate constants and diffusion coefficients. Other tests are possible: the variation of the rate constant with oxygen pressure, the effect of minor amounts of alloying elements on the oxidation rate, and so forth.

2 TRANSPORT BY PROCESSES OTHER THAN BULK DIFFUSION

The idea that in some cases transport may involve other mechanisms than bulk diffusion is a relatively old one. Evans[6] summarized the early investigations under the general heading "movement through pores", but included grain-boundary and dislocation diffusion within this. He expressly considered the situation where an easy diffusion path may become blocked as a result of growth of new oxide and commented that this may result in a decrease in oxidation rate with increasing temperature. Gulbransen[7] accounted for negative evaluations of the activation entropy of parabolic oxidation by proposing diffusion of reactants along boundaries of oxide crystallites. Evans presented two kinetic models, one in which the pores simply become blocked, and one in which compressive stresses arising from the blocking of one pore also block all those within a certain distance of the first: with some simplifying assumptions, these result in an asymptotic rate equation and a direct logarithmic equation, respectively.

Kofstad[2] also discussed the possibility of short-circuit processes of various kinds, and pointed out that if the density of short-circuit paths remains constant, a parabolic rate law will again be observed; the existence of a parabolic rate law is thus no guarantee that transport of the reactants is by bulk diffusion.

Smeltzer, Haering, and Kirkaldy[8] remarked that many metals oxidize too rapidly to be explained by lattice diffusion of the reactants through the oxide at relatively low and intermediate temperatures, and presented a phenomenological model based on the idea that the flux through the oxide consisted of two parallel terms: one representing bulk diffusion as in the Wagner Model and the other representing diffusion through low resistance paths of temporarily decreasing density, in the manner postulated by Davies et al.[9] The model was concerned with oxidation in which the oxygen ion was mobile, and the fraction of available oxygen sites within the array of short-circuit paths f varied with time by a first-order rate equation:

$$-\frac{df}{dt} = kf \qquad (5)$$

or

$$f(t) = f^{\circ} \exp(-kt) \quad . \qquad (6)$$

Now, following Hart[10], the effective diffusion coefficient is a weighted average of the two separate coefficients:

$$D_{eff} = D_b\,(1 - f) + D_{sc}\,f \quad , \tag{7}$$

where D_b is the bulk diffusion coefficient and D_{sc} is the short-circuit diffusion coefficient. Then,

$$\frac{dx}{dt} = \Omega\left[\,D_b\left\{1\text{-}f^{\circ}\,\exp(-kt)\right\} + D_{sc}\,f^{\circ}\,\exp(-kt)\,\right]\frac{\Delta c}{x}\ , \tag{8}$$

where Ω is the volume of oxide per oxygen ion, Δc is the difference in oxygen-ion concentrations at the two interfaces, and x is the film thickness. In the absence of the short-circuit term, this equation integrates to give a parabolic rate law, with

$$K_p = 2\,\Omega\,D_b\,\Delta c \quad , \tag{9}$$

and Eq. (8) can be integrated to

$$x^2 = K_p\left[\,t + \frac{f^{\circ}\,\alpha}{k}\left\{1\text{-}\exp(-kt)\right\}\right] \quad , \tag{10}$$

where $\alpha = D_{sc}/D_b$. Figure 1 shows the agreement between Eq. (10) and the experimental results of Kofstad et al.[11] for the oxidation of titanium

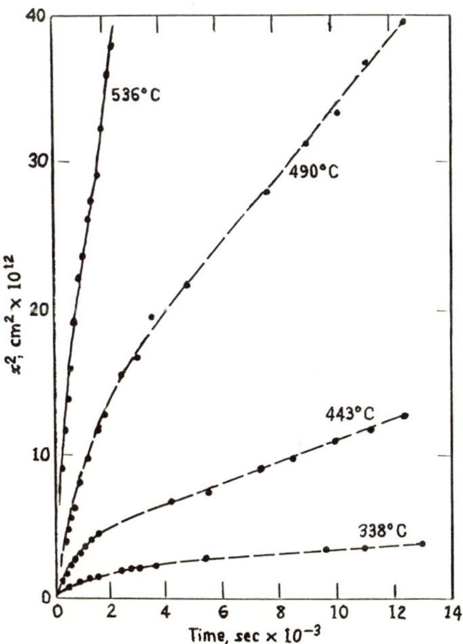

Fig. 1. Oxidation of titanium. Experimental points from Kofstad et al.[11]; dashed lines calculated from Eq. (10).[8]

in the range 338 to 536°C. Similar agreement was obtained for zirconium and hafnium. Since the equation contains a number of adjustable parameters, the real test is the physical reasonableness of the values that must be assigned to them. Smeltzer et al. concluded that some 10 percent of the oxygen sites are in the short-circuit paths, and commented that this "represents very defective material indeed". They postulate that the short-circuit paths may be associated with 10 Å-diameter dislocation pipes at a density of $10^{12}/cm^2$.

3 POSSIBLE SHORT-CIRCUIT PATHS

The kinetic model of Evans, and its extension by Smeltzer, did not expressly consider the nature of the short-circuit path, except insofar as Smeltzer discussed the values of the arbitrary parameters in terms of dislocation pipes or grain-boundary-diffusion paths, and it is convenient therefore to describe the possible paths.

3.1 Pores or Microcracks

The simplest short-circuit path is clearly a fissure through the scale that allows the gas phase transport of the oxidant inwards towards the metal surface. In the late stages of oxidation, large cracks often develop within the scale, but these present no barrier to gaseous transport and are of no special interest from a kinetic viewpoint. Hancock and co-workers[12] have demonstrated in recent years that many oxide scales that appear quite compact and continuous in a conventional metallographic examination may nevertheless contain extensive microcracking. Stringer[13] suggested a kinetic model based on gaseous diffusion through fine pores in TiO_2 to explain a particular type of rapid reaction sometimes observed in the high-temperature oxidation of titanium. Because of the fineness of the pores, a Knudsen flow model was used, producing a general growth law of the form.

$$\frac{dx}{dt} = \frac{n}{M\rho_x} \left(\frac{8}{3}\right) \left(\frac{2}{\pi}\right)^{1/2} \rho_1{}^{1/2}\, p \left\{ \int_0^x \frac{g(y)}{[f(y)]^2}\, dy \right\}^{-1} \tag{11}$$

where ρ_1 is the density of the gas at unit pressure, M is the weight fraction of oxygen in the oxide, ρ_x is the density of the oxide, n is the number of pores per unit area, and p is the external gas pressure (the internal pressure is assumed to be zero). The term in braces is related to the geometry of the pore distribution, and Stringer used a model in which the dilation of the oxide arising from the growth was incorporated. If this is neglected and the pores are assumed to be of constant diameter r_o, Eq.

(11) reduces to the parabolic rate law:

$$x^2 = \frac{16}{3}\left(\frac{\rho_1}{2}\right)^{1/2} \frac{\pi}{M\rho_x} p\, n^{3/2}\, r_o{}^4\, t \qquad (12)$$

The only adjustable parameters are r and n; Stringer assumed that there was approximately one pore per grain in the oxide, and that the radius was of the order of 60 Å, and this gave reasonably good agreement with the experimental observations; in addition, the parabolic rate constant was shown to be linearly dependent on oxygen pressure, as required by Eq. (12). The value of 60 Å for the pore radius resulted from the identification of the pores with screw dislocation cores.

Mrowec[14] suggested that atoms or molecules of sulfur diffused inwards through pores during the sulfidation of binary alloys, and inferred from specific density and specific surface measurements that pores less than 10^5 Å were indeed present. Cox and Johnston[15] postulated a rate-controlling gaseous diffusion of oxygen through a "porous plug" of oxide in the linear oxidation of niobium in the temperature range 350 to 600°C; again, the pressure dependence of the linear rate constant is close to unity. Stringer[16] showed that the presence of nitrogen in the atmosphere retarded the linear oxidation of tantalum, and suggested that this too was consistent with the existence of a "porous plug" effect of the nonprotective oxide. Mrowec[17] has developed his earlier model, showing that when a sulfide film grows by outward diffusion of metal ions, it eventually becomes impossible to maintain contact between scale and metal; a dissociation of the inner sulfide then takes place, the sulfur reacting with the metal to form new sulfide, and the metal diffusing outwards to the surface, again to react with sulfur. This dissociative breakdown takes place more readily at the sulfide grain boundaries, and fissures develop which eventually extend to the surface of the scale, permitting the inward diffusion of gaseous sulfur. Tuck[18] has also suggested that gaseous sulfur-containing mixtures could be transported through oxide scales by a system of continuously opening and closing pores. Birks[19] considers that CO is transported through microcracks in the scale during the simultaneous oxidation and carburization of stabilized steels in CO_2. Howes[20] has examined the interface morphology for a range of iron-chromium alloys and sees some evidence of localized penetration of oxygen through the main oxide, which may possibly involve gaseous diffusion through microfissures. Surman and Castle[21] oxidized steel in steam at temperatures below 570°C to form Fe_3O_4 and found, in agreement with Potter and Mann[22], that inert markers were buried at the middle of the scale, suggesting both inward and outward growth of oxide. The inner layer did not seem porous, or indeed any different from the outer layer, but careful scanning electron microscopy showed that in fact it was porous. They postulated a mechanism involving the gaseous transport of a volatile species through the scale, which was developed from the earlier model of Castle and Masterson[23] for transport in liquid within pores during aqueous corrosion. The experimental results followed an approximately parabolic rate law, and the parabolic rate constant fitted an Arrhenius expression with a frequency factor of 6.1 x 10^5 g^2 m^{-4} sec^{-1} and an activation energy of 34,600 cals/mole. They suggest that three possible transport mechanisms may be considered: solid-state diffusion of iron atoms, solid-state diffusion of oxygen ions, and vapor phase diffusion of volatile $Fe(OH)_2$. These are tested by using Eq. (3), using data for the various terms from the literature; the results are shown in Table II.

Table II. Activation Energies and Frequency Factors for the Oxidation of Iron to Magnetite, Calculated by Surman and Castle[21]

	Frequency Factor, k_o, g^2 m^{-4} sec^{-1}	Activation Energy, Q, cals $mole^{-1}$
Cation diffusion	2.0×10^7	55,000
Anion diffusion	1.2×10^{-7}	17,000
Gas phase transport	$4.8 \times 10^4\, \nu\, T^{1/2} \frac{P_{H_2O}}{P_{H_2}}$	38,900
Experimental	6.1×10^5	34,600

The gas phase transport model is different in character in that the temperature dependence of the diffusion coefficient is small, but the concentration difference term across the scale contains an activation energy which dominates the overall temperature dependence. The frequency factor has the form shown in Table II, where $\bar{\nu}$ is a porosity factor, which Surman and Castle estimate from scanning-electron-microscope pictures to be less than 0.1, and probably in the range 0.001 to 0.1. The water vapor/hydrogen equilibrium ratio cannot be predicted, and values were taken between 1 and 10^3. Within the extremes of these two adjustable parameters, it was possible to straddle the experimental results, as shown in Fig. 2. Recently, Surman[24] has extended the analysis. This restricts the range of partial-pressure ratios that can be taken, the frequency factor becoming 1.25 x 10^4 $\bar{\nu}$ $T^{1/2}$, and within the expected range of $\bar{\nu}$, it is not possible to fit the experimental data, although the correspondence is still very much better than with the solid-state-diffusion models.

Several investigators have shown that the magnetite films formed on iron in hot water or steam are indeed porous.[25]

Recently, Kovacina[26] developed a variety of techniques for studying

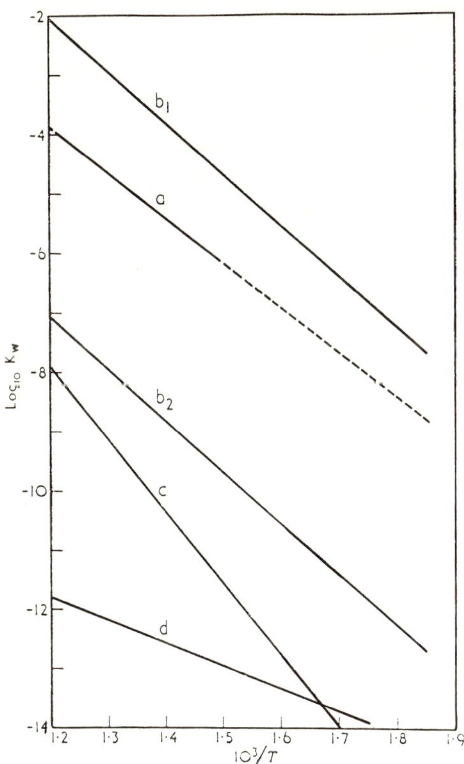

Fig. 2. Oxidation of mild steel in steam, from Surman and Castle[21]. Line (a) represents the best fit to the experimental results of Potter and Mann[22], interpreted in terms of a parabolic rate law. Lines (b_1) and (b_2) are calculated from the vapor transport model, the two lines representing the extremes possible by varying the adjustable parameters. Line (c) is the expected rate of growth if bulk diffusion of iron were rate controlling; Line (d) is the expected rate if oxygen bulk diffusion is controlling.

the porosity of Fe_3O_4 films formed on mild steel in aqueous NaOH at 316°C, and concluded that pores of the order of 0.1 μm (1000 Å) were certainly present and permitted ready penetration of water, dissolved salts, and gases through the oxide film. Hussey and Cohen[27] studied the oxidation of Fe_3O_4 to Fe_2O_3, and observed blistering of the oxide which appeared to become permeable to oxygen by the production of micro-cracks. The permeation increased with increasing oxygen pressure.

Porosity has also been reported in SiO_2 scales: Ing et al.[28] demonstrated the existence of pores and microchannels, and both Ligenza and Spitzer[29] and Deal and Grove[30] have reported that the parabolic rate constant for the oxidation of silicon depended approximately linearly on the oxygen pressure. The diffusion coefficient of oxygen in quartz glass at 1115°C in oxygen pressures between 40 and 710 torr is proportional to oxygen pressure[31]; the diffusion rate is about three times that in crystalline quartz. Williams[32] has also found that the diffusion coefficient varies linearly with oxygen pressure in the range 850 to 1250°C. Kofstad[33] suggests that these results imply that oxygen diffuses as a neutral atom. Gulbransen and Jansson[34] discussed the formation of SiO_2 films on silicon and SiC, and remarked that cracks and pores must be expected, with gas transport possibly causing healing of the holes, at least in the passive region. The existence of a transiton from passive to active behavior in the oxidation of SiC is regarded as a proof of the existence of open pores in the SiO_2 film, with the nature of the oxidation process depending on the magnitude of the pressures of SiO(g) and CO(g) in the pores in relation to the external oxygen pressure. Similarly, in the oxidation of silicon it was concluded that SiO(g) could be transported through the pores in the oxide.

The evidence for the existence of gaseous diffusion through pores as a significant process in oxidation reactions is reasonably convincing as a mass, but any single experiment is less than wholly convincing. Of necessity, the kinetic models contain a number of very considerable simplifications, and usually involve a number of adjustable parameters; the pores required are usually too small to be directly observed. In many cases the arguments are qualitative and based on mechanistic or morphological grounds. Many of the examples given above have been questioned by other investigators, and alternative explanations have been offered. The temperature dependence of the rate constants is difficult to estimate unless a good model for the formation of the pores is available, and it usually is not. Probably the most convincing fact is the existence of one or two reasonably documented examples of parabolic rate constants linearly dependent on the oxygen pressure; it is difficult to think of alternative explanations for these. The increasing weight of observations showing the existence of fine pores and cracks in what were previously regarded as compact scales by the use of the scanning electron microscope and in techniques such as that of Hancock is also significant.

3.2 Grain Boundaries

Several investigators have invoked grain-boundary diffusion to explain anomalies in the oxidation of metals, and this is partly because there is excellent experimental evidence for the existence of accelerated diffusion. This may have the effect of increasing the rate of transport of the species which is mobile in the bulk of the grain, but frequently it has the more dramatic effect of increasing the rate of movement of the species which is slower moving in the bulk so that it becomes the effective mobile species. This then appears as an anomaly between the mobile species as determined by marker experiments in an oxidation experiment, and that determined by a conventional diffusion experiment.

The majority of the direct experimental evidence on grain-boundary diffusion has been obtained for metals, and there have been two comprehensive recent reviews.[35,36] Most theoretical discussions are based on the model of Fisher[37], who treated the boundary as a slot of high-diffusivity material embedded in the relatively low diffusivity matrix. Calculations then involve both the diffusivity within this grain-boundary region, D_{gb}, and the width of the slab, δ. Fisher used a numerical solution, choosing $D_{gb}/Db \sim 10^6$ and $\delta \sim 5 \times 10^{-8}$ cm, to fit experimental results for silver at 478°C. For the particular case of <100> tilt boundaries in silver, Hoffmann and Turnbull[38] showed that grain-boundary diffusivity increased with tilt in the range 9 to 28 degrees, and interpreted this in terms of "pipe" diffusion along the dislocations in the boundary network (see next section). Love and Shewmon[39] examined the case of twist boundaries in silver, again interpreting their results in terms of dislocation pipe diffusion, concluding that the diffusivity along screw dislocations in this case was less than 1/10 that along edges, principally because of a smaller pre-exponential term. Again, they estimate the boundary width as a few Burgers vectors. Bokshtein, Magidson, and Svetlov[40] used a different model, treating the solid as an assembly of spherical grains embedded in a high-diffusivity matrix, and concluded that an estimate of a grain-boundary width of 5×10^{-6} cm quoted from the work of Arkharov was more likely to be correct than was Fisher's.

Mimkes and Wuttig[41] have pointed out that in alkali halides and oxides there are indications that the high-diffusivity paths are of considerable dimensions, and have developed a more complete treatment, appropriate to the circumstance where much of the diffusant can be regarded as being inside the high-diffusivity region. This model was then applied to the experimental results of Wuensch and Vasilos[42] for the diffusion of Ni^{2+} into MgO bicrystals at 1200°C, and it was concluded that the boundary width was 75 ± 5 μm, a very large value indeed.

Irving[43] has considered the role of grain boundaries in growing oxide scales in some detail, applying a simple extension of the Fisher model, due to Le Claire[44], to the problem of oxidation. His formulation allows the effect of grain size to be expressly included. The results show that as the grain size changes so that the grain-boundary effects vary in importance, different apparent rate laws will be obtained, varying from an accurately parabolic law identical with the Wagner expression to a variety of different empirical expressions; if the apparent empirical rate law is written

$$x^n = kt \quad ,$$

then n may have values ranging from about 1.25 to 4; these arise because over several orders of magnitude in time, the true curve changes slope, and the apparent rate law within the two or three orders of magnitude time range used by most investigators may have an intermediate form.

Gibbs[45] has criticized Irving's model on a number of grounds, and in particular has pointed out that in the case of a mobile anion in the boundary, accumulation of stress at the inner surface due to the formation of oxide will eventually reduce the diffusion rate until it becomes equal to that in the bulk. This is similar to the view of Rhines and Wolf[46] who suggested that during the oxidation of nickel at 1000°C, oxygen diffused rapidly into the oxide grain boundaries, reacting with nickel arriving by bulk diffusion through the grains and forming new oxide; this resulted in the development of growth stresses which, it was suggested, would approach a steady-state value.

Enhanced grain-boundary diffusion has been deduced from the difference in diffusivities measured in single crystals and polycrystals. An early and important example is the enhanced diffusion of oxygen observed in polycrystalline Al_2O_3 by Oishi and Kingery[47]; Fig. 3 is taken from Paladino and Coble[48]. Numerous examples are quoted by Kofstad[33];

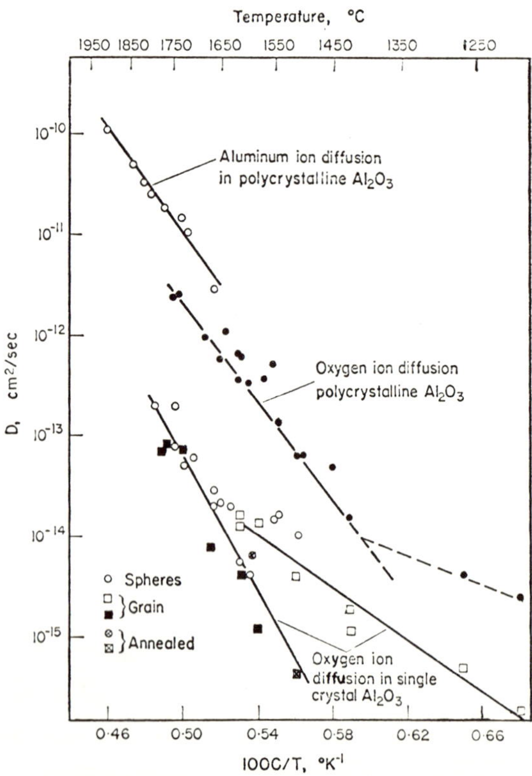

Fig. 3. Aluminum and oxygen diffusion in single crystals and polycrystals of Al_2O_3, showing the enhanced oxygen diffusion in polycrystals, presumably due to grain-boundary migration. The data are from Oishi and Kingery[47]; the figure is taken from Paladino and Coble[48].

Holt[49] found for oxygen in single-crystal BeO at 1300 to 1700°C that $D_0 = 2.95 \times 10^{-5}$ cm^2 sec^{-1}, Q = 68,500 cals mole^{-1}, whereas Austerman et al.[50] using an identical technique on polycrystalline BeO obtained values of 4×10^{-7} and 42,000, respectively. However, Cline and Newkirk[51] found that the more rapid diffusion of beryllium in BeO was insensitive to the presence of grain boundaries. Several authors have reported enhanced diffusion of cation impurities along grain boundaries in MgO.[33] Holt[52] has reported enhanced oxygen diffusion along grain boundaries in CoO. Alcock et al.[53] and Yajima et al.[54] have shown enhanced grain-boundary diffusion of uranium in UO_2. Roberts and Wheeler[55] reported that the diffusion of zinc in ZnO between 800 and 1400°C was predominantly along grain boundaries.

This is by no means a comprehensive list, but it is sufficient to indicate that diffusion of both anions and cations along grain boundaries in oxides is possible at normal temperatures; there is some indication that it is more common for enhanced diffusion of the anion, but this is not certain. Unfortunately, the numerical data are disappointing, because of the extreme irreproducibility of diffusion measurements in oxides. In metals, the activation energy for grain-boundary diffusion is approximately half that for bulk diffusion, but there are some indications that the difference may not be as great in oxides. Equally, it seems probable that the boundary width is greater in oxides. It should be remembered that the segregation of impurities to grain boundaries will produce a region of different chemistry with very different properties; some of the observations may well be due to this rather than to an intrinsic grain-boundary effect.

3.3 Dislocations

Again, the existence of rapid diffusion along dislocation lines has been known for a number of years. The method used to treat the problem is analogous to that used for grain boundaries, except that the dislocation is regarded as a "pipe" of high-diffusivity material. Again, qualitative analysis requires a knowledge of D_d, the diffusivity in the dislocation, and a, the radius of the pipe. An exact solution of the problem for a reasonably sophisticated model has recently been given by Mimkes and Wuttig[56], and applied to the data for tellurium of Ghostagore[57]; the pipe radius is of the order of 1.5 μm, which is surprisingly large. In metals, it is usual to regard the enhanced diffusion as confined to the dislocation core[58], although Baker, Wuttig, and Birnbaum[59] reported values of the order of 10^{-6} cm for nickel and Co-Ni alloys. Baluffi[58] finds these values very hard to rationalize.

For metals, the activation energy is again about half that for bulk diffusion; a frequency factor cannot be quoted without also choosing a value for the total pipe area. It appears that edge dislocations are much more effective than screws.[39]

There is less specific information on dislocation diffusion in oxides. Moore and Williams[60] studied the diffusion of oxygen in ZnO in oxygen at 0.5 atm in the range 1000 to 1300°C, and obtained a (corrected) frequency factor of 2.75×10^{13} cm^2 sec^{-1} and an activation energy of 176,000 cals $mole^{-1}$; D was proportional to the square root of the oxygen pressure. They suggested that the very large preexponential factor implied dislocation pipe diffusion, but in this case the activation energy seems very large. Kofstad[33] remarked that the pressure dependence suggested a diffusion involving neutral oxygen atoms.

4 EXAMPLES OF OXIDATION PROCESSES APPARENTLY INVOLVING SHORT-CIRCUIT TRANSPORT

4.1 Oxidation of Iron

Tallman and Gulbransen[61] have studied the growth of αFe_2O_3 whiskers on iron in dry oxygen at 400 to 500°C. Diffraction experiments showed that there was an axial screw dislocation lying along each whisker, and it was proposed that the wisker grows by rapid diffusion of iron along the dislocation to the tip of the whisker.[62] The same authors[63] have studied the formation of bladelike platelet αFe_2O_3 growths formed on iron in a water vapor atmosphere containing a trace of oxygen at 450°C, and have shown that the blade face plane is a twin boundary, the two halves of the blade being mirrored in this plane. They commented that the growth rates of both the whiskers and the plates require diffusion rates between 10^4 and 10^6 higher than can be accounted for by bulk diffusion coefficients derived from sintering experiments, and conclude that there is rapid diffusion of iron along the twin boundaries to the blade tip. Rapid diffusion along twin boundaries is a little surprising, since the registry is very good; Balluffi[58] has expressed doubt at the claim by Baker et al.[59] that there is accelerated diffusion along stacking faults, and the cases would seem to be equivalent. However, in a later paper, Tallman and Gulbransen[64] determined the crystallography of similar boundaries in plates formed by oxidizing unannealed iron in dry oxygen at 400°C, from which it appears that the interface is only partly coherent. It appears to be a twist boundary with an array of screw dislocations; and the fact that screw dislocations are not very effective short-circuit paths in metals does not, of course, imply that the same is true in oxides.

Caplan, Sproule, and Hussey[65] have oxidized both annealed and cold-worked high-purity iron in oxygen at 1 atm pressure at 500 and

550°C. The apparent parabolic rate constant for the oxidation of the cold-worked material has a high initial value which drops rapidly, approaching a constant after about 20 hours (Fig. 4); it appears that diffusion through the oxide is initially rapid. One possibility is that lattice mismatch in the cold-worked metal and polygonization during heating generates an oxide overgrowth of fine crystallite size in which the boundaries act as short-circuit diffusion paths. Recrystallization and grain growth in the oxide later eliminate these; or possibly the oxide formed after the metal has recovered contains fewer short-circuit paths. Hussey and Cohen[66] oxidized iron at low pressure, and considered that the results implied rapid transport of cations across the Fe_3O_4 layer, facilitated by passage along preferred paths, which may or may not be grain boundaries. Again, the rate diminished at longer times.

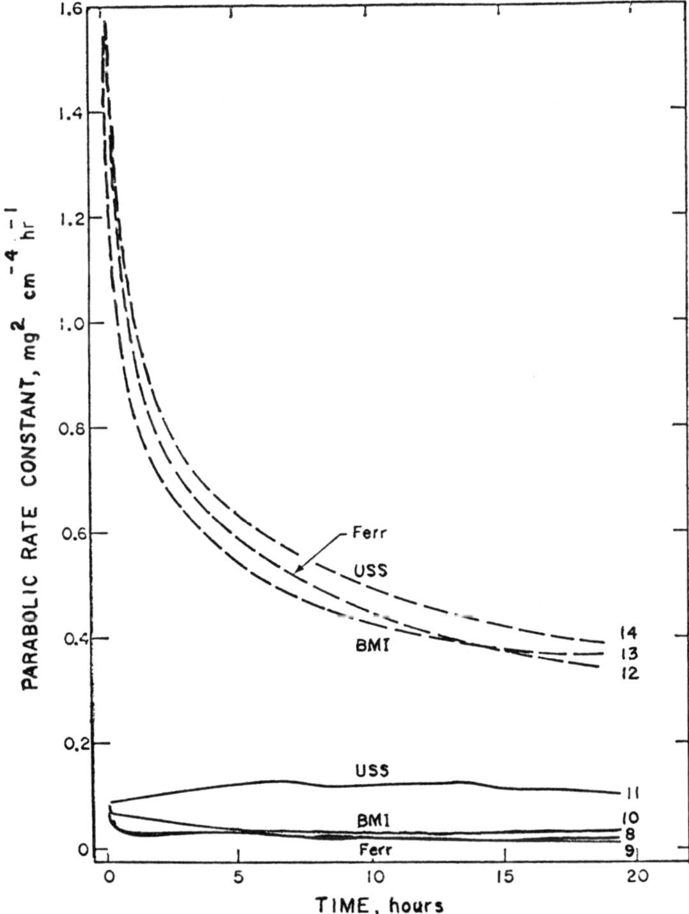

Fig. 4. Oxidation of iron in oxygen at 550°C, from Caplan et al.[65]; plotted in terms of the "instantaneous parabolic rate constant", 2w dw/dt. Solid lines – annealed specimens; broken lines – cold-worked specimens.

Channning and Graham[67] have used Mössbauer spectroscopy to study the iron oxidation process at 450 to 550°C. The activation energy for the growth of αFe_2O_3 on Fe_3O_4 was 49 kcal mole^{-1}, and that for iron transport through the Fe_3O_4 was 31 kcal mole^{-1}; the authors considered that because of the low activation energies, and the rapid growth rates observed, transport must involve short-circuit paths. Birchenall[68] has discussed the problem of the growth rate of αFe_2O_3 in some detail, pointing out that if the iron is mobile, the observed bulk diffusion coefficients are several orders of magnitude too small to explain the rate.

4.2 Oxidation of Nickel

Perrow, Smeltzer, and Ham[69] examined the NiO films formed in nickel at 500 and 600°C in oxygen at 400 torr using transmission electron microscopy, and calculated dislocation densities of the order of 5×10^{10} cm^{-2}, in reasonable agreement with the value of 10^{11} cm^{-2} estimated if the short-circuit model of Smeltzer et al.[8] is to apply to nickel. On morphological grounds, they concluded that the greater part of the measured weight gain is due to short circuit (presumably dislocation pipe) diffusion. Later this work was extended by Perrow, Smeltzer, and Embury[70], who concluded that the most important defect was crystallite boundaries, and that as the film thickened these were removed by recrystallization and grain growth. A simple kinetic model was constructed similar to that of Smeltzer et al.[8] and applied to the oxidation data; it was concluded that Q_{gb}/Q_b was approximately 0.6, assuming a crystallite size of 100 Å and a boundary width δ of 10 Å; the ratio of the diffusion coefficients lay in the range 10^2 to 10^5, which is not unreasonable.

Cathcart, Peterson, and Sparks[71] have examined NiO films 25 to 500 Å thick grown on various single crystal surfaces of nickel at temperatures between 400 and 600°C, using an X-ray technique to measure the mosaic spread in the oxide. They concluded that the films on all the crystal faces studied contained large numbers of short-circuit diffusion paths, and following their similar work on copper[72], related the varying types and densities of the defects to the anisotropy of the oxidation rate. This anisotropy in the oxidation rate of a cubic metal forming a cubic oxide is a most powerful argument for the existence of short-circuit transport, and the success of Cathcart and co-workers in rationalizing the anisotropy in terms of the detailed defect structure is highly significant.

Herchl et al.[73] have also studied the growth of nickel oxide scales on single-crystal faces of nickel in the temperature range 500 to 800°C to form much thicker oxides than those of Cathcart et al: 0.7 to 14 μm. The kinetics could be satisfactorily interpreted by including the short-circuit term, modified to allow for a decrease in the available paths due to grain growth.

Berry and Paidassi[74] have oxidized a number of grades of nickel over the temperature range 600 to 1400°C. Above 1000°C, the activation energy for the parabolic oxidation is 49 to 57 kcal $mole^{-1}$; below 1000°C, the value is 24 kcal $mole^{-1}$. This compares with a value for bulk cation diffusion in NiO of 61 kcal $mole^{-1}$ [75], and while it is unwise to attach too much importance to activation-energy determinations, these results are certainly consistent with a short-circuit mechanism.

4.3 Oxidation of Zirconium

Cox and Pemsler[76] have measured the diffusion of oxygen in growing oxide films, and were able to distinguish the much greater penetration due to short-circuit paths and the relatively small penetration due to bulk diffusion; at the temperature used (400 to 500°C), $D_{sc} \approx 10^4 D_b$.

Madeyski, Poulton, and Smeltzer[77] have studied the parabolic oxidation of zirconium in the range 750 to 950°C, and have determined the diffusivity of oxygen in the scale. Oxygen migration occurred largely by short-circuit diffusion, and it was shown that lattice diffusion is approximately 10^4 slower than the short-circuit path. The parabolic rate constant was calculated using both the bulk diffusion coefficient and the short circuit-diffusion, and reasonable agreement was obtained; with bulk diffusion alone the rates were as much as an order of magnitude too small (Fig. 5). Lightstone and Pemsler[78] have also suggested that there is a large density of potential short-circuit paths in thick oxides, and that tracer

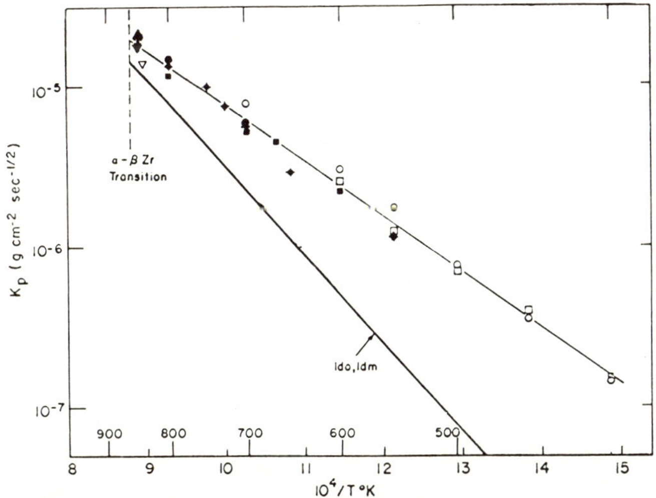

Fig. 5. Oxidation of alpha zirconium, from Madeyski et al.[77] The various points are experimental results from the literature (see ref. 77 for the sources). The heavy line ido, ldm is calculated assuming lattice diffusion is the sole mechanism of oxygen transport in the scale and in the metal.

measurements are consistent with this. However, Kofstad[33] has pointed out that these analyses all depend on an assumption that anion diffusion is rate controlling; if the scale exhibits ionic conductivity, so that electron transport is rate determining, it may be possible to explain the results without having to invoke short-circuit paths.

4.4 Oxidation of Chromium, and the Growth of Cr_2O_3 on Alloys

There is a wide variation in the oxidation rates of chromium reported in the literature, which has been reviewed briefly by Stringer et al.[79] However, at first sight there seems no problem; chromium is known[33] to diffuse more rapidly than oxygen in Cr_2O_3, and marker experiments indicate that chromium is the mobile species in the oxide. The activation energy for the oxidation is 40 kcals mole^{-1}, and values for chromium diffusion in Cr_2O_3 lie in the range 40 to 100 kcals mole^{-1}, with the lower value corresponding to the better specimens. Birchenall[68] comments that the rate of oxidation is consistent with the diffusion coefficients.

However, the presence of inert oxide dispersions in chromium[79,80] or in alloys forming Cr_2O_3 oxides[81] reduces the rate of oxidation very markedly. Stringer et al.[81] have pointed out that it is extremely difficult to envisage a mechanism that would result in slowing down a bulk diffusion process, and therefore concluded that in spite of the apparent agreements listed above chromium normally migrates through Cr_2O_3 layers by a short-circuit mechanism. Paidassi, Cadiou, and Berry[82] had earlier proposed that below 1100°C a short-circuit process was involved, because of the small activation energy they obtained (29.2 kcals mole^{-1}); above 1100°C the activation energy is 58.5 kcals mole^{-1} and it was assumed that bulk diffusion became dominant.

Furthermore, marker experiments, and morphological studies have shown that oxygen can diffuse through Cr_2O_3 scales grown on alloys containing inert oxide dispersions[79-81], and it was suggested that this occurred by a grain-boundary process.

Clearly, further work is needed on this technically important reaction.

4.5 The Growth of Al_2O_3 on Alloys

Pure aluminum melts at too low a temperature, and Al_2O_3 grows too slowly, to allow any reasonable information to be derived from the oxidation of the pure metal. However, the growth of Al_2O_3 scales at elevated temperatures on a number of alloys has been studied. Pettit[83] has shown that a continuous Al_2O_3 layer forms on Ni-25 wt % Al at temperatures in the range 900 to 1300°C in 0.1 atm oxygen. The rate of growth of the oxide is slow, but the parabolic rate constant is only three orders of magnitude slower than that for Cr_2O_3 on pure chromium at 1000°C. For single crystals, the difference in cation diffusion rates would be six or more orders of magnitude. Marker experiments show that, in fact, oxygen appears to be mobile in the scale[84], and following Oishi and Kingery it might be concluded that this is an example of short-circuit diffusion at grain boundaries. However, it should be remembered that in Oishi and Kingery's experiments, the diffusion rate of oxygen became only comparable with that of aluminum in the polycrystalline sample, and did not exceed it. The grain size of their specimens was 20 to 30 μm, and the grain size in Al_2O_3 scales is typically much smaller than this, often less than 1 μm, so the rate might well be further enhanced.

The activation energy for the oxidation is 65 kcal mole^{-1} which is just over half that for the diffusion of aluminum in Al_2O_3; there is no wholly satisfactory value for oxygen diffusion, but Oishi and Kingery obtained a value of 57.6 kcal mole^{-1} for their polycrystalline samples. Clearly, an accelerated transport process is operating, but as yet the growth of Al_2O_3 scales is not wholly understood.

5 SUMMARY

There are clearly a number of transport mechanisms which can operate in the growth of oxides on metals in addition to bulk diffusion. There is reasonably good evidence that accelerated grain-boundary diffusion is possible in oxides. There are some anomalies in the interpretation of the oxidation kinetics of some metals, which might be resolved by considering short-circuit transport processes.

This is as far as one can go at the moment, since there are few examples where it is possible to say this oxidation certainly involves short-circuit diffusion; the difficulties in understanding the complex oxidation process, the inherent simplifications introduced by the experimental techniques, and the irreproducibility often encountered prevent complete certainty. Nevertheless, there is a sufficient weight of evidence to suggest that, in fact, short-circuit diffusion processes are operative to some extent in oxidation reactions, and that their possible effect must be considered in any investigation.

REFERENCES

1. Kubaschewski, O., and Hopkins, B. E., *Oxidation of Metals and Alloys,* 2nd Edition, Butterworths, London (1967).
2. Kofstad, P., *High Temperature Oxidation of Metals,* Wiley, New York (1966).

3. Wagner, C., *Atom Movements,* American Society for Metals, Cleveland (1951), p. 153.
4. Carter, R. E., and Richardson, F. D., *Trans. AIME,* **203,** (1955), p. 336.
5. Moore, W. J., and Selikson, B., *J. Chem. Phys.,* **19,** 1539 (1951); **20,** 927 (1952).
6. Evans, U. R., *Corrosion and Oxidation of Metals,* Edward Arnold, Ltd., London (1960).
7. Gulbransen, E. A., *Ann. New York Acad. Sci.,* **58,** 721 (1954).
8. Smeltzer, W. W., Haéring, R. R., and Kirkaldy, J. S., *Acta. Met.,* **9,** 880 (1961).
9. Davis, D. E., Evans, V. R., and Agar, J. N., *Proc. Roy. Soc.,* **A225,** 443 (1954).
10. Hart, E. W., *Acta. Met.,* **5,** 597 (1957).
11. Kofstad, P., Hauffe, K., and Kjollesdal, H., *Acta. Chim. Scand.,* **12,** 239 (1958).
12. See for example Hancock, P., in AGARD Conference Proceedings No. 120 "High Temperature Corrosion of Aerospace Alloys", Stringer, J., Jaffee, R. I., and Kearns, T. F. (Eds.), 1973, p. 117.
13. Stringer, J., *Acta. Met.,* **8,** 758, 810 (1960).
14. Mrowec, S., *Arch. Hutnicwa,* **6,** 327 (1961).
15. Cox, B., and Johnston, T., *Trans. AIME,* **227,** 36 (1963).
16. Stringer, J., *J. Electrochem. Soc.,* **112,** 1083 (1965).
17. Mrowec, S., *Arch. Hutnicwa,* **15,** 231 (1970).
18. Tuck, C. W., *Anti-Corrosion Methods and Materials,* **15,** 4 (1968).
19. Birks, N., *Brit. Corr. J.,* **3,** 56 (1968).
20. Howes, V. R., *Corrosion Sci.,* **10,** 99 (1970).
21. Surman, P. L., and Castle, J. E., *Corrosion Sci.,* **9,** 771 (1969).
22. Potter, E. C., and Mann, G.M.W., 2nd Int. Congr. Metal Corrosion, New York (1963).
23. Castle, J. E., and Masterson, H. G., *Corrosion Sci.,* **6,** 93 (1966).
24. Surman, P. L., *Corrosion Sci.,* **13,** 113 (1973).
25. Adams, A. M., Field, E. M., Holmes, D. R., and Stanley, R. C., 2nd Inter. Congr., Metal Corrosion, New York (1963).
26. Kovacina, T. A., Naval Research Laboratory Report 6788, December, 1968.
27. Hussey, R. J., and Cohen, M., *Corrosion Sci.,* **11,** 713 (1971).
28. Ing, S. W., Jr., Morrosion, R. E., and Sandor, J. E., *J. Electrochem. Soc.,* **109,** 221 (1962).
29. Ligenza, J. R., and Spitzer, W. G., *J. Phys. Chem. Solids,* **14,** 131 (1960).
30. Deal, B. E., and Grove, A. S., *J. Appl. Phys.,* **36,** 3770 (1965).
31. Haul, R., and Dümbgen, G., *Z. Elektrochem.,* **66,** 636 (1962).
32. Williams, E. L., *J. Amer. Ceram. Soc.,* **48,** 190 (1965).
33. Kofstad, P., *Stoichiometry, Diffusion, and Electrical Conductivity in Binary Metal Oxides,* Wiley, New York (1972).
34. Gulbransen, E. A., and Jannson, S. A., *Oxidation of Metals,* **4,** 181 (1972).
35. Hassner, A., *Diffusion in Metalischen Werkstoffen* (1970), p. 41.
36. Gjostein, N. A., "Physicochemical Measurements in Metal Research", *Techniques of Metal Research,* **4,** 405 (1970).
37. Fisher, J. C.,*J. Appl. Phys.,* **22,** 74 (1951).
38. Hoffman, R. E., and Turnbull, D., *J. Appl. Phys.,* **22,** 634 (1951); Turnbull, D., and Hoffmann, R. E., *Acta. Met.,* **2,** 419 (1954); Hoffmann, R. E., *Acta. Met.,* **4,** 97 (1956).
39. Love, G., and Shewmon, P. G., *Acta. Met.,* **11,** 899 (1963).
40. Bokshtein, B. S., Magidson, I. A., and Svetlov, I. L., *Fiz. Metal. Metalloved.,* **6,** 1040 (1958).
41. Mimkes, J., and Wuttig, M., *J. Appl. Phys.,* **41,** 3205 (1970).
42. Wuensch, R. J., and Vasilos, T., *J. Amer. Ceram. Soc.,* **47,** 63 (1964).
43. Irving, B. A., *Corrosion Sci.,* **5,** 471 (1965).
44. Le Claire, A. D., *Brit. J. Appl. Phys.,* **14,** 351 (1963).
45. Gibbs, G. B., *Corrosion Sci.,* **7,** 165 (1967).
46. Rhines, F. N., and Wolf, J. S., *Met. Trans.,* **1,** 1701 (1970).
47. Oishi, Y., and Kingery, W. D., *J. Chem. Phys.,* **33,** 480 (1960).
48. Paladino, A. E., and Coble, R. L., *J. Amer. Ceram. Soc.,* **46,** 133 (1963).
49. Holt, J. B., *J. Nucl. Mater.,* **11,** 107 (1961).
50. Austerman, S. B., Meyer, R. A., and Swartbout, D. G., Atomics International Lab. Report NAA-Sr-6427 (September, 1961); quoted in Ref. 49.
51. Cline, C. F., and Newkirk, H. W., *J. Chem. Phys.,* **49,** 3496 (1968).

52. Holt, J. B., *Proc. Brit. Ceram. Soc.*, **11**, 157 (1967).
53. Hawkins, R. J., and Alcock, C. B., *J. Nucl. Mater.*, **26**, 112 (1968).
54. Yajima, S., Furuya, H., and Hirai, T., *J. Nucl. Mater.*, **20**, 162 (1966).
55. Roberts, J. P., and Wheeler, C., *Phil. Mag.*, **2**, 570 (1957).
56. Mimkes, J., and Wuttig, M., *Phys. Rev.*, **B2**, 1619 (1970).
57. Ghoshtagore, G. N., *Phys. Rev.*, **155**, 589, 603 (1967).
58. Baluffi, R. W., *Phys. Stat. Solidi*, **42**, 11 (1970).
59. Baker, C., Wuttig, M., and Birnbaum, H. K., *Jap. Inst. Met. Proc. Intern. Conf. on Strength of Metal and Alloys*, Tokyo, September, 1967; Suppl. *Trans. Japan Inst. Metal*, **9**, 268 (1968).
60. Moore, W. J., and Williams, E. L., *Farady Soc. Disc.*, **28**, 86 (1959).
61. Tallman, R. L., and Gubransen, E. A., *J. Electrochem. Soc.*, **114**, 1227 (1967).
62. Gulbransen, E. A., *Mem. Sci. Rev. Met.*, **62**, 253 (1965).
63. Tallman, R. L., and Gulbransen, E. A., *J. Electrochem. Soc.*, **115**, 770 (1968).
64. Tallman, R. L., and Gulbransen, E. A., *J. Electrochem. Soc.*, **117**, 250 (1970).
65. Caplan, D., Sproule, G. I., and Hussey, R. J., *Corrosion Sci.*, **10**, 9 (1970).
66. Hussey, R. J., and Cohen, M., *Corrosion Sci.*, **11**, 699 (1971).
67. Channing, D. A., and Graham, M. J., *Corrosion Sci.*, **12**, 271 (1972).
68. Birchenall, C. E., in "Mass Transport in Oxides", Wachtman, J. B., and Franklin, A. D. (Eds.), *NBS Special Publication 296* (1968), p. 119.
69. Perrow, J. M., Smeltzer, W. W., and Ham, R. K., *Acta. Met.*, **15**, 577 (1967).
70. Perrow, J. M., Smeltzer, W. W., and Embury, J. D., *Acta. Met.*, **16**, 1209 (1968).
71. Cathcart, J. V., Petersen, G. F., and Sparks, C. J., Jr., *J. Electrochem. Soc.*, **116**, 664 (1969).
72. Cathcart, J. V., Peterson, G. F., and Sparks, C. J., Jr., *Mem. Sci. Rev. Met.*, **62**, 11 (1965); *Proc. Sagamore Conference on Physical and Chemical Characteristics of Surfaces and Interfaces*, University of Syracuse Press, Syracuse, New York (1967).
73. Herchl, R., Khoi, N. N., Homma, T., and Smeltzer, W. W., *Oxidation of Metal*, **4**, 35 (1972).
74. Berry, L., and Paidassi, J., *Compt. Rend. Acad. Sci. Paris*, **262**, 1353 (1966).
75. Volpe, M. L., and Reddy, J., *J. Chem. Phys.*, **53**, 1117 (1970).
76. Cox, B., and Pemsler, J. P., *J. Nucl. Mater.*, **28**, 73 (1968).
77. Madeyski, A., Poulton, D. J., and Smeltzer, W. W., *Acta. Met.*, **17**, 579 (1969).
78. Lighstone, J. B., and Pemsler, J. P., in "Mass Transport in Oxides", Wachtman, J. B., and Franklin, A. D. (Eds.), *NBS Special Publication 296* (1968); *Reactivity of Solids* (1969), p. 615.
79. Stringer, J., Hed, A. Z., Wallwork, G. R., and Wilcox, B. A., *Corrosion Sci.*, **12**, 625 (1972).
80. Wright, I. G., and Stringer, J., *Metallography*, **6**, 65 (1973).
81. Stringer, J., Wilcox, B. A., and Jaffee, R. I., *Oxidation of Metals*, **5**, 11 (1972); Stringer, J., and Wright, I. G., *ibid.*, p. 59.
82. Paidassi, J., Cadiou, L., and Berry, L., *Compt. Rend.*, **266**, 413 (1968).
83. Pettit, F. S., *Trans. AIME*, **239**, 1297 (1967).
84. Tien, J. K., and Pettit, F. S., *Met. Trans.*, **3**, 1587 (1972).

DISCUSSION on Paper by John Stringer

SHEWMON: One can coarsen the grain size of a protective oxide layer by raising the temperature briefly. Has anyone measured the increase in the low-temperature oxidation rate that results from coarsening the oxide grain size with a brief high-temperature excursion? Also might not stresses in the oxide be relieved by rapid diffusion along the metal/oxide interface?

STRINGER: Where such experiments have been made, the intention was to determine the effect of a relaxation of the defect gradient rather than to investigate the effects of short-circuit diffusion on oxidation rates.

The concept of rapid diffusion along the metal/oxide interface is entirely feasible but often overlooked. The particular model (due to Rhines and Wolfe) for stress generation in NiO growing on nickel is, however, slightly different. In this case, oxide formation takes place in the oxide grain boundary itself, from oxygen diffusing along the grain boundary and nickel diffusing through the oxide grain body. The stress set up by the formation of this oxide cannot be relieved by diffusion along the metal/oxide interface.

GRAHAM: I would like to add one other oxide, silica, to your list of Cr_2O_3 and Al_2O_3 as oxides where an understanding of the short-circuit transport is of high technological importance.

STRINGER: There is indeed much evidence to suggest that silica grows by transport through pores, rather than through the more conventional short-circuit paths.

KOFSTAD: Prof. Stringer has shown a simple relationship between the parabolic rate constant and the self-diffusion coefficient for the faster moving ionic species in the oxide scale [$K_p = (1+p)D$]. It should be emphasized that this relationship applies only to an ideal limiting case for scales of metal-deficient oxides. For oxygen-deficient scales, the relationship is not equally simple. Also, in comparison of parabolic rate constants and diffusion coefficients in the latter case, the diffusion coefficients to be used are for diffusion in the oxide at the metal/oxide interface. This serves to emphasize that the "correct" diffusion coefficients have to be used in the above comparisons.

JORGENSEN: In the marker experiment showing oxygen is transported faster than is aluminum during oxidation, which phase of alumina occurred in the oxide scale and what was the grain size?

STRINGER: At the temperatures of interest, the oxide is α-alumina, with a grain size of approximately 1 μm. Unlike the columnar-grained scales shown earlier by Prof. Smeltzer, the grain structure of these α-alumina scales tends to be equiaxed.

HAUFFE: The rate law which considers also the pore formation in the growing oxide layer is rather complicated and usually useless because it cannot be evaluated. Therefore, it is not recommendable to go further in this direction. It is much more reasonable to elaborate on alloying elements which enter the oxide layer and decrease the pore formation, as this has been done for instance by Jorgensen to improve the sintering of Y_2O_3 by adding of small amounts of ThO_2.

STRINGER: If the mathematical analysis is good, it helps to identify the process of scale growth, and may give some idea of the kind of pores that form. The porosity associated with short-circuit transport is exceedingly fine, and it may be difficult to induce densification in such oxides, especially when the porosity results from the growth mechanism.

DEFECTS AND IMPURITIES INFLUENCES ON PHASE TRANSFORMATIONS IN RARE-EARTH OXIDES

P.E. Caro and G. Schiffmacher

Laboratoire des Terres Rares CNRS
Bellevue, France

C. Boulesteix and Ch. Loier

Institute d'Optique
Orsay, France

R. Portier

ENSCP
Paris, France

ABSTRACT

Several types of defects are observed on thin films (up to 2500 Å) of pure rare-earth oxides of type A-Ln_2O_3 and B-Ln_2O_3. They are (a) structural steps connected with the layered character of the structures which contain the $(LnO)_n^{n+}$ entities parallel to the thin-film surface, (b) reciprocal twins in B-Ln_2O_3 of type I [twin plane (313)] and of type II [twin axis (132)] whose combination yields microtwinned zones; (c) planar defects; (d) dislocations; and possibly (e) antiphase boundaries. The twinning in B-Ln_2O_3 is directly seen, under the electron microscope, securing the syntaxy conditions between the two phases when the A → B-Ln_2O_3 or B → A-Ln_2O_3 transition takes place. A small amount of yttrium in A-Nd_2O_3 induces the formation of domains of B-Nd_2O_3. The transformation is, from X-ray investigation, complete at 10 atomic percent yttrium, and consequently the B-Nd_2O_3 phase is stabilized. An electron microscopy investigation of thin films of mixed oxides containing up to 10 atomic percent yttrium suggests the possibility that microdomains of C-Ln_2O_3 are operative in the transition from A to B-Nd_2O_3, which can be produced in the pure phase by a mechanical stress. A well-defined epitaxy between C and B-Ln_2O_3 was determined in the mixed oxides thin films.

1 INTRODUCTION

It is well known that the rare-earth sesquioxides Ln_2O_3 exhibit, depending on temperature and atomic number, three main crystallographic structures: type A-Ln_2O_3 hexagonal, type B-Ln_2O_3 monoclinic, and type C-Ln_2O_3 cubic. The three polymorphs can be described from the point of view of oxygen coordination.[1] It is then apparent that A and B-Ln_2O_3 are layered structures involving a complex framework $(LnO)_n^{n+}$ (C_3), made of OLn_4 tetrahedra sharing edges with a ternary symmetry (a Ln-O-O-Ln slice of a fluorite structure LnO_2 perpendicular to [111]). A slight displacement of the $(LnO)_n^{n+}$ layers with respect to each other in A type along the {1100} hexagonal directions yields the B-type Ln_2O_3. The $(LnO)_n^{n+}$ entities can be recognized in the structures of a score of other rare-earth compounds (the oxysalts). On the contrary, C-type Ln_2O_3 can be seen as a three-dimensional framework of OLn_4 tetrahedra sharing four edges out of six. The situation in the rare-earth oxides can be topologically compared with the one in the layered silicates [$(Si_2O_5)_n^{2n-}$ sheets] on one hand, and the various forms of silica on the other hand.[2]

It is quite easy to prepare thin films of the rare-earth oxides with thicknesses up to 2500 Å by direct oxidation of thin, evaporated metal films. Those samples can be observed within the electron microscope and they thereby provide a direct view of the active structural elements, as, for example, extended defects. It seems possible to translate the results from thin films to bulk materials because essentially the same results are inferred from the observation of thin flakes from powdered or high-temperature sintered samples.

The material from the oxidized metal is usually the low-temperature form of the oxide, that is C-Ln_2O_3 (including lanthanum). It has very small crystals which immediately recrystallize into large crystals of the high-temperature (A- or B-Ln_2O_3), or large crystals of C type, when submitted to the electron beam. When the A type is obtained, it has *always* a definite orientation: the $\underline{c}$ hexagonal axis is perpendicular to the thin-film surface (i.e., the $(LnO)_n^{n+}$ layers are parallel to the thin-film plane). When the film is not too thick, the B type crystals similarly have the $(20\bar{1})$ plane parallel to the thin-film plane; here again the $(LnO)_n^{n+}$ layers are parallel to the film surface. The C type has a more erratic behavior, but a large number of crystals have their (111) plane parallel to the film surface. Consequently, the ternary (or pseudo ternary for B type) axis is generally perpendicular to the film surface. In the last 2 years, we have identified numerous structural extended defects within the pure phases. They are especially spectacular in B type crystals.

2 TYPE OF DEFECTS IN RARE-EARTH OXIDE THIN FILMS

2.1 Structural Steps

In A- and B-Ln_2O_3, the layered behavior is apparent from the thin-film orientation habits. However, types of defects can be identified which are directly connected to it. They are structural steps, and they appear on the images (Fig. 1) as:

Type 1, a complex lattice of curves

Type 2, lines with definite crystallographic orientations [010], $[1\bar{3}2]$, [132] for B-Ln_2O_3, <10.0> for A-Ln_2O_3

Type 3, isolated hexagon-like figures whose contours have the same orientations as the lines above.

These aspects have been linked (see ref. 3 for details) to differences in thicknesses of the films. That is, the surface is not regular but made of a succession of plateaus separated by cliffs whose slopes are the faces of the OLn_4 tetrahedra not parallel to the film surface: $(0\bar{2}3)$, $(\underline{201})$, (023) in B-Ln_2O_3 with $(20\bar{1})$ as the plateau's surface; (02.3), $(20.\bar{3})$, $(\bar{2}2.3)$ in A-Ln_2O_3 with (00.1) as the plateau's surface. The height of the steps can be 100 Å. Some of the hexagonal contours in B-Ln_2O_3 are indeed quite high stepped towers (we call them "ziggurats"). The formation of the "ziggurats" proceeds from the transition from C to B-Ln_2O_3. It is a dynamic process that can be directly observed and recorded on cine film.

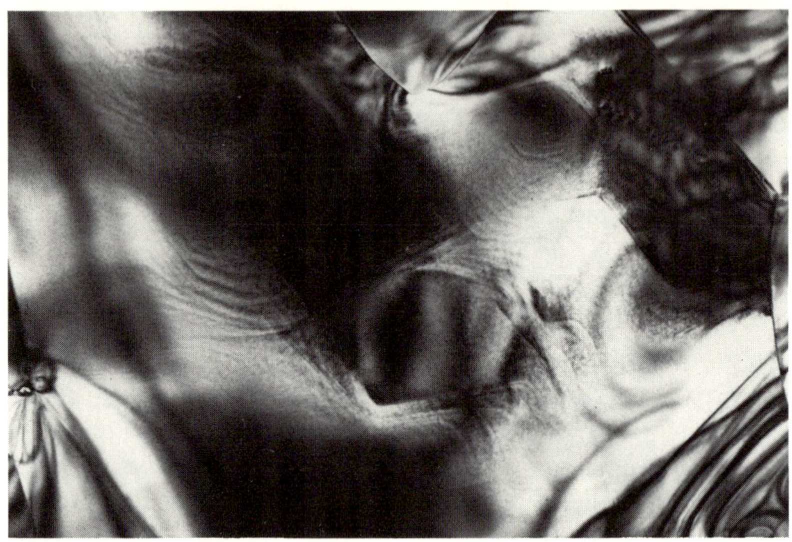

Fig. 1. Structural steps and "ziggurats" on a crystal of B-Gd_2O_3; note the series of dislocations on the lower left side. (X24,000)

When the steps are curved, a moderate annealing breaks them in type 2 linear segments.

2.2 Twins

The monoclinic structure can be derived from the hexagonal one by a deformation of a haxagonal layers together with a 3 degree inclination of the $\underline{c}$ axis. Some symmetry elements of the A type become pseudo-symmetry elements of the B type. The three hexagonal {110} and the axis perpendicular to them become two {313} planes and one (020) in the B type. The {313} are pseudo-symmetry elements of the B type; they are twin elements, with twin axes <132> which are pseudo perpendicular to them. On the contrary, (020) is a symmetry plane in the monoclinic structure.

The interesting consequence is the formation of two reciprocal systems of twins. There is a twin of type I with a twin plane (313) and a reciprocal twin of type II with a twin axis [132]. The other system is a twin with a plane $(\bar{3}13)$ together with a twin whose axis is $[1\bar{3}2]$. It is quite easy to recognize the two types of twins on the images. When the electron beam is along the $[10\bar{1}]$ direction, a twin boundary of type I appears as a thin line, because $[10\bar{1}]$ is within the contact plane, the two twinned crystals being symmetrical with respect to (313). On the other hand a twin boundary of type II is broad and shows fringes. The rationnal element of the twin is the binary <132> axis, which is within the $(20\bar{1})$ plane of the foil[4] (Fig. 2).

The simultaneous occurrence of those two modes of twinning is a rather rare one in a given material. As a result, spectacular images are observed when the twin systems interpenetrate[5] (Fig. 3). When twin bands belonging to nonreciprocal types intersect, there are in the common part microtwins of type I or II with tapered ends within one or the other band. Also, when bands of type II meet a reciprocal band of type I, one observes the formation of tapered twins at the end of the type II bands. A complete study of the several types of twin interactions in B type thin films has just been completed by C. Loier.[6]

The twins in the B-type crystals are mechanical ones. They are associated with the stresses produced when the narrowly localized electron beam is displaced through the thin films to create the large, high-temperature phase crystals; for instance, a tear originating in a single crystal induces the appearance of a twin. But the same twins also are created by annealing as a result of the recrystallization process. Twinning is also currently observed on flakes from bulk samples. We can sequently think that it is an inherent property of the B phase.

Fig. 2. Twins of type I and type II in B-Sm_2O_3. (X50,000)

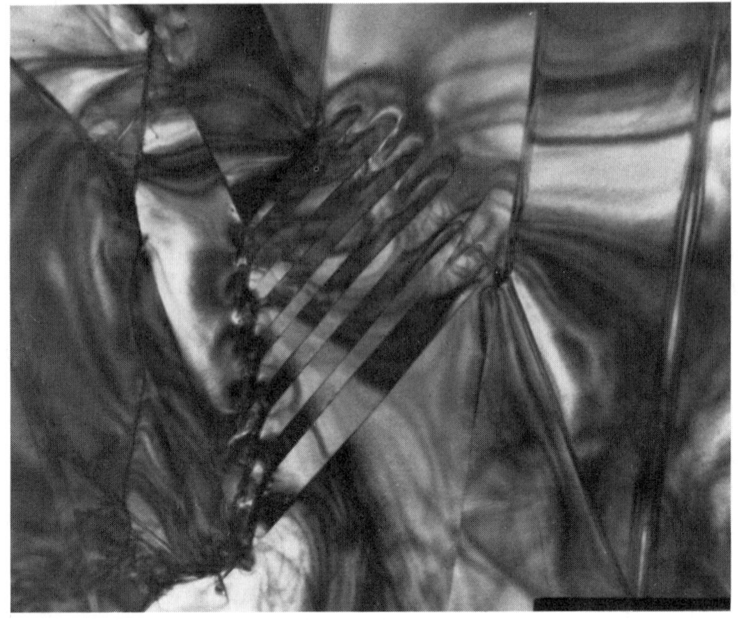

a

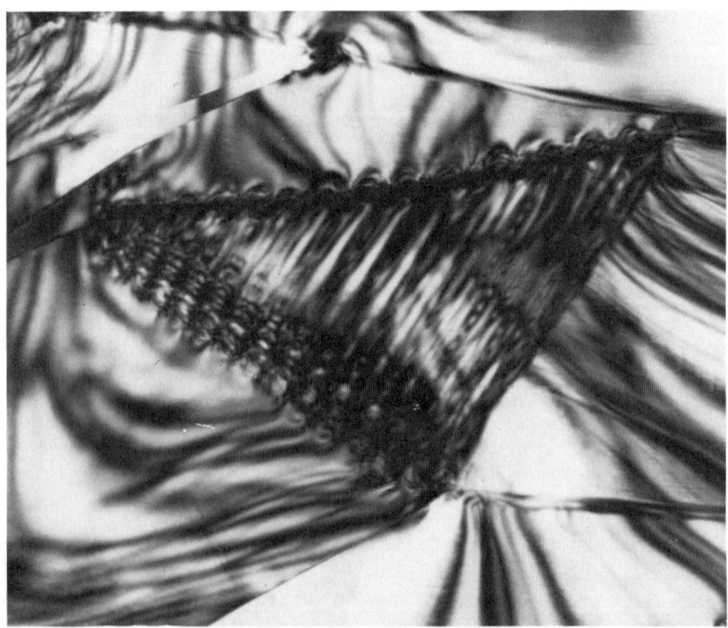

b

Fig. 3. Microtwins (a) of type I in B-Gd_2O_3 and (b) of type II in B-Sm_2O_3. (X20,000)

2.3 Planar Defects

There are several types of such defects in B-Ln_2O_3. They can be observed:

- In large thin crystals in the vicinity of a hole in the film
- Between two twin boundaries which make either a twin band or a tapered twin.
- At the extremity of very narrow twins. (Fig. 4.)

When a planar defect ends in the middle of a crystal, it terminates by a partial dislocation.

The planar defects are presently under investigation by R. Portier at the ENSCP-Paris.

Fig. 4. Tapered twins terminated by a planar defect ending on a partial dislocation- B-Sm_2O_3 (X40,000); courtesy of R. Portier.

2.4 Dislocations

If the crystals of B-Ln_2O_3 in thin films are strongly twinned, there are comparatively few dislocations, contrary to what is observed in metals. In films with thicknesses smaller than 1000 Å, the dislocations are extremely rare and are perpendicular to the surface of the film. In thicker films (1600 Å), it is possible to create and keep, by rapid quenching, more numerous dislocations. They then have a longer length, and also the crystals sometimes have another orientation than $(20\bar{1})$. Then one can observe

- Piled-up groups of equidistant dislocations within a single crystal. They have as Burgers vector $\vec{b}$ = [010]. They are perfect dislocations, with a strong edge character.
- Twin dislocations in twin boundaries. They are imperfect dislocations, and from the formation of tapered twins their edge character can be inferred.
- Dissociated dislocations, separated by a faulty zone (Fig. 5). Dislocations in the vicinity of planar defects are to be linked with this last type.

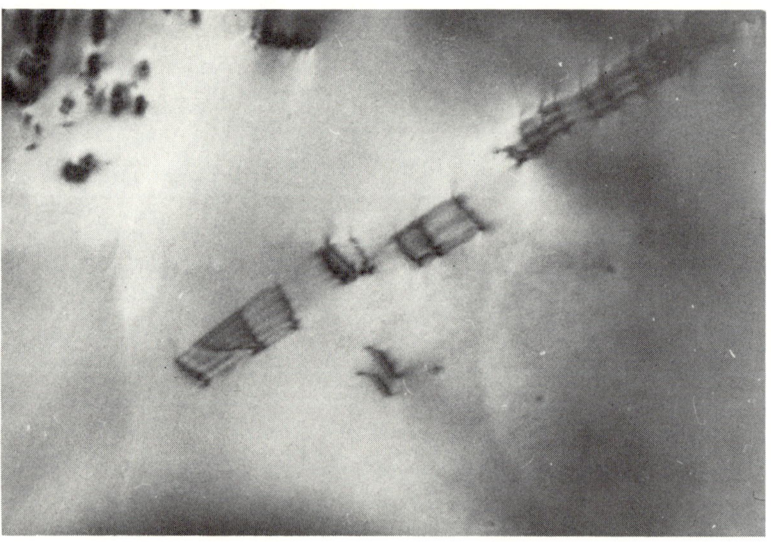

Fig. 5. Dissociated dislocations in B-Sm_2O_3. (X80,000)

2.5 Antiphase Boundaries

The monoclinic unit cell has a volume which is three times the one of the hexagonal unit cell. This is because there are three different crystallographic sites for the rare earths in B type instead of one for the A type. This is also true for the "ionic" oxygens, the ones that are between the $(LnO)_n^{n+}$ layers. For both, one goes from one site to another by a vector translation of $\left[\frac{132}{6}\right]$ or -$\left[\frac{132}{6}\right]$. The existence of three antiphase domains linked together by those translations is then possible. Corresponding antiphase boundaries were recently suspected (Boulesteix and Portier), but problems involving the residual contrast are, at the present time, not fully solved, possibly because the translation vectors are not exactly equal to the values mentioned above. Those values seem complicated but, in fact, the monoclinic unit cell is not the smallest one. There is a triclinic one and, within this frame the above vectors become $\left[\frac{111}{3}\right]$ and -$\left[\frac{111}{3}\right]$.

3 INTERACTION OF DEFECTS AND PHASE TRANSITIONS

When a thin film of Sm_2O_3 is strongly heated within the electron microscope, crystals of the A phase, the high-temperature form, are created. When the crystals are cooled one can observe the appearance of arrow-shaped crystals made of several twinned B-type crystals which split off the arrow (Fig. 6). The arrows grow in a jerky way from the coolest part at the expense of the A type crystal. When the cooling is completed the B type crystals exhibit parallel twin boundaries and twin intersections. The respective orientations of the A and B phases are the same as the one which has been observed when the formation of the B phase is induced by a local thermal shock in pure A-Nd_2O_3 or -Pr_2O_3.[7] In that case the B phase created from the stresses disappears within minutes. In Sm_2O_3, the B phase is the stable one. There is a definite epitaxy between A- and B-Ln_2O_3, but the necessary conditions can be achieved only through twinning in the B phase. In Nd_2O_3, on one side of an arrow there is an epitaxy between (11.0) hexagonal and (313) monoclinic, and on the other side, after a type II twinning, there is an epitaxy between (424) monoclinic and ($3\bar{1}$.0) hexagonal.

In Sm_2O_3 we have observed B-phase arrows with one type I twin securing an epitaxy between approximately (535) monoclinic and (21.0) hexagonal, but there are also arrows with two twins type I and type II at 30 degrees from each other; then the monoclinic plane of contact is (424) on both sides of the arrow, which has then an overall angle of 60 degrees securing epitaxy with the hexagonal {21.0}. By growing, the several arrows end into a B-type crystal with parallel twin boundaries. (See Fig. 6.)

A- and B-Ln_2O_3 are then in a three-dimensional syntaxy, because the layered structures have their packing axis distinct but parallel.

When a C → B transition is directly observed in the electron microscope, for instance on Dy_2O_3, a germ of the B type grows more easily at the expense of a large cubic crystal if ($20\bar{1}$) and (111) of B are parallel, respectively, to (111) and (110) of C. The reversibility of the transition depends on the thickness of the film and the atomic number. It is not reversible up to holmium for films with thicknesses smaller than 1000 Å,

Fig. 6. Arrows of B-Sm_2O_3 with twins of type I and II at 30 degrees from each other in epitaxy on a crystal of A-Sm_2O_3. (X30,000)

but can be reversible at dysprosium, for thicknesses greater than 1000 Å. The free-energy change between the phases is very small[8]; consequently, defect and surface energy may decide, according to the history of a particular sample, the temperature at which the change is taking place as well as the reversibility. This is the reason for the debate on the temperature at which the C → B transition takes place.[9] Recently, Coutures and collaborators[10] have observed the B phase for all rare earths including yttrium, by quenching the oxide vapor. Electron-microscope examination of the resulting powder shows that the crystals are very small with thicknesses of 30 Å or so, that is, a few unit cells. The formation of the C phase can then be understood as a process which keeps the oxygen atoms as far apart as possible, but still within an OLn_4 tetrahedra linkage scheme. However, when the crystal is small, the B-type phase can occur, because then the oxygen atoms can relax through the crystal boundaries the stresses induced by the lanthanide contraction within the rigid framework of OLn_4 tetrahedra in $(LnO)_n^{n+}$.

4 STABILIZED Nd_2O_3-B PHASE BY YTTRIUM ADDITION

An X-ray and optical investigation of the binary system A- Nd_2O_3 $\frac{1}{M}$ C-Y_2O_3[11] shows that a very small addition of yttrium oxide to A-Nd_2O_3 induces the appearance of a B-type phase involving neodymium. This can be seen in the 1000-ppm-yttrium atomic-concentration range from optical studies because of the additional electronic-absorption lines from the three sites of neodymium in the B type, and is clearly detected from the appearance of B-phase X-ray diffraction lines at 0.5 percent yttrium. X-ray diffraction lines for the A phase disappear at 10 percent yttrium, and the C-type lines appear at 20 percent; they are alone from 20 percent and up yttrium. This is clearly a case of the stablization of an otherwise unstable phase by addition of another element.

We have prepared alloys of neodymium and yttrium by coevaporation of the two metals and we oxidized them according to the usual process for pure metals. The yttrium concentration range studied was between 2 and 10 percent from X-ray fluorescence determinations. Electron-microscope examination of the thin films so produced indicated that the material was roughly comparable to the bulk samples; that is, within the range of concentration, they exhibited a large number of crystals of B-type Ln_2O_3 with the same characteristics as those of B-type oxides for pure metals.

We thought the system might be interesting from the standpoint of getting a direct view of the solubility process of the yttrium. The optical data indicates that yttrium does not replace neodymium in the A-type C_{3v} site. On the other hand, yttrium B phase is not known. The studies on the pure phases have shown that neodymium B-type domains can be created by a stress, so there was the possibility that yttrium oxide segregates as C-type oxide, and by some kind of epitaxial relationship combined with the lanthanide contraction, induces the necessary stress, yielding B-type domains out of the A phase. Such a model is suggested by the general possibilities offered by the syntaxies observed in several rare-earths compounds involving structural layered entities like the $(LnO)_n^{n+}$.[12]

The examination of the binary thin films shows that a very large number of defects like the one described in the first part of this paper are present. The general appearance of the images is, however, far more complex than for a pure B phase (Fig. 7). Microtwinning is, for instance, easily recognized (Fig. 8).

Important differences are noticeable however with the pure oxides. First, even for 10 percent Y_2O_3, there are a few rather large crystals which belong to A-type Ln_2O_3. Second, side by side with the heavily twinned B-type zones, there are areas which are untwinned (Fig. 9) and which can

Fig. 7. Image of a thin film of Nd_2O_3 – 7 percent Y_2O_3. (X20,000)

exhibit pure C-phase as well as pure B-phase diffraction patterns, and also a superposition of the two but always with a well-defined orientation. The image (Fig. 10) of the zones giving both diffraction patterns (Fig. 11) has a peculiar striated feature. Pure C-type diffraction patterns are rare and seem to correspond to very small areas. There is apparently no clear contrast on the images separating the particular (B, B and C, or C) diffracting zones. When the two patterns appear together, the electron beam is along $[10\bar{1}]$ for B-Ln_2O_3 and along $[11\bar{2}]$ for C-Ln_2O_3; the epitaxial relationship is $(4\bar{4}0)$ cubic parallel to (020) monoclinic and (222) cubic parallel to (202) monoclinic. (See the indexation of the Fig. 11 pattern in Fig. 12.)

Moreover, when an impure stock of neodymium oxide is used to prepare the metal thin films (by thorium reduction see[7]), the behavior of the oxidized film is very strange. When the sample is heated to grow the large crystals, one observes, if the film is thick, a migration of material towards the hot zones, which results in the formation within the A-type crystals of an inclusion-like feature which exhibits the diffraction patterns of both B- and C-Ln_2O_3 with the same epitaxial relationship as defined above.

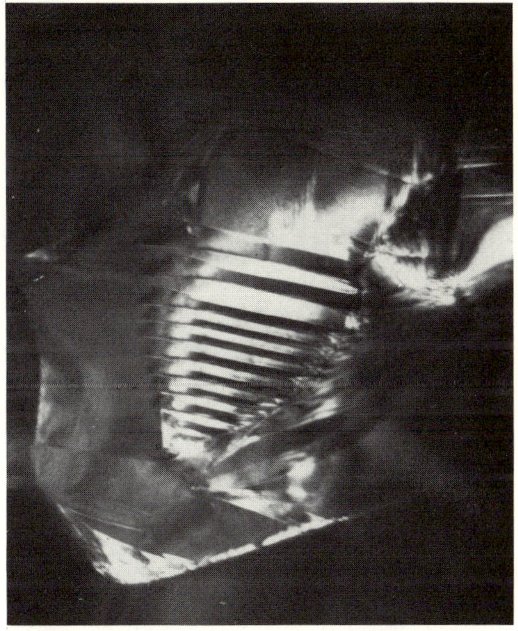

a

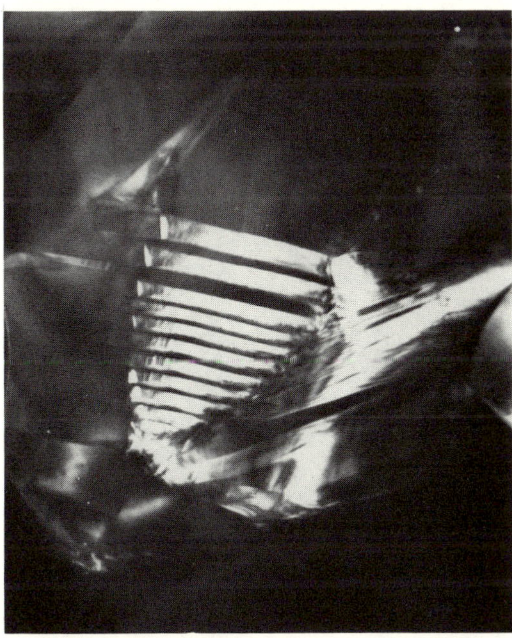

b

Fig. 8. Microtwinning in Nd_2O_3 – 7 percent Y_2O_3 a and b dark-field images on each component of the twin. (X30,000)

Fig. 9. Image of a thin film of Nd_2O_3 – 10 percent Y_2O_3 showing contact between twinned zones and striated nontwinned zones. Cubic diffraction patterns are obtained on the clear striated zones at the upper left. (X20,000)

Fig. 10. Image of a nontwinned zone in Nd_2O_3 – 10 percent Y_2O_3. (X30,000)

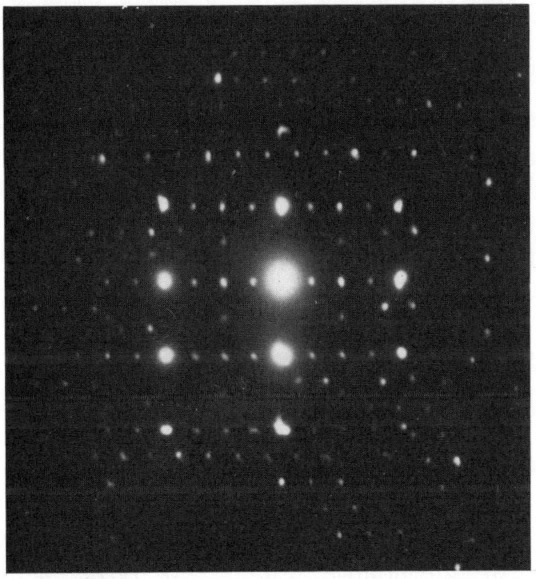

Fig. 11. Diffraction pattern of the clear striated zone in Fig. 10 (on the extinction contour).

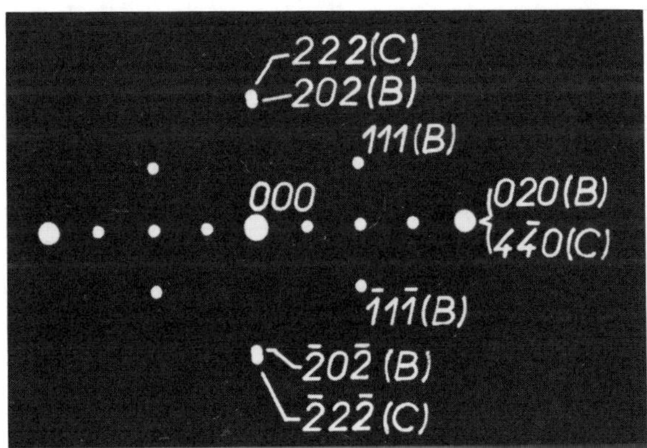

Fig. 12. Indexation of the pattern in Fig. 11, showing the epitaxy between B- and C-Ln_2O_3.

We were able to make few experiments with the Jeol energy nondispersive X-ray spectrometer (EDAX system) on binary thin films. The analysis indicated so far that there was no yttrium in the A-type crystals, a little in the strongly twinned B-type crystals, and a large amount in areas which exhibit C-type diffraction patterns.

There is consequently a chance that C-type microdomains possibly associated with yttrium segregation exists in the thin films studied. However we must recognize that the preparation does not secure the homogeneity of the thin films, and also that segregation can further occur because of temperature effects under the electron beam within the microscope. Strong heating of the material results in large crystals without defects (A type?) which on cooling reverse to the strongly twinned and defect-rigged material exhibited in the photographs in a very fast way.

It is quite important that the presence of small domains of C phase with a well-defined epitaxy to B phase can indeed be derived from the diffraction patterns. The question is one of solubility in the solid state. Solubility can occur when one atom replaces another one in its local site, i.e., a point-to-point substitution. Neodymium for instance can be seen from the displacement towards the blue of the ${}^4I_{9/2}(Z_1) \rightarrow {}^2P_{1/2}$ transition absorption line at 4.2 K, to replace lanthanum in A-La_2O_3.[11] But another possibility is an insertion, ordered or not, of a small block of one structure into another if proper epitaxial conditions can exist. The inserted block may be limited to a few atomic planes; in that case it will not be detected by X-ray. This second type of solubility may be defined as "platonician" versus the point-to-point solubility defined as "democritian" (thereby resuscitating a very old debate in science)[12]. There is no doubt that the epitaxies needed for a platonician-like solubility do occur in the rare-earth oxides binary system under investigation. However the matter of which atom is in which phase can be definitely settled only by using local structural probe methods like the optical ones, which may provide answers when enough information will have been collected on the crystal-field energy-levels scheme in well-defined geometrical sites in pure compounds. The search may be facilitated by the fact that pure rare-earth thin films as well as the mixed-rare-earth ones yield very easily lattice fringes on a high-resolution electron microscope, thereby showing the arrangement of atomic planes. Such a lattice-fringes image of a tapered twin is shown in Fig. 13.

REFERENCES

1. Caro, P. E., *J. Less Common Metals,* **16**, 367 (1968).
2. Caro, P. E., *NBS Special Publication 364,* "Solid State Chemistry", 367-383 (1972).
3. Portier, R., Fayard, M., Boulesteix, C., and Caro, P. E., *Mat. Res. Bull.,* 8, 693 (1973).
4. Boulesteix, C., Caro, P. E., Loier, Ch., and Portier, R., *Phys. Stat. Sol.* (a) **11**: 771 (1972).
5. Boulesteix, C., and Loier, Ch., *Phys. Stat. Sol.* (in print).
6. Loier, Ch., Thesis, University of Orsay-France (1973).
7. Boulesteix, C., Caro, P. E., Gasgnier, M., La Blanchetais, C. Henry, and Schiffmacher, G., *Acta Cryst.,* Sect. A, **27**, 552 (1971).
8. Fitzgibbon, G. C., Pavone, O., and Holley, C. E., Jr., *Proc. Eighth Rare Earth Research Conf.,* Reno, Nevada, (1970), Vol. I, p. 224.
9. Brauer, G., *Proc. Sixth Rare Earth Research Conf.,* Gatlinburg, Tennesee, (1967), p. 385.

10. Coutures, J. P., Coutures, J., Renard, R., and Benezech, G., Compt. Rend. Acad. Sci. Paris, **275C,** 1203 (1972).
11. Caro, P. E., Faucher, M., Derouet, J., and Alves, M. G., *Rev. Chim. Min.*, **10**, 317 (1973).
12. Caro, P. E., *J. Solid. State Chem.,* **6**, 396 (1973).

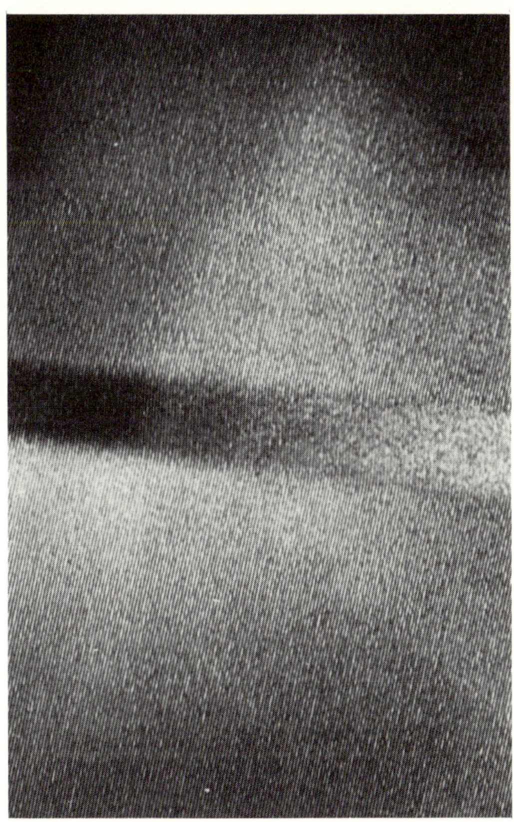

Fig. 13. High-resolution image on a B-phase crystal in Nd_2O_3 – 7 percent Y_2O_3; interference with two tilted beams, planes (111), d = 3.25 Å ; tapered type I twin band. (X500,000)

INTERDIFFUSION AND ASSOCIATION PHENOMENA IN SOME OXIDE SYSTEMS

V. S. Stubican and W. J. Minford

Department Material Sciences
The Pennsylvania State University
University Park, Pa.

ABSTRACT

The mechanism of interdiffusion in systems $AO\text{-}B_2O_3$ is discussed. The experimental data obtained on interdiffusion of B^{3+} ions (e.g., Cr^{3+} and Al^{3+}) in an AO compound (e.g., MgO and NiO) show that the relationship $\tilde{D} \simeq N_B$ is only an approximation which implies that no association of defects occurs. This relationship was observed in systems Cr_2O_3-MgO and Cr_2O_3-NiO at temperatures higher than 1300°C. In the systems where the association energy of defects (ΔF) is high, the association effects cannot be neglected even at high temperatures or low concentration of B^{3+} ions.

Using Lidiard's theory of diffusion by impurity-vacancy pairs, ΔF for the $[Al^{\cdot}_{Ni}V''_{Ni}]'$ pairs was calculated to range from 6.5 to 9.0 kcal/mole between 1789°C and 1540°C, which is considerably higher than ΔF for the $[Cr^{\cdot}_{Ni}V''_{Ni}]'$ pairs previously calculated to range from 0.6 to 3.0 kcal/mole between 1300°C and 1000°C.

From interdiffusion studies, the activation energy of motion (ΔH_m) for B^{3+} ions can be obtained if the concentration of vacancies is impurity controlled and if ΔF is small.

1 INTRODUCTION

Many trivalent ions form partial solid solution with magnesium oxide or nickel oxide. Such systems are of interest in that they allow systematic investigation of the interdiffusion of trivalent ions in the sodium chloride-type lattice in oxide systems. The interdiffusion of trivalent ions in AO oxides will introduce cation vacancies so that, by the choice of appropriate experimental conditions, the influence of the vacancy concentration on the interdiffusion can be experimentally determined and theoretically explained.

Considerable information is available in the literature on the interdiffusion process between divalent oxides[1-4] and, in some cases, the obtained interdiffusion coefficients could be directly related to the tracer diffusion coefficients by using an analysis by Darken[4,5].

It is the purpose of this paper to evaluate the experimental data obtained on interdiffusion in $AO\text{-}B_2O_3$ systems and to discuss the mechanism of interdiffusion.

2 EXPERIMENTAL DATA

Those systems that have been studied include $MgO\text{-}Cr_2O_3$[6], $MgO\text{-}Al_2O_3$[7] $NiO\text{-}Cr_2O_3$[8], $NiO\text{-}Al_2O_3$[9], $MgO\text{-}Fe_2O_3$[2], and $MgO\text{-}Ga_2O_3$[10]. The results in the last two systems are not complete or conclusive, so that in the further discussion only the first four systems will be analyzed.

The typical reaction that occurs in these systems is that between AO and B_2O_3 to form AB_2O_4 spinel, which could be stoichiometric or nonstoichiometric. Furthermore, a diffusion profile of B^{3+} ions in AO results. It has been established that in a majority of such systems, the reaction mechanism is the counterdiffusion of cations through the rigid oxygen lattice.[11]

If the diffusion couple of AO and B_2O_3 is considered after a diffusion-anneal treatment, the schematical concentration profile is as shown in Fig. 1. For the case where B_2O_3 is soluble in spinel, e.g., Al_2O_3 in $NiAl_2O_4$, the concentration of B^{3+} ion changes along the spinel layer. The interdiffusion coefficients ($\tilde{D}$) were calculated from the concentration profiles using Wagner's formula[12]:

$$\tilde{D}(N_2^*) = \frac{(N_2{}^+ - N_2{}^-)V_m(N_2{}^*)}{2t(\partial N_2/\partial x)_{x=x^*}} \left[(1-Y^*)\int_{-\infty}^{x^*} \frac{Y}{V_m}\, dx + Y^* \int_{x^*}^{+\infty} \frac{1-Y}{V_m}\, dx \right], \tag{1}$$

where $N_2{}^+$ and $N_2{}^-$ are the initial mole fractions of the binary diffusion couple, $N_2{}^*$ is the particular mole fraction of Component 2 of interest, x* is the distance at which $N_2 = N_2{}^*$, $V_m(N_2{}^*)$ is the molar volume of the components at concentration $N_2{}^*$, Y is defined as $(N_2-N_2{}^-)/(N_2{}^+-N_2{}^-)$, and Y* is $(N_2{}^*-N_2{}^-)/(N_2{}^+-N_2{}^-)$. If the diffusion couple initially consists of pure components, the auxiliary variable Y may be set equal to the mole fraction N_2.

This formula takes into consideration the moving interface between AO and AB_2O_4 spinel, as well as the change in the molar volume of AO with the composition. The evaluation of the integrals in Eq. 1 can be accomplished quite easily if the concentration profile is plotted in terms of Y/V_m.

The variation of the interdiffusion coefficients with the mole fraction of trivalent ions is shown in Figs. 2-5. An apparent linear relationship between $\tilde{D}$ and the concentration of Cr^{3+} ions in MgO and NiO is observed at high temperatures. However, interdiffusion coefficients in systems Al_2O_3-MgO and Al_2O_3-NiO show considerable deviation from linearity if plotted against the concentration of trivalent ions. This is particularly true for the system Al_2O_3-NiO (Fig. 5) where at low concentration of Al^{3+} ions, the interdiffusion coefficients in the NiO increase with increasing amounts of Al^{3+} ions, but a high concentration, the interdiffusion coefficients level off to a constant value.

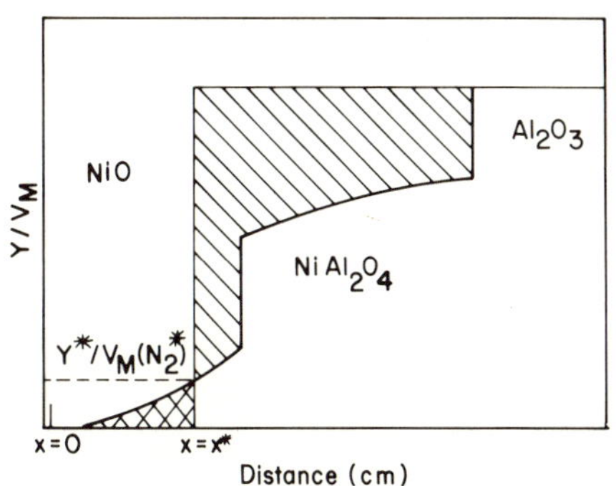

Fig. 1. Concentration (Y/V_m) of Al^{3+} ions versus distance for the diffusion couple NiO/Al_2O_3 after a diffusion anneal; hatched areas are needed for the integrals in Eq. (1).

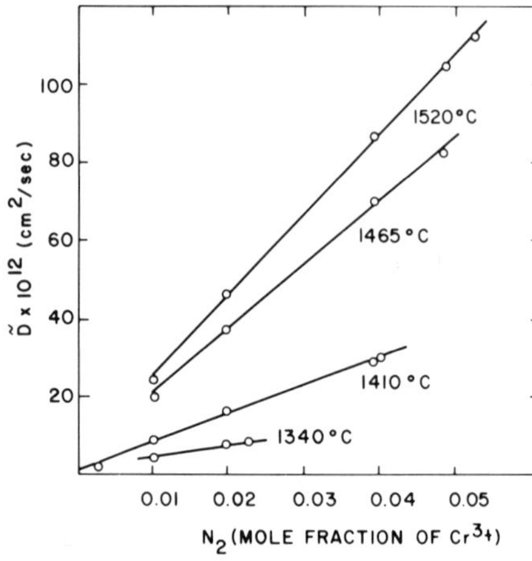

Fig. 2. Variation of $\widetilde{D}$ in MgO with the mole fraction of Cr^{3+} ions.

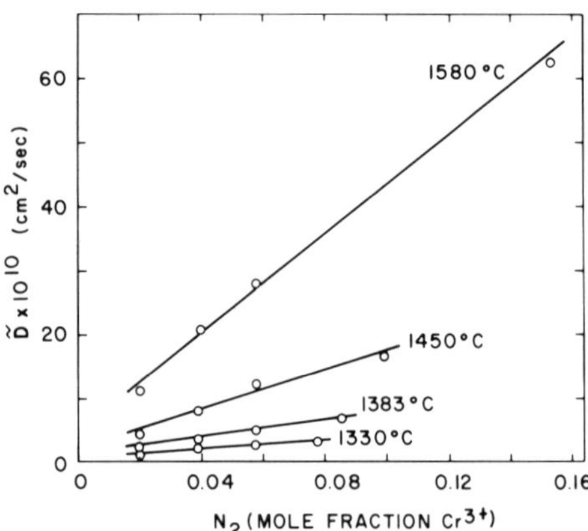

Fig. 3. Variation of $\widetilde{D}$ in NiO with the mole fraction of Cr^{3+} ions.

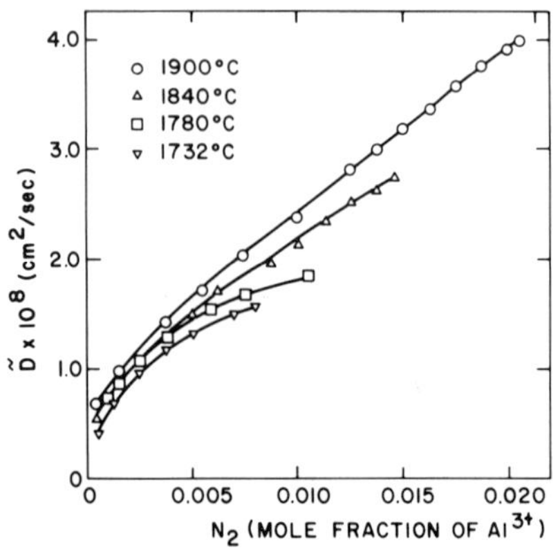

Fig. 4. Variation of $\widetilde{D}$ in MgO with the mole fraction of Al^{3+} ions.

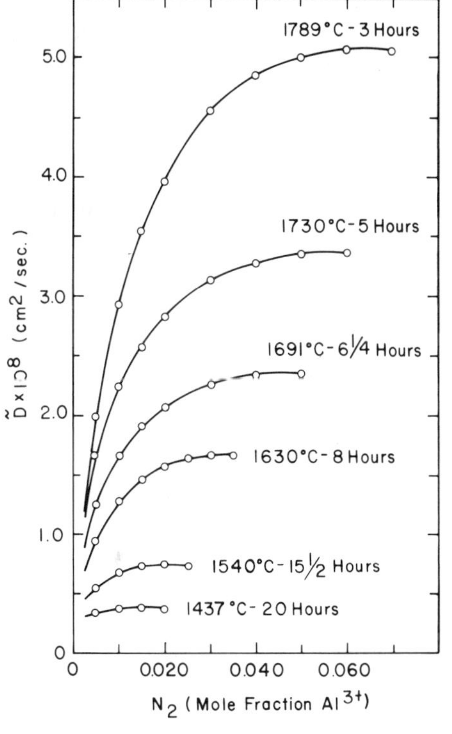

Fig. 5. Variation of $\widetilde{D}$ in NiO with the mole fraction of Al^{3+} ions.

3 DISCUSSION

If an AO compound is equilibrated with the spinel AB_2O_4, B^{3+} ions will diffuse into the AO lattice until the equilibrium concentration at the solubility limit is established. The net flux (j) from the coupling of the diffusion species is then equal to:

$$j = \frac{3D_B \cdot D_v}{(D_B + 2D_v)V_m} \text{ x grad } N_B \quad , \tag{2}$$

where D_B and D_v refer to the self-diffusion coefficients of the B^{3+} ions and vacancies, respectively.

Since B^{3+} ions move via vacancies, and if no association of defects occurs, and $D_v >> D_B$, then D_B is proportional to the local cation-vacancy concentration. The interdiffusion coefficient $\widetilde{D}$ then has the form[13]

$$\widetilde{D} = \frac{3}{2} D_B^o \frac{N_B}{N_B^o} \tag{3}$$

The superscript o refers to the solubility limit and D_B^o is the self-diffusion coefficient of ion B^{3+} in an AO oxide in equilibrium with a AB_2O_4 spinel. From this treatment it is concluded that the interdiffusion coefficient depends linearly on the mole fraction of B^{3+} ions in an AO oxide. However, this treatment assumes that no association of defects occurs.

It was also previously shown[7] that the measured interdiffusion coefficient is equal to 3/2 of the self-diffusion coefficient of ion B^{3+} in an AO oxide:

$$\tilde{D} = 3/2\ D_B \quad . \tag{4}$$

These conclusions appear to agree with the experimental results shown in Figs. 2 and 3. To interpret the results in Figs. 4 and 5, the formation of the complexes between impurity ions B^{3+} and vacancies should be taken into account. This becomes evident if one tries to plot the natural logarithm of the interdiffusion coefficient versus reciprocal temperature for constant concentrations of B^{3+} ions, as is shown in Fig. 6 for the system Al_2O_3-NiO. The activation energy then varies from 37 to 44 kcal/mole for 5×10^{-3} to 15×10^{-3} mole fraction Al^{3+} ions, respectively. The change in activation energy indicates that the number of cations contributing to diffusion is changing with temperature at a specific concentration of Al^{3+} ions.

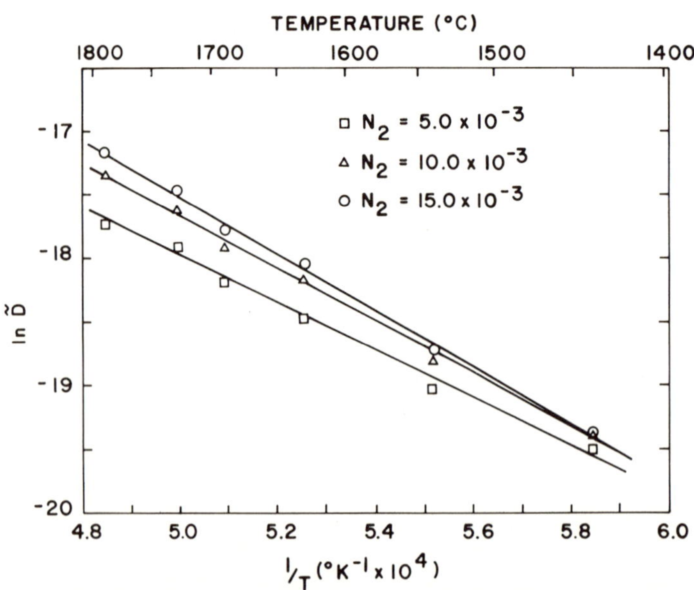

Fig. 6. Ln $\tilde{D}$ versus 1/T for the interdiffusion of aluminum ions in NiO at constant mole fraction of Al^{3+} ions.

For the case treated here, two simple types of complexes may be present, a complex of a vacancy and trivalent ion $[B_A^{\cdot} V_A'']'$ and a three-member complex of one vacancy with two metal ions $[B_A^{\cdot} V_A'' B_A^{\cdot}]$. These types of complexes have been detected by the electron-spin-resonance method in Cr^{3+}-doped MgO[14]. At sufficiently high temperatures, $[B_A^{\cdot} V_A'']'$ pairs, along with unassociated impurities and vacancies, predominate.

According to Lidiard[15], when the complexes are regarded as the diffusion species, the impurity diffusion D(c) is related to the diffusion of the complexes as follows:

$$D(c) = D_p \frac{d(pc)}{dc} \quad , \tag{5}$$

where D_p is the diffusion coefficient of the complex, p is degree of association, and c is total concentration of B^{3+} ions in the AO lattice. For the case of diffusion of trivalent ions in a divalent oxide, p can be expressed as[16]

$$p = \frac{3}{4} + \frac{1}{2Ac} - \left(\frac{1}{16} + \frac{3}{4(Ac)} + \frac{1}{4(Ac)^2} \right)^{1/2} \quad , \tag{6}$$

where $A = 12 \exp\left(\frac{\Delta F}{RT}\right)$.

The free energy of association of the complex, ΔF, is defined as the work gained in bringing a vacancy from a particular distant point to a particular nearest-neighbor position of an impurity ion at constant temperature and pressure. The interdiffusion coefficient will be equal or proportional to the concentration-dependent diffusion coefficient of the impurity ion if the assumption is made that the frequency of B^{3+} ion to vacancy jump is much less than the frequency of A^{2+} ion to vacancy jump. By successive approximation, the experimental values of $\tilde{D}$ versus c for the system Al_2O_3-NiO (Fig. 5) at a particular temperature were used in Eqs. (5) and (6) to obtain values of ΔF and D_p for $[Al_{Ni}^{\cdot} V_{Ni}'']'$ pairs. The values obtained are summarized in Table I. In Fig. 7, the experimental points are compared with the calculated curves. The good agreement shown in these results supports the theory of impurity diffusion by means of impurity-vacancy complexes. Equation (6) was then used to calculate the number of $[Al_{Ni}^{\cdot} V_{Ni}'']'$ pairs at each temperature as a function of aluminum concentration, as shown in Fig. 8. Using these data, the temperature dependence of $\tilde{D}$ could be plotted at fixed concentrations of aluminum-vacancy pairs as shown in Fig. 9. The activation energy for aluminum-ion vacancy exchange (ΔH) calculated from these data ranges from 46 to 50±3 kcal/mole. The slight increase in ΔH with increasing concentrations of pairs may be due to the presence of $[Al_{Ni}^{\cdot} V_{Ni}'' Al_{Ni}^{\cdot}]$

complexes at high concentration of aluminum ions, which was not taken into consideration in the interpretation of the results.

Table I. Results of the Analysis of the Diffusion Profiles of Aluminum in NiO

Temperature, °C	1540	1630	1691	1730	1789
D_p (theor.), 10^{-8} cm²/sec	1.8	3.8	5.3	7.6	11.5
ΔF, kcal/mole	9.0±0.5	8.5±0.5	8.0±0.5	7.5±0.5	6.5±0.5

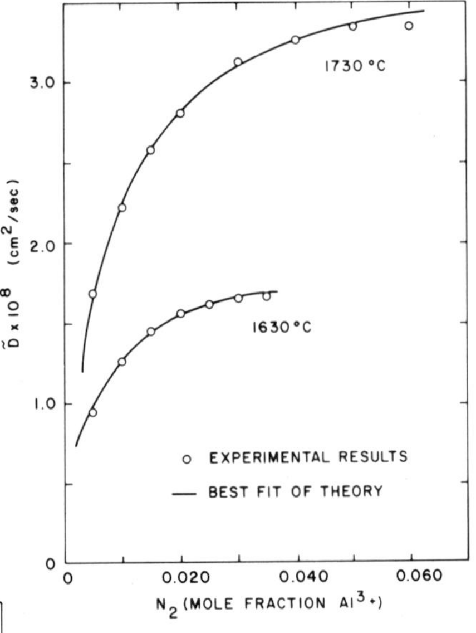

Fig. 7. Variation of $\tilde{D}$ in NiO with the mole fraction of Al^{3+} ions at 1630°C and 1730°C; points represent experimental results; solid curves give the best theoretical fit.

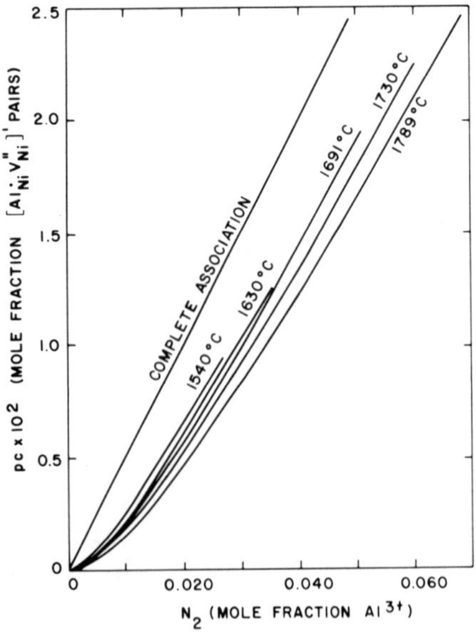

Fig. 8.CConcentration of $[Al^{\cdot}_{Ni}V''_{Ni}]'$ complexes in NiO as a function of temperature and aluminum-ion concentration.

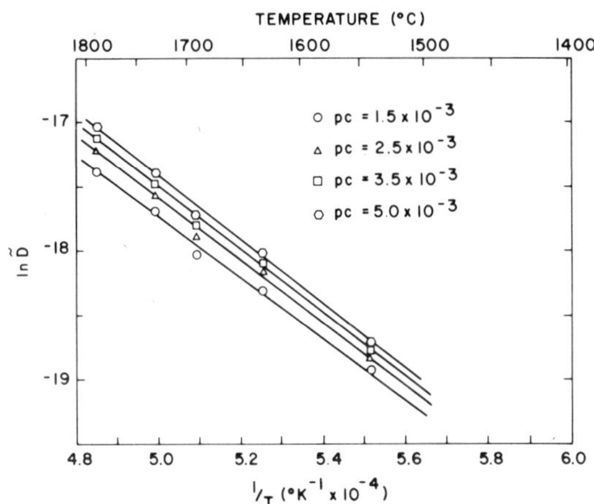

Fig. 9. Ln $\tilde{D}$ versus 1/T for the interdiffusion of aluminum ions in NiO at constant mole fraction of impurity-vacancy pairs (pc).

Perkins and Rapp[16] have studied concentration-dependent diffusion of the Cr^{3+} ion in NiO in the temperature range 1000 to 1300°C. They found that at these low temperatures, the interdiffusion coefficients do not vary linearly with the concentration of Cr^{3+} ions as found by Greskovich[8] who has studied interdiffusion in the same system between 1330 to 1580°C. Using Lidiard's theory, Perkins and Rapp[16] have calculated that the energy of association of chromium-vacancy pairs (ΔF) in NiO was 3 kcal/mole at 1000°C and 0.6 kcal/mole at 1300°C. These are considerably smaller values than those of 9 to 6.5 kcal/mole calculated for ΔF for aluminum-vacancy pairs in NiO[9] in the temperature range 1540 to 1790°C. It seems that the linear dependence of interdiffusion coefficients on the concentration of impurity ions observed in some systems (e.g., Cr_2O_3-MgO and Cr_2O_3-NiO) at relatively high temperatures is due to the low energy of association between the trivalent ions and vacancies. However, even in these systems, at relatively low temperatures (800 to 1300°C) and high concentrations of trivalent ions, the association effects should be taken into account. With systems where ΔF is high (e.g., Al_2O_3-NiO), the influence of the association of trivalent ions and vacancies cannot be neglected even at high temperatures or low concentrations of trivalent ions.

From the interdiffusion experiments, the activation energy of motion (ΔH_m) can be obtained if the concentration of vacancies is impurity controlled and if ΔF is small. The natural logarithm of interdiffusion coefficients for the system Cr_2O_3-MgO as a function of reciprocal temperature is plotted in Fig. 10. The calculated value for ΔH is 67±2 kcal/mole.[6] Tagai et al.[17] obtained ΔH = 68 kcal/mole for the diffusion of

Cr^{3+} in MgO. In both cases, the activation energy obtained is ΔH_m, as most of the existing diffusion data for MgO is characteristic of extrinsic behavior. This is not surprising if one takes into consideration that the energy of formation of Schottky defects in MgO was estimated to be ΔH_f = 5±1 ev/mole.[18,19]

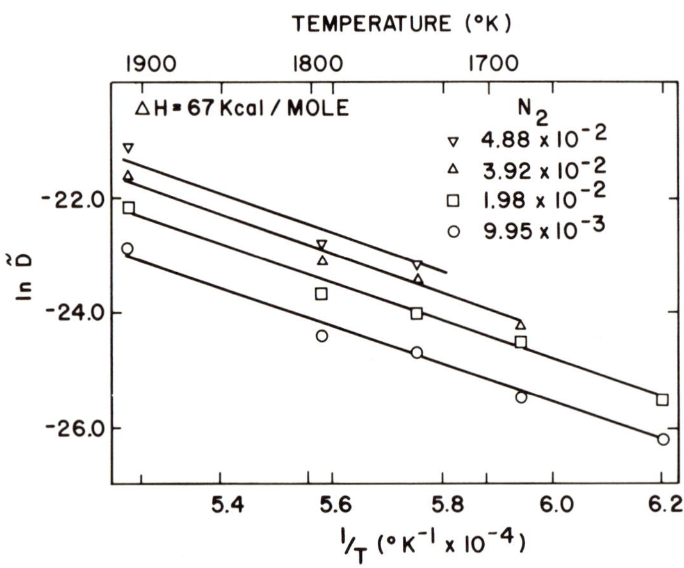

Fig. 10. Ln $\tilde{D}$ versus 1/T for the interdiffusion of chromium ions in MgO; N_2 is the mole fraction of Cr^{3+} ions.

The activation energy for the diffusion of Cr^{51} in NiO was determined to be 67.4±1.1 kcal/mole.[10] From the interdiffusion studies, ΔH_m was found to be 55 kcal/mole.[8] Since in the pure NiO, the concentration of cation vacancies at constant P_{O_2} is thermally controlled, the ΔH of 67.4 kcal/mole includes the activation energy for the formation of cation vacancies, chromium ion-vacancy binding energy, and the activation energy for motion.[20]

ACKNOWLEDGMENT

Part of this work was sponsored by the Office of Naval Research, under Contract Number N0001467-A-0385-0011.

REFERENCES

1. Rigby, E. B., and Cutler, I. B., *J. Am. Ceram. Soc.*, **48**, 95 (1965).
2. Blank, S. L., and Pask, J. A., *J. Am. Ceram. Soc.*, **52**, 669 (1969).
3. Appel, M., and Pask, J. A., *J. Am. Ceram. Soc.*, **54**, 152 (1971).
4. Wei, G.C.T., and Wuensch, B. J., *J. Am. Ceram. Soc.*, **56**, 562 (1973).
5. Levin, R. L., and Wagner, J. B., Jr., *Trans. Met. Soc., Am. Inst. Min. Met. Engrs.*, **233**, 159 (1965).
6. Greskovich, C., and Stubican, V. S., *J. Phys. Chem. Solids*, **30**, 909 (1968).
7. Whitney, W. P., and Stubican, V. S., *J. Phys. Chem. Solids*, **32**, 305 (1971).
8. Greskovich, C., *J. Am. Ceram. Soc.*, **53**, 498 (1970).
9. Minford, W. J., and Stubican, V. S., *J. Am. Ceram. Soc.*, in press.
10. Katz, G., Ph.D. Thesis, The Pennsylvania State University, 1965.
11. Koch, E., and Wagner, C., *Z. Physik. Chem.*, **B34**, 317 (1936).
12. Wagner, C., *Acta Met.*, **17**, 99 (1969).
13. Wagner, C., *J. Chem. Phys.*, **18**, 1227 (1950).
14. Wertz, J. E., in "Mass Transport in Oxides", J. B. Wachtman, Jr., and A. D. Franklin (Eds.), *NBS Special Publication*, **296**, p. 11 (1968).
15. Lidiard, A. B., "Ionic Conductivity", *Encyclopedia of Physics*, Vol. XX, W. Flugge (Ed.), Springer Verlag, Berlin (1957), p. 298.
16. Perkins, R. A., and Rapp, R. A., *Met. Trans.*, **4**, 193 (1973).
17. Tagai, H., Twai, S., Tseki, T., and Saho, M., *Radex Rundschau*, **4**, 577 (1965).
18. Yamashita, J., and Kurosawa, T., *J. Phys. Soc. Japan*, **9**, 944 (1954).
19. Lidiard, A. B., *Proc. Brit. Ceram. Soc.*, **9**, 1 (1967).
20. Chen, W. K., Peterson, N. L., and Robinson, L. C., *J. Phys. Chem. Solids*, in press.

IONIC CONDUCTION IN OXIDES AT LOW TEMPERATURES

Robert A. Huggins

Stanford University
Stanford, California

ABSTRACT

Unusually large values of cationic conductivity have recently been found in some solid oxides at relatively low temperatures. This behavior is related to the presence of crystallographic tunnels within which certain ions can be very mobile.

Prominent examples of such fast-ion conductors are materials of the beta alumina family, which have a two-dimensional tunnel network. The special characteristics of the structure of these materials are discussed, and both ionic and electronic transport measurements involving single crystals containing a group of monovalent cations are reported. Results obtained from measurements of the ionic conductivity over an unusually wide range of temperature by the use of appropriate mixed conductors as non-polarizing solid electrodes are compared with data obtained by the use of other techniques.

A theoretical approach is outlined that explains the unusually high mobility and low activation energies observed in some materials with tunnel structures, as well as the large sensitivity to the size of the mobile ions within the crystallographic tunnels. Additional results obtained from these calculations are that the transport path of lowest energy is typically not along the centerline of the tunnel, and equilibrium sites are not necessarily in the centers of void polyhedra.

1 INTRODUCTION

There are several groups of materials in which mass transport occurs with unusual rapidity, sometimes at surprisingly low temperatures. Some of these, such as a group of silver and copper salts, have been known for some time to have very high values of ionic conductivity in their high-temperature forms, with a sharp decrease of several orders of magnitude accompanying a phase change at lower temperatures. Two significant changes have occurred in recent years. Several silver-conducting materials have been found in which the highly conducting phase persists to below ambient temperatures. In addition, a number of other materials have been found in which ions other than silver and copper are highly mobile.

The possibility of the technological use of such materials, called fast ion conductors, has caused a significant amount of attention to be focused upon this area, with efforts being expended upon the exploration for new materials with desirable properties as well as upon the reasons for the unusual characteristics of fast ion conductors which cause their ionic transport parameters to be so much greater than is typical for more commonly studied materials.

It is now recognized that there are several characteristics of the fast-ion conductors which are substantially different from those found in the simple metals and ionic conductors which have received most scientific attention in the past. These differences are a consequence of crystallographic structures that are not close packed but, instead, contain ion-sized tunnels or passageways composed of connected void polyhedra within which the mobile ions reside and move. Typically, the potentially available sites within these tunnels are only dilutely occupied and, at least in some cases, the distribution of ions among these available sites is random rather than ordered. The openness of such structures sometimes produces a tendency for structural instability at lower temperatures, and this results in phase transformations to structures of higher density.

Although all of the early fast-ion conductors were halides and other simple salts, more recent work has shown that similar phenomena can also occur in some oxide systems at low temperatures. In the next section, the presently most prominent example, viz., oxides of the beta alumina family, are discussed in some detail.

2 MATERIALS OF THE BETA-ALUMINA FAMILY

A great deal of attention has recently been given to materials of the beta-alumina family, which are oxides with the nominal formula $MAl_{11}O_{17}$, where M can be any of a number of different monovalent cations. These materials exhibit unusually large values of ionic conductivity for silver and some of the alkali-metal ions over a wide range of temperature. As a result, they are expected to have important practical applications in devices such as high-performance batteries, and developmental programs aimed in this direction are being undertaken in a number of countries.

Although materials of the beta-alumina type had long been known, and the basic crystallographic structure determined some time ago[1,2], intense recent interest arose as a result of the important work at the Ford Motor Company[3-5]. They not only called attention to the unusual physical properties of these materials, but also described[6] how it is possible to use sodium beta alumina as a solid electrolyte with liquid sodium and sodium polysulfide phases in the electrode compartments to produce a new type of high-energy-density battery.

There are actually several different structural types in this family. In addition to beta alumina itself, which more properly, should be written as $M_{1+x}Al_{11-x/3}O_{17}$, the other major phase is generally called β'', and has the nominal formula MAl_5O_8. There are substantial ranges of composition in both cases, and M can be any of a long list of monovalent cations, even Ga^+ or NO^+.

These beta-alumina structures are composed of blocks (generally four oxygen layers thick) of Al_2O_3 with the γ-alumina (spinel) structure. These are connected by bridging layers containing only M^+ ions and oxide ions. These bridging layers are not close packed, and the oxide ions are arranged to form a hexagonal array of parallel tunnels in which the M^+ ions reside. The cation-site occupation within the tunnels is quite dilute; all the sites would be filled if x were unity, but x values are typically 0.1 to 0.3. In contrast to normal γ alumina, where aluminum ions reside on only two-thirds of the cation sites of the spinel structure, all these sites are occupied in the so-called γ blocks of beta alumina. This structure is illustrated schematically in Fig. 1.

The primary difference between the β and β'' phases lies in the manner in which the γ alumina blocks, in which the oxygens have a close-packed cubic arrangement, are stacked upon one another. In the β structure, the bridging layers which contain the M^+ ions are mirror planes, so that the overall structure has hexagonal symmetry. In the case in which the M^+ ion is sodium, the lattice parameters are $a_o = 5.59$ and $c_o = 22.53$ A. The distance between bridging layers is therefore about 11.27 A. In the

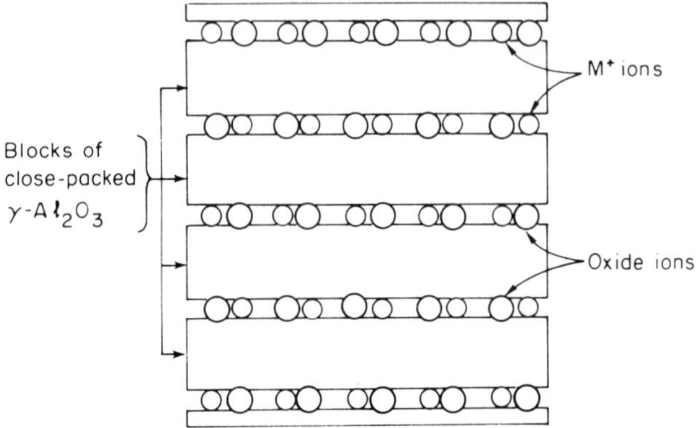

Fig. 1. Schematic representation of the structure of beta alumina showing γ alumina blocks separated by bridging layers containing M^+ and oxide ions.

case of the β'' structure[7-10], the distance between bridging layers is about the same, but the c_o value is approximately 33.9 A and the symmetry is rhombohedral. The spinel blocks in β'' are related to each other by rotation about a threefold screw axis normal to the bridging layers, whereas in the β case, the screw axis is twofold.

In addition to β and β'', there have been reports[11,12] of two related structures, designated as β''' and β''''. The structure of β''' is reported to be similar to that of β alumina, except that the γ blocks contain six layers of oxygen atoms. The bridging layers are thus 15.9 A apart. The β'''' structure is rhombohedral, similar to β'', but again has six layers of oxygen in the γ blocks.

Yamaguchi and Suzuki[8] reported another phase intermediate between β and β'', which they labelled β'. However, it evidently differs from the β structure only with respect to the positions deduced for the sodium ions in the bridging layer. As a result, it has become generally assumed that their results related to a stoichiometric variation of the normal β phase, rather than to a different phase.

It has been found[12] that the presence of MgO or Li_2O stabilizes the β'' structure, and much of the work done on this phase has involved samples containing modest amounts of one or the other of these constituents. This structure contains some 50 percent more M^+ ions in the bridging layers than does the β phase; presumably, the magnesium ions replace some of the aluminum ions in the γ blocks, most probably along the midplane.

The arrangement of the ions within the bridging layer in the β structure is indicated in Fig. 2. It can be seen that the oxygen ions occupy one-fourth of the positions that would be filled in a close-packed plane. The M^+ ions reside in the hexagonally disposed tunnels that permeate this

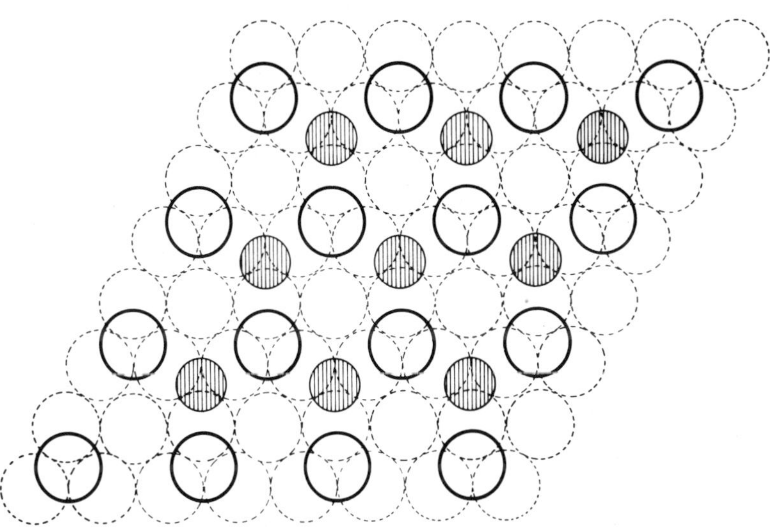

Fig. 2. Arrangement of ions in the bridging layer of beta alumina, assuming ideal stoichiometry. Actual materials have excess cations. Oxide ions in adjacent close-packed plane of γ alumina structure are indicated by dashed lines, oxide ions in bridging layer by heavy circles, and mobile cations by smaller cross-hatched circles.

oxygen array. The normal M^+ ion positions are octahedrally surrounded by oxygens above and below the bridging layer. An alternative and evidently higher energy type of site is also present, in which the M^+ ions lie between oxygen pairs. It was mentioned earlier that samples of beta alumina typically contain excess M^+ ions – that is, more M^+ ions than can be accommodated by filling only the preferred sites in the bridging layer. X-ray diffraction results by Roth[13] on silver beta alumina indicated that the excess silver ions occupy these alternative sites. On the other hand, work by Peters et al.[14] on sodium beta alumina appeared to show that excess sodium is accommodated in a different way. In this case, the M^+ ions occupy octahedrally coordinated sites directly in line between pairs of oxygen ions in the bridging layer. Obviously, if an M^+ ion were in such a position it would have to dislodge one of the nearby ions from a normal position to a similar nearby in-line site.

Thus, the addition of one extra M^+ ion can be accommodated either by its residing upon an alternative site or by the occupation of a pair of in-line sites at the expense of one ion upon a normal site. However, this situation is complicated even more by the suggestion[13,15,16] that deviations from stoichiometry might be accomplished by the presence of a domain-type structure in the bridging layers, with the filling of excess sites at the domain walls.

The mobile ion in the bridging layer can be exchanged with a number of other cations by immersion in appropriate molten salts[3] or by electrochemical pumping[10]. There is also evidence that water intercalates into these structures[5,17,18], although it is not clear whether the water is present as neutral H_2O molecules or as H_3O+ ions. In any case, its presence impedes the transport of the other cations present.

Information on the self-diffusion of several different ions in the beta-alumina structure was presented by Yao and Kummer[3] and Radzilowski et al.[4] by the use of radiotracer techniques. The dielectric-loss method was also used by Radzilowski et al.[4] to evaluate ionic motion in polycrystalline β alumina containing various M^+ ions. In some cases they found quite good agreement between these results, which relate to ionic motion at relatively low temperatures, and data obtained from the radiotracer experiments, whereas in others there was a considerable discrepancy in the apparent activation energy.

Direct measurements of the ionic conductivity of members of the beta-alumina family using ion-blocking electrodes and alternating-current techniques have been reported by a number of authors. However, the results have typically shown a frequency dependence, and comparison among the results of experiments by different investigators has been disappointing. A major step forward was the introduction[19] of the use of mixed conductors, in which the M^+ ion has a high chemical-diffusion coefficient, as ionically reversible electrodes. By using this technique, frequency-independent values of the ionic conductivity could be obtained over a wide range of temperature. Using properly selected tungsten-bronze phases as electrodes, reliable data have been obtained for the ionic conductivity of single crystals of beta alumina containing silver[20], sodium[19] and potassium, thallium, and lithium[21] from over 800°C to well below ambient temperature. The results of these measurements of the conductivity are linear over an unusually wide temperature range.

Information acquired by the use of these three methods, radiotracer self-diffusion, dielectric loss, and ionic conductivity, should be consistent, as they all measure ionic transport. This is illustrated for sodium beta alumina in Fig. 3. Ionic-conductivity data calculated from the radiotracer and dielectric-loss data measured by the Ford group[3-5] are plotted along with the directly measured single-crystal-conductivity values obtained at Stanford[19-21]. The values extracted from the radiotracer diffusion data by use of the Nernst-Einstein relation are expected to be somewhat lower than the true conductivity values because of correlation effects. From the magnitude of this difference, the correlation factor D_T/D_σ can be obtained. This allows conclusions to be drawn concerning the detailed jump mechanism involved in the transport process.

Similar data for beta alumina containing silver and potassium in the bridging layer are shown in Figs. 4 and 5. In the case of silver beta alumina, the conductivity data were not extended below ambient

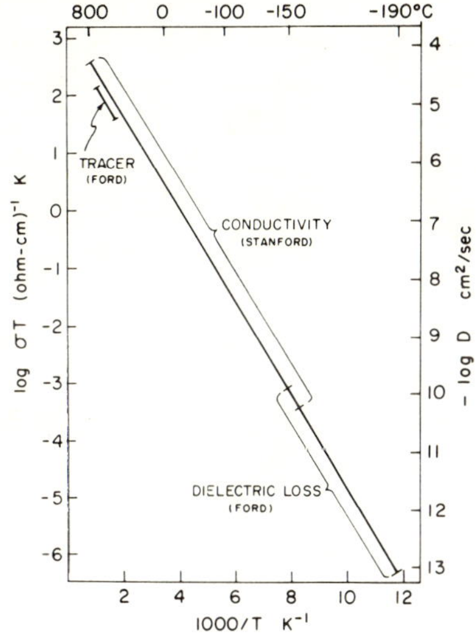

Fig. 3. Temperature dependence of the ionic conductivity and self-diffusion coefficient of sodium beta alumina measured by the use of three different techniques.[21]

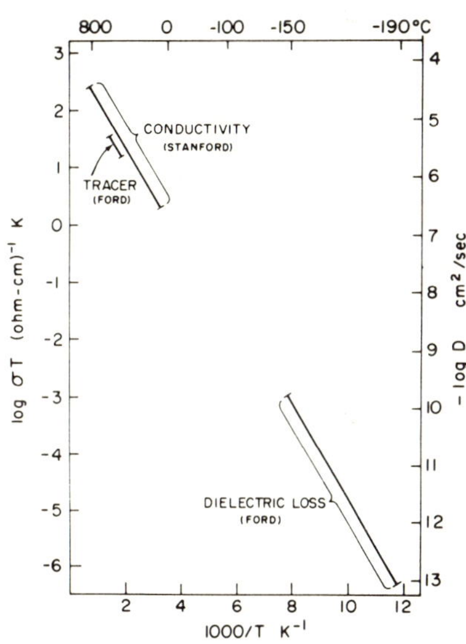

Fig. 4. Temperature dependence of the ionic conductivity and self-diffusion coefficient of silver beta alumina.[21]

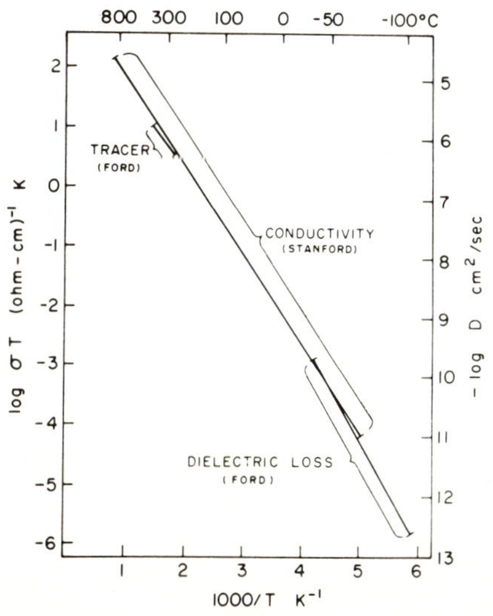

Fig. 5. Temperature dependence of the ionic conductivity and self-diffusion coefficient of potassium beta alumina.[21]

temperature. It can be seen that although the slopes of the radiotracer and dielectric-loss data are quite different in the case of potassium beta alumina, their overall magnitudes are reasonably consistent, and they are tied together quite well by the direct conductivity measurements. Similar comparisons of results obtained by the use of these three methods in the cases of thallium and lithium beta alumina can be found in the paper by Whittingham and Huggins.[21]

The lithium-containing material is evidently somewhat different from the others. The directly measured conductivity data were found to fall upon two connecting straight lines, meeting at about 180°C. The conductivity and radiotracer data essentially coincide in the high-temperature region. The correlation factor is thus unity in this region, implying that the very small lithium ion is moving primarily by a direct interstitial mechanism in this temperature range. Because of its lower temperature dependence, another process, perhaps the normal interstitialcy mechanism, evidently dominates the conductivity at lower temperatures. A further complication observed in lithium-substituted beta alumina is that annealing at temperatures above 800°C causes some structural change – perhaps diffusion of the lithium into the γ blocks – that results in a decrease of the low-temperature conductivity.

The ionic-conductivity data obtained by the reversible-electrode single-crystal method for these five different M^+ ions in the beta-alumina structure are compiled in Table 1, assuming an Arrhenius relation of the form $\sigma = (\sigma_0/T) \exp(-\Delta H/RT)$. Also shown, where available, are the experimental values of the correlation factor. For silver, sodium, and thallium beta alumina these were found to be very close to the theoretical value of 0.6 calculated[20] for a noncolinear interstitialcy (indirect inter-

Table I. Ionic Conductivity of Single-Crystal Beta Alumina Containing Different Mobile Ions

Ion	Temperature Range, °C	σ_0, (ohm-cm)$^{-1}$ K	Activation Energy, kJ/mole	Conductivity at 25°C, (ohm-cm)$^{-1}$	Correlation Factor (D_T/D_σ)	Ionic Radius, A (CN = 6)
Ag	25 → 800	1.6×10^3	16.5	6.7×10^{-3}	0.61	1.15
Na	–150 → 820	2.4×10^3	15.8	1.4×10^{-2}	0.61	1.02
K	– 70 → 820	1.5×10^3	28.4	6.5×10^{-5}	–	1.38
Tl	– 20 → 800	6.8×10^2	34.3	2.2×10^{-6}	0.58	1.50
Li	180 → 800	9.7×10^3	35.8	–	1.0	0.74
Li	–100 → 180	5.4×10^1	18.0	1.3×10^{-4}	–	0.74

stitial) mechanism in the beta-alumina structure. Values of ionic radii were taken from Shannon and Prewitt[22].

It is generally considered that the ionic conductivity of the β'' phase is higher than that found for the β materials, although well-defined and reproducible data for different ions in this structure are not yet available. Some values for sodium β'' samples can be found in the papers by Kummer[5] and Whittingham and Huggins[21].

From the data obtained from these transport measurements, it appears that there is an optimum ionic radius for achieving the maximum mobility in the tunnels within the beta-alumina structure, with both smaller and larger ions slower than those of intermediate size. The importance of this relative size factor was very nicely confirmed by experiments reported by Radzilowski and Kummer[23] which showed a linear change in the ionic conductivity of ion-substituted beta alumina with hydrostatic pressure. At higher pressures, when the tunnel size was reduced, the conductivity of lithium β alumina increased, whereas that of potassium β alumina decreased and that of sodium β alumina was essentially unchanged.

Because of the interest in potential use of beta aluminas as solid electrolytes, it is important to know the electronic transference number, i.e., the fraction of the total conductivity carried by electronic, as distinct from ionic, species. This has been measured in the one case of single-crystal silver beta alumina[20] using the Wagner asymmetric polarization method[24] from 555 to 790°C and over a wide range of oxygen partial pressure. The value of the electronic transference number was found to be about 3×10^{-5} at 750°C and 10^{-6} at 560°C. It was also found that below about 750°C, changes in the ambient oxygen partial pressure did not appreciably influence the electronic conductivity. Presumably, the structure is effectively "frozen" below that temperature. To the extent that one can extrapolate these data, it can be predicted that the beta aluminas should have extremely low values of electronic conductivity at lower temperatures.

3 THEORETICAL CONSIDERATIONS

There are three principal features of the experimental data relating to fast ion conductors that differentiate them from normal ionic conductors and deserve special attention. First, the absolute magnitude of the ionic conductivity can be unusually high, approaching the theoretical limit for a simple jump model in some fast ionic conductors. In the case of some mixed conductors, it appears that ions can be transported even more rapidly, which perhaps indicates some sort of multiple-step jump mechanism.

The second unusual experimental characteristic of fast ionic conductors is the small temperature dependence of the conductivity, which can be interpreted as a low activation energy for motion, as well as a temperature-independent mobile carrier concentration.

The third unique feature is the very small value of σ_o found in these materials, which can be shown[25] to imply either an especially low value of attempt frequency or a very low value of the entropy of migration.

In addition, it is quite apparent that this type of behavior is found only in materials whose crystal structures contain ion-sized tunnels which are sparsely occupied by ions of appropriate size. Ions either too small or too large have greatly reduced mobility.

Initial steps have recently been taken[26] to develop an explicit atomistic model for the transport of ions through crystal structures containing tunnels. The general method initiated by Born and Mayer[27] was used in which the total interaction energy between an ion i and all the surrounding j lattice ions is expressed as the sum of three types of terms, electrostatic point-charge (Coulombic) interactions, dipolar polarization interactions, and overlap repulsion between closed-shell ions. That is, the total energy is written as

$$E_{Total} = E_{Coulombic} + E_{Polarization} + E_{Repulsion} \quad ,$$

where

$$E_{Coulombic} = e^2 \Sigma_j \frac{q_1 q_j}{r}$$

$$E_{Polarization} = -\frac{e^2}{2} \Sigma_j \frac{\alpha_j q_i}{r^4}$$

$$E_{Repulsion} = b \Sigma_j \exp\left[(r_i + r_j - r_{ij})/\rho\right] \quad ,$$

and q_i and q_i are the charges, α_j the polarizability, and r_i and r_j the repulsion radii. b and ρ are constants, and r_{ij} is the distance between the test cation and the jth lattice ion.

With the help of a digital computer, the interaction energy between a test cation placed at any arbitrary position within the crystal structure and the rest of the lattice was calculated, assuming that all the other ions remain fixed in position.

This type of calculation was carried out for the case of the α-AgI structure, in which the cations reside within tunnels which run through the centers of the faces of the body-centered cubic anion lattice. The total energy was found as a function of the cation position in the tunnels, assuming a number of different values of the cation radius. In addition to a series of points along the centerline of the tunnel, a three-dimensional raster of off-center positions was also investigated.

By use of this method, the minimum energy path through the tunnel was found for a series of values of the cation radius. In the case of small cations, there is a symmetrical pair of preferred paths which deviate toward the nearest-neighbor anions lying along the tunnel wall, owing to the large influence of the polarization energy term. For large cations, just the opposite is true, and a pair of equivalent paths are preferred which deviate from the centerline away from the nearby anions, owing to the relatively greater importance of the repulsion term in the total energy. Ions of intermediate size were found to follow compromise minimum energy paths.

In addition to showing the influence of ionic size upon the location of the minimum energy path, these calculations gave information about the variation in potential energy during translation along this preferred transport path for cations of different radii. These results are illustrated in Fig. 6 for a series of different cationic sizes. It is seen that there is, as expected, a periodic variation in energy along the transport path. The amplitude increases as the size becomes more extreme in either direction, as both small and large cations encounter substantial energy barriers owing to the dominance of one or another of the terms in the total energy. In the case of cations of intermediate size, these factors tend to balance each other, and the energy varies less as the ion moves through the tunnel.

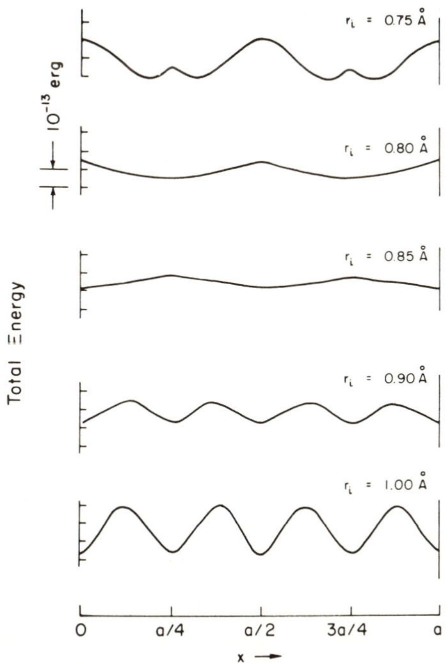

Fig. 6. Variation of potential energy along preferred transport path within unit cell of α-AgI structure for cations of various sizes.[26]

If one interprets the difference between maxima and minima in energy along the preferred transport path as an activation energy for motion, it is clear why ions of intermediate size should have the greatest mobility in materials whose structures contain such tunnels. The influence of cationic size upon the activation energy determined in this way is shown in Fig. 7.

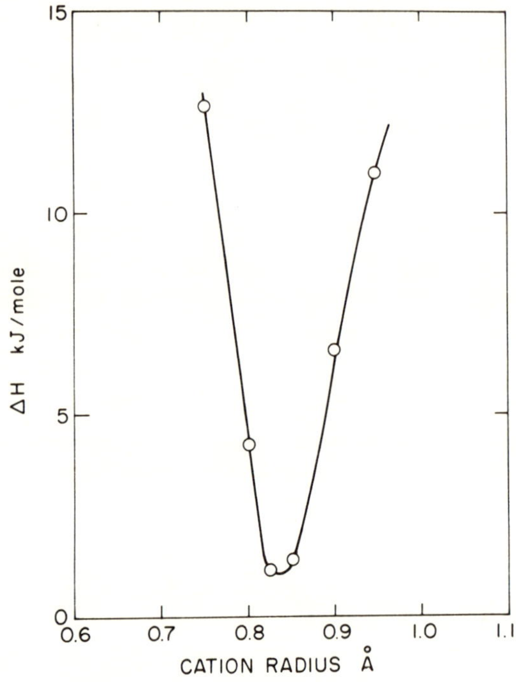

Fig. 7. Calculated influence of cation radius upon activation energy for migration in the α-AgI structure.[26]

Although several important simplifications were made to facilitate this calculation, including the assumptions of no cation-cation interaction, negligible cationic polarization, and a rigid anion lattice, the results are quite reasonable.

They also indicate that the potential wells in which the cations will tend to reside are broad and shallow, leading to an unusually low vibrational frequency, as inferred from the especially small values of σ_0 found experimentally for fast ion conductors. This is also in accord with the observations of unusually large values of the Debye-Waller factor, which is a measure of the mean thermal displacement of ions from their lowest energy sites, in the α-AgI structure.[28] Large Debye-Waller factors have also been found in other fast ion conductors[29,30] and other tunnel-structure materials that may well have high cationic mobility.[31,32]

One other result from this calculation is the indication of a shift in preferred sites within the tunnels for cations of different radii. It appears that the small cations will prefer threefold coordinated positions, while the larger ones would have lower values of energy at fourfold coordinated positions in this structure.

4 CONCLUDING REMARKS

Recent work has demonstrated that some oxides can exhibit fast ion conduction at remarkably low temperatures. One example, the beta-alumina family, in which several monovalent cations are quite mobile, shows excellent promise for use as a solid electrolyte over wide ranges of temperature and thermodynamic conditions.

Although understanding of the basis for the unusually high mobility of ions in such materials is still rather rudimentary, sufficient progress has been made to indicate the primary structural requisites. As a result, it is reasonable to conclude that this type of behavior is not just restricted to a special small group of presently known materials. Instead, it should be found in a number of others in which the necessary criteria can be met. This should result in the opening up of many new opportunities in solid-state electrochemistry.

ACKNOWLEDGMENTS

Work in this area within the Solid State Electrochemistry Group at Stanford has been supported by the Office of Naval Research, The Environmental Protection Agency, the National Science Foundation, and the Advanced Research Projects Agency (through Stanford's Center for Materials Research).

REFERENCES

1. Bragg, W. L., Gottfried, C., and West, J., *Z. Krist.*, **77**, 255 (1931).
2. Beevers, C. A., and Ross, M.A.S., *Z. Krist.*, **97**, 59 (1937).
3. Yao, Y.F.Y., and Kummer, J. T., *J. Inorg. Nucl. Chem.*, **29**, 2453 (1967).
4. Radzilowski, R. H., Yao, Y. F., and Kummer, J. T., *J. Appl. Phys.*, **40**, 4716 (1969).
5. Kummer, J. T., *Prog. Solid State Chem.*, **7**, 141 (1972).
6. Weber, N., and Kummer, J. T., *Proc. Ann. Power Sources Conf.*, **21**, 37 (1967).
7. Thery, J., and Briancon, D., *Rev. Hautes Temp. Refract.*, **1**, 221 (1964).
8. Yamaguchi, G., and Suzuki, K., *Bull. Chem. Soc. Japan*, **41**, 93 (1968).
9. Bettman, M., and Peters, C. R., *J. Phys. Chem.*, **73**, 1774 (1969).
10. Whittingham, M. S., Helliwell, R. W., and Huggins, R. A., *U.S. Gov't. Res. and Devel. Rept.*, **69**, 158 (1969).
11. Bettman, M., and Terner, L., *J. Inorg. Chem.*, **10**, 1442 (1971).

12. Weber, N., and Venero, A., Presented at Annual Meeting of American Ceramic Society, Philadelphia, 1970.
13. Roth, W. L., *J. Solid State Chem.*, **4**, 60 (1972).
14. Peters, C., Bettman, M., Moore, J., and Glick, M., *Acta Cryst.*, **B27**, 1826 (1971).
15. Gool, W. van, in *Fast Ion Transport in Solids*, W. van Gool (Ed.), North-Holland Pub. Co., Amsterdam (1973), p. 201.
16. Gool, W. van, and Bottelberghs, P. H., *J. Solid State Chem.*, **7**, 59 (1973).
17. Saalfeld, H., Matthies, H., and Datta, S. K., *Ber. Deut. Keram. Ges.*, **45**, 212 (1968).
18. Kline, D., Story, H. S., and Roth, W. L., *J. Chem. Phys.*, **57**, 5180 (1972).
19. Whittingham, M. S., and Huggins, R. A., *J. Chem. Phys.*, **54**, 414 (1971).
20. Whittingham, M. S., and Huggins, R. A., *J. Electrochem. Soc.*, **118**, 1 (1971).
21. Whittingham, M. S., and Huggins, R. A., *Solid State Chemistry*, R. S. Roth and S. J. Schneider (Eds.), Nat. Bur. Stand. Special Publication 364 p. 139 (1972).
22. Shannon, R. D., and Prewitt, C. T., *Acta Cryst.*, **B25**, 925 (1969).
23. Radzilowski, R. H., and Kummer, J. T., *J. Electrochem. Soc.*, **118**, 714 (1971).
24. Wagner, C., *Z. Elektrochem.*, **60**, 4 (1956).
25. Huggins, R. A., *Diffusion in Solids: Recent Developments*, A. S. Nowick and J. J. Burton (Eds.), Academic Press (1974).
26. Flygare, W. H., and Huggins, R. A., *J. Phys. Chem. Solids*, **34**, 1199 (1973).
27. Born, M., and Mayer, J. E., *Z. Physik.*, **75**, 1 (1932).
28. Hoshino, S., *J. Phys. Soc. Japan*, **12**, 315 (1957).
29. Krug, J., and Sieg, L., *Z. Naturforsch.*, **7a**, 369 (1952).
30. Hoshino, S., *J. Phys. Soc. Japan*, **10**, 197 (1955).
31. Mumme, W. G., and Wadsley, A. D., *Acta Cryst.*, **23**, 754 (1967).
32. Mumme, W. G., and Reid, A. F., *Acta Cryst.*, **24**, 625 (1968).

DISCUSSION on Paper by R. A. Huggins

FRANKLIN: In the lithium analog of β-Al_2O_3, the dielectric relaxation data (Ford) at low temperatures lies well below the dc conductivity measured with reversible electrodes. A possible explanation might be that in the Ford material there were impurity cations such as Na^+ or K^+, that would effectively block the paths for the Li^+ ions, confining them to relatively short segments. Only by going to higher frequencies would one expect to measure the conductivity of freely moving Li^+ ions.

WELCH: In these calculations you apparently have moved one cation in the field of all the other ions being fixed. Have you considered the possibility of cooperative motion of cations?

HUGGINS: The assumptions made were that the anion lattice is fixed in space and time, and that a test cation is moved through it, placed in different positions. The charge of all the other cations has been spread uniformly through the crystal to achieve charge balance, but the cations were not placed in fixed positions. We have not yet put cooperative cation motion into the model.

HONIG: This is a request for clarification. Have you had the opportunity to do calculations for the activation energy of motion of cations through β-alumina, and if so, do the calculated values of activation energy agree with the measured values?

HUGGINS: We first made the calculations for AgI, which has a unit cell of two ions, so that a sufficiently large number of unit cells could be considered to allow an infinite-crystal approximation to be made. It was found that by going nine unit cells in each direction, essentially the same results were obtained as by using a larger number of unit cells. With β alumina and other oxides like Hollandite, the number of ions per unit cell is larger, so that the calculations are longer. We have not yet done these more extensive calculations.

CARO: Is it not possible to consider β-alumina as an ordered intergrowth between a spinel structure and silver or a sodium metal, or a metal-like or salt-like monolayered compound (i.e., a plane or a metallic or ionic conductor)? The structure is very much like classical ordered intergrowths occurring by coherent epitaxy between chemically different structural units, especially very much like the hexagonal ferrites. Another example would be the Sillen phases [intergrowths between the layered (BiO) Cl, and niobium oxides]. Besides that, are there not several varieties of β-alumina?

HUGGINS: Electrochemically β-alumina is not a metal because electronic conduction along or through layers is very slow. The bridging layer has actually a net negative charge, so that the spinel block must have a net positive charge. Therefore, because it involves a single layer, perhaps the intergrowth model is not a suitable structural model in this case. There are two major phases in the beta-alumina family, β alumina itself, which is hexagonal, and β″ alumina, which is orthorhombic.

O'KEEFFE: Have you calculated for the potential energy profiles you presented whether localized ionic states will always be found?

HUGGINS: We calculated potential energy profiles through the structure following the minimum energy path, and found that such paths wander from the centerline. At present, the interpretation of these findings in terms of localized modes is open to conjecture.

GOODENOUGH: We have come to similar conclusions from our studies on the cubic A^+SbO_3 compounds. These structures have interesting tunnels of face-shared octahedra running parallel to the $\langle 111 \rangle$ directions. The conductivity of $NaSbO_3$ is much higher than that of either

$KSbO_3$ or $LiSbO_3$. From structural studies, we found that a Ag^+ ion becomes stabilized in an octahedral-site face, whereas the Na^+ and K^+ ions are stabilized in the octahedral sites and the Li^+ ions are displaced from the centers of the octahedral sites. Moreover, there is a force of reaction on the tunnels, which change their configuration as the position of the ion within the tunnel changes. At the optimum tunnel size, the forces of action and reaction are minimized.

TANNHAUSER: Were these measurements done on single crystals? Are the values anisotropic?

HUGGINS: The conductivity data shown, extending over a large range of temperature, were obtained on single crystals. The conductivity was extremely anisotropic; the ions only move appreciably in the bridging plane. No one has ever measured motion normal to this plane in beta alumina.

ILSCHNER: What can be said about the thermal stability of this material? It occurred to me that your slides showed data up to ~850°C only. What happens to β alumina as you raise the temperature higher and higher?

HUGGINS: The β-alumina phase seems to melt incongruently at about 2000°C. The data were limited to 850°C by the apparatus used, not by the material. Other experiments have been made on polycrystalline potassium β alumina up to 1500°C, and these gave the same activation energies as found here. A surprising feature of these polycrystal materials is that they have comparable (only three to five times worse) conductivity to single crystals.

ILSCHNER: Is β alumina difficult to prepare, or is it readily available?

HUGGINS: Single crystal β alumina is best obtained from commercially available cast bricks (about $50 from Harbison Carborundum Company). Polycrystalline materials are typically made by decomposing sulfates or nitrates in air. The major experimental difficulty is keeping the monovalent cation and its oxide in place – they tend to sublime away, so that a sealed system is required.

HEUER: Can you make good ionic conductors from alkaline-earth-type beta aluminas, i.e., compounds similar to $BaFe_{12}O_{19}$?

HUGGINS: The problem with structures like magnetoplumbite is that the bridging layers become completely filled. However, small amounts of divalent cation oxides (like MgO) can be put into the beta-alumina family.

WELCH: Would you comment on the correlation factor observed from the discrepancy between conductivity and tracer diffusion?

HUGGINS: The correlation factor we calculated for interstitialcy motion is 0.6, which is essentially what has been found experimentally for Ag, Na, and Tl ions in beta alumina. Lithium is different in that in the high-temperature region experiments indicate that the correlation factor is unity. This means that it probably jumps directly from one site to the next rather than pushing its neighbor out of the way, as in the interstitialcy model.

HAGEMARK: Do you know whether anyone has measured the pressure dependence of the ionic conductivity?

HUGGINS: Radzilowski and Kummer made such measurements for K, Na, and Li ions. They found that the diffusion coefficient for sodium ions was virtually independent of pressure and that the diffusion coefficients of potassium and lithium ions were linearly dependent on pressure, viz., that for potassium decreases with increasing pressure, while that for lithium increases with increasing pressure. These results are exactly as expected from the size dependence predicted by our simple model.

JORGENSEN: Polycrystalline beta alumina is usually formed by a sintering process in which minor amounts of MgO are added. Other additions have also been used, such as NiO, CaO, and Y_2O_3. Some of these additions apparently affect the grain-boundary resistivity; for example, it has been reported that Y_2O_3 additions decrease the resistivity attributed to the grain boundaries. Can you elaborate on the effects of additives?

HUGGINS: Additions of MgO and Li_2O are commonly made in the sintering of these materials, and these seem to lead to a slightly different structure called β'' alumina. This has a slightly higher (2 to 3 times) ionic conductivity than β alumina. Data on the use of Y_2O_3 are not yet available in the published literature. I do not believe that the influence of grain boundaries upon transport processes in fast ionic conductors is yet understood.

VINK: Are also other oxides to be expected to show this kind of phenomena. Or is there some reason, e.g., specific ratio's of radii that have to be obeyed, that this can happen only with β-Al_2O_3?

HUGGINS: There has been some work on β-alumina structures in which aluminum has been replaced by gallium. There are also other oxides which show different structures in which rapid ionic transport is possible. One of these is Hollandite, which is basically a titanium oxide with potassium and magnesium additions.

GOODENOUGH: Can you say something about transport in polycrystalline ceramics containing one-dimensional tunnels? It is my impression that Hollandite, which has one-dimensional tunnels, does not exhibit the ionic conductivity anticipated from dielectric-loss measurements. In this connection, when you use the alkali-ion bronzes as reversible electrodes, do you evaporate layers that form a fiber axis, or do you have a thick polycrystalline layer?

HUGGINS: We have recently made conductivity measurements on polycrystalline Hollandite at temperatures up to 800°C, and obtained results which corresponded well with those from dielectric-loss measurements. However, the relative ionic and electronic contributions to the measured conductivity are still being investigated. For β alumina single-crystal measurements, we used a group of thick polycrystalline tungsten bronzes as reversible electrodes. The bronzes were matched with whatever cations were being investigated; for instance, Li-V bronzes were used for Li ions, tetragonal W bronze for Na ions, and hexagonal W bronze for K ions. The latter two had linear tunnels and the first one is a cubic material.

EYRING: John Goodenough suggests that three-dimensional tunnels might be desirable. What has been done to study ionic conduction in, say, sodalite or the many catalyst-type materials?

HUGGINS: If in the three-dimensional tunnels or cagelike structures, the holes are too big, then ionic conductivity is reduced. If the ions are small relative to the structure, they tend to stick to the sides of the tunnels or to the edge of the cages rather than being able to readily move through the windows.

AGENDA DISCUSSION: OXIDATION OF METALS, SOLID-SOLID REACTIONS, SOLID-STATE ELECTROCHEMISTRY

*R. A. Rapp**

The Ohio State University
Columbus, Ohio

*H. C. Graham***

Aerospace Research Laboratories
Wright-Patterson Air Force Base, Ohio

In the oxidation of pure metals at sufficiently high temperatures, Rapp asserted that for at least four metals (copper, cobalt, iron, and nickel), the classical diffusional mechanism is substantiated, i.e., the calculated values of Wagner's parabolic rational rate constants correspond exactly to the experimental parabolic rate constants. However, the scales of these four metals are of no real practical importance, and the oxidation mechanisms of the important protective scales Cr_2O_3, SiO_2, and Al_2O_3 are not well understood. Thereupon, Anderson and Stringer presented arguments based on breaks or interruptions in perfect parabolic oxidation

*Chairman

**Secretary

kinetics which are observed through precise thermogravimetric investigations or through the elastic vibration technique of Hancock. In their opinion, the gross parabolic oxidation behavior of any of the metals must be questioned because of local cracking of the scale, especially at corners and edges of specimens.

Regarding the calculations of the rational parabolic rate constant, Anderson wondered whether the point defects responsible for the process were sufficiently well known. O'Keeffe inquired whether the correlation factor is properly included in the calculation of the parabolic-oxidation-rate constants. Regarding his comments about agreement between calculated and experimental parabolic rate constants, Rapp qualified his statement to cover only oxidation at temperatures above 800°C where short-circuit paths and cracking are minimal. Further, to evaluate Wagner's rational rate constant for parabolic oxidation, a knowledge of the defect structure (type, concentration, and mobility) is not necessary. Only a value for the self-diffusion coefficient, along with the molar volume of the oxide and the thermodynamic quantities are required. Of course, the self-diffusion coefficient differs from the experimentally measured tracer diffusion coefficient by the correlation factor, f: $D_s^* = \frac{1}{f} D_{tr}$. For the oxidation of iron to wustite, Birchenall has shown that the correlation factor need not be included. For the oxidation of cobalt, nickel, and copper, cation-vacancy diffusion with a known correlation factor predominates. Smeltzer added that a correlation factor had not been included for the calculated rate constants in his paper, because it was not clear what value to use.

Smeltzer commented on the "aging" of oxide scales. During isothermal oxidation of a pure metal, one observes a relatively rapid initial parabolic oxidation of the metal. After interrupting the oxidation to anneal in argon, the kinetics at longer times are described by a much smaller parabolic rate constant. This is the aging phenomenon. From his recent studies of the oxidation of pure, polycrystalline nickel specimens at temperatures less than 900°C, Smeltzer concluded that specimen texture can establish the nature of short-circuit diffusion paths in the oxide scale. These paths can be annealed out by crystallite growth and texturing in the scale during aging to result in the second slower parabolic rate. Above 1000°C, the parabolic rate constant is representative of the bulk material and does not depend upon texture, etc. Smeltzer added that oxidation rates have been measured on single-crystal nickel at 500 to 600°C as a function of crystallographic orientation. In cases where very fine crystallites were formed, the observed rates were 10^4 times greater than in cases where very large crystals grow. Then, the crystalline orientation should be considered in the mechanistic interpretation of the kinetics.

With particular regard to the oxidation of iron, nickel, and cobalt, Rapp noted that it is essentially impossible to obtain carbon-free specimens. Further, if these metals were electrodeposited as a cathode, then they contain hydrogen. Upon oxidation, molecules of CO_2 or H_2O are formed at the metal/scale interface. The resulting disruption of the scale and therefore the oxidation kinetics can depend upon the carbon and hydrogen contents in the starting materials.

Kofstad made a point pertinent to the oxidation of metals at low temperatures. For the oxidation of either nickel or cobalt at T $\geqslant$1000°C, the agreement between the rate constants calculated from diffusion data and the experimental values is fairly good – certainly within 10 to 20 percent. At low temperatures, the discrepancy is much greater and one argument for that is short-circuit diffusion. For $Ni_{1-y}O$ at 1000°C and 1 atmosphere oxygen, y equals about 10^{-4}, 100 ppm vacancies. If these data are extrapolated to 500°C, then y is on the order of 10^{-6}, and no metals or oxides are so pure as this. Thus, at low temperatures, one always oxidizes in the extrinsic range of the oxide. Gradients of the various impurities are also present. Perhaps Smeltzer's aging results could represent a redistribution of impurities in the scale. Whether rapid kinetics at low temperatures result from short circuits or from doping must be considered.

Rapp added that the values for defect concentrations at the gas/NiO interface are the maximum for the entire scale. At the low oxygen activity corresponding to the metal-scale interface, the vacancy concentrations are a factor 10^{-3}, or less, times the maximum. Therefore, in p-type scales, the rate is certainly controlled by impurities at the metal/scale interface.

With regard to the cracks and pores responsible for changing the parabolic rate constant, Schmalzried asked whether these cracks are not so small that they cannot be observed. Because the transport of species in the vapor is perhaps 10^{10} higher than volume diffusion in the scale, even an unobservable crack should be sufficient to affect the kinetics. Stringer responded that in the case of chromium oxidation, it makes no difference whether the Cr_2O_3 scale is attached at the metal/scale interface or not; oxidation continues at the same rate, indicating that pores do not act as diffusion barriers. The reason is that the vapor pressure of chromium is sufficiently high to satisfy the diffusional flux through the Cr_2O_3. In Pemsler's studies of zirconium and hafnium oxidation, pores did seem to act as barriers. Stringer suggested that a pore cannot be a barrier to a volume-diffusion process, but rather to some linear diffusion path. In his paper, Stringer referred to the use of sulfur-tracer experiments and oxygen-isotope experiments to establish the role of pores in the transport of vapor species through the scale.

To end the discussion of the oxidation of the "better understood" metals on a philosophical note, Schmalzried wondered whether it was necessary, from a conceptional point of view, to really understand all the many phenomenological details of oxidation. From this attitude, Galilei would probably have never found his laws. So, doesn't one have to distinguish between the basic concept and the many minor disturbances to ideal behavior? While discussion of this question did not arise, probably, those phenomena considered as details or disturbances by part of the group would have been considered as the important mechanistic issues by others.

Rapp summarized an interesting exchange of discussions in the *Journal of the Electrochemical Society* a couple of years ago concerning the nature of the ionic defect in SiO_2 and the oxidation of silicon. By the application of a voltage between a silicon specimen and a platinum electrode attached to the outside of the scale, Jorgensen could effect an increase or decrease in the thermal scaling rate in air. Further, with the specimen at negative potential, he established the existence of a stopping voltage which corresponded to $\Delta G^o_{SiO_2}$. At this voltage, oxidation ceased to progress. Jorgensen concluded that the predominant diffusing species was an oxygen ion. On the other hand, Raleigh said that Jorgensen's explanation was one possibility, but the applied voltage might also effect an electrolytic decomposition of the scale, such that at the stopping voltage, the rate of decomposition could equal the rate of formation. Furthermore, the pressure dependence for the scaling of silicon, $k_P \alpha P^1_{O_2}$, was not consistent with the diffusion of oxygen ions. So, a few years ago, one did not know whether the diffusing species in SiO_2 was charged or uncharged.

Kröger presented a summary of recent unpublished electrochemical studies of SiO_2 by one of his students, Mills. Mills has thermally oxidized very pure silicon, as well as silicon doped with phosphorus and silicon doped with boron, and obtained amorphous films about 5000 Å thick. A Si-Ge alloy was placed as a second electrode against the outside of the film to set up a galvanic cell of a small but known theoretical emf, whereby the activities in Si-Ge alloys are well known. From the measured cell voltages, one can construct a plot of cell voltage E vs log a_{Si} and add the known line of Nernst. At pure silicon, the experimental galvanic cell voltages corresponded to an ionic transference number of about 0.2 to 0.3; as the oxygen activity was increased, t_{ion} approached unity, and at quite high P_{O_2} for the second electrode, the curve again dropped, corresponding to lower values for t_{ion}. Jorgensen's experimental value for t_{ion} was, of course, an average over oxygen activities between air and Si/SiO_2 equilibrium. But Jorgensen's average t_{ion} corresponded to Mills' average. Mills carried out similar experiments for SiO_2 grown on doped silicon. The SiO_2 grown on P-doped silicon is more ionic than that on pure silicon, with slopes closer to Nernst behavior. For B-doped SiO_2, t_{ion} was radically down. This clearly indicates that at least at higher P_{O_2}, hole conduction is important in the material. The species responsible for the ionic conduction were doubly negatively charged oxygen interstitials.

Jorgensen replied that the model just presented by Kröger for defects in SiO_2 does not explain why the permeation constant and the oxidation-rate constant exhibit a P_{O_2}-power dependence of unity. These results have been rationalized previously as resulting from the transport of O_2 molecules. Thus, a problem still exists here. Raleigh would say that one always stops oxidation at the theoretical cell voltage or the decomposition potential, and at the stopping voltage one might have a balance between a flux of neutral species and a decomposition rate.

Kröger did not agree with this, but added that it might not be too far off. A field can only stop ions, and the neutral passing through by other means would not be affected. But at a slightly higher field, one could balance this. Kröger added that his electrochemical measurements do not refute the possibility that a neutral species is responsible for scale growth.

Seltzer remarked that in hot corrosion, Cr_2O_3 scales provide better protection than does Al_2O_3. However, in his paper, Wagner presented data indicating a much higher diffusivity for sulfur in Cr_2O_3 than in Al_2O_3. Seltzer wondered whether these diffusivities are in conflict with our knowledge about hot corrosion. Wagner replied that the data for sulfur diffusion in Cr_2O_3 were conducted using pressed-powder compacts. The alumina data were taken both for pressed-powder compacts and one datum for an alumina single crystal. The diffusivity values for the Al_2O_3 single crystal were 10^3 to 10^5 smaller than that for the polycrystalline material. Work on single crystal Cr_2O_3 is just beginning. The data for D_S in pressed compacts of Cr_2O_3 of 10-μ grain size were higher than those for compacts of Al_2O_3 of 3- and 10-μ grain size, and indeed higher than for NiO of similar porosity.

The discussion then shifted to the consideration of an interesting "special effect" in alloy oxidation. Rapp summarized the effects on oxidation of adding tiny (~100 Å) dispersed oxide particles (2 vol. %) to a metal: (1) the rate of growth of a protective Cr_2O_3 scale on a T.D. (ThO_2 dispersed) Ni-20Cr alloy is about a factor of ten lower than that for an ordinary Ni-20Cr alloy, (2) less chromium in the T.D. alloy is sufficient to establish the Cr_2O_3 protective scale, (3) from the position of the ThO_2 particle markers, the scale on the T.D. alloy seems to grow at the metal/scale interface instead of at the scale/gas interface, (4) the scale adherence is superior for the T.D. alloy. Thus, three of these effects represent significant practical importance for the protection of high-temperature alloys. One would like to understand these effects, and, furthermore, to realize these advantages through some means for alloys which do not contain ThO_2. Because T.D. alloys are very expensive and difficult to fabricate, one would like to add alloying elements to accomplish the same effects. Perhaps a reactive alloying element like titanium or zirconium will internally oxidize to give an equivalent internal oxide particle under the scale and provide these three important advantageous effects.

Stringer added that other particles such as Y_2O_3 and Al_2O_3 are also effective. The effects can also be realized from the presence of yttrium and other alloying additions in the alloy. There is no assurance that 2 vol.% of dispersed oxide is optimum for the improvements. A much finer grain size is stabilized in the scale on a T.D. alloy and these smaller grains result in slower transport. Relative to the effect of ThO_2 particles on the growth of Cr_2O_3 on Ni-Cr alloys, one must remember that there is a very great variation in the growth rate of Cr_2O_3 anyway. Everyone says that it results from impurities, but the evidence is not very convincing.

In response to questions by Cohen, Stringer added that as well as one can see by using back-scattered electrons in a scanning electron microscope, the ThO_2 particles seem to be mostly in the grain boundaries of the scale. Pettit has a model which involves a quite rapid transport of oxygen through the rare-earth-oxide particles. Somehow the ThO_2 particles in the scale seem to be stopping the diffusion process by which chromium ions normally contribute to the growth of Cr_2O_3 scales. The fact that a whole class of materials seem to be effective as particles suggests that they are physical blocks to a transport process.

Nichols noted that in a number of metallic materials under conditions where one thinks Herring-Nabarro creep is being observed, dispersed particles also reduce the rate of creep to below that calculated and otherwise observed. There may well be a connection. No explanation is offered for this effect.

Johnson added that a dispersion of fine oxide particles also accomplishes a reduction in sintering rate. Some say that the particles tie up dislocations, but there are cases where the process is diffusion controlled also. The effect has been sometimes described in terms of the contact angle between the particle and the metal.

Ilschner asked whether it is possible that the dispersed oxide particles interfere with the contact of phases at the metal/scale interface, thereby either inhibiting the transfer of cations to the scale or else allowing a preferential transfer of one alloying component to the scale. In response, Stringer recalled that the vapor pressure of chromium is so high that local loss of contact does not interfere with chromium transfer to the scale. Wright added that the effect of ThO_2 particles in reducing the growth-rate constant of Cr_2O_3 also occurs for the oxidation of pure chromium containing dispersed ThO_2. Further, about 80 percent of the scale is detached from the metal at temperature, and still the oxidation rate is slower than that for the pure metal. The scale grows at the metal/scale interface. Thus, oxygen transport through Cr_2O_3 does occur, but this process is not understood.

At this stage, progress toward explaining the effect of dispersed oxides on oxidation seems dependent upon a better understanding of the transfer of ions across interfaces like grain boundaries. Rapp referred to Huggins' studies of transport in oxides like beta-alumina which exhibit two-dimensional channels. The ionic conductivity of polycrystalline aggregate specimens is about one-fifth that for the favorably oriented single-crystal material. In other words, here is a case where ions jump across interfaces quite easily.

On ionic transfer across interfaces, Schmalzried remarked that from a fundamental point of view, little is known and it is extremely difficult to study solid-solid interfaces. Electrochemical methods should probably be applied in such studies. Some work has been done: Rickert's work on Ag/Ag_2S and Schmalzried's work on $Ag(+Si)/SiO_2$-saturated silica slags. The difficulty is that interfaces are seldom clean and are probably different under static and under dynamic or steady-state conditions.

Jorgensen commented that for β-alumina, calcium oxide additions increased grain-boundary resistivity and yttria decreased the grain-boundary resistivity. If one assumes a model similar to Oriani's model for grain growth, whereby impurities block sites for grain growth, impurities segregate to grain boundaries and are adsorbed there. This has been observed for alumina, and also for the metal lead where one has counteracting impurities, i.e., some impurities which block sites and then other impurities which negate that blocking.

Goodenough wondered whether the resonance technique won't be suitable for investigating impurities in the grain boundary. Wertz replied that the effects for which one is looking are in similar enough electric fields, essentially identical, so that one would have the superposition of the spectra of a large number of identical defects. If one works with a material of well defined symmetry, and if the defects have well defined symmetry, information can be obtained. But as soon as the symmetry around some of these is destroyed, the lines will be smeared and it's really a question of whether one can get useful information from a line position alone. In favorable cases, that is possible.

Caro related his experience that small particles of yttrium oxide in thin films have a large number of essentially two-dimensional particles with faces parallel to the $\{111\}$ plane of the cubic structure. Caro inquired about the shapes, orientations, and distribution of the dispersed oxide particles in scales on T.D. alloys. Wright replied that the effective dispersed oxide particles are spherical. Flat particles of Al_2O_3 were not effective in reducing the oxidation rate. Rapp added that these dispersed oxide particles also end up on the top of a Cr_2O_3 scale upon high-temperature oxidation, because the scale is evaporating as CrO_3 vapor. However, this dust of particles on the surface does not seem to be any sort of a barrier.

Kofstad remarked that Cr_2O_3 scales formed on ThO_2-dispersed alloys were more dense than when thoria was absent. Without the thoria, the higher porosity can provide rapid vapor transport across pores. Further, if the parabolic rate constant for the oxidation of chromium is compared with that calculated from diffusion measurements, there is poor agreement, probably because of rapid vapor transport across the pores. The diffusion coefficient for oxygen in ThO_2 is several orders of magnitude greater than the diffusivity of chromium in Cr_2O_3. Therefore, one should realize appreciable inward growth by oxygen diffusion, and when scales grow at the metal/scale interface, there is less porosity in the scale. Stringer doubted the poor correlation between the calculated and experimental growth constants for Cr_2O_3. Stringer said that these data were recently compared in great detail with reasonable agreement.

Hauffe volunteered two possible methods for the reduction of cracks and pores in a growing oxide layer:

(1) Considering the improvement of the sintering of oxides by the addition of small particles of foreign oxides, one must focus interest toward finding an alloying element for the oxidizing alloy which favors the pore-free sintering of the growing oxide layer. Such a mechanism seems to govern in the oxidation of Ni-Cr alloys containing small amounts of thorium.

(2) One can decrease the pore formation in the growing oxide if a rather noble metal with a sufficient ductility is alloyed as, for instance, has been done by Turkoogan with copper in steel. During oxidation, precipitation of copper is observed at the steel/oxide interface. This principle also works with a tin addition in the oxidation of Zr-Ni alloys. Obviously, the precipitated ductile tin decreases the formation of pores and cracks.

Johnson interjected that possibly the presence of ThO_2 particles could provide an increased volume of space-charge region in the scale; each particle would be surrounded by a space-charged region. Differing mobilities for the chromium and oxygen in the space-charge region could lead to changed scaling characteristics and rates if their relative volume was sufficiently large.

Stringer commented on oxygen transport in the particle-dispersed Cr_2O_3 scale, saying that it is surprising that Al_2O_3 could produce this effect. That a finer grain size corresponds to faster oxygen diffusion has been observed often, but the really surprising point is that the oxidation rate is reduced. Stringer concludes that normally, in alloys free of dispersions, the growth of Cr_2O_3 is controlled by a relatively rapid short-circuit diffusion of chromium. For ThO_2-dispersed alloys, this short-circuit mechanism is blocked and replaced by inward oxygen transport.

In answer to Jorgensen's questions, Stringer states that the ThO_2-dispersed scales are a few microns thick, with particle spacing of about 1000 Å. Jorgensen states that these particles could then block grain-boundary sliding and plastic deformation of the scale.

To conclude this long discussion of the "special effect" afforded to alloy oxidation by oxide dispersions, Stringer added that some (but not all) of the effects are also observed for Al_2O_3 scales.

Huggins wondered why more attention is not paid to the oxidation of refractory metals for which scales would exhibit the extended (organized) defects. Stringer replied that the oxidation of refractory metals is very complicated, scale plasticity is extremely important, and the transport processes in the scales are poorly known. Many aspects are very interesting, such as an extremely high apparent plasticity in Nb_2O_5 (no cracks in a scale several millimeters thick grown at 700°C). The concept of extended defects has not yet been applied in oxidation.

Smeltzer referred to the transition from parabolic to linear scaling, or breakaway oxidation, which is frequently observed in the oxidation of refractory metals. In one case, the onset of linear behavior can correspond to a fracturing in the outer region of an initially protective scale, or otherwise, an inner compact scale can be transformed to an outer nonprotective phase (called chemical conversion). In the oxidation of tungsten, an inner blue W_4O_{11} scale converts to a nonprotective yellow WO_3 layer. For molybdenum, MoO_2 converts to MoO_3. Smeltzer asked whether extended defects are important to these processes. Anderson stated that the fast oxygen diffusion by easy diffusion paths in Nb_2O_5 at 1000°C would establish a very steep oxygen gradient at the base of the scale with little composition variation across most of the scale.

Goodenough questioned whether diffusionless transformations in oxides could contribute to scale plasticity. He referred to oxide plasticity created by antiphase transitions in grain boundaries. Rapp and Stringer agreed that in the oxidation of refractory metals, the cracking of the scales is the real problem. If greater deviations from stoichiometry would result in increased scale plasticity, this favorable effect would outweigh any concomitant increase in diffusion rate. Stringer reported that both Nb_2O_5 and Ta_2O_5 can exhibit high plasticity, which, however, is enormously dependent upon oxygen pressure. This behavior should be related to the stoichiometry and the defects in these oxides.

Anderson has followed the reoxidation of a niobium oxide specimen in the electron microscope at temperatures up to 1000°C. A single crystal of Nb_2O_5 can be initially reduced to $Nb_{12}O_{29}$ and then reoxidized to Nb_2O_5, the specimen remaining as a single crystal throughout. Thus, plasticity for these flexible foils is not surprising. They seem to undergo rather small adjustments in composition even though the volume of the crystal is changing.

Eyring said that even though one has an oxygen gradient for an oxide with coherently intergrown intermediate phases, the oxide does not necessarily fracture. It is possible to take a dioxide of the CaF_2 structure type, reduce it to a C-type sesquioxide, and then reoxidize it to a single crystal. In the intermediate phases, one has multiple twinning and loses the single crystal, but because of the high symmetry of the fluorite structure, the specimen returns to a single crystal and its integrity is not lost.

Schmalzried returned to the "special effect" of a dispersed oxide, and suggested that if the scale grows by a countercurrent of cations and vacancies, then perhaps the dispersed phase is effective as a sink for the vacancies and thereby eliminates void formation at the metal/scale interface. Stringer replied that this explanation has been one of the popular ones, and Pettit currently feels that the particles very probably do act as vacancy sinks, although Stringer still thinks that the weight of the evidence is against it.

Niesz commented that a possibility for increasing the plasticity of a scale is to add a component that would lead to the formation of a low-volume-percent eutectic liquid.

Jorgensen asked whether one would not lose parabolic kinetics and introduce interface-controlled kinetics if vacancies experienced difficulty in annihilation at the metal/scale interface. Stringer replied that vacancies do not control the scaling kinetics, but, rather, the process goes on and one must remove one product of the reaction – the vacancies. But the means of vacancy annihilation could change the mechanism. With the formation of pores from vacancy condensation, vapor transport across the pores could eventually change the rate.

Seltzer returned to the question concerning the use of the concepts of extended defects in explaining the oxidation of refractory metals. It seems that we need some basic principles relating to transport, plasticity, and adherence for these oxides that can be tested by experiment. We know how to change the point defects, but we need suggestions on how to change and control shear structures.

Lacking any response to the challenge of Seltzer, Tannhauser presented a short prepared discussion describing a two-chamber symmetric apparatus for measuring changes in the stoichiometry of an oxide in a CO/CO_2 gas mixture. Schmalzried has used similar apparatus for equilibrations of oxides with pure oxygen whereby changes in total pressure are seen on a differential pressure gauge. To conduct such an experiment in CO/CO_2 mixtures, which might be necessary for an oxide near the phase boundary with the metal, one finds immediately that, in principle, the experiment cannot work because a change in specimen stoichiometry and the corresponding change in CO/CO_2 ratio do not change the total pressure. Therefore, Tannhauser has incorporated an interferometer to detect changes of the index of refraction of the CO/CO_2 mixture as a measure of its composition. The interferometer has an accuracy of about 0.03 of a fringe, corresponding to 5×10^{-7} mole of atomic oxygen. The device is equipped with bellows to circulate the CO/CO_2 gas within the chamber and thereby arrive at a homogeneous composition. The work will be published.

Smeltzer points out that the Wagner-Hauffe doping relations for the growth of binary-compound scales account only for the concentrations of point defects, and do not consider their mobilities. In the analysis for the growth of three-component scales, the steep concentration gradients cause the cations to experience different point-defect concentrations and, therefore, the differences in mobility play a profound role in the distribution of the two cation species. The previous approximation considering only point-defect concentrations is very crude indeed.

Johnson commented on Singer's work on the hollandites; Singer has reported low d-c conductivity even though the conductivity predicted from the dielectric loss should be very high. Huggins restated his preliminary results for the hollandites. He has conducted a-c conductivity measurements on hollandites of potassium, magnesium, and titanium at typical audio frequencies, 10 to 40,000 hertz. The results fell within the scatter band of the dielectric-loss measurements of Singer. Graphite blocking electrodes were used which would require no compensation. These a-c measurements do not require that the ions move great distances and there was no frequency dependence of the conductivity. In general, one should worry about transport in linear tunnels which could be blocked by some species, or by shear or structural imperfections.

Anderson adds that in hollandites and in most linear tunnel structures, the cations do order, so that one would guess that there are deeper potential wells than in structures like β-alumina. Some electron microscopy on barium, magnesium, and titanium hollandites showed that a change in stoichiometry leads to ordering in a superstructure. The mobility might therefore be considerably lowered. These structures should order well in one direction, whereas they can have imperfections in the other two directions.

Welch said that some theoretical calculations on the calcium apatite structure, where there are no tunnels but chains of very mobile fluoride ions, showed that the chains could be blocked by slow-moving oxygen anions. One would get large electrical dipoles influencing the a-c electrical conductivity.

In response to Goodenough's question about the blocking of tunnels with water molecules, Huggins replied that water could be a problem at the start of an experiment, such that only after a couple of weeks of working with a particular specimen do the results become reproducible and consistent. Generally, with regard to fast-ion conductors, cubic materials offer the great advantage that they are not susceptible to blocking of paths in a given direction.

Honig remarked that the complicated migration of ions through tunnel structures required that the ions move through the tunnels in unison. He wondered how this requirement would affect the comparison between the theory and the experimental results. Huggins responded that these materials have only dilutely filled tunnels, i.e., there is about a factor of two or more tunnel sites than ions. Cooperative motion is certainly possible and is being studied. Abnormally high chemical diffusion coefficients, such as 0.47 cm^2/sec for Ag ions in Ag_2S at 200°C reported by Chu and Rickert, is a puzzling aspect.

In response to Rapp's question about the possibility of doping the tunnel structures and change the tunnel occupancy by varying the alkali oxide content, Huggins replied that composition can be varied but the structure is usually fixed through a high-temperature equilibration. At 1000°C and above, for these structures and for β-alumina, transformations and changes in composition occur and could be engineered.

O'Keeffe commented that the high rate of chemical diffusion for Ag_2S is completely reasonable because the activity is a very strong function of the composition and an activity gradient therefore gives a very strong driving force.

In response to questions by Welch, Huggins responded that two-thirds of the tunnel sites are filled in AgI and one-half are filled for β-alumina. An ordering of sodium in the β-alumina has been established by X-ray diffraction, but this does not affect the transport properties importantly.

Smeltzer remarked that a galvanic cell involving the β-alumina electrolyte could possibly be used as a sodium probe in several nonferrous industries such as the lead, tin, and aluminum industries. Hot shortness occurs in aluminum for sodium contents of less than 5 ppm. Smeltzer wondered about the thermodynamic stability of β-alumina in liquid aluminum. In an industrial test in relative pure liquid aluminum at 1000°C, the outside of the β-alumina was reduced to another phase.

In response to a question by Schmalzried, Huggins answered that the dependence of the mobility of sodium ions in β-alumina as a function of the activity of sodium was not known. The problem is that of obtaining equilibrium with respect to other species. Only at temperatures above

800°C is an effect of oxygen pressure on the partial electronic conductivity important, because the structure is frozen in. In response to Rapp, Huggins replied that fuel cells and ion exchange at low temperatures are possible even though the structure is frozen because the monovalent cations are quite mobile at the low temperatures. In reply to Schmalzried, Huggins added that the use of a β-alumina probe can be effective, because of the facility of ionic exchange, only if one is careful to provide only one exchangeable species at a time. For example, mercury is insoluble in β-alumina. Therefore, one can measure the activity of sodium in Hg-Na solutions. In response to Vink, Huggins said that he believed that β-alumina could be doped with chlorine.

Rapp reported that Brook and Lundsgard have used the β-alumina electrolyte in aqueous solutions as a pH meter because the electrolyte apparently accepts protons. However, the ceramic is mechanically unstable in aqueous solutions and cracking is experienced.

Goodenough wondered whether the diameters of channels could be optimized so that the migrating ions would experience minimum polarization, or whether substitution into the structure to accomplish this causes the phase to become metastable. O'Keeffe and Huggins agreed that at low temperatures such a phase would usually order to produce a structure of lower conductivity.

Rapp reported on the LaOF electrolyte with the fluorite structure of Pelloux and Déportes in Grenoble. Because of a phase transformation, the material is difficult to make and cool, but once obtained, it is a good electrolyte. However, the identity of the conducting ion, oxygen or fluorine or both, is ambiguous. As an experiment to establish which species are conducting, Rapp proposed establishing countergradients of fluorine and oxygen across the electrolyte. Upon evaluating the usual flux equations assuming reversible electrodes, the cell voltage should be given by the following equation:

$$E = \frac{RT}{2F} \int_{P'_{F_2}}^{P''_{F_2}} (t_{F^-})\, d \ln P_{F_2} + \frac{RT}{4F} \int_{P'_{O_2}}^{P''_{O_2}} (t_{O^{2-}})\, d \ln P_{O_2}$$

In plotting the cell voltage vs $t_{O^{2-}}$ [assuming that ($t_{O^{2-}}$ plus t_{F^-}) equals unity], the cell voltage should be a linear function of $t_{O^{2-}}$ and the line crosses the abscissa. Thus, the predicted cell voltages for predominant $t_{O^{2-}}$ or predominant t_{F^-} would have opposite polarity. Kröger added that NH_4Cl exhibits conduction by protons, NH_4^+ or Cl^-, and also by electrons.

Caro remarked about EuOF which can be either a very highly disordered compound or else a layered structure close to the fluorite structure. The diffraction lines for the disordered structure are so broad that one can hardly speak of lattice sites. One can also suppose that a full series of intergrown phases are possible. Rapp asked whether there is a rare-earth oxyfloride which is a stable phase at all temperatures. Further, does europium have a greater tendency than lanthanum to change valence upon changes in oxygen activity? Caro responded that EuOF converts at 600°C to another disordered phase.

Huggins mentioned the material KHF_2 which undergoes a phase transition at about 120°C. In the low-temperature structure protons are conducting, and above the transformation, both K^+ and Cl^- ions conduct in a ratio of about one to three. Rapp commented that the AgI electrolyte has been used to measure Ag^+ concentration in aqueous solutions at room temperature.

Nowick referred to the electrical behavior of CaO- and Y_2O_3-doped CeO_2 at relatively low temperatures. This material shows predominantly ionic conduction when close to stoichiometry, with an ionic conductivity that is more than an order of magnitude higher than that of calcia-stabilized ZrO_2 in the range of 400 to 800°C. Further, with platinum paste as electrodes, it showed considerably lower polarization at temperatures near 600°C than that observed for stabilized ZrO_2 at similar temperatures. Accordingly, doped CeO_2 appears to offer promising possibilities as a solid oxygen electrolyte. Rapp suggested that a silver-paste electrode would be better at 400°C.

With these remarks, the session ended. In retrospect of the session, a lively and enthusiastic discussion was enjoyed by everyone. However, perhaps too frequently the discussion dwelt on a very specific topic ("special effect in oxidation") where the literature was poorly known to the majority of the participants. Nevertheless, the participation of even the theoreticians in considering such engineering-type problems was typical of the prevailing spirit of the meeting.

Part Six

CRITICAL ISSUES

CONCLUDING AGENDA DISCUSSION: CRITICAL ISSUES

*A. S. Nowick**

Henry Krumb School of Mines
Columbia University
New York, N.Y.

*G. G. Libowitz***

Materials Research Center
Allied Chemical Corporation
Morristown, N.J.

In order to provide sufficient time for discussion of each of the more critical issues that arose at the Colloquium, it was decided to divide the concluding agenda discussion into five sections, each dealing with a major problem area in oxides.

1 CHARACTERIZATION OF OXIDES

The chairman posed the questions: What constitutes proper characterization of materials? What can we do to attain it? There appeared to

*Chairman
**Secretary

be two separate approaches to answering the first question. One group of investigators views characterization as defining a particular specimen with respect to purity and structure relative to a pure and perfect sample of that material in equilibrium with its ambient. Others take the view that intrinsic properties of a material also must be included in its "characterization".

Kröger considered the latter approach in defining characterization as the determination of properties as related to structure. He discussed the case of point defects in solids, where it is necessary to ascertain which defects are present and what parameters control the formation of defects. Kröger noted that there are about 50 different properties that depend upon the nature and concentration of defects present in the lattice, and one usually studies those properties that are important to a particular investigator or system. Using cadmium telluride as an example, he illustrated how measurements of a particular property, such as diffusion, over a whole range of thermodynamic variables (e.g., partial pressure of components, temperature, concentration) permit deduction as to types, charges, and concentrations of defects present.

Turning to the former definition of characterization (dealing with a particular specimen), the group then attacked the problem of impurities and their determination. The chairman pointed out that a reliable chemical analysis, especially in the 10 to 100-ppm range is difficult to obtain. Honig warned that emission spectroscopy is not too reliable a method, since it frequently does not detect some species of impurities which are present. He suggested mass spectrometric analysis which is reliable for detecting types of impurities, although it is not too precise in measuring amounts present (usually within a factor of two or three). Neutron-activation analysis appears to be another reliable method. Franklin remarked that in the ppm-level range, the relative amounts of impurities present are fairly easily obtained by any given technique. However, absolute amounts are much more difficult to ascertain because of calibration difficulties in each method of analysis. He cited a case of very pure sapphire where very careful measurements by spark-source mass spectrometry yielded results which differed by an order of magnitude or more for several impurities at the low- and sub-ppm level from those obtained by neutron-activation analysis.

Seltzer noted that the detection of impurities at grain boundaries is also important. Caro proposed the use of an electron microscope as an analytical tool for finding localized impurities. By using the microscope as a microprobe, Caro claimed that areas as small as 5 Å^2 could be studied.

The chairman brought up the question of the desirability of setting up a central bank of well-characterized materials to be used as standards. Two processes that could be handled by such a central bank would be purification and crystal growth.

Schmalzried remarked that even well-characterized materials will change during measurement. For example, chemical analyses of a sample before and after some series of measurements may give different values of impurity concentration. In correlating the results of the investigation with impurity content, the question then arises: Does one take the initial analysis, final analysis, or some value in between? Tannhauser agreed, citing the observations of R. W. Vest of Purdue on nickel oxide where, although the impurity level was less than 100 ppm before electrical-conductivity measurements were carried out, the impurity concentration had increased to 500 ppm during the course of the measurements.

A powerful tool in characterizing materials is high-resolution lattice-imaging electron microscopy as was shown in the paper by Cowley and Iijima. The chairman raised the question as to the limitations of this technique. Cowley stated that the limitations of the method can be discussed in terms of three categories:

(1) Specimens. The requirement that the specimen areas used should be of thickness less than about 150 Å for 100-kev electrons may impose severe difficulties for some types of materials. For example, metals will usually not fracture sufficiently well to give good thin edges of crystals. The standard methods of preparation for electron-microscope specimens of metals, by evaporation or electrochemical thinning, usually give unsuitable rounded edges.

The other severe limitation for many types of specimens comes from the radiation damage suffered under the intense electron irradiation of 10^{12} rads. Although metals and semiconductors are more stable, insulators may disintegrate in a very short time. In some cases, radiation damage may offer advantages along with the disadvantages. For example, Fig. 1 shows a bright-field image of the thin edge of a $LiFeO_2$ crystal. Apart from the fine-scale semiperiodic structure due to the short-range ordering, the image shows a mottling with patches 30 to 40 Å in diameter of varying contrast. These patches appear as a result of electron irradiation and have been shown[1] to be nuclei of the $LiFe_5O_8$ phase. Thus, there is a loss of clarity for studies of the disorder in $LiFeO_2$ but a gain in information regarding a phase transition.

(2) Instruments. At present, 3.5 Å is the best resolution obtainable in images which may be interpretable in a more-or-less intuitive way. Finer detail may be observed, but the interpretation is complicated. The 3.5 Å resolution is obtained with the most advanced current commercial 100-kev electron microscopes. Since the crystal must be aligned with respect to the beam with an accuracy of a fraction of a

Fig. 1. Electron micrograph of a thin edge of a $LiFeO_2$ crystal.

degree in two directions to give an image capable of interpretation, a goniometer or tilting specimen stage must be used. At present, the required resolution can be obtained with such a stage with only one or two of the available microscopes.

For oxides based on the ReO_3 structure, the 3.5 Å capability is sufficient to give a useful amount of detail of the arrangement of the metal atoms. For many other materials, including the oxides based on the NaCl structure, the most commonly occurring separation of rows of atoms in projection is about 2 Å. The possible direction to be taken to overcome the current limitation in resolution is evident from the expression for the minimum resolvable distance for interpretable images,

$$d_{min} = 6^{-1/4} \, C_s^{1/4} \, \lambda^{3/4} \quad ,$$

where C_s is the spherical aberration coefficient and λ is the electron wavelength. A decrease in C_s is difficult to achieve and not very effective. A change in wavelength has greater effect. For a 1-Mev electron microscope, the decrease in wavelength will give d_{min} less than 2 Å. However, the cost increase to go from a 100-kev to a 1-Mev microscope is also a factor of 10.

(3) Interpretation. It must be stressed that even under the ideal imaging conditions when the electron micrograph appears to be a direct projection of the crystal "structure", there is no strictly linear relationship between image intensity and any definable distribution (of electron density, particle density, potential, etc.) within the crystal, especially when, as usually happens, the thickness exceeds 10 to 20 Å for medium-heavy atoms.

An excellent example of this is given by the image of a quenched, disordered crystal of Nb_2O_5[2], shown in Fig. 2. Much of the picture shows a structure well ordered in the beam direction but with a

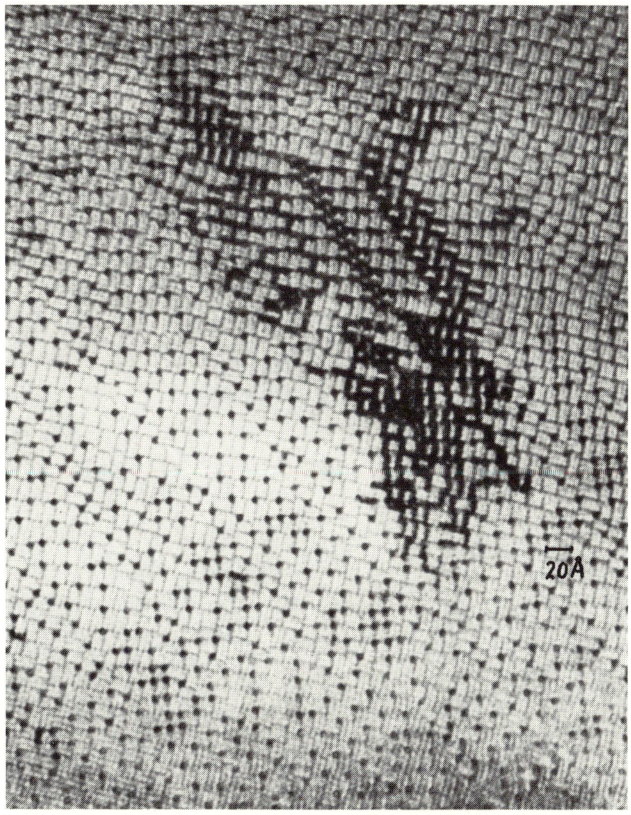

Fig. 2. Electron micrograph of a quenched disordered crystal of Nb_2O_5.

highly disordered superlattice in the two directions perpendicular to the beam, with a clearly resolved configuration of various size blocks of octahedrally coordinated structure and dark spots corresponding to tetrahedrally coordinated metal-atom rows. The extensive dark patches do not, as might be intuitively deduced, represent regions of greater metal-atom concentration. They represent regions where the ordering in the beam direction is interrupted by faults which displace atoms in directions perpendicular to the beam. Therefore, although the number of metal atoms does not increase, there is an increase in the number of metal-atom rows.

Cowley concluded that while some simple rules can be established to assist in the interpretation of such images, the only satisfactory basis for anything beyond a superficial and necessarily uncertain evaluation of images is the application of the known theory (the n-beam dynamical electron-diffraction theory) and computing methods in detailed calculations for model structures. This requires the use of a moderately large computer and a concentrated effort, but satisfactory methods and programs have now been evolved[3,4] and are available.

2 THEORETICAL CALCULATIONS

The chairman raised the question as to how far theory can proceed in predicting structure, particularly with respect to activation energy and extended defect calculations. Lidiard wished to emphasize two points with regard to theoretical calculations. The first is that, given an acceptable model, the properties of defects can be accurately calculated since we are no longer hindered by mathematical approximations. This applies to point defects, clusters, and dislocations. The practical limits depend upon the number of variables one wants to handle or the size of the computer. For real-space calculations, the relaxation of several hundred atoms is perfectly feasible.

Lidiard felt that even very approximate calculations have yielded useful physical ideas. It is important to realize that there is no simple mechanical way of adding to one's knowledge of a system. One always must be thinking physically in relation to the problem.

The second point emphasized by Lidiard is that present models are largely empirical and, therefore, one of the first tasks is to improve the models used for calculations. Ionic materials are on much stronger ground than transition metals because they can be described in terms of interaction between individual ions.

With respect to activation energies, Lidiard stated that it is certainly possible to calculate them, although there has been a tendency to underestimate the values. He suggested that this may be because the models were pushed too far.

Calculations for dislocations are as practical as those for point defects and are being carried out at Harwell. The results should elucidate the interactions between dislocations and point defects. As for extended defects such as crystallographic shear planes, it should be recognized that interactions are describable in macroscopic terms through strain fields. To look more carefully at the detailed structure of extended defects, one must consider suitable atomic models.

In response to a question by Cohen concerning the ability of theory to predict defect structures, Lidiard stressed that theoretical calculations are used as a guide to the interpretation of defect models. However, it is now possible to handle quite complicated defect structures so that the limitation is merely what one considers to be a reasonable range of models to analyze.

Kröger felt that a basic question to be answered by theory with respect to nonstoichiometric compounds is why single point defects occur in some systems, complex defects in others, and crystallographic shear planes and block structures in still other systems. In Lidiard's opinion, it should be possible to deal with this question by utilizing models of surfaces, internal as well as external.

Welch pointed out that while theoretical calculations have been very successful for strongly ionic materials, the situation for essentially covalent crystals having some degree of ionicity is not very good. For example, for such compounds as CdTe or FeO, there are no models available for defect calculations having anywhere near the same degree of reliability as for MgO.

Since there is a recent proposal that there is significant occupation of tetrahedral sites by cations in NiO and CoO, Rapp raised the question as to whether theory can predict the oxidation state of these cations. Because of the need to maintain electroneutrality, he believes that the divalent ion rather than the smaller trivalent ion should be present in the tetrahedral sites. Rapp pointed out that in spinels, both divalent and trivalent cations can occupy tetrahedral sites. Goodenough noted that theories of site-preference energies have been proposed, and although much of this work requires reexamination because the degree of covalency was not sufficiently considered in the crystal-field calculations, some reasonably good rules have evolved. For example, a Cr^{+3} or Ni^{+2} ion cannot be easily put into a tetrahedral site because any ion with a d^3 or d^8 configuration has strong octahedral-site preference as a result of crystal-field splitting. In spinels, the only time Ni^{+2} is found in a tetrahedral site is when Cr^{+3} ions are in the octahedral sites. Goodenough suggests that spin-resonance studies can be used to determine the charge and site occupancy of ions.

3 ELECTRONIC PROPERTIES

In response to a request by the chairman, Goodenough presented some ideas relating electronic properties to transport properties of oxides. He reminded the audience that transition-metal and rare-earth oxides contain filled anionic p bands separated from an empty cationic s band by a large energy gap, and the number of outer d and/or f electrons per transition-metal or rare-earth ion is represented by the formal valence state of the cation, since the Fermi energy (or chemical potential) falls in the energy gap between anion p bands and cation s bands.

Because the anion p states are stable bonding states, it is not energetically favorable to introduce holes into these bands. Consequently, Goodenough suggested that holes introduced by oxidation will become trapped at deep acceptor levels by formation of $(O_2)^=$ ions. Formation of molecular orbitals at a complex anion allows stabilization of occupied states at the expense of unoccupied states. Goodenough pointed out that these complex $(O_2)^=$ ions may be mobile as one of the oxygen atoms of the pair makes and breaks bonds with successive host atoms. Thus, this process represents the transport of neutral oxygen in the lattice. However, since the initial oxidation requires an exceptionally high oxygen pressure because the stability of the anion p bands makes it difficult to force the introduction of holes into these bands, Goodenough suggested that a more likely event would be the introduction of holes into dangling bonds at external or internal surfaces. $(O_2)^=$ ions at surfaces should lead to rapid diffusion of neutral oxygen. Similarly, neutral oxygen might form at and move along certain dislocation lines. Goodenough believes that such a mechanism could account for enhanced oxygen diffusion along grain-boundary surfaces.

Stringer agreed that $(O_2)^=$ ions may explain rapid movement in grain boundaries. He wondered whether one could use an analogous argument for cations in grain boundaries, and also, whether it would be possible to have a situation where in one particular oxide, grain boundaries would short circuit one species while dislocations within the grains short circuited another species. Goodenough stated that dangling bonds and associated surface states would favor $(O_2)^=$ ions on surfaces, while dislocations may favor some other diffusing species, so that this latter situation may very well exist.

Kröger remarked that the $(O_2)^=$ species suggested by Goodenough should not necessarily be restricted to surfaces since it may occur inside crystals. It is the species one would expect to be formed when a neutral oxygen interstitial, O_i^x, is present. Similar species occur in sulfides, selenides, and tellurides, where self-diffusion experiments indicate that, at low metal activities, neutral chalcogen interstitials are the main diffusing species.

The second topic raised by Goodenough pertained to electronic conduction. He pointed out that cations having closed d^{10} cores may have s orbitals that are sufficiently stable to accept electrons. Goodenough noted that this fact is important for the development of transparent conductors for optical filters, display devices, and other devices because of the high mobility of the carriers in the s states. The itinerant n-type conductors $In_{2-x}Sn_xO_3$, $Sn_{1-x}Sb_xO_2$, and $SnO_{2-x}F_x$ are examples of transparent conductors under development. Oxygen vacancies and cation interstitials also can be introduced into such oxides to make them both metallic and good ionic conductors.

Goodenough observed that transition-metal and rare-earth oxides may be excellent electronic insulators over a range of oxygen partial pressures, provided there are no d-state or f-state manifolds in the energy gap between the top of the anion p bands and the bottom of the cation s bands. However, if this condition is not fulfilled, the cations readily assume multiple valence states, and introduction of native defects (cation vacancies, anion vacancies, and cation interstitials) is compensated by mixed valences on the cation array. This leads to nonstoichiometry and the possibility of wide phase fields in which extended defects may form. Compounds with mixed cation valence and ion vacancies are both electronic and ionic conductors.

Goodenough continued by discussing the importance of the chemical potential, which must remain constant in a system in the absence of an electric-field gradient. Matching the chemical potential across a metal-insulator interface may require insulator-band bending induced by a trapping of surface or space charges at the boundary of the insulator. Whether the charges so trapped are provided by electrons, ions, or vacancies depends upon the relative ease with which these species can be created and moved to the boundary. If d or f states are available to the Fermi energy, the necessary charge carriers can be easily created. Similarly, if impurities collect at the grain boundaries, their charge may be compensated by either electrons or native defects. Mass transport can be enhanced by this effect only if native defects compensate the charges.

The chemical potential is also important when considering clusters and extended defects. Goodenough used wustite as an illustrative example. Because the $Fe^{2+}(d^6)$ manifold has a narrow band width, the Fermi surface is pinned within this manifold at a nearly constant energy over the wide variation of $[Fe^{3+}]/[Fe]$ ratios across the FeO_x phase field. Therefore, inhomogeneous concentrations of Fe^{3+} ions do not cause the chemical potential to create a phase boundary between the cluster and the FeO rock-salt matrix. Nevertheless, Fe^{3+}-ion clustering does introduce a distinguishable structural element, and the concentration of the cluster structure increases relative to the rock-salt structure across the phase field. The Koch-Cohen structural element consists of a face-centered cube of O^{2-} ions, four iron atoms in the tetrahedral-site positions of zinc blende within this cube, and octahedral-site cation vacancies at the cube center and the centers of the 12 cube edges. The chemical formula $Fe_{1-\delta}O = Fe^{2+}_{1-3\delta}Fe^{3+}_{2\delta}O$ should be modified to reflect this structural information. The tetrahedral-site ions may be assumed to be Fe^{3+} ions, and they are placed outside the brackets in the following formula in which $x = \delta/9$:

$$Fe^{3+}_{4x}\left[Fe^{2+}_{1-27x}Fe^{3+}_{14x}\square_{13x}\right]O \; .$$

For small-polaron conduction, the Seebeck coefficient is $\alpha = \frac{k}{e}\left\{A + \ln[(1-c)/c]\right\} \approx \frac{k}{e}\ln[(1-c)/c]$, where c is the concentration of mobile holes (Fe^{3+} ions in wustite) on the sites available to the polarons. The transport term A can generally be neglected in a zero-order approximation, which leaves a temperature-independent expression having no adjustable parameters. Therefore, it provides a particularly useful test of atomic models, since it had been shown[5,6] that at elevated temperatures the Seebeck coefficient in $Fe_{1-\delta}O$ changes sign at about $\delta = 0.08$ to 0.09. The simplest model would make all the iron atoms available to the mobile charge carriers. Since the Seebeck coefficient should change sign at $c = 1/2$, this model gives a critical value of $\delta_c = 0.2$ at which $\alpha = 0$, which is outside the phase field. If the tetrahedral-site cations are considered to be always Fe^{3+} ions, this is increased to $\delta_c = 9/41$. Therefore, the observation of a change in sign of the Seebeck coefficient within the phase field would seem to require rather special constraints on the charge carriers if the Seebeck coefficient is temperature independent. For example, the cubic cluster consisting of four tetrahedral Fe^{3+} ions, twelve octahedral-site vacancies, and four O^{2-} ions carries an effective 4+ charge that must repel any Fe^{3+} ions from the nearest-neighbor cation sites. This observation enlarges the cluster to include Fe^{2+} ions at all 42 nearest-neighbor cation sites. If, as above, we now take all the cluster atoms outside the brackets in one chemical formula, we have

$$Fe^{3+}_{4x}Fe^{2+}_{42x}\square_{12x}\left(Fe^{2+}_{1-69x}Fe^{3+}_{14x}\square_{x}\right)O \; ,$$

where $x = \delta/9$. At the sites outside the cluster, i.e., those in the brackets of the formula, $c = 1/2$ occurs for $\delta_c = 0.109$. This value approaches the observed value, which seems to indicate the validity of the assumption that the octahedral-site Fe^{3+} ions outside of a Koch-Cohen cluster are prevented by electrostatic forces from occupying nearest-neighbor cation sites. This fact enlarges the effective size of the cluster from that determined by X-ray diffraction. Cluster-cluster interactions could further reduce δ_c. Also, at high enough temperatures, electrons are excited to the 4s band, and the much more mobile 4s electrons would give an important negative contribution to the Seebeck coefficient, thereby reducing δ_c for a particular temperature.

Stubican commented that elastic-strain energy should be considered in disordered structures since the free-energy equation has a term for strain energy. He reported that in wustite a peak in elastic energy is present when clusters are formed.

The final topic discussed by Goodenough concerned the relationship between oxidation mechanisms and electronic structure. He suggested that the oxidation mechanism of the metals iron, cobalt, and nickel may differ from that of titanium and vanadium. In the latter two metals, oxidation may take place by diffusion of oxygen into the bulk metal as interstitial oxygen, while in the case of iron, cobalt, and nickel, the metal diffuses into the surface oxide layer. Goodenough proposed the following justification for this suggestion.

The transition metals are characterized by overlapping s and d bands. Introduction of an oxygen interstitial would create a strong interaction between the σ-bonding orbitals at the oxygen and nearest-neighbor metal atoms. This interaction would cause a splitting of energy levels, the primarily oxygen 2p states becoming stabilized relative to the Fermi energy, E_F, and the metal s and d_σ states being raised above E_F. In the heavier metals like iron, cobalt, and nickel, the number of occupied metal s and d_σ states that are raised above E_F would exceed the number of stabilized anion states that can accept their electrons. In this case, interstitial oxygen would introduce additional antibonding d electrons which reduce the metal-metal bonding. Consequently, the formation of oxygen interstitials is unlikely in these metals. However, in the lighter elements, such as titanium and vanadium, where the d bands are less than half-filled, any electrons added to the d bands enter bonding states and enhance the metal-metal bonding, thus increasing the likelihood of oxygen interstitials. Therefore, a critical electron/atom ratio, z_c, can be defined in the interval $5 < z_c < 6$ where the mechanism of oxide formation will change (note $z = 5$ for vanadium and 6 for chromium). Goodenough pointed out that there is a marked change in the character of oxides in this interval. For example, VO is a semimetal without spontaneous magnetization, whereas MnO is an antiferromagnetic insulator; V_2O_3 exhibits a first-order semiconductor-metal transition at its Néel temperature, whereas Cr_2O_3 is an antiferromagnetic insulator. Goodenough concluded his discussion by asking the question: Has anyone explored the influence of electron/atom ratio on the oxidation mechanisms?

Stringer recalled the work of Bryant[7] in which additions of molybdenum, rhenium, and ruthenium to niobium all reduce the solubility of oxygen in the alloy, with essentially zero solubility being reached at electron/atom ratios of about 5.75. He suggested that these results may be taken as additional evidence for Goodenough's hypotheses.

4 ROLE OF EXTENDED DEFECTS IN MASS TRANSPORT

The chairman noted that this Colloquium had not really delved into the question of how atom transport takes place in structures containing extended defects, and asked Professor Anderson to open a discussion of this topic.

Anderson first stated that the concept of point defects being present in thermodynamic equilibrium in all crystals is unquestioned, and that, as long as the defect concentration is small, point-defect formulation is perfectly valid. The essential feature of the point-defect description, that there should be an unambiguous one-to-one relation between deviation from stoichiometry and the point-defect concentration, is completely destroyed when a tightly bound defect complex, possibly formed by considerable local reconstruction of the host crystal structure, replaces the isolated interstitial atom or vacancy.

He noted that the acute problem arises with materials such as wustite, for which kinetic measurements, oxygen pressure dependence, etc., can be reproduced by a point-defect description despite the fact that the structural evidence for defect complexes which are formed by reconstructive processes is now beyond doubt. While it is true that too few nonstoichiometric systems have been studied by diffraction methods under equilibrium conditions at high temperatures, he stated that where direct evidence is available, there is essential accord between the high-temperature structures and those deduced from quenched specimens. For wustite, the evidence is direct. Cohen's X-ray work, in accord with the neutron diffraction study of Cheetham et al.[8] shows that the ratio of vacant octahedral cation sites to occupied tetrahedral sites, averaged over all defects, across the whole wustite phase field, is close to that of the Koch-Cohen cluster found in the metastable superstructure. This ratio is quite sensitive to changes in defect structure, and it follows that if the complexes undergo dissociation or other transformations, they do so in such a way as to keep their statistical composition almost unchanged. Anderson suggested that one possible mechanism for cation diffusion is vacancy migration through the continuous matrix; a second (and by intuition, probably the principal mode of diffusion) is by migration of the defect clusters themselves. These clusters are in dynamic equilibrium with point defects, and one can formulate schemes whereby vacant cluster sites are eliminated at one side and created at the other, so that the cluster moves by way of a series of distorted configurations. If the cluster dissociates – for example, into units with a tetrahedral cation surrounded by four empty octahedral sites – the migration process may be facilitated. He stated that Greenwood and Howe have given some consideration to these questions, and their work is continuing, but that there is a need to formulate and to test mechanisms that are applicable to cluster migration. Formation, dissociation, and the unit steps in migration of clusters can all be represented by the quasi-chemical reaction formalism. The constraints are that the average defect structure fits the observed vacancy:tetrahedral cation ratio, and that the overall equilibrium with oxygen fits the measured pressure dependence of the diffusion properties.

Anderson went on to point out that migration properties in oxides with planar defects, and in crystallographic shear phases (which, he emphasized, are not defect structures), are little understood. Little is known about the mechanism by which crystallographic shear (CS) planes are introduced or eliminated. In the reduction of TiO_2 or WO_3, it is probable that a CS plane is nucleated by minor rearrangement in the surface layers and grows inward along its length. A CS plane that does not cross the full width of a crystal is necessarily bounded by, or terminated by, a dislocation and it may be that the growing CS plane is energetically favored as a locus of reaction. The transport properties and electronic properties of CS phases are likely to be highly anisotropic, and since well-ordered crystals of the more reduced CS structure oxides of vanadium and titanium can now be prepared, he expressed the hope that their transport properties will be studied. The thermodynamics of the titanium oxide system have been established by the recent work of Merritt and Hyde[9], so the conditions for measuring both cation diffusion and oxygen diffusion are known.

Information on this subject, Anderson felt, is relevant for the oxidation and corrosion behavior of titanium, niobium, tantalum, etc. He claimed that the existing evidence indicates that transport in the rutile structure, the ReO_3 structure and its derived block structures, and the Ta_2O_5-like structures is facile (and uncontrollable) both for cations and for anions. In all these structures, the oxygen atoms are either in or near close packing, but there is evidence for rather easy penetration through channels between coordination polyhedra which are typically about 2.5 Å across. He noted that Iijima had shown the form that an interstitial-oxygen-defect complex may take in the block structure of $Nb_{12}O_{29}$. Also, work by Dr. Browne at Oxford has shown that oxygen penetrates into the interior of $Nb_{12}O_{29}$ crystals and effects a homogeneous transformation (rapidly at 350 to 400°C and measurably at much lower temperatures) into a new polymorph of Nb_2O_5. In this reaction, the

Iijima complex may well be an initial step. Anderson felt that less is known about the structure of Ta_2O_5, which is related to that of the low-temperature form of Nb_2O_5, but the oxygen atoms in this structure are known to be rather mobile. He noted that there have been several examples (discussed at this colloquium) of oxygen migrating easily in a close-packed anion sublattice. Since, in this class of oxides, the structures are conducive to oxygen migration, metal oxidation can take place at the oxide/metal interface. He concluded, therefore, that one is forced into a rather pessimistic view of the possibilities for control and inhibition of the oxidation reaction.

At this point there ensued general discussion of the topic of diffusion in structures with extended defects. O'Keeffe reported on some experiments on diffusion of various cations in rutile. He stated that although only semiquantitative observations were available, he was able to group the cations into two classes. In the first class of cations are those that diffuse rapidly parallel to the c axis through channels of octahedral sites in the structure, and therefore show a very large anisotropy in the diffusion coefficient. In the second class are those that themselves form shear structures, and these show diffusion which is quite isotropic. Presumably, the latter cases involve cooperative jumps of atoms in shear planes, those jumps having comparable components both parallel and perpendicular to the c axis. The best data for this latter group are for Cr^{3+} diffusion. Kröger asked whether point defects were not still required to move foreign atoms, even though shear planes could move by cooperative motions. Anderson suggested that once two Cr^{3+} ions come side by side, they are in a position to nucleate a shear plane. When further Cr^{3+} ions have joined and aggregated into a large planar singularity, this shear plane will resist deformation and, therefore, further diffusion of these Cr^{3+} ions could occur only through highly cooperative movements. Kröger then asked how the chromium atoms enter the rutile structure, to which Anderson replied that for low concentrations ($\sim 10^{-4}$), point defects would be involved and that it was only for higher concentrations that such shear planes would be created.

The discussion then turned to the role of extended defects in oxidation. Smeltzer felt that shear planes in an oxide might provide channels for rapid migration and wondered whether a metal A can be doped with an impurity B which forms a highly stable oxide (e.g., aluminum, silicon, or chromium) in such a way that, after some degree of formation of AO, enough of B would diffuse through this oxide to form a protective BO oxide at the surface.

Kofstad then noted that oxidation is not usually carried out at such high temperatures that the resulting shear structures become highly ordered. In fact, he noted that oxidation of titanium is usually carried out between 500 and 1000°C while, as Anderson has pointed out, temperatures of at least 1200°C were required for ordering. At these lower temperatures, the shear planes are highly disordered, in the sense of having variable spacings and lying along several equivalent orientations. The question he raised, then, was whether such a highly disordered system of planar defects can behave effectively as if point defects were present. Anderson doubted this, on the grounds that planar singularities cannot be regarded as point defects, but Goodenough asked why point defects could not still provide the rate-limiting step. It was clear that there was a need for more detailed models to evaluate such a situation, which apparently have not yet appeared.

The discussion then took another direction when Goodenough, noting that Anderson's earlier presentation suggested that there is both a strong repulsive force and weak attractive force between shear planes, inquired into the origin of such forces. In particular, he suggested that when quasi-free electrons are present, creation of a periodic structure could introduce an energy discontinuity at the Fermi surface. This phenomenon could provide an electronic origin for the periodicity.

Welch raised the question of whether extended defects could be treated by the methods of statistical mechanics (including configurational entropy) like point defects, or whether they were more like dislocations. Lidiard pointed out that because extended defects cannot move about, their configurational entropy is so small relative to the enthalpy of formation that it can be neglected. In this sense, then, shear planes were like dislocations, although, as Anderson pointed out, the shear planes are in thermodynamic equilibrium while dislocations are not.

5 SOLID STATE REACTIONS AND TECHNOLOGICAL PROCESSES

The final question for this discussion was: Can we describe solid-solid reactions in terms of our knowledge of defects and their mobilities and how useful can this knowledge be for application to technological processes? The chairman called on Professor Schmalzried to comment on these questions.

In answer to the first part of the question, Schmalzried said that, in principle, the answer is definitely "yes", both for quasi-binary and for more complex reactions. However in a ternary or quaternary system, the situation is more complicated in that more possibilities for defects may arise and, thermodynamically, more chemical potentials must be specified. The fact is that, while there are several good models for disorder in binary systems, there are almost none for ternary and none at all for higher systems. The problem, however, is not one of principle, but of application.

Schmalzried pointed out, however, that some ternary systems may be considerably simplified for all practical purposes. Thus, suppose that we have a ternary oxide ABO_2, and we choose as variables the oxygen activity (or partial pressure, p_{O_2}) and, say, the activity of the oxide AO. If the free energy of formation of ABO_2 from the elements were about 200 kcal/mole while that of ABO_2 from AO and BO were about 5 kcal/mole (such numbers can be quite typical), then it follows that the stability of the variability range with respect to p_{O_2} is much higher than with respect to the AO activity. Stated differently, we can vary p_{O_2} over orders of magnitude and thereby change the defect concentration, but by changing the activity of AO considerably we will only create a second phase. Thus, at least to a zero order of approximation, we may forget about the dependence of defect concentrations and also transport coefficients on any component activity other than that of O_2.

Now Schmalzried turned to consideration of a reaction in an n-component system and noted that it requires calculating the reaction path in n-dimensional space. This requires a knowledge of all but one of the partial molar energetic quantities for this system as well as the mobilities of all components or the corresponding significant defects. Such information is, in practice, seldom available. But a further complication occurs if, in addition, the system leaves the single-phase field and moves into a multiple-phase field, for then one must consider nucleation processes which control the morphology. All this leads to complications, as, for example, when cracks are developed in an oxide scale. The question then arises as to whether it is economical to look for worked-out solutions to complex technological problems (i.e., the second part of the question posed at the beginning of this section). Schmalzried then stated his view that it is better to obtain a really clear understanding of the fundamental concepts in limiting cases and to leave the handling of complex problems, such as that of making a better adherent scale, to the technicians. The concepts derived from the limiting cases can then serve to guide the judgments made with regard to the complex technological problems.

Finally, Schmalzried turned to another topic which had come up earlier in this Colloquium but was not discussed in depth, namely the role of interfaces, particularly grain boundaries, in reactions. He emphasized the need to know more about interfaces in order to understand reactions. However, he pointed out, it is one thing to characterize a clean surface, prepared in ultrahigh vacuum and still another to use that surface, for example, in the role of a catalyst. He felt that the same contrast is true for grain boundaries, where, for example, in a conductivity experiment, the boundaries will contain impurities and defects attached to it. The question, therefore, is how do we characterize and find out about an interface not just in its static state, but, rather, under conditions of actual operation.

He noted that in the study of interfaces, there is one tool whose potentiality has not been adequately exploited, namely, the electrokinetic approach. He suggested that work which electrochemists have carried out to study the transfer through an interface in solid-liquid systems could be done in a similar way in solid-solid systems with the aid of solid electrolytes. It requires the observation of responses using pulse kinetics or high frequencies to observe kinetic behavior which takes place across interfaces.

At this point time ran out, so that there was no opportunity for a general discussion of this most important topic.

From a review of all of the topics and attendant problems in oxide research discussed at the Colloquium, it appears that the most perplexing question emerging from the conference is: How does one correlate our knowledge of the structural aspects of extended defects in oxides with observations on transport properties? Although structural studies have established that extended defects such as clusters, crystallographic shear planes, or intergrowth block structures are the predominant defects (rather than point defects) in many oxides of interest, there is, at present, no promising model which can be used to explain diffusion and electrical properties in oxides in terms of extended defects. Perhaps the state of the field is most aptly described by the following "poem" submitted by an anonymous attendee at the conclusion of the conference:

For many years defects were shaped like a point
And nicely experiment with theory was joined
Then Anderson said defects form a phase
And everything we knew turned back into a haze.

REFERENCES

1. Allpress, J. G., *J. Mater. Sci.*, **6**, 313 (1971).
2. Iijima, S., *Acta Cryst.*, **A29**, 18 (1973).
3. O'Keeffe, M. A., *Acta Cryst.*, **A29**, 389 (1973).
4. Fejes, P. L., Ph.D. Thesis, Arizona State University (1973); Goodman, P., and Moodie, A. F., *Acta Cryst.*, **A230**, 280 (1974).
5. Tannhauser, D. S., *J. Phys. Chem. Solids*, **23**, 25 (1962); Bransky, I., and Tannhauser, D. S., *Trans. AIME*, **239**, 75 (1967).
6. Geiger, G. H., Sevin, R. L., and Wagner, J. B., Jr., *J. Phys. Chem. Solids*, **27**, 947 (1966); Hillegas, W. J., and Wagner, J. B., Jr., *Phys. Lett.*, **25A**, 742 (1967).
7. Bryant, R. T., *J. Less-Common Metals*, **4**, 62 (1962).
8. Cheetham, A. K., Fender, B.E.F., and Taylor, R. I., *J. Phys. C: Solid State Phys.*, **4**, 2160 (1971).
9. Merritt, R. R., and Hyde, B. G., *Phil. Trans. Roy. Soc. (London)*, **274**, 627 (1973).

AUTHOR INDEX

SUBJECT INDEX

CHEMICAL INDEX